AXIOMATIC DESIGN

The MIT-Pappalardo Series in Mechanical Engineering

This book is published under the MIT-Pappalardo Series in Mechanical Engineering. This series was established by the generous endowment created at MIT by Neil and Jane Pappalardo. Mr. Neil Pappalardo is a graduate of MIT who started one of the first software companies—Medical Information Technology, Inc. (MEDITECH)—32 years ago and has built it into the leading medical information technology company in the United States. Mrs. Jane Pappalardo is a graduate of Boston University. Both Neil and Jane believe in the importance of education, arts, and public service—they are exceptional people.

The faculty members of the Department of Mechanical Engineering at MIT are most grateful to Mr. and Mrs. Neil Pappalardo. They have given not only a chair for senior professorship, a wonderful teaching laboratory, and this book series to the MIT Department of Mechanical Engineering but also their time and guidance to all parts of MIT. They have contributed a great deal to shape the highest ranked mechanical engineering department in the country. Their confidence and faith in the faculty, students, and staff have done much to strengthen the infrastructure, morale, and continuing promise of the Department.

A number of valuable books, especially a series of integrated undergraduate textbooks by 14 designated professors, are being written for this series. Timing made this book the first.

It has been the pleasure of the MIT editors to work with Oxford University Press in undertaking this series. Our special thanks go to Barbara Wasserman, Senior Vice President of OUP, and Peter Gordon, Executive Editor. Their vision for an outstanding book series and our need to provide the most effective teaching and learning materials have made this series a successful undertaking. We hope that this series will be one of the best of its kind in the years to come.

Rohan C. Abeyaratne
Nam P. Suh
MIT Editors of the Series

AXIOMATIC DESIGN

ADVANCES AND APPLICATIONS

Nam Pyo Suh
Massachusetts Institute of Technology

New York ■ Oxford
OXFORD UNIVERSITY PRESS
2001

Oxford University Press, Inc., publishes works that further Oxford University's
objective of excellence in research, scholarship, and education.

Oxford   New York
Auckland   Cape Town   Dar es Salaam   Hong Kong   Karachi
Kuala Lumpur   Madrid   Melbourne   Mexico City   Nairobi
New Delhi   Shanghai   Taipei   Toronto

With offices in
Argentina   Austria   Brazil   Chile   Czech Republic   France   Greece
Guatemala   Hungary   Italy   Japan   Poland   Portugal   Singapore
South Korea   Switzerland   Thailand   Turkey   Ukraine   Vietnam

Copyright © 2001 by Oxford University Press, Inc.

Published by Oxford University Press, Inc.
198 Madison Avenue, New York, New York, 10016
http://www.oup.com

Oxford is a registered trademark of Oxford University Press

Library of Congress Cataloging-in-Publication Data

Suh, Nam P., 1936–
    Axiomatic design : advances and applications / Nam Pyo Suh.
        p.   cm.—(The Oxford series on advanced manufacturing)
    Includes bibliographical references and index.
    ISBN 978-0-19-513466-7 (alk. paper)
    1. Design, Industrial—Methodology.   2. Axiomatic set theory.
    I. Title.   II. Series.

TS171.4. S84   2000
620'.0042—dc21   00-040635

Printed in the United States of America

To our children and grandchildren

Contents

Preface

Engineering consists of synthesis and analysis, which mutually reinforce each other in a feedback loop. Rational synthesis makes analysis simple, and rigorous analysis of a synthesized engineering system provides the means for improving and optimizing the system through fundamental understanding. Prior to the middle of the nineteenth century, engineering in the United States was done by people without formal engineering education because there were no engineering schools except the military academy. To satisfy human needs, people invented new products and improved existing technologies with great ingenuity and skill—all based on their innate capability, apprenticeship, and experience.

Starting in the middle of the nineteenth century, many engineering schools were established to generate engineers with a strong foundation in science, mathematics, and engineering. One of the goals of some—not all—of those schools was to introduce scientific principles into engineering to make engineering practice more rigorous and reliable. This need to improve engineering systems by applying mathematics and scientific principles to designed systems dominated much of the development of enginering schools in the twentieth century. These schools have made a significant contribution to the creation of the technology-dominated society of today.

The term "engineering science" has been used to describe the subdiscipline of engineering that deals with the application of scientific principles and rigorous mathematical tools to various engineering problems and systems. Engineering science typically implies analysis. One of the underlying assumptions in engineering has been that the synthesis aspect of engineering systems can best be done either heuristically or empirically, as there were no scientific principles that could be applied to the synthesis process. This kind of thinking can be traced to the prevalent approach used in science—reductionism. In science, the driving force has been to understand how nature works, which is done by subdividing a problem until it can be clearly analyzed and understood. However, this reductionism cannot be extended to the synthesis process, because synthesis requires a different way of thinking and different principles.

Design, which is fundamental to engineering, encompasses both synthesis and analysis. Human ingenuity will continue to play an important role in design, but to be more effective in educating engineering students and in achieving industrial tasks, design must be made more scientific. For this to happen, the synthesis process also needs scientific principles. We need both "engineering science for analysis" and "engineering science for synthesis." Just as physical science, information science, and biological science provide a basic scientific foundation to engineering, the discipline of design will become stronger if it, too, is based on scientific principles and paradigms. Then design, which is the ultimate goal of engineering endeavors, can enrich both intellectual and educational endeavors in addition to improving the efficacy of the industrial development process.

In the 10 years since the publication of *The Principles of Design*, the first treatise on axiomatic design, design education, research, and practice have evolved considerably. Many universities in the United States have infused design education into all of their engineering disciplines. As a result, we are now seeing a number of active researchers in design, as well as journals and conferences on design. The corporate world has also begun to recognize the need for basic approaches to design. Because of keen competition, industries have realized that ad hoc design processes can be unreliable, costly, and risky. Such undisciplined practices have resulted in missed schedules, cost overruns, and failed products, while the practice of fixing problems and symptoms of poor design has led to low quality and unreliability as well as high costs of hardware, software, and systems. To overcome these problems, some corporations have actively trained and recruited engineers and designers with a fundamental understanding of basic design principles.

Despite this progress, much more needs to be done to bring the design field to a level at which rational scientific knowledge can reinforce experience-based design expertise. Only through continued efforts can we continue to raise the quality of industrial products and minimize the potential damage of poor designs to society in general. Unfortunately, the culture at some corporations is still dominated by the view that it is possible to design based on experience, improve products through the testing of prototypes, and debug software until a workable solution is developed. We must replace this ill-conceived notion with another— make correct design decisions from the outset, thus eliminating the need for extensive testing and debugging, as well as unforeseen product failures in service. Universities can do their part by giving students the fundamental knowledge of design that can be used in all scientific and technological processes.

This book is an outgrowth of teaching axiomatic design to graduate students at MIT and engineers in many industrial firms in countries throughout the world. Indeed, these students and engineers contributed a great deal to the richness of axiomatic design by applying the theory to their many new design tasks. Nearly all of the examples given in this book were generated during the course of teaching the subject and, in many cases, were solved by the students in the class or as term projects. They enriched the author's experience and enhanced his understanding of the subject matter.

This book is substantially different from *The Principles of Design* in both its content and approach. The first was written as a monograph, whereas this one is intended more as a textbook for advanced college and graduate students and professional engineers. Given the advances made in the past decade, this book also includes newly acquired knowledge and case studies. After covering the fundamental principles of axiomatic design in its first three chapters, it deals with the design of systems, software, materials and materials processing, manufacturing systems, and products. Although this book requires no prerequisite reading, some students of design may still want to refer to the first book for clarification.

From the author's experience in teaching this subject to many engineers and students, it has become clear that axiomatic design is not an easy subject to learn, much less master, without some effort—perhaps because of the conceptual nature of the subject. To truly understand axiomatic design, students must put theory into practice by applying the basic principles to many problems and many design tasks. In some ways, the difficulty of learning axiomatic design may be similar to learning thermodynamics. However, once students understand axiomatic design, they are rewarded with the ability to produce products, manufacturing systems, software, and new processes with both ease and elegance. By designing correctly, a number of engineering problems, such as stability and convergence, become less of an issue or extremely simplified, and the need for extensive modeling and analysis for optimization is also reduced.

Chapter 1 provides an introduction to the key concepts of axiomatic design theory. Chapter 2 considers robustness of design by examining designs that involve only one functional requirement, which can be dealt with using only the information axiom. Chapter 3 presents a number of principles and their implications for design, which are often overlooked by those who design purely based on experience. It covers the fundamental issues involved in designing when there are multiple functional requirements that must be satisfied at the same time. Chapter 4 builds on the materials presented in Chapters 1 through 3 to develop methods of designing complicated systems. It presents many new ideas, such as methods of decomposing functional requirements and design parameters to build a hierarchical structure and the flow chart that represents the system architecture. Because most engineering tasks deal with systems, Chapter 4 should be important to many engineers. In addition, the fundamental concepts of system design presented in Chapter 4 serve as a launching pad for the discussion in subsequent chapters.

Chapter 5, which focuses on software design, demonstrates how easily software can be designed systematically using axiomatic design theory. A special method of creating software—known as axiomatic design of object-oriented software systems (ADo-oSS)—should become a new paradigm for software design. Chapter 6 is devoted to manufacturing systems. It clarifies the roles of "push" and "pull" types of manufacturing systems by showing how they can be designed rigorously. Chapter 7 discusses how axiomatic design can be used to transform customer needs into the development of new materials and materials-processing techniques. Chapter 8, on product design, illustrates how product design can be done systematically. Finally, Chapter 9 deals with the issue of complexity. It presents a theory that postulates and explains four different kinds of complexities—time-independent real and imaginary complexities and time-dependent combinatorial and periodic complexities. The understanding of these four complexities may have a profound influence on how one can view design tasks and natural phenomena.

The author has had the unique privilege of developing axiomatic design theory in the comfortable and supportive surroundings of MIT—an intellectual haven—as a tenured faculty member. The author is particularly appreciative of the MIT culture, which embraces ideas that challenge well-accepted notions. It is extremely important for all universities to nurture new ideas and theories. However, sometimes this is difficult to do in an environment in which young professors have to demonstrate what they can do in relatively short periods of time to earn tenured appointments—and especially if their work challenges the established concepts of powerful senior professors. Even in academia, those who follow the established theories and notions are more protected and promoted than those who advance ideas orthogonal to the entrenched theories accepted by their contemporaries. The temptation of senior faculty to "clone" young faculty in their own image is one of the most serious threats to any great institution.

The MIT Department of Mechanical Engineering, even among MIT's outstanding departments, is unique and exceptionally strong. It has rich talents in many diverse areas, ranging from classical physics to design, manufacturing, mechanics, bioengineering, nonlinear optics, computing, materials, thermal and fluid sciences, networking, control, and precision engineering. During the past decade, it has worked toward three goals: to significantly improve undergraduate education, to transform mechanical engineering disciplines, and to conduct research that makes a meaningful impact on the scientific knowledge base and technology. The Department has developed a new curriculum that provides the right context for learning. It has been transforming the discipline of mechanical engineering from one based primarily on physics to one based on physics, information science, and biology. Its research goal has been emphasizing the two ends (in contrast to the middle) of the research spectrum—basic research and technology innovation—where the impact made on humanity and fundamental knowledge tends to be the greatest. The development of axiomatic design has been accomplished with the hope of making contributions at these two ends of the research spectrum.

This book was written in the belief that it could serve as a small stepping stone in the ever-evolving intellectual science of design. If this book can enlighten some bright young minds in learning design, if it can help industrial firms become more efficient and effective, and, ultimately, if it can benefit humanity, it will have achieved more than its goal.

Nam Pyo Suh
Sudbury, Massachusetts

Acknowledgments

We are all part of a continuum—in genealogy, time, space, and history. Some of us are more fortunate than others in that we have been blessed by family, friends, colleagues, and a greater community of people who have nurtured our minds and spirits—either intentionally or unintentionally. I am one of those fortunate people whose professional and personal lives have benefited by a great wealth of outstanding people.

There are too many to list them all, but several people must be mentioned in a random order: Nathan H. Cook, Milton C. Shaw, Herbert H. Richardson, Ernest Rabinowicz, John Hollick, Water L. Abel, Ralph E. Cross, Alex d'Arbeloff, Elmer Schwartz, Neil Pappalardo, Papken Der Torossian, George N. Hatsopoulos, Edwin H. B. Pratt, Sung-Kyu Kim, Sung-Joon Huh, Woo-Choong Kim, Erich Bloch, B. J. Park, and Matt Pallaver. Although some of them may not even remember me, they have all affected my work, career, and life in many different ways. One was a teacher in elementary school, one was a teacher in high school, one was a headmaster in a prep-school, two were mentors in universities, two were my bosses in industry and generously paid for my doctoral education, and others are friends and colleagues in the professional and corporate world. To all of them, I am most grateful.

What makes the teaching profession so unique is the inseparable linkage—at least in the mind of the teacher—it establishes between the teacher and students. My former students—in tribology, design, manufacturing, and materials processing—were all outstanding. My first graduate student in axiomatic design was Jim Rinderle, followed by Len Albano, Shinya Sekimoto, Mats Nordlund, Derrick Tate, K. Kaneshige, Jack Smith, Sun-Jae Kim, Hiroshi Igata, Y. Yasuhara, Tae-Sik Lee, Jin Pyung Chung, Yun Kang, and John Szatkowski. David Wilson and Yasuo Suga were also important members of the axiomatic design research team. My students in other fields, including Sang-Gook Kim, Dan Baldwin, Chul Bum Park, Vipin Kumar, Turker Oktay, Sung Won Cha, Jonathan Colton, Jason Melvin, Amir Torkaman, and Yoddhojit Sanyal, applied axiomatic design to their projects. I was also most fortunate to have outstanding postdoctorate researchers in Sung-Hee Do and Hod Lipson. All of them made their own unique contributions to the field. Many visiting professors—G. Sohlenius, H. Nakazawa, Susan Finger, Kyung-Jin Park, and Karim Ker—made the research in this field exciting, stimulating, and rewarding. I will cherish a long-standing friendship I have developed with Gunnar Sohlenius, first through axiomatic design. I also enjoyed my interaction with David Cochran, who has done much to develop a firm foundation for design of manufacturing systems based on axiomatic design.

A chance encounter with Hilario L. "Larry" Oh at the NIST Symposium on Quality Control has created a long-term professional relationship and friendship. He taught axiomatic design with me at MIT and in many industrial firms. He made many important contributions to axiomatic design and robust design and solved many important industrial problems, some of which are described in this book.

I was most fortunate to meet Carol A. Vale—an outstanding statistician and a superb editor. She has gone through the entire manuscript, making it more readable, offering many valuable suggestions for changes and additions, and finding all the mistakes I made. I am greatly indebted to her. I am also grateful to Peter Gordon, Executive Editor of Oxford University Press, who made many helpful editorial comments and suggestions. I am pleased to acknowledge that G. Sohlenius, Jon Colton, Milton Shaw, Tae-Sik Lee, and Anna Thornton also read some of the chapters during the early phase of the writing of this book and provided useful comments. Karen Shapiro of Oxford University Press assumed the editorial responsibility for publication of this book.

To teach axiomatic design at MIT has always been a challenge. MIT students are always inquisitive, smart, and hardworking. Challenging their minds and imagination is no easy task. I am grateful to the students who took the course, because they made the task of writing this book a pleasure. I enjoyed teaching the graduate-level axiomatic design course with several former students and colleagues. In particular, I would like to thank Larry Oh, Mats Nordlund, Leonard Albano, and Hod Lipson for their valuable inputs to the course.

Many companies worldwide were kind enough to invite me into their midst to teach axiomatic design to their engineers. I owe much to those who have provided strong support and made teaching at these firms an enjoyable experience. I am especially indebted to Papken Der Torossian, William Hightower, Jeff Kowalski, Dan Cote, Boris Lipkin, and Reese Reynolds at Silicon Valley Group, Inc., Jin-Kyun Kim of the Institute for Advanced Engineering of Daewoo Group, Larry Smith, Carol Vale, and Nathan Soderborg of Ford Motor Company, and Billy Frederickson and Mats Nordlund of Saab Dynamics.

Pamela McCarthy of MIT has given all the support I needed to make the writing of this book possible. She is one of those exceptional persons who go beyond the call of her duty to make things happen and get the job done well—always with a smile. It has been my good fortune to work with her in this task.

This book was written while I headed the Department of Mechanical Engineering at MIT—the highest rated department of its kind in the United States. In many ways, it was a period of transition for the Department, having moved into many new exciting intellectual and educational arenas. I have been most fortunate to work with so many outstanding colleagues who are leaders in their disciplines. They are very competitive and accomplished people, but at the same time, I have found in each of my colleagues a sense of fairness, dedication, humility, and humanity. They make up a strong intellectual community of scholars, which would be difficult to replicate elsewhere. I was particularly fortunate to work with Rohan Abeyaratne and Bora Mikic in leading the department. I am grateful to all of my colleagues for their collegiality.

In shaping and molding my life and personality, there is no one who deserves more credit and thanks than my wife, Young Ja. She has been my inspiration and moral pillar. She has been more than a partner for life—her love has sustained me and nourished me. She has sacrificed much for me, tolerated my shortcomings, and given her best to raise four wonderful daughters who are now serious professionals in their own right. She has now undertaken an additional assignment. She is busy chasing after grandchildren, Kristian, Nicholas, Madeleine, and Henry, to be sure that they, too, grow up and pursue their dreams. She has made our family a happy and thriving one. I am indeed fortunate. Finally, our four daughters and two sons-in-law have been exceptional and most loving—I am perhaps the proudest father in the world.

Nam P. Suh
Sudbury, Massachusetts

On the CIRP Design Book Series

CIRP (International Institution for Production Engineering Research), an international academy created in 1950 after the Second World War demolished the production capability of European nations, works to promote the advances in production engineering through research and education. Today CIRP is made up of as many as 15 full members from each member country and currently has about 250 full members and 250 corresponding members from industrialized countries across the globe.

The CIRP Design Book Series was established by a contract between Oxford University Press and CIRP to promote design research and education worldwide. This book, *Axiomatic Design: Advances and Applications,* was originally undertaken with the CIRP Design Book Series in mind. However, after the MIT-Pappalardo Series in Mechanical Engineering was established, Oxford University Press decided to publish it under both series.

Torsten Kjellberg
For CIRP
Stockholm, Sweden

1 Introduction to Axiomatic Design

1.1 INTRODUCTION

This introductory chapter contains a complete outline of the basic concepts of axiomatic design, with many examples. Because it is easier to illustrate the concept with mechanical case studies, many of the examples in this chapter are mechanical in nature. However, the fundamental ideas of axiomatic design are applicable to software design, organizational design, system design, material design, manufacturing system and process design, and others. In later chapters, the design of these other types of systems and processes based on axiomatic design is presented with examples.

1.2 CURRENT STATE OF DESIGN PRACTICE

In the 300 years since the start of the Industrial Revolution, science and technology have reached an amazing level at an ever-accelerating rate. Humans have walked on the moon, are sending space probes to other planets, and are about to reduce human physiology into basic building blocks of DNA molecules. Memory chips can store, in a single chip, the entire base of scientific knowledge that was available a hundred years ago. These chips are manufactured using equipment that can make microscopic electronic components built on small integrated circuit (IC) chips. These are incredible scientific

1

and technological breakthroughs. Despite these and other brilliant achievements, we are surrounded by many technological and societal problems that have been created through poor design practice.

Some design errors are major problems that have been well publicized. There are also many "small" problems that simply inconvenience or aggravate consumers. All bad designs or products—large or small—can be dangerous, cost money, limit the usefulness of products, or delay the introduction of new products. Many products are recalled. The warranty cost of some products is a significant percentage of the selling price. Poorly designed equipment requires maintenance and wastes valuable time, while some failures result in loss of property and even lives. Furthermore, many poorly conceived and designed development projects suffer from major delays, cost overruns, and, in some cases, complete failure. Although some of these failures are caused by a lack of scientific knowledge, a majority of these problems arise because of poor design of the product, process, software, and systems.

One reason so many design mistakes are being made today is that design is being done empirically on a trial-and-error basis. This problem is not confined to any one country or any one company. It exists everywhere. Universities throughout the world have not given their engineering students generalized, codified, and systematic knowledge in design. Instead, design has been treated as a subject that is not amenable to scientific treatment. Consequently, design has depended on intuitive and innate reasoning rather than on rigorous scientific study. One of the biggest challenges of the design field is to overcome this acceptance of design as a subject in the arts rather than the arts and sciences. Fortunately, the field of design does not have to remain at this stage of empiricism. Just as many fields of technology have gone through similar stages of development, the field of design, too, will evolve into a true discipline with scientific bases. This book, which follows *The Principles of Design* (Suh, 1990), presents an expanded treatise on a scientific foundation for design.

1.3 WHO ARE THE DESIGNERS? HOW DO WE DESIGN? WHAT IS DESIGN?

Are you a designer? Is the mayor of your hometown a designer? If not, should he be? Who performs design activities in your organization? Design has been defined in a variety of ways depending on the specific context and/or the field of interest. Mechanical engineers often design products, and therefore when they say *design*, they typically refer to product design. Manufacturing engineers, on the other hand, think of design in terms of new manufacturing processes and systems (i.e., factories and manufacturing cells). To electrical engineers, design means developing analog or digital circuits, communications systems, and computer hardware, while system architects perceive design in terms of technical or organizational systems where many parts must work together to yield a system that achieves the intended goals.

Although some software engineers think that their primary job is to write computer code, it is difficult for them to produce good software unless they first design the architecture of the software. Business managers design organizations to achieve organizational goals. On a smaller scale, interior designers select and arrange furniture and decorative items to create the right mood for a house or a building. Even the mayor of your hometown must design an effective and efficient government organization and develop strategic plans to achieve his vision for the city.

All of these are design activities, although the contents of these activities and the knowledge required to achieve the design goals are field specific. Although each field utilizes different databases and different design practices, they share many design characteristics. What is common in all these activities is that the designers must do the following:

1. *Know* or understand their *customers' needs.*
2. *Define the problem* they must solve to satisfy the needs.
3. *Conceptualize the solution through synthesis.*
4. Perform *analysis* to optimize the proposed solution.
5. *Check the resulting design solution* to see if it meets the original customer needs.

A definition of design. Design is an interplay between *what* we want to achieve and *how* we want to achieve it. This interplay (or mapping) is illustrated in Figure 1.1. Therefore, a rigorous design approach must begin with an explicit statement of "*what* we want to achieve" and end with a clear description of "*how* we will achieve it." Once we understand the customer's needs, this understanding must be transformed into a minimum set of specifications, which will be defined later as functional requirements[1] (FRs), that adequately describe "*what* we want to achieve" to satisfy the customers' needs. The descriptor of "*how* to achieve it" may be in the form of design parameters (DPs).

Often designers find that the precise description of "what we want to achieve" is a difficult task. Many designers deliberately leave their design goals implicit rather than explicit and then start working on design solutions even before they have clearly defined their design goals. They measure their success by comparing their design with the implicit design goals that they had in mind, which may or may not be what the customer wanted. They spend a great deal of time improving and iterating the design until the design solution and "what they had in mind" converge, which is a time-consuming process at best. To be efficient and to generate the design that meets the perceived needs, the designer must specifically state the design goals in terms of "what we want to achieve" and begin the design process. Iterations between "what" and "how" are necessary, but each iteration loop must redefine the "what" clearly.

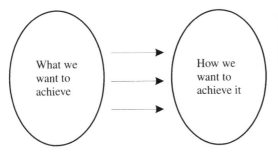

Figure 1.1 A definition of design as a mapping between "what we want to achieve" and "how we want to achieve it."

---

[1] Functional requirements (FRs) are defined as the minimum set of independent requirements that completely characterizes the design goals. The definitions of key terms are given in Section 1.7.2.

EXAMPLE 1.1 Refrigerator Door Design

Consider the refrigerator door design shown in Figure E1.1. Is it a good design? Each time this question is asked, we get many different kinds of answers. Some say the door is not a good design because it is inconvenient for people who are right handed. Some say it is a good design. However, the question posed cannot be answered without asking: What are the design goals (i.e., functional requirements) for the door design?

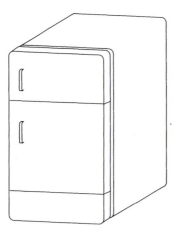

Figure E1.1 Vertically hung refrigerator door.

If the purpose of the door is to provide access to what is inside the refrigerator, then the door performs that function. Therefore, the door design is a good design. On the other hand, if the functional requirements of the door are to (1) provide access to the food in the refrigerator and (2) minimize energy consumption, then the door is a poor design because each time the door is opened, cold air inside the refrigerator is replaced by warm outside air, thus wasting energy. The door should have been designed differently to satisfy these two functional requirements. Consider some alternative designs given the two FRs stated above.

An important lesson to remember! The important lesson of the above example is that you must think in terms of *functions* the product (or software, system, process, or organization) must perform. It is necessary to learn to think *functionally* in designing products, processes, software, organizations, business plans, and policies.

How should we design? In the past, many engineers have designed their products (or processes, systems, etc.) iteratively, empirically, and intuitively, based on years of experience, cleverness, or creativity, and involving much trial and error. Although experience is important because it generates knowledge and information about practical design, experiential knowledge alone is not sufficient, as it is not always reliable, especially when the context of the application changes. Experience must be augmented by systematic knowledge of design, or vice versa.

Although design is an old discipline, it has been slow to organize design knowledge by generalizing, codifying, and systematizing so that engineers/designers can be taught to be good designers. Such codified and generalized knowledge shortens the lead time for developing good design solutions by allowing correct decisions to be made quickly and the first time around. A purpose of axiomatic design is to augment the designer's experience by providing the underlying principles, theories, and methodologies so that they can fully utilize their creativity.

How does human creativity affect the design process? The word "creativity" has been used to describe the human activity that results in ingenious, unpredictable, or unforeseen results (e.g., new products, processes, and systems) that satisfy the needs of society or human aspirations. In this context, creative "solutions" are discovered or derived by inspiration and/or perspiration, often without ever defining specifically what one sets out to create. This creative "spark" or "revelation" may occur because the brain is a huge information storage and processing device that can store data and synthesize solutions through the use of associative memory, pattern recognition, digestion and recombination of diverse facts, and permutations of events.

Sometimes the word "creativity" has been used in a mysterious sense, when we do not understand the process or the logic involved in a given intellectual endeavor (e.g., arts and music), and yet the result of the effort is intellectually, emotionally, or aesthetically appealing and acceptable. A subject is always mysterious when it relies on an implicit thought process that cannot be stated explicitly and explained for others to understand and that can be learned only through experience, apprenticeship, or trial and error. Design has been one of these mysteries, but we must overcome this intellectual and mental barrier.

Design will always benefit when inspiration, creativity, and imagination play a role, but we must augment this process by understanding the nature of human minds and by the development of basic knowledge. Design must become a principle-based subject. The subject of design should attain the same level of intellectual understanding that fields such as thermodynamics and mechanics enjoy. Knowledge from design and other fields should converge and form a continuum of knowledge, rather than remaining as the disparate islands of knowledge that characterize the current situation. Indeed, understanding the design process is one of the most challenging of intellectual quests. The real beneficiaries of structured design knowledge will be humankind, society, industry, and the world.

1.4 WHAT IS THE ULTIMATE GOAL OF AXIOMATIC DESIGN?

Can the field of design be scientific? The ultimate goal of axiomatic design is to establish a scientific basis for design and to improve design activities by providing the designer with a theoretical foundation based on logical and rational thought processes and tools. The goal of axiomatic design is manifold: to make human designers more creative, to reduce the random search process, to minimize the iterative trial-and-error process, to determine the best designs among those proposed, and to endow the computer with creative power through the creation of a scientific base for the design field.

Can we enhance creativity through axiomatic design?[2] Axiomatic design enhances creativity. It demands the clear formulation of design objectives through the establishment of functional requirements (FRs) and constraints (Cs). It provides criteria for good and bad

---

[2] This question was investigated independently by Caldenfors (1998), who showed through a systematic study that students who were taught axiomatic design theory, albeit briefly, performed better than students who were taught other methods.

design decisions to help in eliminating bad ideas as early as possible, enabling designers to concentrate on promising ideas. It also formalizes the decomposition process that enables a systematic flow from creation of concepts to detailed designs.

Can industry benefit through the establishment of a scientific basis for design? In addition to the intellectual reasons given for developing the scientific basis for design, there are also practical reasons. The current design process is both resource intensive and ineffective. Industrial competitiveness demands that industrial firms have strong technical capability in design. These firms are under pressure to shorten the lead time for the introduction of new products, lower their manufacturing cost, improve the quality and reliability of their products, and satisfy the required functions most effectively. The greatest impact on all of these industrial needs rests on the quality and timeliness of developing design solutions. To achieve these practical goals, we must augment the essential ingredients of outstanding designers—knowledge, imagination, experience, and hard work—with science.

Today computers are used in the design field, but they are used primarily for graphic representation, solid modeling, product modeling, and optimization of design solutions. Because computers are becoming ever more powerful and cheaper—lower memory cost, faster number crunching, and smaller physical size—designers should make use of computers as an information storage device and as a design-enhancement tool to augment human capability. This can be done through codification and generalization of design knowledge. The ultimate outcome of design research may be a thinking design machine that should be able to let computers design products (Suh and Sekimoto, 1990).

The modern Renaissance period of the design field is here! The field of design is undergoing an intellectual Renaissance—from the notion that design can be learned only from experience to the idea that it may be amenable to systematic and scientific treatment to enhance the creativity and the experiential elements of design knowledge. This intellectual Renaissance is possible because good design decisions are not as random as they appear to be but are a result of systematic reasoning, the essence of which can be captured and generalized to enhance the design process.

How can axiomatic design help other fields? Axiomatic design (AD) can help many fields of engineering, management, and other intellectual fields. In Chapters 3 and 6, we use AD to solve the scheduling problem more effectively. This type of problem used to be the sole domain of operations research. AD can also make the task of controlling a system—until now, the domain of control theory—simple, reliable, and robust. The lesson is the following:

> *Hardware, software, and systems must be designed right to be controllable, reliable, manufacturable, productive, and otherwise achieve their goals. The performance of poorly designed hardware, software, and systems can rarely be improved through subsequent corrective actions.*

1.5 ROLE OF AXIOMS IN DEVELOPMENT OF SCIENCE AND TECHNOLOGY: A HISTORICAL PERSPECTIVE

History indicates that the Greeks were the first to use axioms. Perhaps the oldest example of the use of axioms is Euclid's geometry, which was created to measure distances. These axioms have had powerful effects on creating the modern mathematical base for manufacturing, navigation, and nearly all fields of science and technology.

Axioms are truths that cannot be derived but for which there are no counterexamples or exceptions. Many fields of science and technology owe their advances to the development and existence of axioms. They have gone through the transition from experience-based practices to the use of scientific theories and methodologies that are based on axioms. In that sense, axiomatic design follows a historical trend in science and technology development.

The role of axioms in natural science. Axioms have played a major role in developing natural science, which includes fields such as physics, chemistry, and biology. These fields deal with energy, matter, living organisms, and their transformations and interrelations. In these fields, axioms were more easily accepted because the predictions that were made based on the axioms could be compared with quantitative and objectively measurable natural phenomena.

Some of the conservation laws, such as the first and second laws of thermodynamics, are axioms. The first law of thermodynamics, which states that energy is conserved, is believed to be true because no observations or measurements contradict either the law or predictions based on the law. It defines the universal concept of energy for all sorts of diverse situations and matter. Similarly, the second law of thermodynamics was not derived. It is a generalization of the commonly observed fact that no net mechanical work could be done by a heat engine unless it exchanges heat with two other bodies. It is also an axiom in that it is believed to be a universal truth for which there are no counterexamples or exceptions. Based on the second law of thermodynamics, the concept of entropy could be derived. The predictions made using the second law of thermodynamics have been found to be consistent with the measurements of natural phenomena. No attempt to violate the second law of thermodynamics has been successful. For example, no one, to date, has devised a perpetual motion machine.

The scientific field of thermodynamics was born as a result of many people attempting to generalize how "good steam engines" work. Before the field of thermodynamics emerged, many people might have said that the steam engine was too complicated to explain and that it could be designed only by experienced, ingenious designers and through trial-and-error processes. Indeed, that might be the reason that the Newcomen engine, which was invented in 1705, had been used for 64 years to pump water out of mines before James Watt realized its shortcomings and invented the Watt engine in 1769. The invention of Watt's steam engine eventually led to the Industrial Revolution. Although thermodynamic axioms were not available to help James Watt, these technological developments gave impetus for the development of the science known as thermodynamics.

Even the law of conservation of matter is an axiom. All measurements seem to support this conservation law, although most engineering measurements do not take into account the mass change associated with transport of photons, electrons, and/or ions from and to the mass during the measurement.

Sir Isaac Newton (1642–1727) changed the world of science and technology by the work he did in the 18 months between the ages of 23 and 25. He developed differential and integral calculus, determined the nature of colors, and calculated the gravitational force that holds the moon in its orbit based on his axioms for mechanics. Newtonian mechanics laid the foundation for modern science and engineering.

Newton formulated three laws or axioms of mechanics. The first law states that if there is no force acting on a body, it will remain at rest or move with constant velocity in a straight line. The second law states that the product of mass and acceleration is equal to the force acting on the body. The third law states that the force that one body exerts on another must always be equal in magnitude and opposite in direction to the force that the second body exerts on the first. These were axioms. Newton's laws established the concept of force.

To prove the validity of these laws or axioms, Newton applied all three of his laws to the motion of planets around the sun. In fact, he needed all three axioms to predict Kepler's three laws of planetary motion, which were developed by Kepler based on his own measurements of the planetary motions as well as those of Tycho Brahe. The measurements gave the radial acceleration of planets, from which Newton could determine the centripetal force acting between the sun and the planets. His third law implied that the same force is acting on the sun and the planets but in the opposite direction. Newton predicted Kepler's three laws of planetary motion based on his own three laws. From this work, he could determine the gravitational force acting between two masses. Newton's three laws are universally accepted because they predict observed natural phenomena and physical measurements. Before Newton explained Kepler's laws with his law for the gravitational force between masses and his more general laws of motion, the motion of planets and other objects had been a mystery.

Although we cannot derive Newton's laws, neither can we come up with exceptions or counterexamples to them. In other words, Newton's laws are axioms because they could not be derived or proven, but there were no exceptions or counterexamples until Einstein advanced the theory of relativity, which has placed a limit on Newton's laws. We now know that Newton's laws are valid when the relativistic effects are small. Einstein's theory of relativity is also an axiom that overcomes the shortcomings of Newton's laws when the relativistic effects are important. Einstein's theory of relativity is also based on an axiom that the speed of light does not depend on the choice of the coordinate system.

These examples show that the development of natural science has been possible because of the advent of important axioms or laws that could generalize the behavior of nature. The validity of these axioms is tested by comparing the theoretical predictions of given phenomena with experimental measurements, by testing hypotheses based on these axioms, and by analysis of observed phenomena using the axioms.

Natural science is not built solely on these axioms. It is also built on other natural laws, which are often based on experimental observations and measurements, such as Maxwell's equations, Boyle's law, and the Stefan–Boltzmann equations. Natural science is also built on discoveries such as the structure of DNA and the periodic nature of elements. All these laws, discoveries, and axioms have provided the foundation for science, engineering, and technology.

How does technology influence science? It seems that axioms played a key role when technology led to the development of science. Is that true? It is well known that scientific discoveries have led to the development of various technologies. The modern biotechnology revolution owes its existence to the identification of the DNA structure by Crick and Watson (Watson, 1969). However, the converse is also true but less well known; technology often has preceded and led to the establishment of scientific fields. Thermodynamics is a well-known example; so is information theory, which was created through an attempt to systematize telecommunications technology. Information theory now finds applications in many fields of science. Axiomatic design is similarly an example of how the technology of design has led to the science of design.

How were the design axioms created? The design axioms were created by identifying the common elements that are present in all good designs. The following questions were asked:

- How did I make such a big improvement in a process?
- How did I create the process?
- What are the common elements in good designs?

Once the common elements could be stated, they were reduced to two axioms through a logical reasoning process. The historical account of how the design axioms were developed in the mid-1970s is given in *The Principles of Design* (Suh, 1990).

1.6 AXIOMATIC APPROACH VERSUS ALGORITHMIC APPROACH

Is there any other way of approaching the subject of design? There are two ways to deal with design: axiomatic and algorithmic. In an ideal world, the development of knowledge should proceed from axioms to algorithms to tools. However, in recent years many algorithms have been advanced ad hoc without the benefit of basic principles.

In purely algorithmic design, we try to identify or prescribe the design process, so in the end, the process will lead to a design embodiment that satisfies the design goals. Generally, the algorithmic approach is founded on the notion that the best way of advancing the design field is to understand the design process by following the best design practice. Algorithmic design is ad hoc for specific situations. For example, design for assembly (DFA) and design for manufacturability (DFM) techniques are algorithmic methods. It is difficult to come up with design algorithms for all situations, especially at the highest conceptual level, and it may not be desirable. Algorithms are generally useful at the level of detail design, i.e., design for assembly, because they are manageable.

Algorithmic methods can be divided into several categories:

- pattern recognition
- associative memory
- analogy
- experientially based prescription
- extrapolation/interpolation
- selection based on probability

Some of these techniques can be effective if the design has to satisfy only one functional requirement, but when many functional requirements must be satisfied at the same time, they are less effective. Axioms provide the boundaries within which these algorithms are valid, in addition to providing the general principles.

The axiomatic approach to any subject begins with a different premise: that there are generalizable principles that govern the underlying behavior of the system being investigated. The axiomatic approach is based on the *abstraction* of good design decisions and processes. As stated earlier, axioms are general principles or self-evident truths that cannot be derived or proven to be true, but for which there are no counterexamples or exceptions. Axioms generate new abstract concepts, such as force, energy, and entropy that are results of Newton's laws and thermodynamic laws.

The axiomatic approach to design is powerful and will have many ramifications because of the generalizability of axioms, from which corollaries and theorems can be derived. These theorems and corollaries can be used as design rules that prescribe precisely the bounds of their validity because they are based on axioms. Design axioms apply to many different kinds of problems and issues, as shown in this book.

What is the relationship between design process and design axioms? In many fields of learning, both the *process* of how something is done and the *abstraction* that can generalize the underlying principles are equally important. For example, when we teach small children the notion of numbers, we do it by counting our fingers—by counting from thumb to pinkie, for example. This is done to teach the *process* of counting, which has the flavor of being algorithmic. However, if we keep starting the process of counting using our thumb and ending up with the pinkie, the child may think that thumb is called "one" and pinkie is called "five." Therefore, we use toes and other objects to teach the notion of numbers through abstraction of the counting process. In design, we also need to do both. We need to teach both the process and the abstracted concept, both the theory of what is a good design and how to develop good designs.

1.7 AXIOMATIC DESIGN FRAMEWORK

1.7.1 The Concept of Domains

The design world consists of four domains. What are domains? Design involves an interplay between "what we want to achieve" and "how we choose to satisfy the need (i.e., the what)." To systematize the thought process involved in this interplay, the concept of *domains* that create demarcation lines between four different kinds of design activities provides an important foundation of axiomatic design.

The world of design is made up of four domains: the *customer domain,* the *functional domain*, the *physical domain*, and the *process domain*. The domain structure is illustrated schematically in Figure 1.2. The domain on the left relative to the domain on the right represents "what we want to achieve," whereas the domain on the right represents the design solution, "how we propose to satisfy the requirements specified in the left domain."

The *customer domain* is characterized by the needs (or attributes) that the customer is looking for in a product or process or systems or materials. In the *functional domain*, the customer needs are specified in terms of *functional requirements* (FRs) and *constraints* (Cs). To satisfy the specified FRs, we conceive *design parameters* (DPs) in the *physical domain*. Finally, to produce the product specified in terms of DPs, we develop a process that is characterized by *process variables* (PVs) in the *process domain*.

For example, a customer in the semiconductor industry needs to coat the surface of a silicon wafer with photoresist. This assessment of customer need is done in the customer domain. Based on this need, the engineers in an equipment company establish the FRs in terms of thickness and uniformity and also the Cs in terms of a tolerable level of contaminant particles, production rate, and cost. Of course, the engineers must seek input from their customers during this process. This formulation of the FRs is done in the functional domain. Then the equipment designer, based on experimental data and past experience, must conceive a design solution and identify the important DPs in the

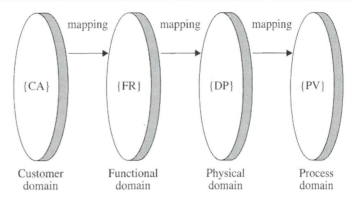

Figure 1.2 Four domains of the design world. The {x} are the characteristic vectors of each domain. During the design process we go from the domain on the left to the domain on the right. The process is iterative in the sense that the designer can go back to the domain on the left based on the ideas generated in the right domain.

physical domain. The designer might choose to coat the surface with photoresist by spraying it onto the disk (for example, one DP could be a sprayer) and then control the thickness by spinning the disk at high speed to make use of centrifugal force (another DP could be a spinner). Then the manufacturing engineer in the process domain must conceive the means of manufacturing the equipment, specifying PVs, such as the means of controlling the delivery of the photoresist, rotational speed, temperature, etc., that can provide the DPs.

In mechanical engineering we usually think of design in terms of product design and often hardware design. However, engineers also deal with other equally important designs, such as software, manufacturing processes, systems, and organizations. All designers go through similar thought processes regardless of the specific goal of their effort, although some believe that their design is unique and different from the designs of everyone else.

In materials science, the design goal is to develop materials with certain properties (i.e., FRs). This is done through the design of microstructures (i.e., DPs) to satisfy these FRs, and through the development of material processing methods (i.e., PVs) to create the desired microstructures. In business, after business goals {FRs} are established, the next task is to design the business structure, organization, and strategy {DPs} to meet the business goals, and then to find human and financial resources {PVs} to staff and operate the business. Similarly, universities must define the mission of their institutions (i.e., FRs), design their organizations effectively to have an efficient educational and research enterprise (i.e., DPs), and deal with human and financial resource issues (i.e., PVs). In the case of the U. S. government, the President must define the right set of FRs, design the government organization and programs (DPs) to achieve these FRs, and secure the resources necessary to get the job done (PVs), subject to the constraints imposed by the Constitution and Congress. In all organizational design, the process domain represents the human and financial resources.

Table 1.1 shows how all these seemingly different design tasks in many different fields can be described in terms of the four design domains. In the case of *product* design, the customer domain consists of the needs or attributes that the customer is looking for in a product. The functional domain consists of FRs, often defined as engineering specifications

TABLE 1.1 Characteristics of the Four Domains of the Design World for Various Designs: Manufacturing, Materials, Software, Organizations, Systems, and Business

	Customer Domain {CA}	Functional Domain {FR}	Physical Domain {DP}	Process Domain {PV}
Manufacturing	Attributes that customers desire	Functional requirements specified for the product	Physical variables that can satisfy functional requirements	Process variables that can control design parameters
Materials	Desired performance	Required properties	Microstructure	Processes
Software	Attributes desired in the software	Output specification of program codes	Input variables Algorithms Modules Program codes	Subroutines Machine codes Compilers Modules
Organizations	Customer satisfaction	Functions of the organization	Programs Offices Activities	People and other resources to support programs
Systems	Attributes desired of the overall system	Functional requirements of the system	Machines Components Subcomponents	Resources (human, financial, etc.)
Business	ROI	Business goals	Business structure	Human and financial resources

and constraints. The physical domain is the domain in which the key DPs are chosen to satisfy the FRs. Finally, the process domain specifies the manufacturing PVs that can produce the DPs.

The design issues in the organizational design of an academic department are illustrated in terms of the design domains in Table 1.2.

All designs fit into these four domains. Therefore, all design activities, be they product design or software design, can be generalized in terms of the same principles. Because of this logical structure of the design world, the generalized design principles can be applied to all design applications and we can consider all design issues that arise in the four domains systematically and, if necessary, concurrently.

1.7.2 Definitions

Is it important that we adhere to certain definitions of key words? Before proceeding with any further discussion of axiomatic design, it is important to provide the definition of a few key words discussed in the preceding section, as axioms are valid only within the bounds established by the definitions of these key terms. Just as the words *heat* and *work* have unique meanings in thermodynamics, which are different from those of their daily usage, so is the case with key words used in axiomatic design. The definitions are as follows:

Axiom: Self-evident truth or fundamental truth for which there are no counterexamples or exceptions. An axiom cannot be derived from other laws or principles of nature.

Corollary: Inference derived from axioms or from propositions that follow from axioms or from other propositions that have been proven.

TABLE 1.2 Four Domains of an Academic Department

Customer Domain	*Functional Domain*	*Physical Domain*	*Process Domain*
CA₁: Customer Satisfaction	**FR₁: Quality**	**DP₁: Programs**	**PV₁: Academic People**
CA₁₁ Undergraduates	FR₁₁ Provide quality undergraduate education	DP₁₁ Undergraduate program	PV₁₁ Strong involvement of faculty
CA₁₂ Graduates	FR₁₂ Provide quality graduate education	DP₁₂ Graduate program	PV₁₂ Academically strong graduate students
CA₁₃ Research Sponsors	FR₁₃ Conduct trend-setting quality research	DP₁₃ Research organization	PV₁₃ Strong faculty
CA₁₄ Public (society at large)	FR₁₄ Promote active participation in public activities	DP₁₄ Service function	PV₁₄ Active support of external activities of faculty
CA₂: Cash Flow	**FR₂: Good Management of Resources**	**DP₂: Administrative Mechanisms**	**PV₂: Administrative People**
CA₂₁ Teaching support	FR₂₁ Use the general fund effectively	DP₂₁ Budget and planning mechanism	PV₂₁ Budget officer
CA₂₂ Research support	FR₂₂ Generate external research support	DP₂₂ Research support infrastructure	PV₂₂ Support staff for research
CA₂₃ Capital investment	FR₂₃ Solicit gifts	DP₂₃ Fund-raising mechanisms	PV₂₃ Department head and faculty fund generators
CA₂₄ Human resource "protection"	FR₂₄ Create chairs, support, etc.	DP₂₄ Incentive system	PV₂₄ Department head and association head
CA₃: Profit	**FR₃: Productivity (Intellectual and Financial)**	**DP₃: Means**	**PV₃: Methods**
CA₃₁ Better teaching paradigms	FR₃₁ Create effective pedagogical tools	DP₃₁ Development of textbooks, videotapes	PV₃₁ Support and reward mechanisms
CA₃₂ Research infrastructure	FR₃₂ Develop labs and centers	DP₃₂ (Better) research organizations	PV₃₂ Establish interdisciplinary research activities
CA₃₃ New inventions and discoveries	FR₃₃ Promote scholarship and creative activities: patents, monographs, prizes and awards	DP₃₃ Active support, promotion, and nomination	PV₃₃ Staff support
CA₃₄ Better tools (equipment/facilities)	FR₃₄ Secure equipment and facilities	DP₃₄ Investment in capital goods	PV₃₄ Fund raising
CA₃₅ Outstanding graduates: captains of industry, researchers, professors, government officials	FR₃₅ Provide mentorship	DP₃₅ Stronger faculty/student interaction	PV₃₅ "Research teams" and commencement of thesis work at sophomore level
CA₄: Growth (Intellectual and Physical)	**FR₄: Innovation**	**DP₄: Environment/ Culture**	**PV₄: Resources**
CA₄₁ Define ME of the twenty-first century	FR₄₁ Create new pedagogical tools and disciplines	DP₄₁ Creative, experimental educational programs	PV₄₁ Faculty time and financial support
CA₄₂ Define engineering of the twenty-first century	FR₄₂ Pioneer new engineering tools, methods, and books	DP₄₂ Active interaction with industry	PV₄₂ "Manufacturing Institute"
CA₄₃ Shape the society of the twenty-first century	FR₄₃ Solve societal problems	DP₄₃ Active interaction with industry, government	PV₄₃ External participation
CA₄₄ Strengthen the human resource of engineering	FR₄₄ Entice minorities and women into engineering	DP₄₄ Special programs	PV₄₄ Financial resources

Theorem: A proposition that is not self-evident but that can be proved from accepted premises or axioms and so is established as a law or principle.

Functional requirement: Functional requirements (FRs) are a minimum set of independent requirements that completely characterizes the functional needs of the product (or software, organizations, systems, etc.) in the functional domain. By definition, each FR is independent of every other FR at the time the FRs are established.

Constraint: Constraints (Cs) are bounds on acceptable solutions. There are two kinds of constraints: *input* constraints and *system* constraints. Input constraints are imposed as part of the design specifications. System constraints are constraints imposed by the system in which the design solution must function.

Design parameter: Design parameters (DPs) are the key physical variables (or other equivalent terms in the case of software design, etc.) in the physical domain that characterize the design that satisfies the specified FRs.

Process variable: Process variables (PVs) are the key variables (or other equivalent term in the case of software design, etc.) in the process domain that characterize the process that can generate the specified DPs.

Most of the key words listed are associated with the Independence Axiom, discussed below. Additional definitions of key words associated with the Information Axiom will be given in a later section. The significance of these definitions should become clearer through the examples given throughout this book.

1.7.3 Mapping from Customer Needs to Functional Requirements

The customer needs or attributes (CAs) desired in a product are sometimes difficult to define or are vaguely defined. Nevertheless we have to do the best we can to understand the customer needs by working with customers to define them. The rule is to ask the right questions to the right customers at the right time.

There are several formal means of rank-ordering the customer needs. A formal methodology called quality function deployment (QFD) for determining CAs was developed by Fuji Xerox and introduced into the United States by Clausing (1994). Toyota introduced the idea of the House of Quality to identify and prioritize the customer needs, relate them to engineering characteristics, benchmark them against competitors' products, establish important engineering characteristics, and select the important areas of improvement. There are other techniques such as the Pugh matrix, the analytical hierarchy process (AHP), the Taguchi loss function, and the weighted sum of product attributes. These methodologies are most useful in improving an existing product, but not very effective in developing innovative products. In the latter case, it is best to remember that the FRs must be defined in a solution-neutral environment.

Notwithstanding the importance of customer input in determining FRs, they cannot always be determined by asking the customers for their preference. Arrow's impossibility theorem states that, in general, the preference indicated by individuals in a group does not have any value in determining the preference of the group as a whole. This is shown in the following example (Hazelrigg, 1998).

EXAMPLE 1.2 Arrow's Impossibility Theorem

Consider the case of having three choices, A, B, and C. Three people were asked to indicate their preference among these three choices. Based on the input from these individuals, can we make a decision as to what the group as a whole prefers?

The answer is "No." The following table lists the preferences indicated by Smith, Kim, and Stein (in Hazelrigg, 1998):

Individuals	Preferences	Choices		
		A vs. B	B vs. C	A vs. C
Smith	A > B > C, A > C	A	B	A
Kim	B > C > A, B > A	B	B	C
Stein	C > A > B, C > B	A	C	C
Group preference		A > B	B > C	C > A

The results show that the group is confused as to what it wants. It prefers A over B, and B over C, but it prefers C over A rather than A over C, as one might have expected from the first two choices.

Once we identify and define the perceived customer needs (or the attributes the customer is looking for in a product), these needs must be translated to FRs. This must be done within a "solution-neutral environment." That means that the FRs must be defined without ever thinking about something that has been already designed or what the design solution should be. If FRs are defined based on an existing design, then we will simply be specifying the FRs of that product, and the result of the design endeavor will be likely to be similar to the existing product, forestalling creative thinking by introducing personal bias.

In the translation of customer needs to FRs, the usage range of the design must be considered, even though the customer will almost never specify such a range. That is, the FRs must be stated with the anticipated environmental variation, customer usage variation, and targeted useful life before degradation as requirements of the design. If such "noise" variables are represented as requirements of the design, then the accommodation to handle them will be an integral property of the design. Thus the design will be inherently robust to such sources of variation, and the optimization process will be significantly simplified.

Industrial firms often use "marketing requirement specification" (MRS)[3] as the product specification document. MRS documents are often very thick, for which the primary inputs are provided by the marketing people. Usually the document is a random mixture of CAs,

---

[3] Sometimes MRS is referred to as "manufacturing requirement statement."

FRs, Cs, DPs, and PVs. When the marketing group specifies DPs and PVs, the design process becomes complicated because they constrain the design options and the designers lose the freedom to come up with innovative design solutions. Marketing input is most useful in establishing CAs, FRs, and Cs, but their prescription of DPs and PVs can constrain the creative process of design.

During the mapping process (for example, going from the functional domain to the physical domain), we must make the right design decisions using the Independence Axiom. When several designs that satisfy the Independence Axiom are available, the Information Axiom can be used to select the best design. When only one FR is to be satisfied by having an acceptable DP, the Independence Axiom is always satisfied and the Information Axiom is the only axiom the one-FR design must satisfy. This is the subject of Chapter 2.

So what are the design axioms? How many design axioms are there? The basic postulate of the axiomatic approach to design is that there are fundamental axioms that govern the design process. Two axioms were identified by examining the common elements that are always present in good designs. They were also identified by examining actions taken during the design stage that resulted in dramatic improvements.[4]

The first axiom is called the *Independence Axiom*. It states that the independence of functional requirements (FRs) must always be maintained, where FRs are defined as *the minimum set of independent requirements* that characterizes the design goals.

The second axiom is called the *Information Axiom*, and it states that among those designs that satisfy the Independence Axiom, the design that has the smallest information content is the best design. To make the design work, information must be supplied—by the user of the design output or by other means—when the information content is finite. Because the information content is defined in terms of probability, the second axiom also states that the design that has the highest probability of success is the best design. Based on these design axioms, we can derive theorems and corollaries. The axioms are formally stated as:

Axiom 1: The Independence Axiom. *Maintain the independence of the functional requirements* (FRs).

Axiom 2: The Information Axiom. *Minimize the information content of the design.*

As many case studies presented in this book show, the performance, robustness, reliability, and functionality of products, processes, software, systems, and organizations are significantly improved when these axioms are satisfied. Conversely, machines and processes that are not working well can be analyzed to determine the causes of their dysfunction or malfunction and to solve the problems based on the design axioms.

1.7.4 The First Axiom: The Independence Axiom

As stated earlier, the FRs are defined as the minimum set of independent requirements that the design must satisfy. A set of FRs is the description of design goals. The Independence Axiom states that when there are two or more FRs, the design solution must be such

---

[4] Suh (1990) gives the history of how the design axioms were created.

that each one of the FRs can be satisfied without affecting the other FRs (*note:* review the refrigerator door example involving two FRs discussed in Example 1.1). That means we have to choose a correct set of DPs to be able to satisfy the FRs and maintain their independence.

The Independence Axiom is often misunderstood. Many people confuse functional independence with physical independence. The Independence Axiom requires that the *functions* of the design be independent from each other, not the physical parts. The second axiom would suggest that physical integration is desirable to reduce the information content if the functional independence can be maintained. This is illustrated using the beverage can as an example.

EXAMPLE 1.3 Beverage Can Design

Consider an aluminum beverage can that contains carbonated drinks. How many FRs must the can satisfy? How many physical parts does it have? What are the DPs? How many DPs are there?

SOLUTION

According to an expert working at one aluminum can manufacturer, there are 12 FRs for the can. Plausible FRs: contain axial and radial pressure, withstand a moderate impact when the can is dropped from a certain height, allow stacking on top of each other, provide easy access to the liquid in the can, minimize the use of aluminum, be printable on the surface, and others. However, these 12 FRs are not satisfied by 12 physical pieces; the aluminum can consists of only three pieces: the body, the lid, and the opener tab (see Figure E1.3). The Independence Axiom requires that the 12 FRs remain independent of one another by proper choice of DPs. The Information Axiom requires that the information content be minimum, which can be done if the DPs can be integrated so that there will not be 12 physical pieces making up the can!

Where are the DPs? To satisfy the Independence Axiom (Theorem 4 of Appendix 1A and further discussed in Chapter 3), there must be at least 12 DPs. Most of the DPs are associated with the geometry of the can: the thickness of the can body, the curvatures at the bottom of the can, the reduced diameter of the can at the top to reduce the material used to make the top lid, the corrugated geometry of the opening tab to increase the stiffness, the small extrusion on the lid to attach the tab, etc.

According to the engineer who improved the can design after taking the axiomatic

Figure E.1.3 Beverage can.

design course at MIT, the aluminum can now has 12 DPs integrated in three physical pieces and satisfies the Independence Axiom.

After we define the FRs, we must conceptualize a solution. When and how does conceptualization take place during the design process? After the FRs are established, the next step in the design process is the conceptualization process, which occurs during the mapping process going from the functional domain to the physical domain.

To go from "what" to "how" (from the functional domain to the physical domain) requires *mapping*, which involves creative conceptual work. After the overall design concept is generated by mapping, we must identify the DPs and complete the mapping process. During this process, we must think of all the different ways of fulfilling each of the FRs by identifying plausible DPs. Sometimes it is convenient to think about a specific DP to satisfy a specific FR, repeating the process until the design is completed. Databases of all kinds (generated through brainstorming, morphological techniques, etc.), analogy from other examples (apparently Thomas Edison's favorite means of invention), extrapolation and interpolation, laws of nature, order-of-magnitude analysis, and reverse engineering (copying somebody else's good idea by examining an existing product) can be used. It is relatively easy to identify a DP for a given FR, but when there are many FRs that we must satisfy, the design task becomes difficult and many designers make mistakes by violating the Independence Axiom. This is the subject of Chapter 3, which is on multi-FR design.

If it is a mapping process, shouldn't we be able to write design equations? The mapping process between the domains can be expressed mathematically in terms of the characteristic vectors that define the design goals and design solutions. At a given level of the design hierarchy, the set of functional requirements that defines the specific design goals constitutes the FR vector in the functional domain. Similarly, the set of design parameters in the physical domain that has been chosen to satisfy the FRs constitutes the DP vector. The relationship between these two vectors can be written as

$$\{FR\} = [A]\{DP\} \tag{1.1}$$

where $[A]$ is called the *design matrix* that characterizes the product design. Equation (1.1) is a design equation for the design of a product. The design matrix is of the following form for a design that has three FRs and three DPs:

$$[A] = \begin{bmatrix} A_{11} & A_{12} & A_{13} \\ A_{21} & A_{22} & A_{23} \\ A_{31} & A_{32} & A_{33} \end{bmatrix} \tag{1.2}$$

When Equation (1.1) is written in a differential form as

$$\{dFR\} = [A]\{dDP\}$$

the elements of the design matrix are given by

$$A_{ij} = \partial FR_i / \partial DP_j$$

With three FRs and three DPs, Equation (1.1) may be written in terms of its elements as

$$\text{FR}_i = \sum_{j=1}^{3} A_{ij}\text{DP}_j$$

or

$$\text{FR}_1 = A_{11}\text{DP}_1 + A_{12}\text{DP}_2 + A_{13}\text{DP}_3$$

$$\text{FR}_2 = A_{21}\text{DP}_1 + A_{22}\text{DP}_2 + A_{23}\text{DP}_3 \qquad (1.3)$$

$$\text{FR}_3 = A_{31}\text{DP}_1 + A_{32}\text{DP}_2 + A_{33}\text{DP}_3$$

In general,

$$\text{FR}_i = \sum_{j=1}^{n} A_{ij}\text{DP}_j$$

where n = the number of DPs.

For a linear design, A_{ij} are constants; for a nonlinear design, A_{ij} are functions of the DPs. There are two special cases of the design matrix: the diagonal matrix and the triangular matrix. In the diagonal matrix, all $A_{ij} = 0$ except those where $i = j$.

$$[A] = \begin{bmatrix} A_{11} & 0 & 0 \\ 0 & A_{22} & 0 \\ 0 & 0 & A_{33} \end{bmatrix} \qquad (1.4)$$

In the lower triangular (LT) matrix, all upper triangular elements are equal to zero, as shown below.

$$[A] = \begin{bmatrix} A_{11} & 0 & 0 \\ A_{21} & A_{22} & 0 \\ A_{31} & A_{32} & A_{33} \end{bmatrix} \qquad (1.5)$$

In the upper triangular (UT) matrix, all lower triangular elements are equal to zero.

For the design of processes involving mapping from the DP vector in the physical domain to the PV vector in the process domain, the design equation may be written as

$$\{DP\} = [B]\{PV\} \qquad (1.6)$$

where $[B]$ is the design matrix that defines the characteristics of the process design and is similar in form to $[A]$.

How do we know whether the Independence Axiom is satisfied? To satisfy the Independence Axiom, the design matrix must be either diagonal or triangular. When the design matrix $[A]$ is diagonal, each of the FRs can be satisfied independently by means of one DP. Such a design is called an *uncoupled* design. When the matrix is triangular, the independence of FRs can be guaranteed if and only if the DPs are determined in a proper sequence. Such a design is called a *decoupled* design. Any other form of the design matrix is called a full matrix and results in a *coupled* design. Therefore, when several FRs must be satisfied, we must develop designs that will enable us to create either a diagonal or a triangular design matrix.

EXAMPLE 1.4 Design Matrix for Refrigerator Door

Reconsider the refrigerator door design discussed in Example 1.1. The FRs[5] of the door are

FR$_1$ =   Provide access to the items stored in the refrigerator.
FR$_2$ =   Minimize energy loss.

Determine the DPs and the design matrix. Show whether the design is uncoupled, decoupled, or coupled.

SOLUTION

The DPs are chosen as

DP$_1$ =   Vertically hung door
DP$_2$ =   Thermal insulation material in the door

The design matrix may be stated as

$$\begin{Bmatrix} FR_1 \\ FR_2 \end{Bmatrix} = \begin{bmatrix} X & 0 \\ X & X \end{bmatrix} \begin{Bmatrix} DP_1 \\ DP_2 \end{Bmatrix}$$

The door design is a decoupled design. Unfortunately, it is still not a good design because the thermal insulation (typically polyurethane foam with freon) put in the door cannot compensate for the cold air that escapes the refrigerator whenever the door is opened.

One solution is to make a horizontally hung door like those found in chest freezers. In this case, the cold air stays inside when the door is opened. The design matrix would then provide an uncoupled design as shown in the following equation.

$$\begin{Bmatrix} FR_1 \\ FR_2 \end{Bmatrix} = \begin{bmatrix} X & 0 \\ 0 & X \end{bmatrix} \begin{Bmatrix} DP_1 \\ DP_2 \end{Bmatrix}$$

The elements of the design matrices [A] and [B] can be either constants or functions. If the matrix elements are constants, the design is linear. If the elements are functions of DPs, the design may be nonlinear. Many examples that illustrate nonlinear designs will be given in later chapters.

Can we use mathematical techniques to transform a full matrix to a diagonal or a triangular matrix? The design matrix is a second-order tensor just as stress, strain, and moment of inertia are. However, there is one big difference between the design matrix and these other second-order tensors. These other tensors can be changed through coordinate transformation to convert any matrix into a diagonal matrix. The diagonal elements of the diagonal matrix are invariant, such as the principal stresses in the case of a stress tensor. However, the coordinate transformation technique cannot be applied to design equations to find the invariant (i.e., the diagonal) matrix, as the design matrix [A] typically involves physical things that are not amenable to coordinate transformation. In other words,

---

[5] It should be noted here that FRs are stated in the imperative starting with verbs, whereas DPs usually are stated with nouns.

we can always mathematically transform any design matrix into a diagonal matrix, but the diagonal elements may not have any physical significance. In the case of software design and organizational design, the concept of coordinate transformation clearly has no mathematical basis.

Furthermore, when the matrix is a full matrix producing a coupled design, we may get a unique solution that gives the right values for FRs, but such a design generates many problems. For example, when one of the FRs is changed, all DPs must be changed. Also whenever the DPs are not exact and deviate from the desired (or set) values, the FRs may not be satisfied. Because most manufacturing processes cannot make exactly identical parts, the system or product may be individually tuned or calibrated. Also in situations in which all the elements of the design matrix cannot be precisely specified, FRs cannot be determined. Furthermore, if one of the DPs changes during the life of the product or process (say by wear), then the machine must be discarded unless all other DPs are changed accordingly.

What are constraints? The design goals are often subject to Cs. Constraints provide bounds on the acceptable design solutions and differ from the FRs in that they do not have to be independent.

There are two kinds of constraints: *input* constraints and *system* constraints. Input constraints are specific to the overall design goals (i.e., all designs that are proposed must satisfy these). System constraints are specific to a given design; they are the result of design decisions made.

The designer often has to specify input constraints at the beginning of the design process, because the designed product (or process or system or software or organization) must satisfy external boundary conditions such as the voltage and the maximum current of the power supply. The environment within which the design must function may also impose many constraints. All of these constraints must be satisfied by all proposed design embodiments regardless of the specific details of the design.

Some constraints are generated because of design decisions made as the design proceeds. All higher level decisions act as constraints at lower levels. For example, if we have chosen to use a diesel engine in a car, all subsequent decisions related to the vehicle must be compatible with this decision. These are system constraints.

Often it is best to treat cost as a constraint. Cost is affected by all design changes, and therefore cost cannot be made independent of other FRs in an uncoupled design. If it is decided that cost must be a functional requirement, then the best we can do is to develop a decoupled design, which also satisfies the Independence Axiom. With cost as a constraint, the design is acceptable as long as the cost does not exceed a set limit. When manufacturing cost and selling price of the product are not related, price can be treated as an FR.

How do we create a design hierarchy through zigzagging? FRs and DPs (as well as PVs) must be decomposed into a hierarchy until we get a complete detailed design or until the design is completed. However, contrary to conventional wisdom on decomposition, they cannot be decomposed by remaining in one domain. One must zigzag between the domains to be able to decompose the FRs, DPs, and PVs. Through this zigzagging we create hierarchies for FRs, DPs, and PVs in each design domain.

Zigzagging to decompose FRs, DPs, and PVs and to create their hierarchies is an important part of axiomatic design. For example, when we wish to design a vehicle, the highest level FRs may be four: FR_1 = go forward, FR_2 = go backward, FR_3 = stop, and FR_4 = turn. We cannot decompose these FRs until we decide on DPs that can satisfy the

FRs by going into (i.e., "zigging") the physical domain. After we decide how we are going to satisfy FR_1 by choosing a DP_1, we can go back (i.e., "zag") to the functional domain and decompose FR_1. For example, FR_1 cannot be decomposed without deciding first in the physical domain "how we propose to go forward." If we choose a horse and buggy as a DP for moving forward, the next layer of FRs will be different from those if an automobile is chosen as the DP for moving forward.

In other words, to create FR, DP, and PV hierarchies, we must map from the domain on the left (the "what" domain) into the domain on the right (the "how" domain), and then come back to the domain on the left to generate the next level in the domain on the right, etc. The decomposition and the design hierarchies will be discussed further in Section 1.7.6.

In many cases, the customer needs or attributes cannot and need not be decomposed, as they are often expressed in terms of the highest level needs. For instance, average customers do not care about the details of automobiles as long as they satisfy their perceived needs for acceleration, steering, braking, traction on slippery roads, appearance, and overall comfort. However, in the future, more products must be customized to satisfy individual tastes and preferences. This can be done by zigzagging between the customer domain and the functional domain so that customers can choose their desired functions among those FRs available at the manufacturer's factories. This is further discussed in Chapter 8.

1.7.5 Ideal Design, Redundant Design, and Coupled Design: A Matter of Relative Numbers of DPs and FRs

So far the discussion of the Independence Axiom has dealt with cases in which the number of FRs and the number of DPs were the same. The coupled design was characterized by the existence of a full design matrix, whereas a diagonal matrix and a triangular matrix characterized the designs that satisfy the Independence Axiom, i.e., the uncoupled design and the decoupled design. The question addressed in this section is: What happens if the number of design parameters is more than or less than the number of functional requirements?

Depending on the relative numbers of DPs and FRs, the design can be classified as coupled, redundant, or ideal.

CASE 1 Number of DPs < Number of FRs: Coupled Design

When the number of DPs is less than the number of FRs, we always have a coupled design. This is stated as Theorem 1, which is given below:

Theorem 1 (Coupling Due to Insufficient Number of DPs)

When the number of DPs is less than the number of FRs, either a coupled design results or the FRs cannot be satisfied.

Proof

Suppose that there are three FRs to be satisfied and the designer has proposed a design with only two DPs. Then, the design equation may be written as

$$\begin{Bmatrix} FR_1 \\ FR_2 \\ FR_3 \end{Bmatrix} = \begin{bmatrix} X & 0 \\ 0 & X \\ A_{31} & A_{32} \end{bmatrix} \begin{Bmatrix} DP_1 \\ DP_2 \end{Bmatrix}$$

If A_{31} and A_{32} are zeros, FR_3 cannot be satisfied. If either A_{31} or A_{32} is not zero, the design is a coupled design.

CASE 2 Number of DPs > Number of FRs: Redundant Design

When there are more DPs than there are FRs, the design is called a redundant design. A redundant design may or may not violate the Independence Axiom as illustrated below.

Consider the following two-dimensional case:

$$\begin{Bmatrix} FR_1 \\ FR_2 \end{Bmatrix} = \begin{bmatrix} A_{11} & 0 & A_{13} & A_{14} & A_{15} \\ A_{21} & A_{22} & 0 & A_{24} & 0 \end{bmatrix} \begin{Bmatrix} DP_1 \\ DP_2 \\ DP_3 \\ DP_4 \\ DP_5 \end{Bmatrix}$$

This design takes on various characteristics, depending on which design parameters are varied and which ones are fixed. If DP_1 and DP_4 are varied after all other DPs are fixed to control the values of FRs, the design is a coupled design. On the other hand, if we fix the values of DP_1, DP_4, and DP_5, the design behaves like an uncoupled design. If DP_3, DP_4, and DP_5 are fixed, then the design is a decoupled design. If DP_1 and DP_4 are set first, then the design behaves like an uncoupled redundant design. Theorem 3 states this fact as follows:

Theorem 3 (Redundant Design)

When there are more DPs than FRs, the design is either a redundant design or a coupled design.

CASE 3 Number of DPs = Number of FRs: Ideal Design

When the number of FRs is equal to the number of DPs, the design is an ideal design, provided that the Independence Axiom is satisfied. This is stated as Theorem 4.

Theorem 4 (Ideal Design)

In an ideal design, the number of DPs is equal to the number of FRs and the FRs are always maintained independent from each other.

Proof

Theorem 1 showed that if the number of FRs is larger than the number of DPs, the design is coupled. Similarly, Theorem 3 states that when there are more DPs than FRs, the design is either a redundant design or a coupled design. Therefore, the only way a design can satisfy the Independence Axiom at all times is when the number of FRs is equal to the number of DPs and the Independence Axiom is satisfied. Any design with an equal number of FRs and DPs and whose design matrix is either a diagonal or a triangular matrix always satisfies the Independence Axiom and thus is an ideal design. In these designs, coupling cannot occur.

Many other theorems and corollaries are presented in Appendix 1A. They may be used as design rules for specific cases.

1.7.6 Examples Involving Decoupling of Coupled Designs[6]

Two examples of coupled designs that were improved by decoupling are given in this section. One is a historical case, which led to the Industrial Revolution, and the other was motivated by a case study worked out by engineers of an aircraft company as part of their exercise in learning axiomatic design. The engineers at the aircraft company solved a long-standing problem, simplifying the manufacturing process and eliminating many problems associated with the original process.

EXAMPLE 1.5 Newcomen Steam Engine vs. Watt Engine

Figure E1.5 shows a schematic drawing of the Newcomen engine, which was invented in 1705. This engine was used to pump water out of coal mines. Recently, this design was examined in terms of the Independence Axiom by Thomas (1995). We have modified that examination further for the purpose of this example.

The Newcomen engine works by injecting steam into a cylinder to push a piston outward to lower the piston in the water pump in the mine. After the piston in the water pump reaches the lowest position, it is raised when the steam in the cylinder is condensed to create a vacuum inside the cylinder and pull the piston inward. This work done by the piston/cylinder attached to the boiler pumps the water out of the mine. The steam in the cylinder is condensed by injecting cold water into the cylinder. During the subsequent cycle, the steam is injected into the cylinder, which raises the temperature of the cylinder and drives the condensed water out of the cylinder before the steam can fully expand and raise the piston again, repeating the cycle.

The attribute the customer is looking for is to "pump water out of the mine." The FRs of the engine at the highest level of this design may be stated as follows:

---

[6] A simplified version of the Axiomatic Design Software developed by Axiomatic Design Software, Inc. can be found on the Internet. This software can simplify the creation of the design matrix, help with the decomposition process, create the final documentation, and assist in making the right design decisions. To download the demo, go to http://www.axiomaticdesign.com/demo and enter: username—acclaro, password—demo, domain—(leave blank). The demo version is identical to the commercial version with three exceptions: (1) Windows only (the commercial version, written in Java, runs on any platform supporting Java 1.1.8), (2) single-user only (the commercial version supports multiuser), and (3) maximum of 15 FRs (there is no software limit in the commercial version). The read me file contains instructions for installation and use.

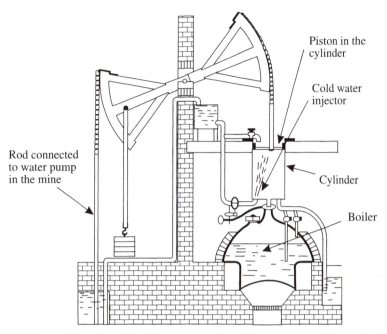

Piston in the cylinder

Cold water injector

Rod connected to water pump in the mine

Cylinder

Boiler

Figure E1.5 Schematic drawing of the Newcomen engine. The water pump in the coal mine is not shown.

$FR_1 =$ Extend the piston.
$FR_2 =$ Contract the piston by creating a vacuum in the cylinder.

The DPs are

$DP_1 =$ Pressure of the steam
$DP_2 =$ Vacuum in the cylinder/piston by condensation of the steam

The design equation may be written as

$$\begin{Bmatrix} FR_1 \\ FR_2 \end{Bmatrix} = \begin{bmatrix} X & 0 \\ 0 & X \end{bmatrix} \begin{Bmatrix} DP_1 \\ DP_2 \end{Bmatrix} \tag{a}$$

Subsequent design decisions we make at lower levels of decomposition must be consistent with the decision represented by Equation (a), i.e., the lower level decisions must maintain the off-diagonal elements as zero. FR_1 may be decomposed as

$FR_{11} =$ Generate the steam.
$FR_{12} =$ Inject the steam.
$FR_{13} =$ Expand the steam and move the piston outward.

FR_2 may be decomposed as

$FR_{21} =$ Condense the steam.
$FR_{22} =$ Move the piston inward.
$FR_{23} =$ Discharge the condensate.

The design parameters are

$DP_{11} =$ Boiler
$DP_{12} =$ Valve
$DP_{13} =$ Steam

$DP_{21} =$ Cold water spray
$DP_{22} =$ Pressure difference caused by condensation
$DP_{23} =$ Discharge valve

The design matrices are

$$
\left\{ \begin{array}{c} FR_{11} \\ FR_{12} \\ FR_{13} \end{array} \right\} = \begin{bmatrix} X & 0 & 0 \\ X & X & X \\ 0 & 0 & X \end{bmatrix} \left\{ \begin{array}{c} DP_{11} \\ DP_{12} \\ DP_{13} \end{array} \right\}
$$

$$
\left\{ \begin{array}{c} FR_{21} \\ FR_{22} \\ FR_{23} \end{array} \right\} = \begin{bmatrix} X & 0 & 0 \\ X & X & 0 \\ 0 & 0 & X \end{bmatrix} \left\{ \begin{array}{c} DP_{21} \\ DP_{22} \\ DP_{23} \end{array} \right\}
$$

To realize the design expressed by Equation (a), none of the $\{DP_{2x}\}$ should affect $\{FR_{1x}\}$ and similarly none of the $\{DP_{1x}\}$ should affect $\{FR_{2x}\}$. However, DP_{21} affects FR_{13}. Also, DP_{13} affects FR_{21} and to a lesser extent, FR_{22}. As a result of this coupling, DP_1 affects both FR_1 and FR_2 because the steam has to heat the cylinder and the piston before the injected steam can expand in the cylinder. Similarly, DP_2 affects both FR_1 and FR_2 because when the cold water is sprayed around the cylinder, the cylinder and the piston have to be cooled before the steam inside the cylinder can be condensed. Therefore, the design matrix is neither diagonal nor triangular. The original intention of the designer expressed by Equation (a) cannot be realized. The Newcomen engine is a coupled design. The efficiency of the engine is low with long cycle times.

The master matrix that shows the two levels of decomposition is shown below.

		DP$_1$			DP$_2$		
		DP_{11}	DP_{12}	DP_{13}	DP_{21}	DP_{22}	DP_{23}
FR_1	FR_{11}	X	0	0	0	0	0
	FR_{12}	X	X	X	0	0	0
	FR_{13}	0	0	X	X	0	0
FR_2	FR_{21}	0	0	X	X	0	0
	FR_{22}	0	0	0	X	X	0
	FR_{23}	0	0	0	0	0	X

This coupled design can be uncoupled by creating a separate condenser elsewhere, in which the steam ejected from the cylinder can be condensed. This is the invention James Watt made in 1769, 64 years after the invention of the Newcomen engine. The engine was further improved by Watt and others, changing it from a single-stroke to a double-stroke engine, for example.

The Watt engine was successful because it was an uncoupled design that had a higher efficiency. In Watt's engine, the FRs can be satisfied independently. It is interesting to note that James Watt's first patent application was based on the idea of separating the function of steam injection from the condensation function to increase efficiency and shorten the cycle time. Watt came out with an uncoupled design without the benefit of the Independence Axiom. The hope is that the explicit statement of the design axioms will enable an ordinary engineer to do what James Watt did in a much shorter period of time. Indeed, many inventions made by engineers after learning axiomatic design demonstrate that this is often the case.

Some 100 years after the invention of the steam engine, because of its importance as a motive source for power, the science of thermodynamics was established. It was done through generalization of the experience gained working with effective steam engines, which is now known as the second law of thermodynamics. It is interesting to note that we may now invoke the second law of thermodynamics and the Independence Axiom to come up with the solution James Watt determined empirically. After the advent of the second law of thermodynamics, the first law of thermodynamics was established. Since then, thermodynamics has impacted all scientific and technological endeavors of humankind.

The following example illustrates the importance of the theorem for ideal design (Theorem 4) in addition to showing how to develop a design concept for a given set of FRs.

EXAMPLE 1.6 Shaping of Hydraulic Tubes

Tubes must be bent to complex shapes in many applications (e.g., aircraft) without changing the circular cross-sectional shape of the tube. This is a particularly difficult job when the tube is made of titanium. Titanium has a hexagonal close packed (hcp) structure so that its mechanical properties are nonisotropic, and it cannot be bent repeatedly because it will fracture.

When an aircraft company tried to bend titanium tubes into complex shapes, it found that the round cross-sectional shape could not be maintained. To prevent this distortion of the cross-sectional shape, they created a special mandrel (i.e., a flexible wire with many thin spacer disks whose diameter was equal to the inside diameter of the tube). The spacer disks were symmetrically mounted on the wire through a hole at the center of the disks. This tool was inserted through the tube to be bent. It was difficult to remove the wire with disks from the tube after the bending operation, and the engineers found that the disks scratched the inner surface of the tube. So they applied a lubricant, which made the removal of the wire easier. However, this solution to the problem created additional operations— they had to clean the lubricant from the inside of the tube with a solvent and then dispose of the solvent.

The engineer in charge of this project happened to take the axiomatic design course specially offered at his company. As part of his term project, he solved the tube-bending problem. The solution was so successful that the company made his solution proprietary to his company. Therefore, the solution presented here to illustrate the design procedure has been obtained independently without the benefit of the engineer's design (Suh, 1995a).

To design a machine and a process that can achieve the task, the functional requirements can be formally stated as:

$FR_1 =$ Bend a titanium tube to prescribed curvatures.
$FR_2 =$ Maintain the circular cross section of the bent tube.

Theorem 4 (see Appendix of Chapter 1 and Chapter 3) states that in an ideal design, the number of DPs is equal to the number of FRs. To come up with an acceptable solution according to this theorem, we must look for a design with two DPs.

The mechanical concept that can do the job is schematically shown in Figure E1.6 for a two-dimensional bending case.

It consists of a set of matching rollers with semicircular grooves on their periphery. These "bending" rollers can counterrotate at different speeds and move relative to each other to control the bending as shown in Figure E1.6. A second set of "feed" rollers, which counterrotate at the same speed, feed the straight tube feedstock into the bending rollers. The centers of these two bending rollers are fixed with respect to each other and the contact point of the bending rollers can rotate about a fixed point. As the tubes are bent around the rollers, the cross-sectional shape will tend to change to a noncircular shape. The deformation of the cross section is prevented by the semicircular cam profile machined on the periphery of the bending rollers. [It may be necessary to make the groove profile slightly oval shaped at the top and bottom of the groove to prevent buckling on the compression side.]

The DPs for this design are

$DP_1 =$ Differential rotation of the bending rollers to bend the tube
$DP_2 =$ The profile of the grooves on the periphery of the bending rollers

The kinematics of the roller motion needs to be determined. To bend the tube, one of the bending rollers must rotate faster than the other. In this case, the tube will be bent around the slower roller. The forward speed of the tube is determined by the average speed of the two bending rollers. The motion of these rollers can be controlled digitally using stepper motors.

The design is an uncoupled design because each of the DPs affects only one FR. Is this the best design? The only way this question can be answered is to develop alternate designs that satisfy the FRs and Cs and the Independence Axiom. Then we need to compute the information content of the proposed designs to select the best design among them.

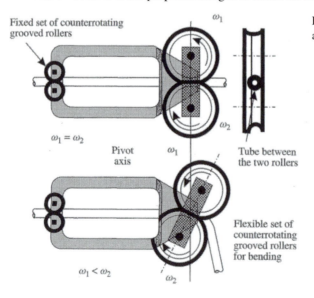

Figure E1.6 Tube-bending apparatus.

If the design is acceptable so far, we need to decompose the FRs to form the next-level FRs (e.g., FR_{11} and FR_{12} for FR_1, based on the chosen DP_1) and then map them into the physical domain to determine the next-level DPs. This process can go on until the design is completed. However, it will not be done here. How would you decompose FR_1 and DP_1?

Another design solution that can achieve the desired goal might be to fill the tube with incompressible material, such as a low melting point metal that can be solidified in the tube before bending the tube. After bending, the metal can be molten and removed from the tube. Is this a better solution than the use of the grooved rollers?

The preceding example illustrates the power of Theorem 4 (Ideal Design). From the beginning of the design process, we knew that we should seek only two DPs as there were only two FRs. This kind of discipline in the design process channeled the thought process. The redundant design originally used by the airplane company made the entire process unnecessarily complicated and probably more costly, in addition to creating secondary problems, such as the disposal of solvents.

1.7.7 Decomposition, Zigzagging, and Hierarchy

Why should we decompose the characteristic vectors FRs, DPs, and PVs? In the preceding example on tube bending, the design was terminated when we mapped from the functional domain to the physical domain at the highest level. The highest level conceptual design provided enough design information at the conceptual level that we knew we would have a successful design when fully implemented. To complete the design of the tube-bending machine, however, detailed designs for the bending mechanism, groove shape, and servo-control, etc. must be developed by decomposing the highest level FRs and DPs. This decomposition process must proceed layer by layer until the design reaches the final stage, creating a design that can be implemented. Through this decomposition process, we establish hierarchies of FRs, DPs, and PVs, which are a representation of the design architecture. (System architecture is further discussed in Chapter 4 in relation to system design.)

In the case of the Newcomen engine, we decomposed the highest level FR and the highest level DP to the second level. To complete the design, we need to decompose the FRs and DPs into lower level FRs and DPs, respectively, until the design is completed so that it can be implemented. The highest level design equation for the Newcomen engine was given as

$$\begin{Bmatrix} FR_1 \\ FR_2 \end{Bmatrix} = \begin{bmatrix} X & 0 \\ 0 & X \end{bmatrix} \begin{Bmatrix} DP_1 \\ DP_2 \end{Bmatrix}$$

When the design details are missing at the highest level of design, the design equation represents the design *intent*. We must decompose the highest level design to develop design details that can be implemented. As we decompose the highest level design, the lower level design decisions must be consistent with the highest level design intent. Unfortunately, the second level design for the Newcomen engine created coupling, which was not consistent with the original design intent. This coupled Newcomen engine, which was inefficient and slow, was nevertheless used as there were no other better alternatives until the invention of a decoupled engine by James Watt.

When the Independence Axiom is violated by design decisions made, we should go back and redesign rather than proceeding with a flawed design.

How do we decompose FRs and DPs? To decompose FR and DP characteristic vectors, we must zigzag between the domains. That is, we start out in the "what" domain and go to the "how" domain. This is illustrated in Figure 1.3. From an FR in the functional domain, we go to the physical domain to conceptualize a design and determine its corresponding DP. Then we come back to the functional domain to create FR_1 and FR_2 at the next level that collectively satisfies the highest level FR. FR_1 and FR_2 are the FRs for the highest level DP. Then we go to the physical domain to find DP_1 and DP_2 by conceptualizing a design at this level, which satisfy FR_1 and FR_2, respectively. This process of decomposition is pursued until the FR can be satisfied without further decomposition when all of the branches reach the final state. The final state is indicated by thick boxes in Figure 1.3, which are called the "leaf" or "leaves."

At each stage of decomposition, how do we know that we have made the right design decisions? To be sure that we have made the right design decision, we must write down the design equation—{FRs} = [A]{DPs}—at each level of decomposition. For example, in the case shown in Figure 1.3, after FR and DP are decomposed into FR_1, FR_2, DP_1, and DP_2, we must write down the design equation to indicate our design *intent* at this level. At this high level of the design process, we can only state our design intent, as we have not yet developed the lower level detailed designs. We know that the design must be either uncoupled or decoupled and, therefore, the intended design must have either a diagonal or a triangular matrix. Suppose that the designer wanted to have a decoupled design represented by the design equation

$$\begin{Bmatrix} FR_1 \\ FR_2 \end{Bmatrix} = \begin{bmatrix} X & 0 \\ X & X \end{bmatrix} \begin{Bmatrix} DP_1 \\ DP_2 \end{Bmatrix}$$

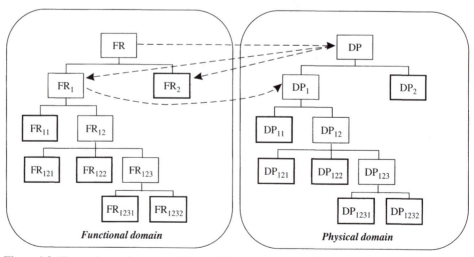

Figure 1.3 Zigzagging to decompose FRs and DPs in the functional and the physical domains and to create the FR and DP hierarchies. Boxes with thick lines represent "leaves" that do not require further decomposition.

Because design details are unknown at this stage of the design process, the triangular matrix represents the design *intent*. All subsequent lower level design decisions must be consistent with this high-level design decision. The consistency of all lower level design decisions can be checked by constructing the full design matrix, which was illustrated in Example 1.5 for the Newcomen engine.

Through the design decomposition process, the designer is transforming design intent into realizable design details. At the highest level of the design process, the designer develops the design concept based on the available knowledge; the designer develops design *intent*. To complete the detailed design, the FR and DP vectors must be decomposed to the lowest level of FRs and DPs, i.e., leaf-level FRs and DPs. Throughout the decomposition process, the designer is transforming the design *intent* expressed by the higher level design matrices into realizable detailed designs given by the lowest level design matrices. At each level of decomposition, the design decisions made must be consistent with all higher level design decisions that were already made. That is, if the highest level design matrix is a diagonal matrix, all lower level decisions must not make— either intentionally or inadvertently—the off-diagonal elements of the highest level design matrix nonzeroes. To check this fidelity and consistency of design decisions, the full design matrix must be constructed by combining all lower level design matrices into a single master matrix.

How does this design process affect inventions and innovations? As a designer tries to develop detailed designs that do not violate the original design intent, the designer may find that existing technologies cannot be used. Then, the designer may develop a new technology that can achieve the original design goals. This process of recognizing the shortcomings of existing technologies and/or designs often leads to inventions and innovations. When a coupled design is replaced by an uncoupled or a decoupled design, major improvements can be made. These novel solutions often constitute inventions or innovations.

What is the current state of design practice as far as the decomposition process is concerned? In some large organizations, there exists a "division for engineering specification" that is charged with creating FRs at all levels. The major task of the division is to develop functional requirements or specifications for their products. These divisions are typically organized so that they have to create FRs at all levels without zigzagging, i.e., by remaining only in the functional or physical domain. As should be quite obvious by now, these divisions cannot do their job right, since FRs cannot be decomposed by remaining in one domain, i.e., without zigzagging. Thus, when designers/engineers are forced to work in such an organization, they often develop FRs or specifications by thinking of an already existing design, which results in respecifying that which already exists.

To decompose FRs and DPs, the designer must zigzag. For example, suppose you want to design a vehicle that satisfies the following four FRs: go forward, go backward, stop, and turn. We cannot decompose these FRs unless we first conceptualize DPs that can satisfy these highest level FRs. If we decide to use an electric motor as a DP to satisfy the FR of moving forward, the decomposed FRs at the next level would be quite different from those that would have resulted had we chosen gas turbines as the DP. Therefore, when we define the FRs in a solution-neutral environment, we have to "zig" to the physical domain, and after proper DPs are chosen, we have to "zag" to the functional domain for further decomposition. Organizations that have created a division for the specific task of specifying FRs at all levels without zigzagging between the domains will not get the results they are looking for and miss important opportunities for innovation.

The process of mapping, zigzagging, and decomposition is illustrated in the following example.

EXAMPLE 1.7 Refrigerator Design

Historically, humankind has had the need to preserve food. Now consumers demand an electrical appliance that can preserve food for an extended time. The typical solution is to freeze food for long-term preservation and to keep some food at a cold temperature without freezing for short-term preservation. These needs can be formally stated in terms of two functional requirements:

$FR_1 =$ Freeze food for long-term preservation.
$FR_2 =$ Maintain food at cold temperature for short-term preservation.

To satisfy these two FRs, a refrigerator with two compartments is designed. Two DPs for this refrigerator may be stated as

$DP_1 =$ The freezer section
$DP_2 =$ The chiller (i.e., refrigerator) section

To satisfy FR_1 and FR_2, the freezer section should affect only the food to be frozen and the chiller (i.e., refrigerator) section should affect only the food to be chilled without freezing. In this case, the design matrix should be diagonal. When we decompose these FRs and DPs, we must be sure that the lower level FRs and DPs are consistent with this highest level design decision.

Having chosen DP_1, we can now decompose FR_1 as

$FR_{11} =$ Control temperature of the freezer section in the range of $-18°C \pm 2°C$.
$FR_{12} =$ Maintain a uniform temperature throughout the freezer section at the preset temperature.
$FR_{13} =$ Control humidity of the freezer section to relative humidity of 50%.

Similarly, based on the choice of DP_2 made, FR_2 may be decomposed as

$FR_{21} =$ Control the temperature of the chiller section in the range of 2°C to 3°C.
$FR_{22} =$ Maintain a uniform temperature throughout the chiller section within 0.5°C of the preset temperature.

To satisfy the second-level FRs (FR_{11}, etc.), we have to conceive a design and identify DPs that can satisfy the FRs at this level of decomposition. Just as FR_1 and FR_2 were independent of each other through the choice of proper DP_1 and DP_2, we must now ensure that FRs at this second level are independent of each other. Furthermore, the choice of the lower level DPs must be consistent with the higher level design matrix, i.e., they must not compromise the independence of FR_1 and FR_2.

The requirements of the freezer section can be satisfied by (1) pumping chilled air into the freezer section, (2) circulating the chilled air uniformly throughout the freezer section, and (3) monitoring the returning air for temperature and moisture in such a way that the temperature is controlled independently of the moisture content of the air. Then, the second-level DPs may be chosen as

DP_{11} = Sensor/compressor system that turns the compressor on (off) when the air temperature is higher (lower) than the set temperature in the freezer section

DP_{12} = Air circulation system that blows air into the freezer section and circulates it uniformly throughout the freezer section at all times

DP_{13} = Condenser that condenses the moisture in the returned air when its dew point is exceeded

Then the design equation may be written as

$$\begin{Bmatrix} FR_{12} \\ FR_{11} \\ FR_{13} \end{Bmatrix} = \begin{bmatrix} X & 0 & 0 \\ X & X & 0 \\ X & 0 & X \end{bmatrix} \begin{Bmatrix} DP_{12} \\ DP_{11} \\ DP_{13} \end{Bmatrix} \tag{a}$$

Equation (a) indicates that the design is a decoupled design. However, it should be noted that we had to change the order of FRs and DPs to make it into a triangular matrix—the fan must be turned on first (or concurrently) before turning on the compressor.

We can now design the chiller section where the food has to be kept in the range of 2°C to 3°C. Here again, we may circulate the chilled air throughout the chiller section and turn on the compressor when the temperature of the returned air is out of the preset range. To have a uniform temperature distribution we should design a fan/vent combination to ensure good circulation of air in the chiller section. The design parameters are

DP_{21} = Sensor/compressor system that turns the compressor on (off) when the air temperature is higher (lower) than the set temperature in the chiller section

DP_{22} = Air circulation system that blows air into the chiller section and circulates it uniformly throughout the chiller section at all times

The design matrix for the $\{FR_{21}, FR_{22}\} - \{DP_{21}, DP_{22}\}$ relationship is triangular as shown below.

$$\begin{Bmatrix} FR_{22} \\ FR_{21} \end{Bmatrix} = \begin{bmatrix} X & 0 \\ X & X \end{bmatrix} \begin{Bmatrix} DP_{22} \\ DP_{21} \end{Bmatrix}$$

The master design matrix that shows these two levels of decomposition is shown in Table E1.7.

TABLE E1.7 Master Design Matrix Showing Two Layers of Decomposition

		DP_1			DP_2	
		DP_{12}	DP_{11}	DP_{13}	DP_{22}	DP_{21}
FR_1	FR_{12}	X	0	0	0	0
	FR_{11}	X	X	0	0	0
	FR_{13}	X	0	X	0	0
FR_2	FR_{22}	0	0	0	X	0
	FR_{21}	0	0	0	X	X

This master matrix shows that the decomposition is consistent with the original design intent.

One of the design questions to be answered here is whether the same compressor and the same fan can be used to satisfy the set $\{FR_{11}, FR_{12}, FR_{13}\}$ and the set $\{FR_{21}, FR_{22}\}$ to minimize the information content without compromising their independence. Clearly we can use one compressor and two fans to satisfy the above design. DP_{11} and DP_{21} are sensor/compressor systems so that the compressor will be on when either one of these two sensors is on. But the fan (represented by DP_{12} and DP_{22}) will not be on unless the temperature of its respective section goes outside of its set range. This fact can be more clearly represented if we decompose FR_{11}–DP_{11} and FR_{21}–DP_{21} to the next level.

The design with two fans and one compressor that satisfies the specified functional requirements and the Independence Axiom is shown in Figures E1.7.a and E1.7.b.

One can propose various specific design alternatives and evaluate these options in terms of the Independence Axiom. If a design allows the satisfaction of these FRs independently, then the design is acceptable for the set of specified FRs. If the designer cannot come up with a good design that satisfies the Independence Axiom, the designer may end up compromising the FRs by eliminating some of them or by giving a much larger range for temperature control, moisture control, etc., to satisfy the Independence Axiom.

It is interesting to compare this new design with a conventional design.

Conventional Refrigerator

A conventional freezer/refrigerator design uses one compressor and one fan that turns on when the temperature of the freezer section is higher than the set temperature. The chiller section is cooled by controlling the opening of the damper between the two sections as shown in Figures E1.7.c and E1.7.d (see Lee et al., 1994). Therefore, the chiller-section temperature cannot be controlled independently from that of the freezer

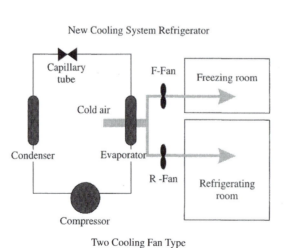

Figure E1.7.a A schematic drawing of a new refrigerator design. (Courtesy of Daewoo Electronic Co.)

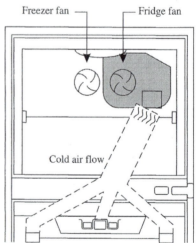

Figure E1.7.b A cold-air circulation system with many vents. (Courtesy of Daewoo Electronic Co.)

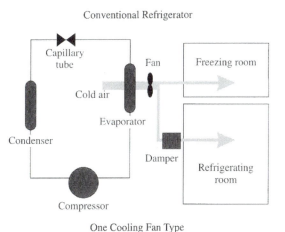

Conventional Refrigerator

One Cooling Fan Type

Figure E1.7.c Schematic drawing of a conventional refrigerator. (Courtesy of Daewoo Electronic Co.)

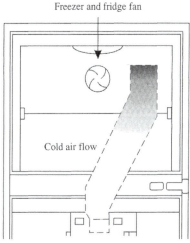

Figure E1.7.d Cold-air circulation in a conventional refrigerator. (Courtesy of Daewoo Electronic Co.)

section. It is a coupled design in that opening the damper when the temperature of the chiller section is greater than 3°C does not satisfy FR_{21}, unless the freezer section turns on the compressor and the fan; in this case, we do not have a DP. The damper is effective only when the temperature goes below 2°C, not when the temperature is over 3°C.

Daewoo Refrigerator

Recently, one company has improved the preservation of food in their chiller section by adding an additional fan so as to control the temperature of the chiller section more effectively, as shown in Figures E1.7.a and E1.7.b (Lee et al., 1994). However, they found out that they did not have to turn on the compressor—only the fan of the chiller section—when the chiller section had to be cooled. It turned out that the evaporator was sufficiently cold and had a large enough thermal inertia to cool the air being pumped into the chiller section even during the period when the compressor was off. This is a better design as it satisfies the Independence Axiom, and therefore food stored in the chiller section stays fresh longer.

The designers of this new refrigerator found that the temperature fluctuation was within their specification (i.e., FR_{21}), as shown in Figure E1.7.e. This design saves electricity because air can be defrosted as a result of the air flow into the chiller section when the compressor is not operating. This new design also enables the use of a quick refrigeration mode in the chiller section by turning on the chiller-section fan as soon as food is put into it. To cool 100 g of water from 25°C to 10°C, this new design took only 37 minutes vs. 96 minutes in a conventional refrigerator, as shown in Figure E1.7.f (Lee et al., 1994).

This idea of using two fans and uniformly positioned ducts may or may not be the best solution if the FRs can be satisfied independently using only one fan, according to Corollary 3 (see Appendix to this chapter). If there is an alternate design that can satisfy

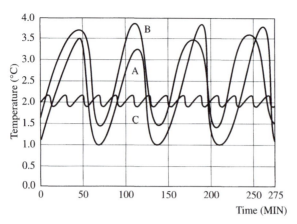

Figure E1.7.e The temperature cycle and the effectiveness of the multi-air-flow system in the new refrigerator (C) versus the old (A and B). (Courtesy of Daewoo Electronic Co.)

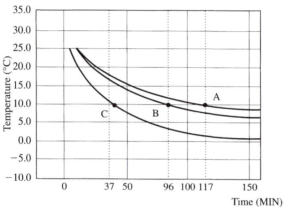

Figure E1.7.f Time required for the refrigerator section to reach the set temperature. The bottom line (C) is for the new design, and A and B are conventional refrigerators. (Courtesy of Daewoo Electronic Co.)

the Independence Axiom, we have to consider the Information Axiom to choose the better of the two designs. Only the detailed calculation of the information content can determine the best design option if one can conceive of designs that use only one fan and yet satisfy the Independence Axiom.

If the design effort produces several designs that are acceptable in terms of the Independence Axiom, we will have to choose the best design among those proposed by invoking the Information Axiom. As explained extensively in a later section, the choice is made by comparing the design range with the system range. The design range is specified by the designer and the system range represents how well a given design (i.e., product or system) can meet the specified functional requirement. The best design is the one that has the minimum information content because it has the highest probability of success. This will be discussed in Section 1.7.8.

What happens if an uncoupled design has more information content than a coupled design? In the preceding example, it was stated that among those designs that satisfy the Independence Axiom, the best design is the one that has the lowest information content. But some may ask: What should be done if there is a coupled design that has a smaller information content than an uncoupled design? Based on years of experience of creating designs or analyzing designs, we have not come across a situation in which a coupled design had lower information content than an uncoupled design. If there is such a case, one go back and look for an uncoupled design that has a lower information content than the coupled design as per Theorem 18.

According to the Independence Axiom, an uncoupled design is always superior to a coupled design, and thus there must always exist an uncoupled or decoupled design that has a smaller information content than that of a coupled design (Theorem 18: The Existence of an Uncoupled or Decoupled Design). In some poor designs that satisfy the Independence Axiom, the information content can be increased unnecessarily by decomposing the FRs and DPs in a complicated way—typically by creating a redundant design in which there are more DPs than FRs. In the examples discussed so far, the number of DPs in the proposed design was always equal to the number of FRs. If this is not done, you may end up with an uncoupled design with a large information content.

We could have come up with designs that have more DPs and FRs, thus inadvertently increasing the information content. As the number of decomposition layers and the number of DPs increase, the likelihood of increasing the information content increases. Many case studies that illustrate this point are given in Suh (1990). For example, the vented compression molding (VCM) of thermal insulation parts reduced the complexity of an existing molding process significantly by reducing the number of process variables and the number of layers of decomposition.

When does analysis come into the picture during the design process? To refine the design, we must model and analyze the proposed design whenever possible. In the preceding three examples, the design matrix was formulated in terms of X and 0. In some cases, it may be sufficient to complete the design using simply X and 0. In many cases, we may take further steps to determine the precise values of design parameters. After the conceptual design is done in terms of X and 0, we need to model the design more precisely to replace the Xs with equations or numbers. Through modeling, we can replace each X with either a constant or a function that involves the DP. We then have a set of equations that relate the FRs to the DPs. This set of equations can be solved separately for uncoupled designs or by following the sequence given by the design matrix for decoupled designs.

1.7.8 Requirements for Concurrent Engineering

When should we bring in manufacturing considerations during the product design stage? After or during product design, we must be sure that the product can be manufactured. Therefore, after certain DPs are chosen, we have to map from the physical domain to the process domain (i.e., process design) by choosing the process variables (PVs). This mapping of process design must also satisfy the Independence Axiom. Sometimes we may simply use existing processes, and at other times we must invent new processes. When existing processes must be used to minimize capital investment in new equipment, the existing process variables must be used and thus they act as constraints in choosing DPs. In developing a product, both the product design and the process design (or selection)

TABLE 1.3 Type of the Concurrent Engineering Matrix [C]

Matrix Type[a]	[A]	[B]	[C] = [A][B]
1. Both diagonal	[\]	[\]	[\]
2. Diagonal × full	[\]	[X]	[X]
3. Diagonal × triangular	[\]	[LT]	[LT]
4. Triangular × triangular	[LT]	[LT]	[LT]
5. Triangular × triangular	[LT]	[UT]	[X]
6. Full × full	[X]	[X]	[X]

[a] *Note that only types 1, 3, and 4 are acceptable from the concurrent engineering point of view.*

must be considered at the same time. This is sometimes called "concurrent engineering" or "simultaneous engineering."

For concurrent engineering to be possible, both the product design represented by Equation (1.1) and the process design represented by Equation (1.6) must satisfy the Independence Axiom. This means that the product design matrix [A] and the process design matrix [B] both must be either diagonal or triangular so that the product of these matrices [C] = [A][B] will be diagonal or triangular. [*Note*: each element is given by $C_{ik} = \Sigma_j(A_{ij}B_{jk})$]. Table 1.3 shows the type of matrix [C] depending on the types of matrices [A] and [B]. For example, to get an uncoupled concurrent design, both matrices must be diagonal. If one is diagonal and the other is triangular, the resulting product of the matrices is triangular. If both [A] and [B] are full triangular matrices, they must be the same kind, either both full [UT] or both full [LT]. If one is [LT] and the other is [UT], the product is a full matrix [X]. Therefore, when [A] and [B] are triangular matrices, either both of them must be upper triangular or both of them must be lower triangular for the manufacturing process to satisfy the independence of functional requirements. This is stated as Theorem 9 (Design for Manufacturability).

When the matrices [LT] and [UT] are not full triangular matrices, the product matrix [C] may still be a triangular matrix, resulting in a decoupled design. Consider, for example, the following case, where [A] is UT and [B] is LT:

$$
\begin{array}{ccc}
[A] & [B] & [C] \\
\begin{bmatrix} X & 0 & 0 \\ 0 & X & X \\ 0 & 0 & X \end{bmatrix} & \begin{bmatrix} X & 0 & 0 \\ 0 & X & 0 \\ X & 0 & X \end{bmatrix} = & \begin{bmatrix} X & 0 & 0 \\ X & X & X \\ X & 0 & X \end{bmatrix}
\end{array}
$$

In this case, it is possible to rearrange the order of the FRs, DPs, and PVs so that both [A] and [B] are [LT], thus making [C] a decoupled design. The original order for [A] is FR_1, FR_2, FR_3. If [A] is reordered as FR_1, FR_3, FR_2, and [B] is similarly reordered, then we have the following:

$$
\begin{array}{ccc}
[A] & [B] & [C] \\
\begin{bmatrix} X & 0 & 0 \\ 0 & X & 0 \\ 0 & X & X \end{bmatrix} & \begin{bmatrix} X & 0 & 0 \\ X & X & 0 \\ 0 & 0 & X \end{bmatrix} = & \begin{bmatrix} X & 0 & 0 \\ X & X & 0 \\ X & X & X \end{bmatrix}
\end{array}
$$

This design is a decoupled design, and therefore concurrent engineering is possible.

1.7.9 The Second Axiom: The Information Axiom

In the preceding sections, the Independence Axiom was discussed and its implications were presented. In this section, we will discuss how we can choose the best design. Even for the same task defined by a given set of FRs, it is likely that different designers will come up with different designs, all of which may be acceptable in terms of the Independence Axiom. Indeed there can be many designs that satisfy a given set of FRs. However, one of these designs is likely to be superior. The Information Axiom provides a quantitative measure of the merits of a given design, and thus it is useful in selecting the best among those designs that are acceptable. In addition, the Information Axiom provides the theoretical basis for design optimization and robust design.

Among the designs that are equally acceptable from the functional point of view, one of these designs may be superior to others in terms of the probability of achieving the design goals as expressed by the functional requirements. The Information Axiom states that the design with the highest probability of success is the best design. Specifically, the Information Axiom may be stated as follows:

Axiom 2: The Information Axiom. *Minimize the information content.*

Information content I_i for a given FR_i is defined in terms of the probability P_i of satisfying FR_i.

$$I_i = \log_2 \frac{1}{P_i} = -\log_2 P_i \tag{1.7}$$

The information is given in units of bits. The logarithmic function is chosen so that the information content will be additive when there are many functional requirements that must be satisfied simultaneously. Either the logarithm based on 2 (with the unit of bits) or the natural logarithm (with the unit of nats) may be used.

In the general case of m FRs, the information content for the entire system I_{sys} is

$$I_{sys} = -\log_2 P_{\{m\}} \tag{1.8}$$

where $P_{\{m\}}$ is the joint probability that all m FRs are satisfied. When all FRs are statistically independent, as is the case for an uncoupled design,

$$P_{\{m\}} = \prod_{i=1}^{m} P_i$$

Then I_{sys} may be expressed as

$$I_{sys} = \sum_{i=1}^{m} I_i = -\sum_{i=1}^{m} \log_2 P_i \tag{1.9}$$

When all FRs are not statistically independent, as is the case for a decoupled design,

$$P_{\{m\}} = \prod_{i=1}^{m} P_{i|\{j\}} \qquad \text{for } \{j\} = \{1, \ldots, i-1\}$$

where $P_{i|\{j\}}$ is the conditional probability of satisfying FR_i given that all other relevant (correlated) $\{FR_j\}_{j=1,\ldots,i-1}$ are also satisfied. In this case, I_{sys} may be expressed as

$$I_{sys} = -\sum_{i=1}^{m} \log_2 P_{i|\{j\}} \qquad \{j\} = \{1, 2, \ldots, i-1\} \qquad (1.10)$$

The Information Axiom states that the design that has the smallest I is the best design, as it requires the least amount of information to achieve the design goals. When all probabilities are equal to 1.0, the information content is zero, and, conversely, the information required is infinite when one or more probabilities are equal to zero. That is, if the probability is small, we must supply more information to satisfy the functional requirements.

A design is called *complex* when its probability of success is low; that is, when the information content required to satisfy the FRs is high. This occurs when the tolerances of FRs for a product (or DPs for a process) are small, requiring high accuracy. This situation also arises when there are many parts, because as the number of parts increases, the likelihood that some of the components do not meet the specified requirements also increases. In this sense, the quantitative measure for complexity is the information content because complex systems may require more information to make the system function. A physically large system is not necessarily complex if the information content is low. Conversely, even a small system can be complex if the information content is high. Therefore, the notion of complexity is tied to the design range for the FRs: the tighter the design range, the more difficult it becomes to satisfy the FRs.

EXAMPLE 1.8 Cutting a Rod to a Length

Suppose we need to cut Rod A to 1 ± 0.000001 m and Rod B to 1 ± 0.1 m. Which has a higher probability of success? How does the probability of success change if the nominal length of the rod is 30 m rather than 1 m?

SOLUTION

The answer depends on the cutting equipment available for the job! However, most engineers with some practical experience would say that the one that has to be cut within 1 μm would be more difficult because the probability of success associated with the smaller tolerance is lower than that associated with the larger tolerance using typical equipment. Therefore, the job with the lower probability of success is more *complex* than the one with higher probability of success.

When the nominal length of the rod is longer, it is more complicated to cut the rod to within the tolerance because the probability of introducing errors increases as the nominal length becomes larger. That is, the total length over which a fixed tolerance must be maintained affects the probability of success. It is much more difficult to achieve the goal as the nominal length increases.

Thus the probability of success is a function of the ratio of the tolerance divided by the nominal length, i.e.,

$$P = f\left(\frac{\text{tolerance}}{\text{nominal length}}\right)$$

The beauty of the above equation is that the designer knows both the nominal length and the tolerance a priori and, therefore, the probability of success can be estimated based on the known ratio. Although we do not know what the function f is, we may approximate it as a linear function in the absence of anything better.[7]

In reality, the probability of success is governed by the intersection of the design range defined by the designer to satisfy the FRs and the ability of the system to produce the part within the specified range. For example, if the design specification for cutting a rod is 1 m plus or minus 1 μm and the available tool (i.e., system) for cutting the rod consists of only a hacksaw, the probability of success will be extremely low. In this case, the information required to achieve the goal would approach infinity. Therefore, this may be called a complex design. On the other hand, if the rod needs to be cut within an accuracy of 10 cm, the hacksaw may be more than adequate, and therefore the information required is close to zero. In this case, the design is simple.

The probability of success can be computed by specifying the *design range (dr)* for the FR and by determining the *system range (sr)* that the proposed design can provide to satisfy the FR. Figure 1.4 illustrates these two ranges graphically.

The vertical axis (the ordinate) is the probability density and the horizontal axis (the abscissa) is either the FR or DP, depending on the mapping domains involved. When the mapping is between the functional domain and the physical domain as in product design, the abscissa is for the FR. When the mapping is between the physical domain and the process domain as in process design, the abscissa is for the DP.

In Figure 1.4, the system pdf is plotted over the system range for the specified FR. The overlap between the design range and system range is called the *common range (cr)*, and this is the only region where the FR is satisfied. Consequently, the area under the system pdf within the common range, A_{cr}, is the design's probability of achieving the specified goal. Then the information content may be expressed as (Suh, 1990)

$$I = \log_2 \frac{1}{A_{cr}} \tag{1.11}$$

Figure 1.4 Design range, system range, common range, and system pdf for a functional requirement.

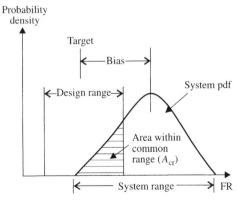

<hr>

[7] This issue was investigated by David R. Wilson (1980).

EXAMPLE 1.9 Cutting of the Rod with a Hacksaw

Let us revisit Example 1.8. We want to cut the rods as specified earlier, but now we know that the equipment available for the job is an ordinary hacksaw. The system pdf is shown in Figure E1.9.

The plot of the system pdf and the design range shows that in the case of cutting Rod B, the system range is completely inside the design range and thus the common range and the system range are the same. Therefore, the probability of success is 1.0, and the information required to fulfill the functional requirement is zero. On the other hand, Rod A has such a tight design range requirement that the common range is almost zero, making the information required approach infinity.

In normal machine shops, the information required is supplied by experienced machinists or toolmakers by carefully measuring the length and making careful cuts, even lapping the part. Because machinists' expertise or skills are limited, the information supplied may not completely compensate for the lack of system capability.

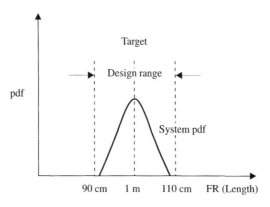

Figure E1.9 The design range and the system pdf for cutting a rod to length.

The Information Axiom is a powerful tool for selecting the best set of DPs when there are many FRs to be satisfied, but should we also use weighting factors? Often design decisions must be made when there are many FRs that must be satisfied at the same time. The Information Axiom provides a powerful criterion for making such decisions without the arbitrary weighting factors used in other decision-making theories. In Equation (1.9), the information content corresponding to each FR is simply summed up with all other terms without a weighting factor for two reasons. First, if we sum up the information terms, each of which has been modified by multiplying with a weighting factor, the total information content no longer represents the total probability (Homework 1.1). Second, the intention of the designer and the importance assigned to each FR by the designer are represented by the design range. If the design ranges for all the FRs are precisely specified and if every specified FR is satisfied within its design range, the goal of the design is fully satisfied. Then there is no need for rank ordering or giving weighting factors to FRs, as the design range specifies their relative importance. The following example illustrates the point.

EXAMPLE 1.10 Buying a House

Professor Sandra Wade of Boston College is planning to buy a new house. She and her husband decided that the following are the four important functional requirements the house must satisfy:

$FR_1 =$ Commuting time for Prof. Wade must be in the range of 15 to 30 minutes.
$FR_2 =$ The quality of the high school must be good, i.e., more than 65% of the high school graduates must go to reputable colleges.
$FR_3 =$ The quality of air must be good over 340 days a year.
$FR_4 =$ The price of the house must be reasonable, i.e., a four bedroom house with 3000 square feet of heated space must be less than $650,000.

They looked around towns A, B, and C and collected the following data with the help of realtors:

Town	FR_1 = Commute time [min]	FR_2 = Quality of schools [%]	FR_3 = Quality of air [days]	FR_4 = Price [$]
A	20 to 40	50 to 70	300 to 320	450k to 550k
B	20 to 30	50 to 75	340 to 350	450k to 650k
C	25 to 45	50 to 80	350 and up	600k to 800k

Which town meets the requirements of the Wade family the best?

SOLUTION

The FRs specify the design range. The system range is given by the above table. Using these design and system ranges, the information content for each FR in each town can be computed using Equation (1.7). Figures E1.10.a and E1.10.b illustrate the overlap (i.e., common range) between the design range and the system range for FR_1 and FR_2 of Town A.

The information content of Town A is infinite as it cannot satisfy FR_3, i.e., the design range and the system range do not overlap. The information content for Towns B and C is computed using Equation (1.9) as follows:

Town	I_1 (bits)	I_2 (bits)	I_3 (bits)	I_4 (bits)	ΣI (bits)
A	1.0	2.0	Infinite	0	Infinite
B	0	1.32	0	0	1.32
C	2.0	1.0	0	2.0	5.0

The information associated with buying a house in Town B is 1.32, whereas for Town C it is 5.0. In Town A, Professor Wade is not likely to find a house that satisfies her needs unless she is willing to change her specifications. The best town in which Professor Wade should buy her house is Town B. In this example, Town A cannot meet her FR for clean

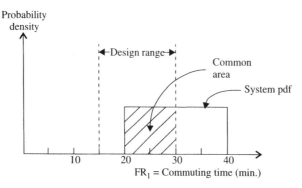

Figure E1.10.a Probability distribution of commuting time.

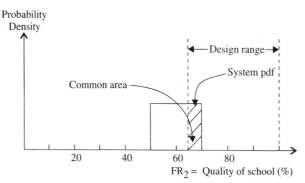

Figure E1.10.b Probability distribution of the quality of schools.

air. In Town C, she will find it more difficult to find houses close to her job (so she can commute in the time allowed) and in the price range she is willing to pay.

If she thinks certain FRs are very important, she can express their importance by means of the specification of the design range without using a weighting factor. For example, if the quality of education is important for her, she can specify higher values for SAT scores or the fraction of students who go to the top 10 universities in the country.

When there is only one FR, the Independence Axiom is always satisfied when there is an appropriate DP that satisfies the FR. In the one-FR case, the only task left is the selection of the right values for the design matrix and the DP to come up with a robust design based on the Information Axiom. In the case of one-FR nonlinear design, various optimization techniques have been advanced when the task is to find a maximum or minimum of an objective function. However, when there is more than one FR, some of these optimization techniques do not work.

To develop a design with more than one FR, we must first develop a design that is either uncoupled or decoupled. If the design is uncoupled, each FR can be satisfied and the optimum points for all FRs can be found because each FR is controlled by its corresponding DP. If the design is decoupled, the FRs must be satisfied following a set sequence, which is further discussed in Chapter 3. The Information Axiom provides a metric that enables us to measure the information content and thus be able to judge a superior design.

1.7.10 Reduction of the Information Content: Robust Design

The ultimate goal of design is to reduce the additional information required to make the system function as designed, i.e., minimize the information content as stated by the Information Axiom. To achieve this goal, the design must be able to accommodate large variations in design parameters and process variables and yet still satisfy the functional requirements. Such a design is called a *robust* design.

To achieve a robust design, the variance of the system must be small and the bias must be eliminated to make the system range lie inside the design range, thus reducing the information content to zero (see Figure 1.4). The bias can be eliminated if the design satisfies the Independence Axiom. There are four different ways of reducing the variance of a design if the design satisfies the Independence Axiom.

1.7.10.1 Elimination of Bias

In Figure 1.4, the target value of the FR is shown at the middle of the design range. The distance between the target value and the mean of the system pdf is called *bias*. To have an acceptable design, the bias associated with each FR should be very small or zero. That is, the mean of the system pdf should be equal to the target value inside the design range.

How can we eliminate the bias? What are the prerequisites for eliminating the bias? In a one-FR design, the bias can be reduced or eliminated by changing the appropriate DP, because the DP controls this FR and we do not have to worry about its effect on other FRs. Therefore, it is easy to eliminate the bias when there is only one FR.

When there is more than one FR to be satisfied, we may not be able to eliminate the bias unless the design satisfies the Independence Axiom. If the design is coupled, each time a DP is changed to eliminate the bias for a given FR, the bias for the other FRs changes also, making the design uncontrollable. If the design is uncoupled, the design matrix is diagonal and the bias associated with each FR can be changed independently as if the design were a one-FR design. When the design is decoupled, the bias for all FRs can be eliminated by following the sequence dictated by the triangular matrix.

1.7.10.2 Reduction of Variance

What is variance? What causes variance? How do we control it? How is it related to redundant design? Variance is a statistical measure of the variability of a pdf. Variability is caused by a number of factors, such as noise, coupling, environment, and random variations in design parameters. In a multi-FR design, the prerequisite for variance reduction is the satisfaction of the Independence Axiom. In all situations, the variance must be minimized. The variation can be reduced in a few specific situations discussed below.

a. Reduction of the Information Content through Reduction of Stiffness. Suppose there is only one FR that is related to its DP as

$$FR_1 = (A_{11})DP_1 \tag{1.12}$$

In a linear design, the allowable tolerance for DP_1, given the specified design range for FR_1, depends on the magnitude of A_{11}, i.e., the "stiffness." As shown in Figure 1.5, the smaller the stiffness, A_{11}, the larger the allowable tolerance of DP_1. However, there is a lower bound for stiffness, which is discussed in Chapter 2.

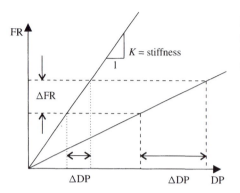

Figure 1.5 Allowable variation of DP as a function of stiffness. For a specified ΔFR, the allowable variation of DP increases with a decrease in the stiffness, A_{11}.

EXAMPLE 1.11 Automobile Wheel Cover (Hubcap)

Consider the wheel cover (hubcap) for a passenger automobile shown in Figure E1.11.a. Sheet metal is pressed to make a decorative cover to hide the lug nuts that hold the rim of the wheel assembly onto the car. The design is simple. Holes are punched in the rim of the wheel and metallic clip springs are welded on the wheel cover. To attach the cover to the rim, the springs welded on the cover are pushed into the holes in the rim. The springs deflect and snap in the holes. The interference between the spring and the hole keeps the cover attached to the rim. The interference fit must be just right to prevent the cover from falling off the wheel when the car goes over a bump and at the same time to make the mounting of the rim easy when the tires are replaced. Tests showed that the force for retention and installation must be 34 N ± 4 N. That is, the design range is from 30 N to 38 N. However, because of slight misplacement of the spring during welding and the wear of punching dies

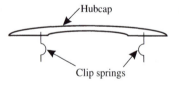

Figure E1.11.a A hubcap for automobiles.

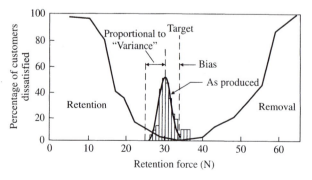

Figure E1.11.b Percentage of parts versus retention force.

during fabrication of the rim, it was found that the force is not always in this range. Figure E1.11.b shows the force distribution (Oh, 1988).

SOLUTION

According to the Information Axiom, the system range of the existing design is broader than the design range; some of the rims are outside of the design range and thus are not acceptable from a quality-control point of view. As this design involves only one FR, it is easy to change the bias and bring the center of the system range into the design range. The DP is the deflection of the spring that is determined by the relative radial position of the spring with respect to the punched-hole location. The design equation may be written as

$$FR_1 = \text{Retention force} = A_{11}(DP_1)$$

$$= (\text{stiffness of spring}) \, \Delta(r_{\text{rim}} - r_{\text{spring}})$$

To reduce the system range, which is associated with the variance, so that the system range is inside the design range, we need to use springs with softer stiffness, as shown in Figure E1.11.c. The design with a stiffer spring requires tighter control of the interference between the hole and the spring than does a design with a softer spring. When the stiffer spring was used, the welding operation, the wear of the dies, and the positioning of the spring clip during welding all had to be controlled within such a tight tolerance that the production operation could not consistently manufacture to the specification. By simply using softer spring clips, the production problem was eliminated without any additional changes in the manufacturing operation.

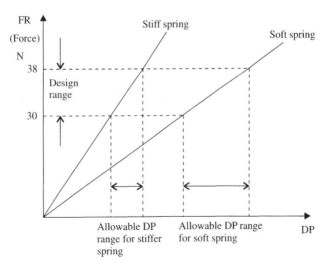

Figure E1.11.c Spring force as a function of the required deflection of the spring. The deflection is determined by the position of the spring relative to the location of the hole in the rim.

b. Reduction of the Information Content through the Design of a System That Is Immune to Variations. When the stiffness shown in Figure 1.5 is zero, the system will be completely insensitive to variation in DP. If the goal is to vary the FR by changing the DP, the stiffness must be large enough to allow control of the FR, although from the robustness point of view, low stiffness is desired. When there are many DPs that affect a given FR, design should be done so that the FR will be "immune" to variation of all these other DPs except the one specific DP chosen to control the FR. In the case of nonlinear design, we should search for such a design window where this condition is satisfied.

The variance is the statistical measure of the spread of the distribution of the output. If a number of DPs are affecting an FR, the total variance of the FR is equal to the sum of the separate variances of the DPs when these DPs are statistically independent.

Often the variation in the system range may be due to many factors that affect the FR. Consider the one-FR design problem. The designer might have created a redundant design as follows:

$$FR = f(DP_a, DP_b, DP_c)$$

or

$$FR = A_a DP_a + A_b DP_b + A_c DP_c \tag{1.13}$$

where A_a, A_b, and A_c are coefficients and the DPs are design parameters that affect the FR. In this case, variation in FR can be introduced by any uncontrolled variation in all coefficients and DPs. The variance can be reduced by creating the design so that the FR is not sensitive to (or is immune to) DP_b and DP_c changes, which can be done either if A_b and A_c are small or if DP_b and DP_c are fixed so that they remain constant. In this case, because FR would be a function of only DP_a, FR can be controlled by changing DP_a. In this case, the only source of variation is the random variation of A_a.

Now consider the case of the multi-FR design given by

$$\begin{Bmatrix} FR_1 \\ FR_2 \\ FR_3 \end{Bmatrix} = \begin{bmatrix} A_{11} & 0 & 0 \\ 0 & A_{22} & 0 \\ 0 & 0 & A_{33} \end{bmatrix} \begin{Bmatrix} DP_1 \\ DP_2 \\ DP_3 \end{Bmatrix} \tag{1.14}$$

In this ideal design with a diagonal design matrix, the variance will be minimized if the random variation in A_{11}, A_{22}, and A_{33} can be eliminated. It should be noted that any error in DPs would contribute to the variance and the bias. Therefore, the coefficients A_{11}, A_{22}, and A_{33} should be small, but large enough to exceed the required signal-to-noise ratio. This subject will be discussed more extensively in Chapters 2 and 3.

c. Reduction of the Information Content by Fixing the Values of Extra DPs. When the design is a redundant design, the variance of the FRs can be reduced by identifying the key DPs and preventing the extra DPs from variation, i.e., fixing the values of these extra DPs.

Consider a multi-FR design given by

$$\begin{Bmatrix} FR_1 \\ FR_2 \\ FR_3 \end{Bmatrix} = \begin{bmatrix} A_{11} & 0 & 0 & A_{14} & A_{15} & 0 \\ 0 & A_{22} & 0 & 0 & A_{25} & A_{26} \\ 0 & 0 & A_{33} & A_{34} & 0 & A_{36} \end{bmatrix} \begin{Bmatrix} DP_1 \\ DP_2 \\ DP_3 \\ DP_4 \\ DP_5 \\ DP_6 \end{Bmatrix} \tag{1.15}$$

Equation (1.15) represents a redundant design because there are more design parameters than functional requirements. The task now is to reduce the information content of this redundant design. The first thing we have to do is to find a way to make the design represented by Equation (1.15) an ideal, uncoupled design as shown by Equation (1.14). This can be done in two different ways: by fixing DP_4, DP_5, and DP_6 so that they do not act as design parameters or by making the coefficients associated with these DPs equal to zero. Fixing DP_4, DP_5, and DP_6 also will minimize the variance in the FRs due to any variation of these three DPs. The variation can also be reduced by setting A_{14}, A_{15}, A_{25}, A_{26}, A_{34}, and A_{36} to zero so that the FRs will be immune to changes in DP_4, DP_5, and DP_6. If the design matrix were different from the one shown above, other appropriate design elements should be made zero or other appropriate DPs should be fixed to reduce the variance of the FRs.

d. Reduction of Information Content by Minimizing the Random Variation of DPs and PVs. One way of reducing the variance of the FRs is to reduce the random variation of input parameters as they contribute to the total random variation of the FRs. The variance of FR may be expressed as

$$\sigma_{FR_i}^2 = \sum_{j=1}^{n} A_{ij}^2 \sigma_{DP_j}^2 + 2 \sum_{j=1}^{n} \sum_{k=1}^{j-1} A_{ij} A_{ik} \, \text{Cov} \, (DP_j, DP_k) \tag{1.16}$$

By reducing the variance of any of the DP_j, we can reduce the contributions to the variance of FR_i. Moreover, if some of the DPs are independent of one another, the relevant covariance terms disappear from Equation (1.16), further reducing the contributions to the variance of FR_i.

e. Reduction of the Information Content by Compensation. The one-FR design given by Equation (1.13) was a redundant design, having three DPs rather than one. For the design given by Equation (1.13), we could satisfy the FR with only one DP, as per Theorem 4 (Ideal Design). Therefore, the best solution for dealing with random variations—noise—for one-FR design is to eliminate the unnecessary DPs and lower the stiffness of the one DP that has been selected to satisfy the FR. However, there may be situations where a given redundant design must be made to work.

Suppose that we have to work with less than an ideal design represented by Equation (1.1), and that the design cannot be made "immune" to random variations by having low stiffness, because the coefficients associated with the redundant DPs cannot be made sufficiently small. In this case, the effect of random variations of the extra DPs on FR can be eliminated by "compensating" for it through the adjustment of the selected DP.

In Equation (1.13), suppose that the following is true:

$$A_a \gg A_b \quad \text{and} \quad A_a \gg A_c$$

Then, we should choose DP_a as the chosen DP and try to minimize the effect of the variations of DP_b and DP_c on FR. The random variations will be represented as δDP_b and δDP_c. If we want to change FR from one state to another state, which is represented by ΩFR, it can be done by changing DP_a by ΩDP_a. For this change of state of FR, Equation (1.13) may be written as

$$\Omega FR = A_a \Omega DP_a + \sum_{i=\text{extra terms}} A_i \, \delta DP_i \tag{1.17}$$

If the allowable random variation of FR, i.e., the design range of FR, is represented as ΔFR, the random noise term represented by the second term of the RHS of Equation (1.17) can be compensated by adjusting DP_a. The necessary adjustment ΔDP_a to compensate for the random variation is given by

$$\Delta DP_a = \frac{\Delta FR - \sum_{i=\text{extra terms}} A_i\, \delta DP_i}{A_a} \tag{1.18}$$

In Equation (1.18), if the noise term is larger than the allowable tolerance of FR, we have to look for a new design by choosing new DPs.

This means of compensating for the random error can be done with multi-FR designs as well as with one-FR designs, if the Independence Axiom is satisfied by the multi-FR design. This kind of compensation scheme can be employed to eliminate the effect of the random variations introduced during manufacturing. Such an example is given later in Example 2.6 (Van Seat Assembly).

It is clear from Equation (1.16) why it is easier to reduce the information content for uncoupled designs because only one DP contributes to the variance of FR_i and there are no covariance terms.

f. Reduction of the Information Content by Increasing the Design Range. In some special cases, the design range can be increased without jeopardizing the design goals. The system range may then be inside the design range. This can be illustrated by revisiting Example 1.10 (Buying a House).

EXAMPLE 1.12 Reevaluation of the Decision on Purchase of a House

Professor Wade decided to eliminate the possibility of buying a house in Town A because the design range for air quality (FR_3) could not be satisfied by Town A. The design range called for acceptable air quality for at least 340 days a year, but the town had good air quality only for between 300 and 320 days a year. Therefore, the design range and the system range did not overlap.

After having evaluated the effect of air quality on health, Professor Wade decided that her original design range on air quality was too stringent. Therefore, she has changed FR_3: the air quality must be good over only 300 days a year. The design range has been expanded by lowering the minimum number of acceptable days to 300 days. Now the information associated with air quality for Town A is zero as the system range is completely inside the design range. Even then, Town B is a better town to look for a house.

1.7.11 Reduction of the Information Content through Integration of DPs

In the preceding section, a means of reducing the information content of a design by making the system range fit inside the design range was presented. This technique is normally called "robust design." There is another equally significant means of reducing the information content—through integration of DPs in a single physical part without compromising the independence of FRs. In this way, the information content can be made small by reducing the likelihood of introducing errors when many physical parts are assembled or by making the manufacturing operation simple.

A good example of DP integration is the beverage can, which was illustrated in Figure E1.3. There were 12 DPs, but only three physical pieces. Another example is a can and bottle opener that must open bottles and cans, but not at the same time. In this case, the DP that opens the bottle and the DP that opens the can (by punching a triangular opening on the lid of the can) may be integrated in the same steel sheet stock—the can opener at one end and the bottle opener at the other end [Figure 3.3 in *The Principles of Design* (Suh, 1990)].

When the design has been achieved by decomposing FRs and DPs to many levels, the integration of DPs can be done in the physical domain. In this case, only the leaf-level DPs of each branch need to be integrated, as higher level DPs are made up of the leaf-level DPs.

To create a system, all physical parts that contain the leaf-level DPs must be integrated into a physically functioning system. This system-level integration must be done from the viewpoint of minimizing information content. As of now, there is no automatic means of assembling a system without human intervention.

1.7.12 Designing with Incomplete Information

During design, we encounter situations in which the necessary knowledge about the proposed design is insufficient and thus design must be executed in the absence of complete information. The basic questions are as follows:

- Under what circumstances can design decisions be made in the absence of sufficient information?
- What are the most essential kinds of information for making design decisions?

These questions will be explored in this section.

Throughout the design process, the designer collects, manipulates, creates, classifies, transforms, and transmits information. Information in design assumes a variety of different forms—knowledge, databases, causality, paradigms, etc. The information necessary to design must be distinguished from the *information content* we need to minimize as required by the Information Axiom. Information is not as specific as the *information content*, which was specifically defined as a function of the probability of satisfying the FR in terms of design range and system range [see Equations (1.7) and (1.8)].

For example, in mapping from the CAs of the customer domain to the FRs of the functional domain, the information needed is in the form of customer preference, potential FRs, and the relationship between the CAs and the FRs. Similarly, information is needed when FRs are mapped into the physical domain and when the DPs are mapped into the process domain.

The information we need is indicated by the design equations. First, we need information on the characteristic vectors (i.e., what they are, etc.). Given an FR, the most appropriate DP must be chosen, the likelihood of which increases with the size of the library of DPs that satisfies the FR. Similarly, given a DP, the more PVs we have, the more options we will have. Once DPs and PVs are chosen, information must be available on the elements of the design matrix, which define the relationship between "what we want to achieve" and "how we want to achieve it."

One of the central issues in the design process is to determine the minimum information necessary and sufficient for making design decisions given a set of DPs for a given set of FRs. The necessary information depends on whether the proposed design satisfies the Independence Axiom. In the case of a coupled design, which violates the Independence Axiom, all the information associated with all elements of the design matrix is required. That is, in the case of coupled designs, design cannot be done rationally without complete information.

a. Information Required for an Uncoupled Design. Consider an ideal uncoupled design that satisfies the Independence Axiom and consists of three FRs. For this uncoupled design, which is the simplest case, the design equation may be written as

$$\begin{Bmatrix} FR_1 \\ FR_2 \\ FR_3 \end{Bmatrix} = \begin{bmatrix} A_{11} & 0 & 0 \\ 0 & A_{22} & 0 \\ 0 & 0 & A_{33} \end{bmatrix} \begin{Bmatrix} DP_1 \\ DP_2 \\ DP_3 \end{Bmatrix} \tag{1.19}$$

A_{11}, A_{22}, and A_{33} relate FRs to DPs. They are constants in the case of linear design, whereas in the case of nonlinear design, A_{11} is a function of DP_1, etc. To proceed with this design, we must know the diagonal elements. Therefore, the minimum information required is the information associated with the diagonal elements. The information required for the uncoupled case is less than that for the coupled case because the off-diagonal elements are zeros.

b. Information Required for a Decoupled Design. Again consider the three-FR case, but this time the design is a decoupled design given by the following design equation:

$$\begin{Bmatrix} FR_1 \\ FR_2 \\ FR_3 \end{Bmatrix} = \begin{bmatrix} A_{11} & 0 & 0 \\ A_{21} & A_{22} & 0 \\ A_{31} & A_{32} & A_{33} \end{bmatrix} \begin{Bmatrix} DP_1 \\ DP_2 \\ DP_3 \end{Bmatrix} \tag{1.20}$$

As in the case of the uncoupled design given by Equation (1.19), we need to know the diagonal elements A_{ii}. It will also be desirable to know the off-diagonal elements A_{ij}. However, information on the off-diagonal elements may not be required to satisfy the given set of FRs with a given set of DPs. We can proceed with the design if the diagonal elements are known and if the magnitudes of the off-diagonal elements are smaller than those of the diagonal elements, i.e., $A_{ii} > A_{ij}$. This can be done because the value of FR_1 can be set first, and then the value of FR_2 can be set by varying the value of DP_2, regardless of the value of A_{21}. When DP_2 is chosen, we must be certain that it does not affect FR_1, but it is not necessary that any information for A_{21} be available if DP_2 has the dominant effect on FR_2, i.e., $A_{22} > A_{21}$. Similarly, as long as DP_3 does not affect FR_1 and FR_2, the design can be completed, even if we do not have any information on A_{31} and A_{32}. This is the only case when design can proceed in the absence of complete information. This is stated as Theorem 17.

Suppose that the upper triangular elements are not quite equal to zero but have very small values as shown in Equation (1.21):

$$\begin{Bmatrix} FR_1 \\ FR_2 \\ FR_3 \end{Bmatrix} = \begin{bmatrix} A_{11} & a_{12} & a_{13} \\ A_{21} & A_{22} & a_{23} \\ A_{31} & A_{32} & A_{33} \end{bmatrix} \begin{Bmatrix} DP_1 \\ DP_2 \\ DP_3 \end{Bmatrix} \tag{1.21}$$

The absolute magnitudes of the elements a_{ij} are much smaller than those of A_{ij}, i.e., $|a_{ij}| \ll |A_{ij}|$. In this case, FR$_1$ will still be affected by large state changes of DP$_2$ and DP$_3$ and this effect may not be negligible because

$$\Omega FR_1 = A_{11} \Omega DP_1 + a_{12} \Omega DP_2 + a_{13} \Omega DP_3 \tag{1.22}$$

where Ω signifies a large change in the value of FR$_i$ due to large state changes in the DPs. In this case, we must compensate for the effect of the DP state changes if the design range of FR$_i$ is smaller than the variability caused by these state changes.

1.8 COMMON MISTAKES MADE BY DESIGNERS

- *Coupling Due to Insufficient Number of DPs (Theorem 1)*

Designers do not recognize that they have a coupled design and try to make it work by a brute-force approach. Coupled designs are created when the number of DPs selected to satisfy a given set of FRs is less than the number of FRs (or when the number of PVs selected to satisfy a given set of DPs is less than the number of DPs). In an ideal design, the number of FRs and the number of DPs are the same (Theorem 4).

- *Not Recognizing a Decoupled Design*

Although a decoupled design satisfies the Independence Axiom, one must first recognize that the design is decoupled and then change the DPs or PVs according to the proper sequence. Many designers do not know that they have a decoupled design, so they randomly change DPs in an effort to make things work. The probability of finding the correct design when a decoupled design with a full triangular matrix is randomly searched is the inverse of n factorial ($n!$), where n is the number of FRs. Consequently, the design appears to be dysfunctional.

- *Having More DPs than FRs*

When we have more DPs than FRs, we have a redundant design. In this case, if possible, it is important to fix the extra DPs and create an uncoupled or decoupled design in order to satisfy the Independence Axiom. Otherwise, the redundant design is a coupled design.

- *Not Creating a Robust Design—Not Minimizing the Information Content through Elimination of Bias and Reduction of Variance*

Products can go out of tolerance easily and thereby develop operational problems when the design is not robust. Robust design can accommodate large variation in DPs or PVs, and yet satisfy the FRs. This can be done by reducing "stiffness," bias, and variance as discussed in Sections 1.7.9 and 3.5.

- *Concentrating on Symptoms Rather Than Cause—Importance of Establishing and Concentrating on FRs*

Surprisingly, a large number of designers, engineers, and managers begin the design process without first determining functional requirements. In the absence of well-established FRs, the designer will go through a random process of ideation and will not be able to communicate to and work with others during the design process. When a complete new product, process, software, or system is to be designed, FRs must be established in a solution-neutral environment.

When an existing product is analyzed for its malfunctions, most people concentrate on symptoms rather than concentrating on functions. If a product satisfied its functional requirements well, symptoms that impede functional performance would not have appeared. Therefore, it is imperative that the analysis begin by asking what FRs must be satisfied by the product and then examining how well the design goals are achieved.

For example, an automobile manufacturer found that its hood lock-and-release mechanism was making a strong undesirable sound each time it was activated by opening the hood. Engineers were given the task of eliminating the sound. They immediately began concentrating on how the sound is created and investigating various means of eliminating the sound. What should they have done? They should have examined the relationship between the FRs and the DPs. Once they understood the design matrix, they could develop a way to satisfy the FRs without making the noise.

EXAMPLE 1.13 Hood Lock-and-Release Mechanism

An automobile company found that the lock-and-release mechanism of the trunk lid shown in Figure E1.13.a makes a very undesirable sound. To eliminate the noise, it was suggested that the current design be improved. Analyze the current design. Design an improved lock-and-release mechanism for the trunk lid.

A POSSIBLE SOLUTION

If the functional requirements were properly satisfied, the noise might not have been generated. We will treat the noise level as a constraint. Therefore, it would be a mistake to concentrate on the noise problem from the beginning. We have to ask what the functional requirements are and investigate how the lock-and-release mechanisms should be designed

When hood is released, metallic noise is produced at latch mounted area. (resonance sound of spring, clang)

Release cable is laid in place where the cable can be seen through radiator grille from outside vehicle.

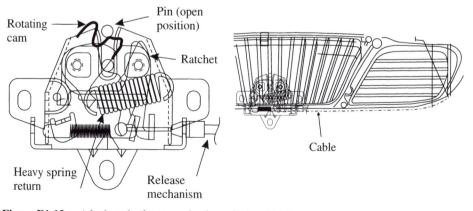

Figure E1.13.a A lock-and-release mechanism of a hood latch system.

to satisfy these FRs through mapping, zigzagging, and decomposition. This process should reveal the source of the noise and the means of reducing the noise. Once an uncoupled or decoupled design is developed, the selected design should be optimized to create a robust design.

The design task is to hold a pin (attached to the hood) in the lock when the hood is closed and to release the pin to an open position when the hood is to be opened. Therefore, the highest level FRs and Cs are

$FR_1 =$ Hold the pin (attached to the hood) in the locked position.
$FR_2 =$ Release the pin from the locked position to an open position.
$C_1 =$ Noise level should not exceed 50 dB.
$C_2 =$ The manufacturing cost cannot be more than the current cost.

Having decided on the FRs, we have to map them into the physical domain by conceiving a design idea that can provide a solution for these high-level FRs. At this stage, we may choose the corresponding DPs as

$DP_1 =$ Mechanical locking mechanism
$DP_2 =$ Release mechanism

Although DP_1 and DP_2 are created without any detailed mechanisms in mind, the original design shown in Figure E1.13.a is consistent with this choice of DPs at the highest level. As we decompose these DPs to lower level DPs, many different mechanisms may be conceived.

For these high-level DPs, the design equation may be expressed as

$$\begin{Bmatrix} FR_1 \\ FR_2 \end{Bmatrix} = \begin{bmatrix} X & 0 \\ 0 & X \end{bmatrix} \begin{Bmatrix} DP_1 \\ DP_2 \end{Bmatrix}$$

This is an uncoupled design. This is the best design at this level of decision making. All subsequent lower level design equations must be consistent with this higher level design equation.

Because DP_1 and DP_2 cannot be implemented without further design details at a lower level of the hierarchy, they must be decomposed. FR_1 may be decomposed to generate the next-level FRs as

$FR_{11} =$ Locate the pin (attached to the hood) at the locked position.
$FR_{12} =$ Lock the pin.

The pin is the part attached to the hood that engages the locking mechanism, which is attached to the engine compartment. Having established FR_{11} and FR_{12}, we have to conceive a design solution at this second level. The following DPs are chosen:

$DP_{11} =$ A cam plate that provides a dead-stop position
$DP_{12} =$ Rotating cam plate with a slot for the pin and a cam profile to engage a spring-loaded ratchet mechanism (to keep the ratchet spring loaded against the cam surface)

It turns out that the original design shown in Figure E1.13.a has these two features and thus satisfies the functional requirements at this level.

It is always a good idea to sketch the DPs at each level as we decompose the FR and DP hierarchies as shown in Figure E1.13.b, which is a new design.

The design equation for the second-level FRs and their corresponding DPs is

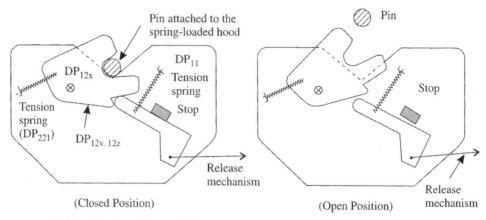

Figure E1.13.b Sketch of DP_{11} and DP_{12}.

$$\begin{Bmatrix} FR_{11} \\ FR_{12} \end{Bmatrix} = \begin{bmatrix} X & 0 \\ 0 & X \end{bmatrix} \begin{Bmatrix} DP_{11} \\ DP_{12} \end{Bmatrix}$$

Again the design is an uncoupled design.

FR_2 may now be decomposed to its next-level FRs. The second-level FRs are established as

$FR_{21} =$ Release the pin
$FR_{22} =$ Put the pin and the rotating disk at the normally open position

The corresponding second-level DPs may be determined as

$DP_{21} =$ Ratchet-removing mechanism
$DP_{22} =$ Spring force at the hinge of the hood to pull the pin and rotate the rotating locking disk out of the locked position (let's put a "spring" on the hood hinge)

The design equation is

$$\begin{Bmatrix} FR_{21} \\ FR_{22} \end{Bmatrix} = \begin{bmatrix} X & 0 \\ 0 & X \end{bmatrix} \begin{Bmatrix} DP_{21} \\ DP_{22} \end{Bmatrix}$$

The original design shown in Figure E1.13.a differs from this proposed design in that it relies on the heavy spring to unlock and push the hood upward. To provide enough energy to accelerate the hood upward, a heavy spring was used. Then a stopper had to be placed on the lock plate to stop the cam. In the proposed design, the equilibrium position of the hood is the "normally open" position because of the spring placed on the hinge of the hood. The original design can also be modified to be consistent with the new design by replacing the heavy spring with a light spring to keep the cam plate in place and by placing a spring at the hinge of the hood. This will clearly eliminate the strong sound that emanates when the latch is opened.

The decomposed FRs of FR_{22} are:

$FR_{221} =$ Put the rotating disk at "normally open" position.
$FR_{222} =$ Put the pin at its "normally open" position.

The original design shown in Figure E1.13.a couples FR_{221} and FR_{222} as the heavy spring does both of these jobs, i.e., there is only one DP. A new proposed design can be as follows:

$DP_{221} =$ Soft spring of the latch (that replaces the heavy spring)

$DP_{222} =$ Equilibrium position determined by the spring force on the hood hinge and the weight of the hood

The design equation is

$$\begin{Bmatrix} FR_{221} \\ FR_{222} \end{Bmatrix} = \begin{bmatrix} X & 0 \\ 0 & X \end{bmatrix} \begin{Bmatrix} DP_{221} \\ DP_{222} \end{Bmatrix}$$

This design is a completely uncoupled design.

It should be noted that we could have chosen a different DP_{22}. Then, FR_{221} and FR_{222} would be quite different from the ones listed above. For example, had we chosen a mechanism that does not use a spring at the hood hinge but rather uses a spring mounted on the latch mechanism just like the original design shown in Figure E1.13.a, the third-level FRs may be stated as

$FR_{221} =$ Accelerate the pin (and hood) out of its closed position.

$FR_{222} =$ Decelerate the pin and the disk and stop at the open position.

$FR_{223} =$ Put the rotating disk at the "normally open" position.

Now we have to choose the appropriate design concept and the design parameters for these third-level FRs (Homework 1.5).

In the case of the original design shown in Figure E1.13.a, FR_{22} was decomposed in a different way. In this original design, the FRs were

$FR_{221} =$ Accelerate the pin (and hood) out of its closed position.

$FR_{222} =$ Put the cam plate at "normally open" position.

The corresponding DPs were

$DP_{221} =$ Heavy spring/cam plate

$DP_{222} =$ Stopper

The design matrix is

$$\begin{Bmatrix} FR_{221} \\ FR_{222} \end{Bmatrix} = \begin{bmatrix} X & 0 \\ X & X \end{bmatrix} \begin{Bmatrix} DP_{221} \\ DP_{222} \end{Bmatrix}$$

The original design is a decoupled design, but noise is made because of the conversion of mechanical energy to acoustic energy by the stopper.

◼ 1.9 COMPARISON OF AXIOMATIC DESIGN WITH OTHER METHODOLOGIES

Questions are often asked about how Axiomatic Design differs from other design methodologies. People who ask these questions have many different methodologies in mind, including statistical process control (SPC) techniques, the Taguchi methodology (Taguchi, 1987), and the Altshuller inventive problem-solving methodology (Altshuller, 1996). The following are offered as general comments:

- Axiomatic design deals with principles and methodologies rather than with algorithms or tools. Based on the two axioms, it derives theorems and corollaries and also develops methodologies based on functional analysis and information minimization, which lead to robust design.
- Axiomatic design is applicable to *all* designs: products, processes, systems, software, organizations, materials, and business plans.
- To be valid, all approaches, including the Taguchi method, must satisfy the design axioms. For example, the Taguchi method is valid only on designs that satisfy the Independence Axiom. So far, there seems to be no contradiction between Altshuller's methodologies and the design axioms. However, Altshuller's original method was used primarily to come up with DPs that are consistent with Cs based on the rule of "contradiction."
- The Taguchi method is instructive on how to make design decisions. It is a method of checking and improving an existing design.
- Both axiomatic design and the Taguchi method lead to robust design for designs that satisfy the Independence Axiom.
- Robust design cannot be accomplished by applying the Taguchi method if the design violates the Independence Axiom. (See the example involving design of an automatic transmission given in Chapter 3.)
- Although many efforts are being made in industry to improve a bad design using optimization techniques, a design that violates the Independence Axiom cannot be improved unless it is first made to satisfy the Independence Axiom. Optimization of a bad design may lead to an optimized bad design or minor improvements. Optimization often implies a trade-off between competing FRs. Designs that satisfy the design axioms do not have to be optimized in the traditional sense.
- In the analytic hierarchy process (AHP), priorities are given to FRs,[8] whereas in axiomatic design theory, there is no need to rank order (or give hierarchical priorities to) FRs. When the design range is precisely defined for all FRs, the design goal is fully specified, and any design that satisfies the FRs within their specified design ranges fully satisfies the design objective. Therefore, there is no need for priorities. When a design cannot be developed that satisfies all of the FRs within their specified design ranges, or when the design is a coupled design, it may be necessary to specify priorities for the FRs, but this approach is not consistent with the Independence Axiom and the Information Axiom.

1.10 SUMMARY

The field of design covered in this book is a broad one that transcends specific engineering fields and encompasses such areas as management and business. In this chapter, the basic concepts of axiomatic design theory are presented as a scientific base for the design field. Axiomatic design is advanced to facilitate the educational process and to improve the design skills of multitudes of people engaged in synthesis in many fields. The basic concepts and methodologies of axiomatic design include the concepts of domains, mapping, the two

---

[8] Dr. Xianyun (Sheldon) Wang and Professor Mitchell M. Tseng of the Hong Kong University of Science and Technology provided critical insights on the issues discussed under the analytical hierarchy process.

design axioms (the Independence Axiom and the Information Axiom), decomposition, hierarchy, and zigzagging.

Several key terms, such as functional requirement (FR), design parameter (DP), and process variable (PV), are carefully defined, as strict adherence to definitions is required in an axiomatic treatise of the subject matter for internal consistency, logical deduction, and mathematical derivation of the resulting relationships. The acceptance of these definitions is a prerequisite in applying the axiomatic principles for design.

Mapping between the domains generates design equations and design matrices. The design equation models the relationship between the design objectives (*what* the design is trying to achieve) and the design features (*how* the design goals are to be satisfied) of a given design. The design matrix describes the relationship between the characteristic vectors of the domains and forms the basis for functional analysis of the design in order to identify acceptable designs. Uncoupled and decoupled designs are shown to satisfy the Independence Axiom and thus are acceptable. Coupled designs do not satisfy the Independence Axiom and thus are unacceptable.

The Independence Axiom states that the functional requirements must always be maintained independent of one another by choosing appropriate design parameters. To be able to satisfy the functional requirements, the designer must always think in terms of FRs before any solution is sought. Robust design is a design that satisfies the functional requirements easily, although large tolerances are given to DPs and PVs. Decomposition of FRs and DPs can be done by zigzagging between the functional and the physical domains.

The Information Axiom deals with information content, the probability of satisfying the FRs, and complexity. Information content is defined in terms of the probability of success and is the additional information required to satisfy the functional requirement. Complexity is related to information content, as it is more difficult to meet the design objectives when the probability of success is low. Computing the information content in a design is facilitated by the notion of the design range and the system range. The design range is specified for each FR by the designer, whereas the system range is the resulting actual performance of the design embodiment.

REFERENCES

Altshuller, G. *And Suddenly the Inventor Appeared*, Technical Innovation Center, Worcester, MA, 1996.

Caldenfors, D. *Top-Down Reasoning in Design Synthesis and Evaluation,* Linkoping Studies in Science and Technology Thesis No. 713, Linkoping University, Sweden, 1998.

Clausing, D. *Total Quality Development*, ASME Press, New York, 1994.

Hazelrigg, G. A. "Theoretical Foundations of Engineering Design," Lecture at MIT, 1998.

Lee, J., Cho, K., and Lee, K. "A New Control System of a Household Refrigerator-Freezer," Presentation at the International Refrigeration Conference at Purdue University, 1994.

Oh, H. L. "Modeling Variation to Enhance Quality in Manufacturing." Presented at the Conference on *Uncertainty in Engineering Design*, NBS, May 10–11, 1988.

Suh, N. P. *The Principles of Design*, Oxford University Press, New York, 1990.

Suh, N. P. "Axiomatic Design of Mechanical Systems," *Journal of Mechanical Design* and *the Journal of Vibration and Acoustics, Transactions of the ASME*, Volume 117, pp. 1–10, June 1995(a).

Suh, N. P. "Design and Operation of Large Systems," *Journal of Manufacturing Systems*, Vol. 14, No. 3, pp 203–213, 1995(b).

Suh, N. P., and Sekimoto, S. "Design of Thinking Design Machine," *Annals of CIRP*, Vol. 1, 1990.

Taguchi, G. *Systems of Engineering Design: Engineering Methods to Optimize Quality and Minimize Cost*, American Supplier Institute, Dearborn, MI, 1987.

Thomas, J. "The Archstand Theory of Design for Information," Ph.D. Thesis, Department of Civil Engineering, MIT, February 1995.

Watson, J. D. *The Double Helix*, Athenaeum, New York, 1969.

Wilson, D. R. "Exploratory Study of Complexity in Axiomatic Design," Ph.D. Thesis, Department of Mechanical Engineering, MIT, 1980.

APPENDIX 1–A Corollaries and Theorems

Some of these theorems are derived in this book as well as in the references given. For those theorems not derived in this book, the readers may consult the original references.

1. COROLLARIES

Corollary 1 (Decoupling of Coupled Designs) Decouple or separate parts or aspects of a solution if FRs are coupled or become interdependent in the designs proposed.

Corollary 2 (Minimization of FRs) Minimize the number of FRs and constraints.

Corollary 3 (Integration of Physical Parts) Integrate design features in a single physical part if the FRs can be independently satisfied in the proposed solution.

Corollary 4 (Use of Standardization) Use standardized or interchangeable parts if the use of these parts is consistent with the FRs and constraints.

Corollary 5 (Use of Symmetry) Use symmetrical shapes and/or components if they are consistent with the FRs and constraints.

Corollary 6 (Largest Design Ranges) Specify the largest allowable design range in stating FRs.

Corollary 7 (Uncoupled Design with Less Information) Seek an uncoupled design that requires less information than coupled designs in satisfying a set of FRs.

Corollary 8 (Effective Reangularity of a Scalar) The effective reangularity R for a scalar coupling "matrix" or element is unity. [*Note*: Reangularity is defined in Suh (1990) and in Chapter 3.]

2. THEOREMS OF GENERAL DESIGN

Theorem 1 (Coupling Due to Insufficient Number of DPs) *When the number of DPs is less than the number of FRs, either a coupled design results or the FRs cannot be satisfied.*

Theorem 2 (Decoupling of Coupled Design)   When a design is coupled because of a larger number of FRs than DPs (i.e., m > n*), it may be decoupled by the addition of new DPs so as to make the number of FRs and DPs equal to each other if a subset of the design matrix containing* n × n *elements constitutes a triangular matrix.*

Theorem 3 (Redundant Design)   When there are more DPs than FRs, the design is either a redundant design or a coupled design.

Theorem 4 (Ideal Design)   In an ideal design, the number of DPs is equal to the number of FRs and the FRs are always maintained independent of each other.

Theorem 5 (Need for New Design)   When a given set of FRs is changed by the addition of a new FR, by substitution of one of the FRs with a new one, or by selection of a completely different set of FRs, the design solution given by the original DPs cannot satisfy the new set of FRs. Consequently, a new design solution must be sought.

Theorem 6 (Path Independence of Uncoupled Design)   The information content of an uncoupled design is independent of the sequence by which the DPs are changed to satisfy the given set of FRs.

Theorem 7 (Path Dependency of Coupled and Decoupled Design)   The information contents of coupled and decoupled designs depend on the sequence by which the DPs are changed to satisfy the given set of FRs.

Theorem 8 (Independence and Design Range)   A design is an uncoupled design when the designer-specified range is greater than

$$\left(\sum_{\substack{i \neq j \\ j=1}}^{n} \frac{\partial FR_i}{\partial DP_j} \Delta DP_j \right)$$

in which case, the nondiagonal elements of the design matrix can be neglected from design consideration.

Theorem 9 (Design for Manufacturability)   For a product to be manufacturable with reliability and robustness, the design matrix for the product, [A] *(which relates the FR vector for the product to the DP vector of the product), times the design matrix for the manufacturing process,* [B] *(which relates the DP vector to the PV vector of the manufacturing process), must yield either a diagonal or a triangular matrix. Consequently, when either* [A] *or* [B] *represents a coupled design, the independence of FRs and robust design cannot be achieved. When they are full triangular matrices, either both of them must be upper triangular or both must be lower triangular for the manufacturing process to satisfy independence of functional requirements.*

Theorem 10 (Modularity of Independence Measures)   Suppose that a design matrix [DM] *can be partitioned into square submatrices that are nonzero only along the main diagonal. Then the reangularity and semangularity for* [DM] *are equal to the product of their corresponding measures for each of the nonzero submatrices.* [*Note:* Chapter 3 defines reangularity and semangularity. See also Suh (1990).]

Theorem 11 (Invariance) Reangularity and semangularity for a design matrix [DM] are invariant under alternative orderings of the FR and DP variables, as long as the orderings preserve the association of each FR with its corresponding DP.

Theorem 12 (Sum of Information) The sum of information for a set of events is also information, provided that proper conditional probabilities are used when the events are not statistically independent.

Theorem 13 (Information Content of the Total System) If each DP is probabilistically independent of other DPs, the information content of the total system is the sum of the information of all individual events associated with the set of FRs that must be satisfied.

Theorem 14 (Information Content of Coupled versus Uncoupled Designs) When the state of FRs is changed from one state to another in the functional domain, the information required for the change is greater for a coupled design than for an uncoupled design.

Theorem 15 (Design–Manufacturing Interface) When the manufacturing system compromises the independence of the FRs of the product, either the design of the product must be modified or a new manufacturing process must be designed and/or used to maintain the independence of the FRs of the products.

Theorem 16 (Equality of Information Content) All information contents that are relevant to the design task are equally important regardless of their physical origin, and no weighting factor should be applied to them.

Theorem 17 (Design in the Absence of Complete Information) Design can proceed even in the absence of complete information only in the case of a decoupled design if the missing information is related to the off-diagonal elements.

Theorem 18 (Existence of an Uncoupled or Decoupled Design) There always exists an uncoupled or decoupled design that has less information than a coupled design.

Theorem 19 (Robustness of Design) An uncoupled design and a decoupled design are more robust than a coupled design in the sense that it is easier to reduce the information content of designs that satisfy the Independence Axiom.

Theorem 20 (Design Range and Coupling) If the design ranges of uncoupled or decoupled designs are tightened, they may become coupled designs. Conversely, if the design ranges of some coupled designs are relaxed, the designs may become either uncoupled or decoupled.

Theorem 21 (Robust Design When the System Has a Nonuniform pdf) If the probability distribution function (pdf) of the FR in the design range is nonuniform, the probability of success is equal to one when the system range is inside the design range.

Theorem 22 (Comparative Robustness of a Decoupled Design) Given the maximum design ranges for a given set of FRs, decoupled designs cannot be as robust as uncoupled designs in that the allowable tolerances for DPs of a decoupled design are less than those of an uncoupled design.

Theorem 23 (Decreasing Robustness of a Decoupled Design)   The allowable tolerance and thus the robustness of a decoupled design with a full triangular matrix diminish with an increase in the number of functional requirements.

Theorem 24 (Optimum Scheduling)   Before a schedule for robot motion or factory scheduling can be optimized, the design of the tasks must be made to satisfy the Independence Axiom by adding decouplers to eliminate coupling. The decouplers may be in the form of a queue or of separate hardware or buffer.

Theorem 25 ("Push" System vs. "Pull" System)   When identical parts are processed through a system, a "push" system can be designed with the use of decouplers to maximize productivity, whereas when irregular parts requiring different operations are processed, a "pull" system is the most effective system.

Theorem 26 (Conversion of a System with Infinite Time-Dependent Combinatorial Complexity to a System with Periodic Complexity)   Uncertainty associated with a design (or a system) can be reduced significantly by changing the design from one of serial combinatorial complexity to one of periodic complexity.

3. THEOREMS RELATED TO DESIGN AND DECOMPOSITION OF LARGE SYSTEMS

Theorem S1 (Decomposition and System Performance)   The decomposition process does not affect the overall performance of the design if the highest level FRs and Cs are satisfied and if the information content is zero, irrespective of the specific decomposition process.

Theorem S2 (Cost of Equivalent Systems)   Two "equivalent" designs can have substantially different cost structures, although they perform the same set of functions and they may even have the same information content.

Theorem S3 (Importance of High-Level Decisions)   The quality of design depends on the selection of FRs and the mapping from domain to domain. Wrong selection of FRs made at the highest levels of design hierarchy cannot be rectified through the lower level design decisions.

Theorem S4 (The Best Design for Large Systems)   The best design for a large flexible system that satisfies m FRs can be chosen among the proposed designs that satisfy the Independence Axiom if the complete set of the subsets of FRs that the large flexible system must satisfy over its life is known a priori.

Theorem S5 (The Need for a Better Design)   When the complete set of the subsets of FRs that a given large flexible system must satisfy over its life is not known a priori, there is no guarantee that a specific design will always have the minimum information content for all possible subsets and thus there is no guarantee that the same design is the best at all times.

Theorem S6 (Improving the Probability of Success)   The probability of choosing the best design for a large flexible system increases as the known subsets of FRs that the system must satisfy approach the complete set that the system is likely to encounter during its life.

Theorem S7 (Infinite Adaptability versus Completeness)   A large flexible system with infinite adaptability (or flexibility) may not represent the best design when the large system is used in a situation in which the complete set of the subsets of FRs that the system must satisfy is known a priori.

Theorem S8 (Complexity of a Large Flexible System)   A large system is not necessarily complex if it has a high probability of satisfying the FRs specified for the system.

Theorem S9 (Quality of Design)   The quality of design of a large flexible system is determined by the quality of the database, the proper selection of FRs, and the mapping process.

4. THEOREMS FOR DESIGN AND OPERATION OF LARGE ORGANIZATIONS (SUH, 1995b)

Theorem M1 (Efficient Business Organization)   In designing large organizations with finite resources, the most efficient organizational design is the one that specifically allows reconfiguration by changing the organizational structure and by having flexible personnel policy when a new set of FRs must be satisfied.

Theorem M2 (Large System with Several Subunits)   When a large system (e.g., organization) consists of several subunits, each unit must satisfy independent subsets of FRs so as to eliminate the possibility of creating a resource-intensive system or a coupled design for the entire system.

Theorem M3 (Homogeneity of Organizational Structure)   The organizational structure at a given level of the hierarchy must be either all functional or product oriented to prevent duplication of effort and coupling.

5. THEOREMS RELATED TO SOFTWARE DESIGN

Theorem Soft 1 (Knowledge Required to Operate an Uncoupled System)   Uncoupled software or hardware systems can be operated without precise knowledge of the design elements (i.e., modules) if the design is truly an uncoupled design and if the FR outputs can be monitored to allow closed-loop control of FRs.

Theorem Soft 2 (Making Correct Decisions in the Absence of Complete Knowledge for a Decoupled Design with Closed-Loop Control)   When the software system is a decoupled design, the FRs can be satisfied by changing the DPs if the design matrix is known to the extent that knowledge about the proper sequence of change is given, even though precise knowledge about the design elements may not be known.

HOMEWORK

1.1 Prove that if each information content term of the right-hand side of Equation (1.9) is multiplied by a weighting factor k_i, the total information content will not be equal to information.

1.2 Consider the design of a hot and cold water tap. The functional requirements are the flow rate and the temperature of the water. If we have a faucet that has one valve for hot water and another valve for cold water, the design is coupled as the temperature and flow rate cannot be controlled independently. We can design an uncoupled faucet that has one knob for temperature control only and another knob for the flow-rate control only. Design such an uncoupled faucet by decomposing the FRs and DPs. Integrate the DPs to reduce the number of parts.

1.3 Professor Smith of the University of Edmonton raised the following question about the water-faucet design. If we take the coupled design (i.e., the design with two valves, one for cold water and the other for hot water) and then add a servo-control mechanism, we may be able to control the flow rate and the temperature independently. Therefore, Professor Smith says that a coupled design is as good as the uncoupled design.

How would you answer Professor Smith's question? Analyze the design proposed by Professor Smith by establishing FRs and DPs, by creating a design hierarchy through zigzagging, and by constructing the design matrices at each level. Is Professor Smith's design a coupled, an uncoupled, or a decoupled design?

1.4 In some design situations, we may find that we have to make design decisions in the absence of sufficient information. In terms of the Independence Axiom and the Information Axiom, explain when and how we can make design decisions even when we do not have sufficient information. What kinds of information can we do without and what kinds of information must we have in design? Illustrate your argument using a design task with three FRs as an example.

1.5 For the latch mechanism discussed in Example 1.13, develop a design solution and state DP_{221}, DP_{222}, and DP_{223}. Sketch your latch-mechanism design.

1.6 Prove Theorem 18, which states that there is always an uncoupled design that has a lower information content than coupled designs.

1.7 Prove Theorems 2, 6, 15, and 16.

1.8 A surgical operating table for hospitals is to be designed. The position of the table must be adjustable along the horizontal and the vertical directions as well as the inclination of the table. Design a mechanism that can satisfy these functional requirements.

If the functional requirements of the table are modified so that the table has to change from one fixed position (i.e., fixed horizontal, vertical, and inclination) to another fixed position, how would you design the mechanism?

1.9 One of the major problems in the automobile business is the warranty cost associated with the weather-strip. It is typically made of extruded rubber to prevent dust, water, and noise from coming into the vehicle. The weather-strip also affects the force required to close the door. One of the problems identified is that the gap between the door and the body can vary from about 10 to 20 mm. Design the weather-strip.

1.10 Compare elements of axiomatic design theory to those of other design methodologies, specifically quality function deployment (QFD), robust design (Taguchi methods), and Pugh concept selection. Where do they agree and where do they differ?

1.11 Investigate how the spark-ignition internal combustion (IC) engine in automobiles works. What are the functional requirements? Is this a good design? Why? Suggest an improved design that is either uncoupled or decoupled.

1.12 With the result of Homework 1.1 in mind, prove Theorem 16 (Equality of Information Content): All information contents that are relevant to the design task are equally important regardless of their physical origin, and no weighting factor should be applied to them.

1.13 The two linear equation sets below describe two designs. Each of the design matrices can be made uncoupled or decoupled depending on whether the variable x is set to 0 or 1, respectively. Analytically compute the probability of success of each of the designs for each value of x (four cases in total). All distributions are uniform. What can you conclude about relationship between information content and coupling?

$$\begin{Bmatrix} -1 < FR_1 < 1 \\ -1 < FR_2 < 1 \end{Bmatrix} = \begin{bmatrix} 1 & 0 \\ x & 1 \end{bmatrix} \begin{Bmatrix} 0 < DP_1 < 2 \\ 0 < DP_2 < 2 \end{Bmatrix}$$

$$\begin{Bmatrix} -1 < FR_1 < 1 \\ 1.5 < FR_2 < 3.5 \end{Bmatrix} = \begin{bmatrix} 1 & 0 \\ x & 1 \end{bmatrix} \begin{Bmatrix} 0 < DP_1 < 2 \\ 0 < DP_2 < 2 \end{Bmatrix}$$

1.14 The equation below describes a design with two DPs and two FRs. The first DP has a uniform distribution and the second DP has a normal distribution around its nominal value. Write a short program (in MATLAB for example) that numerically computes the probability of success of this design, and plot each FR(DP$_1$, DP$_2$). Use the same program to recompute the results for Homework 1.13.

$$\begin{Bmatrix} 2 < FR_1 < 5 \\ 1.5 < FR_2 < 3.5 \end{Bmatrix} = \begin{bmatrix} 0.7 & 1.6 \\ 1.1 & 0.5 \end{bmatrix} \begin{Bmatrix} 1 < DP_1 < 3 \\ DP_2 = 2, \quad \sigma = 0.8 \end{Bmatrix}$$

1.15 Given a system with m independent events with probability of success P_i, prove that the total information content is the sum of individual information contents of these events.

2

One-FR Design, the Information Axiom, and Robust Design

2.1 INTRODUCTION

This chapter presents the theory of axiomatic design when only one functional requirement (FR) is to be satisfied by a proposed design without further decomposition. When the design task is to come up with a design that satisfies one FR with or without constraints, the only relevant axiom is Axiom 2 (the Information Axiom) because Axiom 1 (the Independence Axiom) is automatically satisfied. The Information Axiom requires that we minimize the information content of the design. Information content is measured in terms of the probability of achieving the FR or the information needed to fully satisfy the FR. It is measured by comparing the *design range* specified by the designer with the *system range*.

This chapter considers many implications of the Information Axiom: robust design, integration of parts (assembly), selection of DPs, determination of the best value for a selected DP, decision making in a one-FR design, and manufacturing-process design. The means of developing a robust design is presented, along with many examples. A review of the concept of signal-to-noise ratio in the context of robust design is also included.

◼ 2.2 INTRODUCTION TO ONE-FR DESIGN

One of the key messages of axiomatic design is that you must think of design in terms of functional requirements. *That is, define "what you want to achieve" before you proceed. Are you thinking* functionally? In Chapter 1, the overall concepts and methodology of axiomatic design were presented. Specific concepts introduced in the first chapter may be summarized as follows:

1. Existence of four domains: the customer domain, the functional domain, the physical domain, and the process domain.
2. Understanding customer needs in the customer domain.
3. Mapping the customer needs into the functional domain to determine the functional requirements (FRs) and the constraints (C) in a solution-neutral environment.
4. Mapping from the "what" in the functional domain to the "how" in the physical domain to establish design parameters (DPs) that satisfy the FRs in the case of product design and to establish process variables (PVs) that satisfy the DPs in the case of process design.
5. Need to satisfy the Independence Axiom during the mapping process.
6. Zigzagging between the domains to decompose FRs, DPs, and PVs. For example, after establishing DPs, we must come back to the functional domain to decompose the FRs and then map them in the physical domain again to establish the corresponding next-level DPs, etc.
7. Establishment of the design hierarchy through the decomposition process until the design is completed.
8. Optimization of the uncoupled or decoupled design through the elimination of bias and the reduction of variance of DPs and PVs to develop a *robust design*—a design that can always satisfy the FRs (i.e., 100% of the time) by tolerating large variation in DPs and PVs.
9. When there are many design solutions that satisfy the FRs, choosing the best solution based on the Information Axiom.

Axiom 1: The Independence Axiom. *Maintain the independence of functional requirements.*

The Independence Axiom states that acceptable designs must maintain the independence of the FRs in the functional domain—not the independence of physical things characterized by the DPs of the physical domain[1]— to eliminate coupling of FRs in the functional domain. It was shown that uncoupled and decoupled designs satisfy the Independence Axiom.

Axiom 2: The Information Axiom. *Minimize the information content.*

The Information Axiom states that the best design among those that satisfy the Independence Axiom is the one with the least information content, i.e., the one that

---

[1] Do you remember the beverage-can design of Example 1.3? It had three physical pieces, but satisfied 12 FRs with 12 DPs.

requires the least additional information to achieve the FRs. The design that has the highest probability of achieving the FRs is the best design.

These two axioms form the basis for making the right design decisions, selecting the best design, determining the best design window for the selected design if the design is nonlinear, and creating a robust design. The design must be robust so that we can allow the largest possible tolerances for the DPs and/or PVs when the FRs and their design ranges are specified.

In addition to the Independence Axiom and the Information Axiom, the design must also satisfy constraints, such as cost and geometric bounds imposed by external factors. Then the best design window must be determined for performance, manufacturability, and reliability through modeling of the relationship between the FRs and the DPs to replace each X with either a number or a function, especially when the design is non-linear.

2.2.1 One-FR Design versus Multi-FR Design

What is the difference between the one-FR design and the multi-FR design? There are two kinds of design tasks: one-FR design and multi-FR design. When there is only one FR, the Independence Axiom is always satisfied and therefore coupling is not a design issue. The only axiom we must satisfy is the Information Axiom, which can be done through the *minimization of the information content* associated with satisfying the FR. Therefore, design issues become simpler when only one FR must be satisfied within the bounds established by constraints. The key issue is: "How do you choose an appropriate DP and an appropriate PV so that we can have a robust design?" This is the topic of this chapter.

The necessary and sufficient condition for satisfying the Information Axiom is that the system range of an FR must lie completely within its specified design range. In this case, the probability of success is equal to 1 and the information required is zero. This condition can be satisfied if the bias is very small (ultimately zero) and if the system range (variance) is much smaller than the design range (see Figure 1.4). When the system range is completely inside the design range, the information content is zero.

As discussed in Section 1.7.10.2, there are five possible ways of making the system range lie within the design range and thus of making the information content equal to zero.

1. Reduce the *stiffness* (the coefficient that relates DP to FR) of the system.
2. Make the system totally insensitive (i.e., immune) to random variation of the design parameters or process variables.
3. If the design is redundant, make the extra DPs or PVs constant (i.e., fixed) so that they cannot add to the variance of the system.
4. Compensate for the random variations introduced by extra DPs of a redundant design through the adjustment of the selected DP.
5. Make the design range larger, if it can be done without compromising the functional requirements of the design.

In this chapter, these four different ways of satisfying the Information Axiom will be presented again in a somewhat different context.

Can a one-FR design become a multi-FR problem? Sometimes a one-FR design task becomes a multi-FR design task when the one FR must be decomposed to be able to implement the design task by developing lower level detailed designs. Decomposition enables the development of the next level detailed design of the design hierarchy. When more than one lower level FR must be satisfied, the Independence Axiom must be satisfied when we choose the DPs for the lower level FRs.

The methodology for dealing with a one-FR design can be applied equally well to a multi-FR design task if the design is completely *uncoupled*. If the design is a *decoupled* multi-FR design, we can apply a similar thinking process, but with certain modifications in sequence of changes for FRs. This topic is discussed further in Chapter 3.

2.2.2 Minimization of the Information Content

The Information Axiom states that the design with the highest probability of achieving the given FR is the best design. To satisfy the Information Axiom, we must minimize the information content of a design. This was done in Chapter 1 by making sure that the system range (i.e., the actual variation of the FR of the system) lies inside the specified design range associated with the FR. This simple statement has many implications that will be explored in this chapter.

Robust design is a consequence of the Information Axiom. There are many corollaries related to the Information Axiom: Corollary 3 (Integration of Physical Parts), Corollary 4 (Use of Standardization), Corollary 5 (Use of Symmetry), Corollary 6 (Largest Design Range), and Corollary 7 (Uncoupled Design with Less Information).

Selection of DPs should also be based on the measurement of the information content. The best value for the selected DP can be obtained by minimizing information content. When there are many FRs, the information content can be used as the decision-making tool—it is necessary to select a system with the minimum information content. Manufacturing processes have been determined by measuring the information content (see Suh, 1990).

Information content I was defined in terms of the probability P_i of satisfying a given FR$_i$ as

$$I_i = -\log_2 P_i \tag{2.1}$$

In Chapter 1, it was shown that the probability of success can be computed by specifying the *design range (dr)* for the FR and by determining the *system range (sr)* that the proposed design can provide to satisfy the FR.[2] When the system range is not known a priori, we must rely on precedents or use educated guesswork. Figure 2.1 shows these two ranges graphically.

It was shown that the information content may be expressed as

$$I = \log_2 \left(\frac{1}{A_{cr}} \right) \tag{2.2}$$

where A_{cr} = the area of the system probability density function (pdf) over the common range = probability of success.

---

[2] From a statistical point of view, FR may be considered a *random variable* defined in an FR sample space. FR may be discrete or continuous.

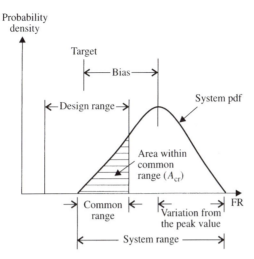

Figure 2.1 Design range, system range, common range, and probability density function (pdf) of a functional requirement.

The probability of achieving FR in the design range may be expressed, if FR is a continuous random variable,[3] as

$$P = \int_{dr^l}^{dr^u} p_s(FR) \, dFR \tag{2.3}$$

where $p_s(FR)$ is the system pdf for FR. Equation (2.3) gives the probability of success by integrating the system pdf over the entire design range (i.e., the lower bound of design range, dr^l, to the upper bound of the design range, dr^u). In Figure 2.1, the shaded area—the area of the common range, A_{cr}—is equal to the probability of success P. Therefore, the information content is equal to

$$I = -\log_2 P = -\log_2 \int_{dr^l}^{dr^u} p_s(FR) \, dFR \tag{2.4}$$

To minimize the information content, we must eliminate the bias and reduce the variance so that the system range will be inside the design range. When the system range is completely inside the design range, the probability of success is always equal to 1.0 and the information content is therefore always equal to zero. The most robust design is the one that always requires zero information to satisfy the FR by being tolerant of variation in DPs and PVs.

EXAMPLE 2.1 Probability and Information Content of a One-FR Design

The design range and the system range of a design with one FR are shown in Figure E2.1. Determine the probability of success and the information content of the design.

---

[3] Definitions and a brief description of probability, expected value, variance, and standard deviation are given in Appendices 2B and 2C. Appendix 2B is for the case of a discrete random variable, and Appendix 2C is for the case of a continuous random variable.

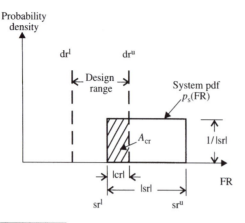

E2.1 Design range and system range of a design with one FR.

SOLUTION

The probability P of achieving the FR is given by A_{cr}, which in the case of this uniform pdf is

$$P = A_{cr} = \int_{sr^l}^{dr^u} p_s(FR)\, dFR = \frac{dr^u - sr^l}{|sr|} = \frac{|cr|}{|sr|}$$

where $|cr|$ is the common range, $|sr|$ is the system range, and sr^l is the lower bound of the system range.

The information content is

$$I = \log_2 \frac{|sr|}{|cr|}$$

In the preceding example, it was assumed that any point in the design range is equally acceptable. However in some situations, one value of the FR may be more desirable than other values of the FR within the design range. For example, the customer needs (or attributes) may be better satisfied if the FR is closer to a target value. The probability of success P and the information content I are still given by Equations (2.3) and (2.4).

Information may be reduced when the design is either uncoupled or decoupled, i.e., when the design satisfies the Independence Axiom. For uncoupled designs, the system range shown in Figure 2.1 can be shifted horizontally by changing the DPs (to shift the system range) until the information content is at a minimum. For decoupled designs, the system range can also be shifted to seek the minimum information point by changing the DPs in the sequence given by the design matrix. The best values of DPs can be obtained by finding where the value of the information reaches its minimum when the following two conditions are satisfied:

$$\sum_{i=1}^{j} \frac{\partial I_j}{\partial DP_i} = 0$$

$$\sum_{i=1}^{j} \frac{\partial^2 I_j}{\partial DP_i^2} > 0 \tag{2.5}$$

For an uncoupled design, the solution to the above two equations can be obtained for each DP without any regard to other DPs, i.e., each term of the series must be equal to zero. In the case of decoupled designs, these equations must be evaluated in the sequence given by the design equation, since the design matrix is triangular.

When the system probability distribution is fully contained within the design range, the probability of success is equal to 1.0. Other system ranges that are inside the design range with their mean (peak) near the target value of the design range may also have probability of success equal to 1.0.

2.3 DESIGN ISSUES FOR THE ONE-FR DESIGN

Do we have to worry about the Independence Axiom in a one-FR design? When there is only one FR, coupling of functional requirements cannot occur, by definition, as coupling refers to the creation of interdependencies among FRs in the functional domain through improper selection of DPs. So with the one-FR design, the remaining design issue becomes the selection of the right DP that will provide the most robust design. That is, we want to find a DP that can always yield a value of the FR within the specified design range without being affected by noise and random variation.

How do we select the best DP in a one-FR design? In theory, any DP that satisfies the FR is a good DP. However, if there are many DPs that can satisfy the FR, as shown in Equation (2.6), we must decide which one of these DPs is the best one to choose.

$$FR_1 \ \$ \ f(DP_1^a, DP_1^b, DP_1^c, \ldots, DP_1^n) \tag{2.6}$$

where $\$$ signifies the fact that any one of the DP_1s in parentheses can satisfy FR_1.

The answer is given by the Information Axiom, i.e., the DP with the minimum information content is the best one. That is, the one with largest A_{cr} provides the best design. This may be stated as

$$\text{Select } DP_1^a \text{ over } DP_1^b \text{ if } (A_{cr})_a > (A_{cr})_b \tag{2.7}$$

When the sr is completely inside the dr, the probability of achieving the FR is equal to 1.0, regardless of the specific system pdf. Therefore, among the set $\{DP_1^a, DP_1^b, DP_1^c, \ldots, DP_1^n\}$, those whose cr = sr are the best DPs if they do not violate the constraints. If cr < sr, the probability of achieving the FR < 1.0; in this case, the design with the largest A_{cr} is the best design.

However, when there are constraints, constraints will limit the acceptable set of DPs because the selected DP should not be in conflict with the constraints. There are many ways of resolving conflicts between DPs and constraints. Altshuller (1988) formalized this process.

Are there optimum values for FR and DP? In *linear* one-FR design, the best design is not related to any maximum or minimum value of the FR. In *nonlinear* one-FR design, although there may be maximum and minimum values for the FR for certain values of the DP, the goal of design is not necessarily to come up with a maximum or a minimum FR value. Rather the goal of the one-FR design is to satisfy the specified FR by means of a specific design parameter DP.

In some fields, the maximum or minimum is associated with the word "optimal," but in axiomatic design, "optimum design" refers to the design that satisfies the FR and constraints with zero information content.

Are the foregoing statements regarding the one-FR design equally valid in multi-FR design? Even when there is more than one FR, the design problem can be treated as if it is a one-FR design task if and only if the design is an uncoupled design. When the design is a decoupled design with many FRs, each FR can be satisfied if the FRs are changed in the right sequence, which is dictated by the governing design equation. These issues are discussed in Chapter 3.

2.4 ONE-FR DESIGN AND INFORMATION CONTENT

The Information Axiom states that the information content should be minimized. Because the information content is a logarithmic function of the probability of success, the best design is the one with zero information content.

2.4.1 One-FR Design with No Constraints

Why and how can the information content be greater than zero? In an ideal one-FR design, we should be able to satisfy the FR without violating any constraints by mapping it to a corresponding DP in the physical domain. In this case, the information content should always be equal to zero. However, when there are random variations and physical limitations in implementing the design intent, the system pdf of the FR can vary over a wide range for a given value of the DP.

Consider a design with one FR and one DP, for which the design equation may be written as

$$\text{FR}_1 = A_{11}\text{DP}_1 \tag{2.8}$$

or in differential form[4] as

$$d\text{FR}_1 = \frac{\partial \text{FR}_1}{\partial \text{DP}_1} d\text{DP}_1 \tag{2.9}$$

The relationship between A_{11} and $(\partial \text{FR}_1/\partial \text{DP}_1)$ is obtained by differentiating Equation (2.8) with respect to DP_1 as

$$\frac{\partial \text{FR}_1}{\partial \text{DP}_1} = A_{11} + \frac{\partial A_{11}}{\partial \text{DP}_1}\text{DP}_1 \tag{2.10}$$

Then Equation (2.9) may be written as

$$d\text{FR}_1 = (A_{11} + \frac{\partial A_{11}}{\partial \text{DP}_1}\text{DP}_1)\, d\text{DP}_1 = \frac{\partial \text{FR}_1}{\partial \text{DP}_1} d\text{DP}_1 \tag{2.11}$$

In the case of a linear design, A_{11} is constant and equal to $(\partial \text{FR}_1/\partial \text{DP}_1)$, i.e.,

$$A_{11} = \frac{\partial \text{FR}_1}{\partial \text{DP}_1} = \text{constant} \tag{2.12}$$

---

[4] This is equivalent to retaining only the first term of the Taylor Series expansion of FR_1 about a point (FR_{1_0}) so that $d\text{FR}_1 = \text{FR}_1 - (\text{FR}_{1_0})$.

In the case of nonlinear design, A_{11} and $(\partial FR_1/\partial DP_1)$ will vary as a function of DP_1.

The system pdf, which covers the system range, may not be fully contained within the design range ΔFR_1. Moreover the mean of the system pdf may differ from the target (FR_{1_o}) as shown in Figure 2.2. For a one-FR design, it is easy to remove the bias by changing the value of DP_1.

To reduce the variance, we must reduce the random variation of DP_1 and the magnitudes of the higher order derivatives of FR_1 with respect to DP_1. The variance can also be reduced by lowering the stiffness of the system, i.e., by making A_{11}, i.e., $(\partial FR_1/\partial DP_1)$, small, as illustrated in Example 2.2. A_{11} must be large enough to have an acceptable signal-to-noise ratio. Several different ways of reducing the bias and the variance are discussed in Section 1.7.10.

The variance σ^2 is the square of the standard deviation. The estimated variance s^2 may be expressed as

$$s^2 = \frac{1}{N-1} \sum_{i=1}^{N} \left(FR_{1_i} - FR_{1_{avg}} \right)^2 \tag{2.13}$$

where FR_{1_i} = the ith value of N measurements of FR_1 and $FR_{1_{avg}}$ = the average value of the N measurements.

When there are many independent contributors to the variance of a system, the total variance of the system is given by

$$s_{tot}^2 = \sum s_i^2 \tag{2.14}$$

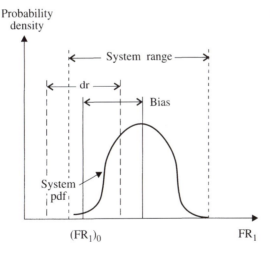

Probability density

Figure 2.2 Design range and system range. The difference between the system mean of FR_1 and the target value (FR_{1_o}) is the bias.

EXAMPLE 2.2 Measurement of Air Velocity

Design an instrument for measuring the velocity of one-dimensional flow of air. The instrument must be able to measure the air velocity within 1% of the absolute air velocity. Specify the random variations you can allow in the key design parameter.

SOLUTION

The design equation may be written as

$$FR_1 = A_{11}DP_1 \tag{a}$$

or in differential form as

$$dFR_1 = \frac{\partial FR_1}{\partial DP_1}dDP_1 \tag{b}$$

FR_1 is "measure the velocity of air." There are many ways we can measure air velocity. Suppose that we decided to predict the air velocity by measuring the stagnation pressure using a Pitot tube shown in Figure E2.2. That is, DP_1 is the relative stagnation pressure $(P_s - P) = \tilde{P}_s$, where P is the ambient pressure and P_s is the absolute stagnation pressure.

From fluid mechanics (Bernoulli's equation), we can find the value of A_{11} that relates the FR (velocity of air = V) to the stagnation pressure P_s and density ρ. The Bernoulli equation may be written as

$$\frac{V^2}{2} + \frac{P}{\rho} = \frac{P_s}{\rho} \tag{c}$$

$$V = \sqrt{\frac{2(P_s - P)}{\rho}} = \sqrt{\frac{2\tilde{P}_s}{\rho}} \tag{d}$$

If we put Equation (d) in the form of the design equation given by Equation (a), we obtain

$$V = \sqrt{\frac{2}{\rho\tilde{P}_s}}\,\tilde{P}_s = f(P_s)\tilde{P}_s \tag{e}$$

where

$$A_{11} = f(P_s) = \sqrt{\frac{2}{\rho\tilde{P}_s}} \tag{f}$$

This is a nonlinear design, as the element A_{11} of the design matrix is a function of DP_1 and varies with DP_1.

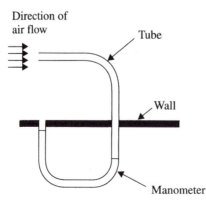

Direction of air flow

Tube

Wall

Manometer

Figure E2.2 A Pitot tube. It is used to determine the velocity of the air stream by measuring the stagnation pressure.

If we differentiate Equation (d) to put the design equation in the form of Equation (b), we obtain the differential

$$\frac{\partial V}{\partial p_s} = \sqrt{\frac{1}{2\rho \tilde{P}_s}} \tag{g}$$

We may also obtain the same result as Equation (g) by differentiating Equation (e).

When we actually make the device and measure V, we find that there is random variation in the measured values of V. If we symbolize a random variation by the use of δ, then δV may be expressed in terms of $\delta \tilde{P}_s$ and δA_{11} as

$$\delta V = \sqrt{\frac{2}{\rho \tilde{P}_s}} \delta \tilde{P}_s + \delta \left(\sqrt{\frac{2}{\rho \tilde{P}_s}} \right) \tilde{P}_s = \sqrt{\frac{1}{2\rho \tilde{P}_s}} \delta \tilde{P}_s \tag{h}$$

Equation (h) states that variation in the value of V is due to the random fluctuations of the instrument represented by δA_{11} and the random variations in the stagnation pressure $\delta \tilde{P}_s$ caused by many possible factors. The allowable fluctuations in the measurement of the stagnation pressure to measure the velocity within 1% accuracy can be obtained by dividing Equation (h) by Equation (e) as

$$\frac{\delta V}{V} = \frac{1}{2} \frac{\delta \tilde{P}_s}{\tilde{P}_s} = 0.01 \tag{i}$$

$$\frac{\delta \tilde{P}_s}{\tilde{P}_s} = 0.02 \tag{j}$$

The allowable error in pressure measurement depends on the absolute magnitude of the stagnation pressure of the air. The higher the pressure, the larger is the allowable random variation of the pressure measurement. The probability of success requires that the total system variation be small enough for the system range to be fully contained within the design range. If there is a bias, it may be due to the fact that the Pitot tube was not located parallel to the direction of the flow, which can be corrected to eliminate the bias. To reduce the variance, the source of the variance must be determined.

2.4.1.1 Robustness through Lower Stiffness

If the specified design range is ΔFR and if the system range is sr(FR), the information content is zero if ΔFR $>$ sr(FR) and the bias b is equal to zero. Then the design is acceptable.

How do we make a design robust? In axiomatic design, robust design is defined as a design that always satisfies the functional requirements, i.e., ΔFR $>$ sr(FR) and $b = 0$, even when there is large random variation in the design parameter δDP.

The specified tolerance ΔDP is determined by the magnitude of A_{11} (i.e., the *stiffness* of the design) and the magnitude of the design range of FR, that is, ΔDP $= \Delta$FR$/A_{11}$. The idea is to make ΔDP as large as possible so that the effect of the random variation of DP on FR is always much smaller than the specified design range ΔFR.

The stiffness of the system should be reduced to enhance the design robustness even when there is random variation. A simple example is now given to illustrate the concept.

EXAMPLE 2.3 Estimating the Height of the Washington Monument

Two groups of 25 high school students—one group from California and the other group from Massachusetts—came to Washington, DC on study tours in October 1995. When they went to see the George Washington Monument, they were asked to estimate its height. They were given tape measures that can measure the length of the shadow of the monument accurately. Then they were asked to eyeball the angle from the end of the shadow to the top of the monument (see Figure E2.3.a). All 25 students in each group estimated the angle by standing at the end of the shadow and looking up to the tip of the monument. The students from California did it at 5 P.M. and the students from Massachusetts did the same at 1 P.M.

The teachers collected the data and determined the average value and the standard deviation of the angles estimated by the students in each group. The average angle estimated by the students from Massachusetts was 75° and that for the California group was 30°. The average value of the angle estimated by the two groups was different because of the different shadow length (at 5:00 P.M. it was much longer than at 1 P.M.). The standard deviation for both groups was about 7°. Which group of students do you believe came closer to estimating the true height of the George Washington Monument? Were they smarter than the other group?

SOLUTION

The functional requirement FR is "measure the height H." Because they did not have any device to measure the angle accurately, the students had to estimate it. The design parameter DP was the angle. The design equation for this one-FR design problem may be expressed as

$$FR = ADP$$

$$H = Af(DP) = L \tan \theta$$

The error in estimating H is given by

$$\Delta H = \left(L \sec^2 \theta\right) \Delta\theta = \frac{H}{\tan \theta} \frac{1}{\cos^2 \theta} \Delta\theta = \frac{2H}{\sin 2\theta} \Delta\theta$$

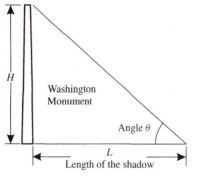

Figure E2.3.a Estimation of the height of the Washington Monument.

The error is proportional to the "stiffness" given by $(2H/\sin 2\theta)$ at a given standard deviation of the estimate $\Delta\theta$. When the sun was setting at 5 P.M., θ was $30°$ and the stiffness was $4H/\sqrt{3}$, but when the sun was high at 1 P.M. $\theta = 75°$ and the stiffness was $(4H)$. Therefore, the estimate made by the students from Massachusetts had a larger error than the estimate made by the students from California. However, because the estimates of the angle were made under different conditions (time of day), the students from California cannot claim to be smarter than the students from Massachusetts.

The estimate gives only the mean of the estimated height, which may be different from the true height, i.e., the bias may not be equal to zero. For example, if the measurement of the shadow is inaccurate—has a systematic error—it will introduce a bias in the estimation of the height of the monument, which can be eliminated by measuring the length of the shadow correctly. Ultimately, we need to calibrate the mean with respect to an absolute measure so as to eliminate a systematic error, if there is any.

One of the key messages of this example is that the "stiffness"—the coefficient that relates the change of FR to the DP variation—should be small to make the design more robust. This is illustrated in Figure E2.3.b.

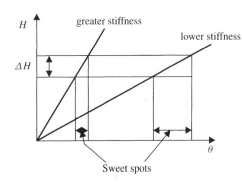

Figure E2.3.b The stiffness given by $(2H/\sin 2\theta)$. The stiffness is greater when the shadow is measured at 1 p.m. than when it is measured at 5 p.m. The system with lower stiffness has a larger sweet spot.

The above example has implications for industrial practice. In the manufacturing industry, statistical process control (SPC) is extensively used to assure quality control. In SPC, careful measurements are made of critical dimensions (sometimes called key characteristics) so that the variation will be within specified bounds. However, this example shows that it is much more important to make the process robust so that the system can tolerate large variation in process parameters than to maintain a tight tolerance to accommodate poor process design.

Other examples of robust design are given in *The Principles of Design* (Suh, 1990). One of the examples is the design of a snap fit for a wheel cover, which is presented in Chapter 1 (Example 1.7) again. The solution was to reduce the stiffness[5] of the spring clips.

Can we make the stiffness infinitesimally small? We stated that when there is no constraint, the FR can be satisfied by choosing an appropriate DP and by making certain

---

[5] The term "stiffness" will be used to describe the ratio of the FR_i to the DP_i—an element of the design matrix that relates FRs to DPs.

that the random variation of the FR is within the specified design range for the FR. Making the stiffness, A_{11}, small can minimize the variation of the FR caused by random variation of the DP. The bias is easily removed by changing the mean value of DP_1.

However, the stiffness cannot be reduced indefinitely, as the signal (i.e., FR_0) must be much larger than the noise (i.e., δFR) to make the signal-to-noise (S/N) ratio larger than the minimum S/N ratio[6] (Suh, 1995).

The signal-to-noise ratio is defined as

$$\eta = 10 \log_{10} \left(\frac{\text{Signal}}{\text{Noise}} \right)^2 \tag{2.15}$$

The minimum S/N ratio may be expressed as the ratio of the desired target output FR_1 to the specified design range as

$$\eta_{\min} = 10 \log_{10} \left(\frac{FR_0}{\Delta FR} \right)^2 \tag{2.16}$$

The actual performance of the system must have an S/N ratio greater than $\eta_{\min}$. The S/N ratio of the real system, i.e., the system range of FR, may be expressed as

$$\eta_{\text{sys}} = 10 \log_{10} \left(\frac{FR_0}{\delta FR} \right)^2 \tag{2.17}$$

where $\delta r(FR_1)$ is the random variation of FR, i.e., noise. To have an acceptable design, η_{sys} must be larger than or equal to $\eta_{\min}$.

In most cases, the system should perform at $\eta_{\text{sys}} = \eta_{\min}$. However, there may be exceptional cases that require $\eta_{\text{sys}} > \eta_{\min}$. They are described below.

2.4.1.2 Stiffness and Response Rate

What is the trade-off between the stiffness and the response characteristics of a system? In some cases, we may need rapid response. However, a robust design with low stiffness may be too slow to respond in time. The time rate of change of FR_1, dFR_1/dt, is related to the time rate of change of DP_1, dDP_1/dt, as

$$\frac{dFR_1}{dt} = A_{11} \frac{dDP_1}{dt} \tag{2.18}$$

In this case, the magnitude of A_{11} must also satisfy the response-time criterion. The smaller the magnitude of A_{11}, the slower is the response rate of dFR_1/dt at a given dDP_1/dt. If the minimum response rate is given by $(dFR_1/dt)_{\min}$, the response criterion that the system must satisfy may be written as

$$\frac{dFR_1}{dt} > \left(\frac{dFR_1}{dt} \right)_c \tag{2.19}$$

To have a rapid response rate, either dDP_1/dt or A_{11} must be large.

---

[6] The signal-to-noise ratio was originally used to measure sound levels and electrical signal levels to characterize the magnitude of the signal relative to background noise. Its unit is decibels.

In summary, design must be optimized with respect to the following three considerations:

1. *Robustness*—The stiffness of the system should be as small as possible.
2. *Signal-to-noise ratio*—The minimum stiffness must be larger than the minimum S/N ratio determined by the desired target output and the given allowable design range, i.e., $FR_0 / \Delta FR$.
3. *Rate of response*—The rate of response of the system may require that the stiffness be greater than that dictated by the minimum S/N ratio in order to meet the desired response rate.

2.4.1.3 Robust Design by Making the System "Immune" to Variation

What is a design that is immune to variation? When FR_1 must remain constant and insensitive to the random variation of DP_1, the desired design solution is the one that will make FR_1 "immune" to the variation of DP_1. This can be done by letting A_{11} (i.e., dFR_1/dDP_1) and all higher order derivatives of FR_1 be equal to zero at the set value of FR_1 and DP_1 even when DP_1 fluctuates about, or drifts from, the set value.

EXAMPLE 2.4 Windshield Wiper—Robust Mounting Design[7]

Windshield wipers of automobiles can make squeaking sounds when the windshield is not sufficiently wet, thus annoying some customers. The goal is to devise a design solution that eliminates squeaking sounds under all conditions.

Figure E2.4.a shows the windshield and the wiper blades. The windshield-wiper assembly consists of the pivot of the wiper blades, itself, and the mounting bracket with three mounting holes. It is supplied by outside vendors. There are three mounting holes punched out on the automobile body. The location of the three holes defines the angle of the pivot of the wiper-blade assembly when it is mounted on the car. When the blade wipes across the windshield, it flips its angle as the direction of the wiper reverses.

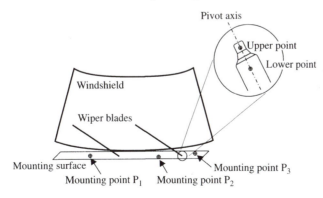

Figure E2.4.a Windshield and windshield-wiper assembly. The wiper blades are mounted on pivots. The wiper-blade assembly is mounted on a car body using mounting holes at P_1, P_2, and P_3, which determine the angle of the pivot axis.

[7] Adapted from Oh (1997).

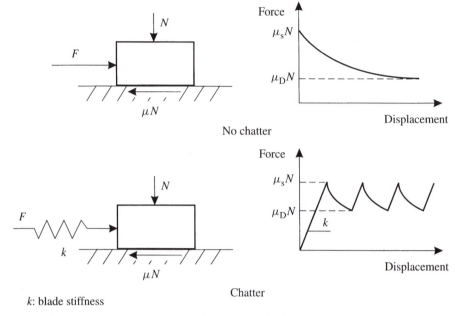

Figure E2.4.b Modeling of chatter as vibration of a mass/spring/friction system. (From Oh, 1997.)

The source of noise is the chatter of the rubber blade due to frictional force and the elastic nature of the blade. The physics of chatter is illustrated in Figure E2.4.b, which shows that force is acting on the end of a spring that is connected to a mass serially. As the force is applied, the spring exerts a force on the mass. As a result of the applied force, the spring will be compressed until the force is greater than the friction force acting on the mass. Then the mass will suddenly accelerate, stretching the spring. Once the mass starts moving, the frictional force will decrease as the spring expands. When the spring force cannot overcome the frictional force again, the mass will come to a stop until the spring force builds up again to overcome the frictional force. This process will repeat and create chatter and noise.

SOLUTION

Figure E2.4.c shows the blade and its angle with respect to the windshield. Normally the blade will be dragged along as shown when it moves from left to right. When the direction of the wiper reverses at the end of the stroke and goes from right to left, the blade will experience the reversal of the loading direction—elastic unloading first, followed by the change in the direction of the applied force. Then the blade may not flip over immediately when the direction of force reverses. The direction of the blade will eventually reverse, flipping over to the left side, when the frictional force is greater than or equal to the force pushing the blade. Then the blade will be dragged along in the opposite direction.

For the blade to flip, the frictional force between the blade and the windshield must be large enough to resist the sliding action due to the tangential component of the force applied to the blade at the pivot. Figure E2.4.d shows a plot of the tangential force as a function of

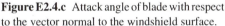

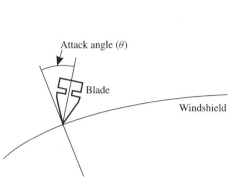

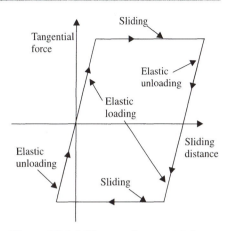

Figure E2.4.c Attack angle of blade with respect to the vector normal to the windshield surface.

Figure E2.4.d History of tangential force acting on the blade as a function of sliding distance.

sliding distance measured from the left end of the blade stroke. If the frictional force is not large enough, the blade will be pushed by the applied force and slide on the glass. This is the condition under which chatter occurs.

The forces acting on the blade are shown in Figure E2.4.e. The condition for flipping is

$$\mu F h \cos \theta > F h \sin \theta$$

or   (a)

$$\mu > \tan \theta$$

where F is the resultant contact force applied to the wiper, h is the height of the blade, and μ is the friction coefficient. Equation (a) states that the attack angle must be smaller than arc tangent of the coefficient of friction to flip the wiper blade.

Figure E2.4.f shows a plot of μ vs. $\tan \theta$, with the design range and the system range. The system distribution for the attack angle is determined by finding the actual attack angle distribution, and the design range is given by the intersection of the nominal friction coefficient and the line m = $\tan \theta$, as shown.

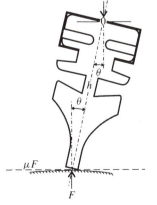

Figure E2.4.e Condition for flipping of the blade: $\mu F(h)$ $(\cos \theta) > Fh(\sin \theta)$ or $\mu > \tan \theta$.

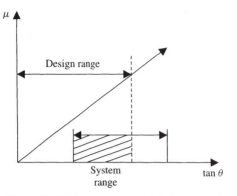
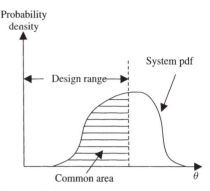

Figure E2.4.f System range and design range in a plot of actual friction force vs. blade angle.

Figure E2.4.g System pdf and design range of the attack angle θ.

Figure E2.4.g shows the probability density vs. the attack angle θ, showing the actual frequency of the attack-angle distribution. The unshaded area outside the common range is where chattering will occur.

This is a one-FR problem. The FR may be stated as

FR =   Control the attack angle θ.

The attack angle θ is a function of the positions of the windshield, the twist arm on which the blade is attached, and the pivot axis on which the twist arm with blade is mounted. Therefore, there are the following three DPs:

DP_1 =   The pivot axis on which the twist arm with the blade is mounted
DP_2 =   The angle of the orientation of the twist arm on which the blade is attached
DP_3 =   The position of the windshield

The design equation may be expressed as

$$\theta = \theta(X_1, X_2, X_3) \tag{b}$$

where X_1, X_2, and X_3 are vectors that define the pivot axis, the position of the twist arm, and the vector normal to the windshield. This is a one-FR design problem. In an ideal one-FR design, we need only one DP. This can be achieved if we can "fix" two of the three DPs. For example, X_2 and X_3 may be "fixed" by proper installation of the windshield and the accurate manufacture of the twist arm. Then, we may assume that the random variation of these two DPs is negligible. Then, Equation (b) may be expressed as

$$\theta = \theta(X_1) \tag{c}$$

Expanding Equation (b) in Taylor's series about its mean positions X_1^*, we obtain

$$\theta = \theta^*(X_1^*, X_2^*, X_3^*) + \frac{\partial \theta}{\partial X_1}(X_1 - X_1^*) \tag{d}$$

where $(\partial \theta / \partial X_1)(X_1 - X_1^*)$ is a dot product of two vectors and is equal to

$$\frac{\partial \theta}{\partial X_1}(X_1 - X_1^*) = \frac{\partial \theta}{\partial x_1}(x_1 - x_1^*) + \frac{\partial \theta}{\partial y_1}(y_1 - y_1^*) + \frac{\partial \theta}{\partial z_1}(z_1 - z_1^*) \tag{e}$$

where x, y, and z are three components of the vector X. The mean shift in the attack angle is given by

$$\theta - \theta^* = \frac{\partial \theta}{\partial X_1}(X_1 - X_1^*) \tag{f}$$

The variance of the mean shift angle $(\Delta \theta)^2$ is

$$(\Delta \theta)^2 = \left(\frac{\partial \theta}{\partial X_1} \Delta X_1 \right)^2 \tag{g}$$

The primary random variation of the pivot axis during assembly is caused by the errors introduced during mounting of the wiper assembly on the car body.

The deviation of the attack angle due to the pivot axis may be expressed in terms of the error in the windshield wiper assembly as received and the errors due to deviations of mounting point coordinates for three holes, which may be expressed as

$$\frac{\partial \theta}{\partial X_1} \Delta X_1 = \frac{\partial \theta}{\partial X_1} \left(\Delta X_1 \bigg|_{\text{as received}} + \sum_{i=1}^{3} \frac{\partial \theta}{\partial M_i} \Delta M_i \right) \tag{h}$$

where M_i are the mounting point coordinates for the three holes, which are given by $P_1(x_1, y_1, z_1)$, $P_2(x_2, y_2, z_2)$, and $P_3(x_3, y_3, z_3)$. What we need to do is to find the best mounting scheme so that the term $[(d\theta/dM_i)\Delta Mi]$ will be small.

One way of achieving this goal is to make the coefficient $(d\theta/dM_i)$ so small that the attack angle is "immune" to the variation of the mounting position ΔM_i. In other words, to minimize the variance $(\Delta \theta)^2$, we must make it immune to the variation of θ with respect to the position coordinates M_i.

The position of the rigid body is determined when six degrees of freedom are controlled. The automobile companies have been doing this by punching out three holes on the sheet metal, as illustrated in Figure E2.4.h, which shows the particular case where the "net" hole—

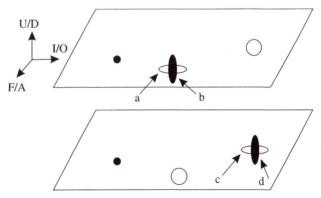

Figure E2.4.h Location of three mounting holes. The solid circle at Position 1 represents a net hole that eliminates the three degrees of freedom (df) of the wiper system; the elongated holes eliminate 2 df (U/D motion and either I/O or F/A motion); the large hole restricts only the vertical motion (U/D) to the sheet metal surface. A large statistical sample indicated that the most robust wiper-system assembly is obtained with the net hole at Position 3. (a) I/O slot at point 2; (b) F/A slot at point 2; (c) I/O slot at point 3; (d) F/A slot at point 3 (courtesy of GM).

shown by a solid circle—is placed at Point 1 with two different possibilities for Points 2 and 3. The net hole on the sheet metal holds the wiper system tightly, thus eliminating three degrees of freedom. The elongated hole allows an in-and-out motion along one direction, thus using another two degrees of freedom, and a loose hole only restricts the vertical motion.[8]

Because it is difficult to determine the sensitivity coefficient of the attack angle $(d\theta/dM_i)$ by analytical means, this automobile company measured it by checking the actual attack angle of 23 cars. From these measurements, they found out that when the "net" hole is at Position 3, the mean shift and the standard deviation of the attack angle were the smallest. Once the number 3 hole is fixed, the sensitivity was not very high, regardless of whether number 1 hole or number 2 hole was the elongated hole, although the results favored making number 1 the elongated hole.

One of the key points of this example is that the variation can be minimized by choosing a system that is "immune" to variations—by making the slope given by $d\mathrm{FR}_1/d\mathrm{DP}_1$ very small.

2.4.2 One-FR Design with Constraints

When there are constraints (Cs), the design must satisfy both FRs and Cs. In general, it is better to ignore the Cs during the first stage of the mapping process from the functional domain to the physical domain or from the physical domain to the process domain, except in special situations. Once appropriate DPs are chosen, we can go back and check whether the Cs are violated. For some Cs, such as cost, the design must be completed before it can be checked against the Cs and improved to satisfy the Cs. However, throughout the design process, designers should normally consider the Cs implicitly when they choose DPs. For example, if one C is cost, designers should look for the most inexpensive material that can do the job rather than choosing a more expensive material. If the vibration of a structure is a C, it would be better to select a rotary compressor than a piston-type compressor.

In choosing a DP, we may find a conflict between the DP and one or more of the Cs. Many experienced designers can resolve this conflict by trial and error or by using a database such as the morphological method. Altshuller's Theory of Inventive Problem Solving (TIPS) is one of the techniques that has been proposed to generate DPs based on the argument of contradiction.

EXAMPLE 2.5 Electric Circuit Breaker Box[9]

An electric circuit breaker has two electrodes in contact as shown in Figure E2.5.a. We want to install a new circuit breaker that can transmit twice the power of the original

---

[8] This is the principle behind the general notion of kinematic coupling. In typical kinematic coupling, three V-grooves with three balls are used to eliminate the six degrees of freedom of motion and position two mating surfaces.

[9] From Nordland (1996).

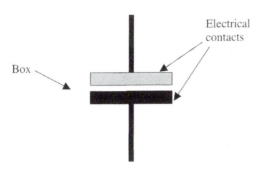

Box

Electrical
contacts

Figure E2.5.a Original electric contact.

design in the same available space. You are asked to design such an electric circuit breaker.

There must be many ways of making a circuit breaker, such as a plasma-based circuit breaker, a mechanical contact type (i.e., the original design), and a conventional fuse. To simplify our task, we will design a mechanical contact type.

One of the constraints is the space available for installation of the new circuit breaker. Another constraint to be considered is the temperature rise of the circuit breaker. It will be assumed that most of the resistance to current flow is the contact resistance at the contact surface rather than in the electrodes.

The design goal (i.e., FR) is to double the power transmitted without increasing the box size and overheating the circuit breaker box. For the given FR of transmitting twice the power of the conventional circuit breaker, we may choose the contact area of the circuit breaker plate as the DP.

One simple way of increasing the power rating is to increase the surface area of the contact. However, there is a conflict between this DP and the C, i.e., volume of the box. If we simply scale up the current design, then the volume will increase. Therefore, we must devise ways of increasing the contact area without increasing the volume.

We can think of several solutions with little effort. A couple of suggestions are given in Figures E2.5.b and E2.5.c, which were generated by Nordlund (1996) using the software

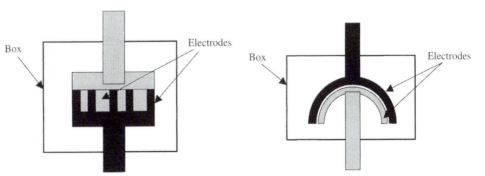

Box

Electrodes

Box

Electrodes

Figure E2.5.b Comb-like structure to increase the contact area.

Figure E2.5.c Hemispheric surface to increase the contact area.

program called the Invention-Machine Principle, which is based on Altshuller's Theory of Inventive Problem Solving (1988).

This extremely simple example illustrates the fact that there are many ways of satisfying an FR even when there are constraints. In many designs, there are several constraints, such as cost, geometry, temperature, state codes governing a particular industry, and many others. Design must be done without violating these constraints.

2.4.3 Nonlinear One-FR Design with Constraints

Regardless of whether the design is linear or nonlinear, after the FR is satisfied by choosing a right DP, the designer must check the design to determine whether it violates any Cs. If the proposed design does violate one or more of the Cs, the choice of DP must be reconsidered.

In nonlinear one-FR design, we have to look for design windows. There are several possibilities: searching for a maximum or minimum of the FR, seeking designs that are "immune" to variation, and finding designs that are most sensitive to the variation of inputs.

For some nonlinear designs, the problem can be posed as an optimization problem of finding a maximum or minimum of an objective function, subject to a set of Cs. Nonlinear one-FR design with Cs can be optimized if the design goal is to find the maximum or the minimum point, which is not always the case. The design equation and Cs may be expressed as

Maximize

$$FR = f(DP^a) \qquad (2.20)$$

Subject to

$$\{C_i(DP^b)\} = 0 \qquad (2.21)$$

$$\{C_j(DP^b)\} > 0$$

where { } indicates a vector consisting of many constraints and DP^b represents design parameters, including DP^a.

When the design objective is to maximize the FR, it is possible to seek a mathematical solution for the maximum or the minimum that satisfies the FR without violating the Cs. To achieve this goal, we may use the calculus of variations, which requires that the Cs be expressed as functions of the DP [see, for example, Hildebrand (1952) and Papalambos and Wilde (1988)]. However, in most cases, we may not be looking for a maximum or a minimum. It is often difficult to obtain analytical solutions, and numerical solutions may have to be used instead. There are also many Cs imposed on the DPs by manufacturing technologies.

Not all nonlinear designs involve finding a maximum or minimum. In many one-FR designs, we must develop a means of satisfying and changing the FR by choosing the right DP. In this case, we may simply look for a robust design window so that the design can be implemented easily, either mathematically or empirically. In the following example, an empirical means of finding the best set of DPs under a highly constrained manufacturing environment is illustrated.

EXAMPLE 2.6 Van Seat Assembly[10]

An automobile company has designed seats for vans in such a way that the entire seat assembly can be removed from the vehicle to allow more room for cargo and other uses. The design of the seat is shown in Figure E2.6.a. The seat is engaged to the pins mounted on the van floor by the use of linkages, a rear latch, and a front leg. To install the seat, the front leg engages the front pin first while the seat is partially folded, and then the seat is lowered to engage the rear pin with the rear latch. When the rear latch hits the pin, the latch opens. When the rear pin is fully engaged in the latch, the latch closes.

These seats must be exchangeable so that selective assembly is not required (i.e., any seat can fit in any position of any van) and the customers can install them without much effort. Depending on the van, there can be as many as five seats, and as the production rate in modern assembly plants is one van per minute, every one of the latches must work perfectly.

When this automobile company started to produce these vans, they found out that 5% of the seats could not be installed without forcing the pins. This resulted in major production delays and poor quality products. To solve this urgent problem, the automobile company put together a team of experts in quality control. If you were in charge of this project, how would you solve the problem?

SOLUTION

The FR of the seat engagement linkage is that the distance between the front leg and the rear latch when the seat engages the pins must be equal to the distance between the pins, which is 340 mm. The linkages shown in Figure E2.6.b, which are L_{12}, L_{14}, L_{23}, L_{24}, L_{27}, L_{37}, L_{45}, L_{46}, L_{56}, and L_{67}, determine the FR. Table E2.6 shows the nominal lengths of the linkages.

The design must be such that the pins go into the front leg and the rear latch without exerting too much force on the jaw of the rear latch. The pin must hit the sweet spot shown

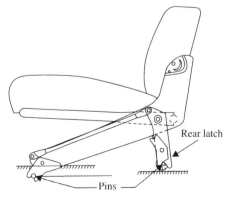

Figure E2.6.a Schematic drawing of a van seat. Seat can be removed and installed easily using a pin/latch mechanism.

Rear latch

Pins

---

[10] This example is created based on the information given in the paper by Hsieh, Oh, and Oh (1990) and from the viewgraphs used in the lecture at MIT by Dr. H. L. Oh (1997). The actual treatment of the problem given in this example deviates from the published paper, although the same data are used.

Table E2.6 Length of Linkages and Sensitivity Analysis

Links	Nominal Length (mm)	Sensitivity (mm/mm)
L_{12}	370.00	3.29
L_{14}	41.43	3.74
L_{23}	134.00	6.32
L_{24}	334.86	1.48
L_{27}	35.75	6.55
L_{37}	162.00	5.94
L_{45}	51.55	11.72
L_{46}	33.50	10.17
L_{56}	83.00	12.06
L_{67}	334.70	3.71

in Figure E2.6.c, which is the location of the radius of gyration, so that the jaw will rotate about the hinge without transmitting any reaction force to the pin of the hinge.

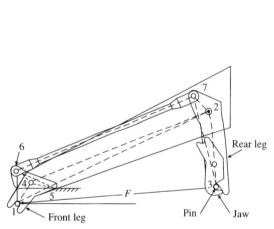

Figure E2.6.b Linkage arrangement of the seat. The sweet spot of the rear latch must be hit so that the jaw will rotate about the hinge.

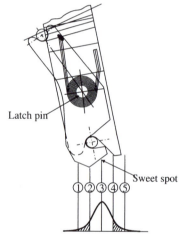

Figure E2.6.c Impact must be made at the sweet spot (3) so that the jaw will simply rotate.

Traditional SPC Solution to Reliability and Quality

The traditional way of solving this kind of problem has been to do the following:

a. *Analyze the linkage to determine the sensitivity of the error.*

Determine $\Delta FR/\Delta L_{12}$, etc. so that we can identify the linkage that must be made very precise to eliminate the major source of error. The sensitivity numbers indicate how much the FR will be off if the error in measuring L_{12} is 1 mm. Table E2.6, which contains both the nominal length and the sensitivity of each linkage, shows that the most sensitive linkages are L_{45}, L_{46}, and L_{56}.

b. *Assess uncertainty through prototyping and measurement.*

The manufacturer of this van measured the distance from the front to rear leg span (FR = F) as shown in Figure E2.6.d.[11] The mean value of FR is determined to be 339.5 mm with a standard deviation of 3.37 mm. Then we can fit the data to a distribution function. If we assume a normal distribution, then the reliability R is given by

$$R = \int_{334}^{346} \frac{1}{\sqrt{2\pi}\,\sigma_F} e^{-(FR-\overline{FR})^2/2\sigma_F}\, d\text{FR} \tag{a}$$

The data plotted in Figure E2.6.d yield a reliability of 95%. The reliability of success is the same as the probability of success.

It should be noticed in Figure E2.6.d that the system range is the normal distribution and the design range is between 334 and 346 cm. The information content for this case is equal to

$$I = \log_2(1/P) = \log_2(1/R) = \log_2(1/0.95) = 0.074 \tag{b}$$

    c. *Develop fixtures and gages to make sure that the critical dimensions are controlled carefully.*

    d. *Hire inspectors to monitor and control the key characteristics using SPC.*

By making sure that the key characteristics remain within certain upper and lower bounds, we can improve reliability. The new data obtained by the use of SPC are shown in Figure E2.6.e. In this prototyping approach, 100% reliability is obtained. However, this is a very expensive way of mass producing the product.

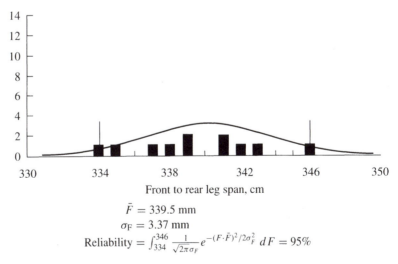

$$\bar{F} = 339.5 \text{ mm}$$
$$\sigma_F = 3.37 \text{ mm}$$
$$\text{Reliability} = \int_{334}^{346} \frac{1}{\sqrt{2\pi}\,\sigma_F} e^{-(F\cdot\bar{F})^2/2\sigma_F^2}\, dF = 95\%$$

Figure E2.6.d Traditional implementation method of trying to assess uncertainty. It is done through prototyping and piloting and then making all linkages accordingly. The test data show that the reliability is only 95%. The bar chart shows the actual measurements and the normal distribution curve is fit to the data.

---

[11] See Appendix 2C for definitions of expected value, variance, and standard deviation for continuous random variables.

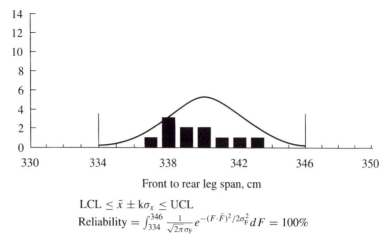

$$\text{LCL} \le \bar{x} \pm k\sigma_x \le \text{UCL}$$
$$\text{Reliability} = \int_{334}^{346} \frac{1}{\sqrt{2\pi}\,\sigma_F} e^{-(F \cdot \bar{F})^2/2\sigma_F^2}\, dF = 100\%$$

Figure E2.6.e The improvement made in reliability by following traditional SPC steps where k = 3. This is a very expensive solution. The bar chart shows the actual measurements and the normal distribution curve is fit to the data.

Unfortunately, these expensive measures do not guarantee that 100% reliability can be achieved even if all the key characteristics are measured, as the assembly process may introduce new errors, in addition to having errors in the measurements. We have to seek a robust manufacturing method to obtain 100% reliability.

Robust Design Solution

The new manufacturing paradigm is to think about how we can achieve the FR, which is to have the pin hit the sweet spot shown in Figure E2.6.c and rotate the jaw about the pivot point, i.e., the latch pin. To achieve this goal, we can use the following steps:

a. If we analyze the forces acting on the jaw, the pin must hit the jaw at (3) such that the jaw will simply rotate about the latch pin. This is similar to the way baseball players must hit the ball. To have no impact forces at the point where the player is holding the baseball bat, the ball must hit the bat at the radius of percussion.

b. Therefore, instead of using the upper and lower bounds for the functional requirement FR, our task is to select the sweet spot and make sure that all the linkages are identical.

The sweet spot FR(x) can be expressed by expanding it in Taylor series about the sweet spot x^{**} as

$$\text{FR}(x) = \text{FR}(x^{**}) + \sum_{i=1}^{n} \frac{\partial \text{FR}}{\partial x_i}(x_i - x^{**}) \tag{c}$$

In terms of the mean value of x, the above equation may be expressed as

$$\text{FR}(x) = \text{FR}(x^{**}) + \sum_{i=1}^{n} \frac{\partial \text{FR}}{\partial x_i}(\bar{x}_i - x^{**}) \tag{d}$$

The variance is given by

$$\sigma_{\text{FR}}^2 = \sum_{i=1}^{n} \left(\frac{\partial \text{FR}}{\partial x_i} \sigma_{x_i} \right)^2 \tag{e}$$

The variance can be minimized if the derivative of FR with respect to DP_i is made small as well as making the variation of $(\text{DP}_i - \text{DP}_i^{**})$ small. In this case, since the FR is a nonlinear function, the sensitivity varies depending on whether the seat is folded or open. In this robust design approach, one would look for design spots that are immune to random variations by evaluating the coefficients and also by minimizing the variance.

From the axiomatic design point of view, this is a simple, straightforward problem in that this is a one-FR design problem that has many extra DPs—all those 10 linkages.

This design has one FR, the front-to-rear leg span, F. This FR is a function of 10 DPs, i.e., 10 linkages. This may be expressed mathematically as

$$\text{FR} = f\,(\text{DP}_1, \text{DP}_2, \ldots, \text{DP}_{10}) \tag{f}$$

where the DPs are the linkages.

This is a case in which the number of DPs exceeds the number of FRs. The random variations (i.e., noise) of these extra DPs cause a random variation of FR, which needs to be brought into the design range.

One possible solution is to fix all extra DPs except one, and vary the one remaining DP to compensate for all the errors made during the manufacturing and assembly operation. The variation of FR may be expressed as

$$\delta \text{FR} = \frac{\partial f}{\partial \text{DP}_i} \Delta \text{DP}_i + \sum_{\substack{j=1 \\ i \neq j}}^{10} \frac{\partial f}{\partial \text{DP}_j} \delta \text{DP}_j \tag{g}$$

The first term of the RHS of Equation (g) represents the compensation effect of the primary DP, which was chosen to vary FR, and the second term is the sum of all the random variations—noise—introduced by extra DPs. What we want to do is to make $\delta \text{FR} = 0$ by compensating error term with ΔDP_i. Since the stiffness of this design is large, especially for linkages L_{45}, L_{46}, and L_{56}, there are only two ways of making δFR equal to zero:

1. Set all the random variations of all DPs equal to zero, which is not realistic because of the variations introduced during manufacturing and assembly.
2. Assemble all the linkages of the seat and fix them except one DP_i [which is equivalent to setting the second term of the RHS of Equation (f) constant], and finally adjust ΔDP_i so that it can compensate for the accumulated errors by setting the first and the second term of the RHS equal to each other. This may be expressed as follows:

$$\frac{\partial f}{\partial \text{DP}_i} \Delta \text{DP}_i = -\sum_{\substack{j=1 \\ i \neq j}}^{10} \frac{\partial f}{\partial \text{DP}_j} \delta \text{DP}_j \tag{h}$$

The automotive company did indeed minimize the variance by assembling all the linkages except one. Then the seat was folded in a vertical position and put in a fixture that fixed the distance F between the front leg and the rear latch. Then, the last linkage, DP_i, was welded in place to satisfy the FR—the distance F between the front leg and the rear latch.

This method of assembly removes the effect of all the random variations introduced by the extra DPs during assembly. In this manner, the FR was satisfied within the specified tolerance—the design range—as shown in Figure E2.6.f. The primary DP_i should be the one that can readily compensate for the sum of the random variations introduced by extra DPs.

Additional Complications Encountered in the Factory

The automobile company found out that the installation of van seats is more complicated than the solution provided in the preceding paragraph. The engineers found out that even after making the distance F constant by welding the last linkage in a fixture, there was a new source of noise. The pins on the van floor introduced additional random variations because they were not installed at exact nominal positions.

The engineers at the automotive company solved this problem by determining the optimum lengths of all 10 linkages. To achieve this optimization, the minimum value of an objective function was obtained. The objective function may be stated as

$$J = \sum_i E[\text{FR}(DP_i, y) - F]^2 \tag{i}$$

where y is the distance between the pins mounted on the van floor, which is a random variable that affects the performance of the van seat assembly, F is the target value of the distance between the pins, and E is the mean value of $[\text{FR}(DP_i, y) - F]^2$. Since y is random, the averaging is done over all DP_i and y. The probability distribution of y must be known to find the minimum value of Equation (i). The engineers assumed it to be Gaussian with mean of F, since the random variation of y was small.

How would you deal with this new source of noise? From the axiomatic design point of view, we may solve this new problem in several different ways. First, we can make the design range, i.e., ΔF, larger, and then come up with a design with a larger sweet spot. Second, we can add another FR to make it into a two-FR problem as

$FR_1 =$ Make the latch mechanism work at $F \pm \Delta F = 340 \pm 2$mm.
$FR_2 =$ Accommodate the random variations in y, δy, over the range of 336 to 344 mm.

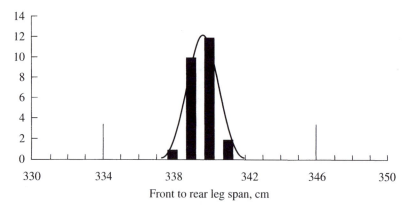

Figure E2.6.f FR distribution when the modern manufacturing method is implemented. Note the narrow distribution of the FR.

The specific values given for FR_2 are based on the measurements made at the automotive company, which found that the random variation—the noise—of y lies within 336 to 344 mm. Since FR_1 has already been solved, the only remaining task is to satisfy FR_2. We must determine DP_2 that will not create coupling between FR_1 and FR_2. What DP_2 would you add to the existing design so as to satisfy FR_2 and the accompanying constraints? It appears that it is not a difficult design task. By introducing this new DP, we can take care of the noise in y by means of a proper design rather than depending on an optimization technique. It should be remembered that the new design should not add to the cost of the van seats. (This is Homework 3.20.)

2.5 ELIMINATION OF BIAS AND REDUCTION OF VARIANCE

As discussed in Section 2.3.1, one of the requirements for making the information content zero is reduction of the bias (shown in Figure 2.1) so that the mean of the system pdf is inside the design range. Because FR is a function of DP, the value of the DP can be changed to eliminate the bias.

The other requirement for making the information content zero is the reduction of variance. As discussed in the preceding sections and in Chapter 1, Section 1.7.10, variance can be reduced by the following means:

a. Decrease the "stiffness" of the system.
b. Minimize random variation of DP and PV.
c. Make the system immune to the variation of DP and PV by lowering the sensitivity of the FR with respect to the DP or the sensitivity of the DP with respect to the PV.
d. If the design has more DPs than the one FR, choose a DP to satisfy the FR and to compensate for the random variations caused by the extra DPs.
e. Make the design range larger.

Examples of (a) are estimating the height of Washington Monument (Example 2.3), hubcap design (Example 1.9), and press fitting of aluminum tube onto a steel shaft (Example 2.7). Examples of (b) and (c) are the windshield wiper design (Example 2.4), and an example of (d) is the van seat assembly (Example 2.6).

2.6 ROBUST DESIGN

The term robust design has been used to mean different things to different people. In the context of axiomatic design, it is defined as the design that satisfies the functional requirements even though the design parameters and the process variables have large tolerances for ease of manufacture and assembly. This definition of robust design was originally established for product design and manufacturing, but as discussed in Chapters 4 through 10, it is equally applicable to systems, software, manufacturing, materials processing, and organizational design.

2.6.1 Determination of Tolerances for Robust Design

As stated in Chapter 1, functional requirements (FRs) are formulated to satisfy customers' needs. The problem formulation is completed when the design ranges for the FRs are

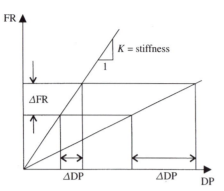

Figure 2.3 Allowable ΔDP for different values of stiffness. The specified tolerance ΔFR can be more easily achieved by lowering the stiffness of the system, thus allowing ΔDP to be larger, which leads to robust design.

specified, along with the constraints. The design ranges for the FRs define what is acceptable for the end-products, whatever they may be. Once the FRs and their design ranges are established, they, in turn, determine the tolerances associated with the DPs and PVs, as shown below.

Consider a one-FR design with a specified design range ΔFR_1. Based on this FR range, the allowable tolerances ΔDP_1 and ΔPV_1 for DP_1 and PV_1, respectively, can be calculated using the design equation. For a one-FR design, the design equation may be written as

$$FR_1 + \Delta FR_1 = A_{11}[DP_1 + \Delta DP_1]$$

$$DP_1 + \Delta DP_1 = B_{11}[PV_1 + \Delta PV_1]$$

or   (2.22)

$$\Delta FR_1 = A_{11}[\Delta DP_1] = (A_{11})(B_{11})\Delta PV_1$$

ΔDP_1 and ΔPV_1 are the *maximum* allowable tolerances for DP_1 and PV_1, respectively.

Equation (2.22) states that the smaller the coefficients A_{11} and B_{11}, the larger are the maximum allowable tolerances ΔDP_1 and ΔPV_1. Therefore, to have a robust design, we must use small coefficients or design a low stiffness system. This was shown graphically in Figure 1.4 in Chapter 1, which is reproduced here as Figure 2.3. However, as discussed in a previous section, the stiffness cannot be made too small—it must exceed the minimum signal-to-noise ratio.

2.6.2 Effect of Noise on FRs in Design and Manufacturing

Unexpected random variations introduced during manufacture and use of a product are called *noise*. Noise may be due to random variations introduced by machining processes, the temperature fluctuations the product is subjected to in use, variations in usage, and other environmental factors, all of which contribute to the random variation of DP. Once we can estimate the variation of $DP(\sigma_{DP_1})$, its effect on the variation of $FR(\sigma_{FR_1})$, can be determined as

$$\sigma_{FR_1} = A_{11}\sigma_{DP_1} = f(a, b, \dots, x)\sigma_{DP_1} \qquad (2.23)$$

Since the system range of FR must be smaller than the specified design range for the FR [i.e., sr(FR) < ΔFR], we have to adjust the value of A_{11}, which may be a function of geometric and other variables, to make it smaller than the maximum value of A_{11}. The minimum value of A_{11} is equal to $\Delta FR_1/\Delta DP_1$.

EXAMPLE 2.7 Joining of Aluminum Tube to Steel Shaft

A part for a machine used in harsh environments is made of a 7075-T6 aluminum tube pressed on a 1020 steel rod as shown in Figure E2.7a. The part must maintain a tight fit in the temperature range of $-30°C$ to $+70°C$. The required interference fit between the rod and the tube is 500 to 1000 psi. The machining accuracy of the mass-production machines selected to make these parts is ±0.001 inch. The radius of the rod is 0.5 inch and the wall thickness of the tube is 0.5 inch. It was found that many of the parts failed in actual service. Determine the cause of the failure. Suggest a robust design of the part so that the functional requirement can always be satisfied.

The properties of these materials are as follows:

	Coefficients of Thermal Expansion	Yield Strength (psi)	$E(x10^6$ psi)	$G(x10^6$ psi)
Aluminum	$25 \times 10^{-6}/°C$	47,000	10.4	3.9
Steel	$15 \times 10^{-6}/°C$	51,000	30.0	11.6

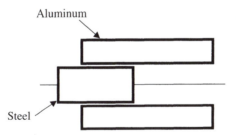

Aluminum

Steel

Figure E2.7a A steel rod is press fit in an aluminum tube.

SOLUTION

This is a one-FR problem and, therefore, the Independence Axiom is always satisfied. The FR is the compressive stress $(\sigma_{rr})_{r=r_o}$ between the steel shaft and the aluminum tube. The DP is the interference fit ϑr (i.e., the difference between the nominal value of the shaft and the nominal value of the cylinder). We have to develop a robust design that can always provide a compressive stress of between 500 and 1000 psi.

The desired nominal value of FR may be set at 750 psi and then the nominal value of DP, ϑr, may be set to correspond to FR $= 750$ psi. Then, we have to choose the dimensions of the part so that FR will be between 500 and 1000 psi when the random variation of DP is $\delta(\vartheta r)$. This has to be done always even though random variation of the interference fit, $\delta(\vartheta r)$, will occur due to the manufacturing variability of the shaft diameter and the inner diameter of the tube and also during service by the temperature fluctuation. The choice of the dimensions, r_o, and the thickness of the tube t must be determined so that they can easily be made using the conventional manufacturing process, which can control the machining dimensions to only within ±0.001. FR must always be satisfied in the temperature range of

$-30°C$ to $+70°C$. The relationship between FR and DP is illustrated in Figure E2.7b. The slope of the curve determines the robustness.

The maximum variation in interference fit due to machining error is 0.002 inch and that due to temperature variation is $[\alpha_{al} - \alpha_{st}]r_o(T_r - T) = 0.00025$ inch, where α_{al} and α_{st} are the coefficients of thermal expansion for aluminum and steel, respectively. T_r and T are the temperature at assembly and the extreme service temperatures ($-30°C$ and $70°C$), respectively. T_r is assumed to be $20°C$. Therefore, the total maximum random variation $\delta(\vartheta r)$ is ± 0.00225 inch.

The design equation is

$$FR_1 = A_{11}DP_1 \tag{a}$$

$$(\sigma_{rr})_{r=r_o} = f(r_o, t)[\vartheta r \pm \delta(\vartheta r)]$$

The function f can be obtained from stress analysis of the tube and the shaft. Equation (a) can be written at the two bounds of the interfacial stress as

$$1,000 \text{ psi} = f(r_o, t)(\vartheta r + 0.00225)$$

$$500 \text{ psi} = f(r_o, t)(\vartheta r - 0.00225) \tag{b}$$

In Equation (b), we have two unknowns: ϑr and $f(r_o, t)$. Solving Equation (b) for these unknowns, we obtain

$$f(r_o, t) = 1.11 \times 10^5 \frac{\text{lb}}{\text{inch}}$$

$$\vartheta r = 0.00675 \text{ inch} \tag{c}$$

It appears that $\vartheta r = 0.00675$ inch may be difficult to achieve when $r_o = 0.5$ inch and $t = 0.5$ inch, because the aluminum tube may yield (i.e., deform plastically). Therefore, we have to determine r_o and t at which the tube will just reach the yield point. The Tresca yield criterion for the aluminum tube is

$$[\sigma_{\theta\theta} - \sigma_{rr}]_{r=r_o} = \sigma_{y,al} = g(r_o, t) \tag{d}$$

Yielding will occur at $r = r_o$, since $\sigma_{\theta\theta}$ is the maximum tensile stress at $r = r_o$ and σ_{rr} is the minimum compressive stress at $r = r_o$. From Equations (c) and (d), we can solve for r_o and t.

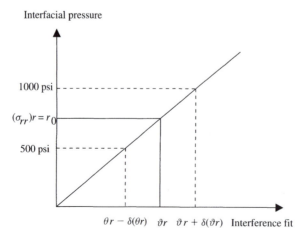

Figure E2.7b Interfacial compressive stress (FR) vs. interference fit (DP).

The functions $f(r_o, t)$ and $g(r_o, t)$ are obtained by analyzing the stress distribution in a thick wall tube.[12]

$$f(r_o, t) = \frac{1}{\dfrac{1}{A} + \dfrac{1}{B}} = 1 \times 10^5 \frac{\text{lb}}{\text{inch}} \tag{e}$$

$$g(r_o, t) = \left[\frac{2b^2}{b^2 - r_0^2}\right] (\sigma_{rr})_{r=r_o} = 47{,}000 \text{ psi} \tag{f}$$

where

$$b = r_o + t \tag{g}$$

$$A = \left(\frac{E_{\text{al}}}{r_o}\right)\left(\frac{r_0^2 + b^2}{b^2 - r_o^2} - \nu_{\text{al}}\right) \tag{h}$$

$$B = \frac{E_{\text{st}}}{r_o(1 - \nu_{\text{st}})} \tag{i}$$

where E_{al} and E_{st} are Young's modulus of aluminum and steel, respectively, and ν is Poisson's ratio of the material described by the subscripts. ν is related to E and G as

$$G = \frac{E}{2(1 + \nu)}$$

Solving Equations (e) and (f) numerically, we obtain $r_o = 3$ inches and $t = 0.1$ inch.

In the foregoing example of designing for the interference fit of the shaft and the tube, the random noises were simply aggregated as the random variation of the chosen DP (i.e., the interference fit Δr). Based on the introduction of noise, δDP, the element of the design matrix, A_{ii}, which consists of the dimensions of the parts and material properties, and the interference fit were determined so as to satisfy the specified FR, the interface pressure. Any other source of noise can be similarly dealt with deterministically rather than by invoking statistical arguments.

2.6.3 Robustness and the Rate of Response in Nonlinear Design

In Section 2.4.1.1, it was shown that the robustness of a linear system could be enhanced by lowering the stiffness of the system, but at the expense of its response rate. As the stiffness is lowered, the response rate is decreased. In this section, we will consider this issue further for the case of nonlinear design.

We can take advantage of nonlinearity in design to improve robustness and response rate by choosing different design points in the nonlinear design space. First, when the design task is to maximize or minimize a functional requirement with constraints, we can find such a maximum or minimum design point. Second, a design window can be found that best satisfies the Independence Axiom in the case of multi-FR design as discussed in Chapter 3.

---

[12] Timoshenko and Goodier (1970). Also see Appendix 2A.

Third, when both the robustness and the response rate of a system must be considered simultaneously, nonlinearity can be used to advantage. In this section, this last aspect of nonlinear design is considered further.

Suppose the relationship between the FR and the DP is as shown in Figure 2.4. At design point a, the design is extremely sensitive to small changes in DP, whereas at design point c, FR is "immune" to the variation of DP. Therefore, the designer should take advantage of the design point c if the FR must not vary when the DP fluctuates. At design point a, the design will respond quickly, but is very sensitive to any variation in DP. Design point b may provide a combination of reasonable robustness and good response-time characteristics.

The response rate of a system is proportional to the slope of the curve shown in Figure 2.4. A system is most likely to respond fastest at a and slowest at c. Therefore, depending on the design task, we can choose different design points by considering the signal-to-noise ratio, the rate of response, and the sensitivity.

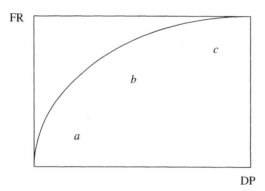

Figure 2.4 Variation of FR as a function of DP in a nonlinear design.

2.7 DESIGN PROCESS

The design process depends on the specific tasks involved and the field of application. Nevertheless we can point out the following important steps involved in a design process:

1. The first step is to understand customers: listening to their concerns and needs, gathering relevant data and past history.

2. The second step is to formulate the FRs and Cs based on an assessment of customer needs. Often, it is best to get inputs from everyone concerned with the proposed product, including marketing, engineering, service, and reliability groups. Whenever possible, the target values of FRs and Cs should be stated.

3. Once the FRs are formulated, the management, marketing, and service groups must concur that indeed the selected FRs are the right ones. If there is disagreement, they should not proceed until they can resolve the issue because the final product will be different, depending on the set of FRs and Cs to be satisfied.

4. Then the designer has to map the FRs into the physical domain by coming up with design ideas and identifying DPs. In a one-FR design, as many plausible DPs as

possible must be considered before discarding any. Sometimes, to generate good ideas for a potential DP, it is necessary to consider extreme cases, contradictory cases, prior examples from other fields, and analogous examples. Sometimes, it is useful to play "what if" games to generate ideas.

5. Once the DP is chosen, the next step is to write the design equation to relate the possible DPs to the FR. In one-FR design, the designer does not have to be concerned about functional independence as coupling is not an issue. Yet the designer must be certain that we are not violating the laws of nature.

6. The proposed design should be compared with constraints to verify that they are not violated.

7. If possible, sketches and drawings should be made to capture the DPs chosen in graphic form at each level of the design hierarchy, in addition to the design equation.

8. To have complete documentation, write down why certain DPs are chosen and why others are not chosen.

9. If the chosen DP is not the final solution and therefore must be further decomposed, go back to the functional domain. Then consider the next-level FRs that are consistent with the DP chosen and the parent FR. When there is more than one FR at this level, which is most likely, make sure that these lower level FRs remain functionally independent by choosing appropriate DPs that do not violate the Independence Axiom.

10. Either during the product design stage or later, consider appropriate manufacturing issues in terms of PVs.

11. At any time during the design process, the designer can change his or her mind and go back and redo the entire design, including modification of FRs, DPs, and PVs.

12. The designer must then go into the implementation stage, including detailing the manufacturing method, schedule, cost, and human resources required.

13. Somewhere in the design process, the design range and the system range should be estimated to determine which DP is a more suitable choice. Sometimes it is nearly impossible to estimate the system range without actual construction of a prototype, but the effort spent estimating the system range can enhance the thinking behind the design. If the design range cannot be established, work with the target value and the acceptable variance of FR, which is equivalent to the design range.

14. After the design is completed, go back to the original customer needs (or attributes) and evaluate the design from the customer's point of view.

15. Benchmarking is a good practice if the product is to be sold competitively with existing products in a given market. It is recommended that benchmarking be done during the later stages of the design process if the goal is to develop an innovative product, but earlier for a more static design.

A systematic development of design following axiomatic design theory may appear to be taking extra time at the early stages of the project, but will save a great deal of time during the execution phase. This kind of practice will guarantee shorter lead time, more reliable products, and lower cost than an experienced-based approach to design. Remember the following:

> *A successful product of any kind is a combined result of macrovision—concept and ideas—and microlevel rational execution that does not leave any decision to random chance.*

2.8 SUMMARY

In this chapter, the issues related to a one-FR design were presented. The importance of identifying customer needs and formulating the correct functional requirement was emphasized. The establishment of the functional requirement can be a lengthy process and take much effort. However, its importance cannot be overemphasized.

Because the one-FR design always satisfies the Independence Axiom, the critical task is to map from the functional domain to the physical domain properly by developing good ideas and identifying potential DPs. Once the DP is selected, the design task involves two things: satisfying the FR within the bounds established by constraints and reducing the information content to zero to satisfy the Information Axiom. This will also lower the cost.

Robust design is defined and the means of creating a robust design are presented with many examples. Robustness and rate of response can be two opposing requirements in some system designs.

One-FR design typically represents a simple design task if the design solution can be found without going through the decomposition process. When the highest level FR must be decomposed, the design task becomes a multi-FR problem, which is the topic of Chapter 3.

REFERENCES

Altshuller, G. S. *Creativity as an Exact Science*: *The Theory of the Solution of Inventive Problems,* Gordon & Breach, Newark, NJ, 1988.

Hildebrand, H. B. *Methods of Applied Mathematics,* Prentice-Hall, Englewood Cliffs, NJ, 1952.

Hsieh, C. C., Oh, K. P., and Oh, H. L. "Design Technique for Minimizing Net Build Errors," *Sensitivity Analysis and Optimization with Numerical Methods,* Vol. 115, pp. 61–69, 1990.

Nordlund, M. "An Information Framework for Engineering Design Based on Axiomatic Design*,"* Doctoral Thesis, Royal Institute of Technology, Stockholm, 1996.

Oh, H. L. Lecture Notes (Viewgraphs) on Robust Design at MIT, 1997.

Papalambros, P., and Wilde, D. J. *Principles of Optimal Design,* Cambridge University Press, New York, 1988.

Ross, S. *A First Course in Probability*, Prentice Hall, Upper Saddle River, NJ, 1997.

Suh, N. P. *The Principles of Design,* Oxford University Press, New York, 1990.

Suh, N. P. "Designing-in of Quality through Axiomatic Design," *IEEE Transactions on Reliability*, Vol. 44, No. 2, pp. 256–264, 1995.

Timoshenko, S. P., and Goodier, J. N. *Theory of Elasticity*, McGraw-Hill, New York, 1970.

APPENDIX 2–A Stress in a Thick Wall Tube[13]

Consider a thick wall tube shown in Figure A.2A.1, which is subjected to both internal and external pressure.

---

[13] From Timoshenko and Goodier (1970).

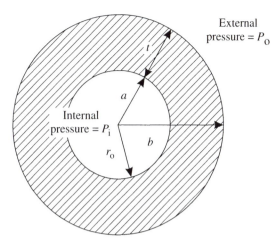

External
pressure = P_0

Figure A.2A.1 Thick wall tube.

$P_0 = $ External pressure
$P_i = $ Internal pressure
$a = r_0$
$b = r_0 + t$

The radial normal stress σ_{rr}, the hoop stress $\sigma_{\theta\theta}$, and the normal strain along the hoop direction $\varepsilon_{\theta\theta}$ are given by

$$\sigma_{rr} = \frac{a^2 b^2}{(b^2 - a^2)r^2}(P_0 - P_i) + \frac{(P_i a^2 - P_0 b^2)}{(b^2 - a^2)} \tag{A.2A.1}$$

$$\sigma_{\theta\theta} = -\frac{a^2 b^2}{(b^2 - a^2)r^2}(P_0 - P_i) + \frac{(P_i a^2 - P_0 b^2)}{(b^2 - a^2)} \tag{A.2A.2}$$

$$\varepsilon_{\theta\theta} = \frac{u}{r} = \frac{\sigma_{\theta\theta} - \nu\sigma_{rr}}{E} \tag{A.2A.3}$$

For the aluminum tube of Example 2.7, the boundary conditions are

$$a = r_0 \tag{A.2A.4}$$

$$b = r_0 + t \tag{A.2A.5}$$

$$P_0 = 0 \tag{A.2A.6}$$

$$P_i = \text{Internal pressure} = P_{\text{int}} \tag{A.2A.7}$$

Equations (A.2A.1) and (A.2A.2) reduce to

$$\sigma_{rr} = \frac{a^2 \left[1 - (b^2/r^2)\right]}{(b^2 - a^2)} P_{\text{int}} \tag{A.2A.8}$$

$$\sigma_{\theta\theta} = \frac{a^2 [1 + (b^2/r^2)]}{(b^2 - a^2)} P_{\text{int}} \tag{A.2A.9}$$

$$(\sigma_{rr})_{r=r_0} = -P_{\text{int}} \tag{A.2A.10}$$

At $r = a$

$$(\sigma_{\theta\theta})_{r=r_o} = P_{\text{int}} \frac{b^2 + a^2}{b^2 - a^2} \tag{A.2A.11}$$

From (A.2A.3), (A.2A.10), and (A.2A.11) we obtain

$$(\sigma_{rr})_{r=r_o} = -A\mu_{\text{al}} \tag{A.2A.12}$$

$$u_{\text{al}} = \frac{(\sigma_{rr})_{r=r_o}}{A} = \frac{P_{\text{int}}}{A} \tag{A.2A.13}$$

where

$$A = \frac{E_{\text{al}}}{a} \frac{b^2 - a^2}{a^2} \left[\frac{1}{(1 - \nu_{\text{al}}) + \dfrac{b^2}{a^2}(1 + \nu_{\text{al}})} \right] \tag{A.2A.14}$$

For the steel rod, we substitute $P_i = 0$, $P_0 = P_{\text{int}}$, $a = 0$, and $b = a$ in (A.2A.1) and (A.2A.2) to obtain

$$[\sigma_{rr}]_{r=r_o} = -P_0 = -P_{\text{int}} \tag{A.2A.15}$$

$$[\sigma_{\theta\theta}]_{r=r_o} = -P_{\text{int}} \tag{A.2A.16}$$

From (A.2A.3), (A.2A.15), and (A.2A.16) we obtain

$$(\sigma_{rr})_{r=r_o} = \frac{E_{\text{st}}}{a(1 - \nu_{\text{st}})} u_{\text{st}} = B u_{\text{st}} \tag{A.2A.17}$$

$$B = \frac{E_{\text{st}}}{a(1 - \nu_{\text{st}})} \tag{A.2A.18}$$

It can be seen from (A.2A.13) that u_{al} is positive (outward radial displacement), whereas from (A.2A.18) it can be seen that u_{st} is negative (inward radial displacement). Thus,

$$\vartheta r = u_{\text{al}} - u_{\text{st}} = \frac{A + B}{AB} P_{\text{int}} \tag{A.2A.19}$$

The pressure at the interface is

$$P_{\text{int}} = \frac{AB}{A + B} \vartheta r = f(a, b)\vartheta r = f(r_o, t)\vartheta r \tag{A.2A.20}$$

where $f(r_o, t)$ is a function that depends on material properties and geometry.

Solving Equation (A.2A.20) for $f(r_o, t)$, we obtain

$$f(r_o, t) = \frac{AB}{A + B} \tag{A.2A.21}$$

Also the difference between the maximum principal stresses is the maximum shear stress, which is given by

$$g(r_o, t) = \lfloor \sigma_{\theta\theta} - \sigma_{rr} \rfloor_{r=r_o} \tag{A.2A.22}$$

From (A.2A.10) and (A.2A.11) we see that

$$g(r_0, t) = (\sigma_{\theta\theta} - \sigma_{rr})_{r=r_0} = P_{\text{int}} \frac{2b^2}{b^2 - r_0^2} = -\frac{2b^2}{b^2 - r_0^2}(\sigma_{rr})_{r=r_0} \quad \text{(A.2A.23)}$$

$$g(r_0, t) = \frac{2b^2}{b^2 - r_0^2}\left(\frac{AB}{A+B}\right)\Delta r \quad \text{(A.2A.24)}$$

APPENDIX 2–B Discrete Random Variables: Expected Value, Variance, and Standard Deviation[14]

If the FR is a discrete random variable (e.g., success vs failure; failure mode A vs. failure mode B vs. failure mode C), the probability function that describes the probability P that a discrete random variable (e.g., number of successes) takes on some specific value is called a probability mass function p.

As with any probability function, the total probability must be 1.0.

$$\sum_{i=1}^{\infty} p(\text{FR}_i) = 1 \quad \text{(A.2B.1)}$$

where FR_i is a specific value of FR.

Expected Value

The expected value of FR, denoted by $E[\text{FR}]$ or μ_{FR}, is defined by

$$E[\text{FR}] = \mu_{\text{FR}} = \sum_i \text{FR}_i \, p(\text{FR}_i) \quad \text{(A.2B.2)}$$

It is estimated from a sample of n observations by

$$\overline{\text{FR}} = \frac{1}{n}\sum_i^n \text{FR}_i \quad \text{(A.2B.3)}$$

Variance

The variance of FR, denoted by $\text{Var}(\text{FR})$ or σ_{FR}^2, is defined as

$$\text{Var}(\text{FR}) = \sigma_{\text{FR}}^2 = E[(\text{FR} - \mu_{\text{FR}})^2] = E(\text{FR})^2 - \mu_{\text{FR}}^2 \quad \text{(A.2B.4)}$$

It is estimated from a sample of n observations by

$$s_{\text{FR}}^2 = \frac{\sum_{i=1}^{n}(\text{FR}_i - \overline{\text{FR}})^2}{n - 1} \quad \text{(A.2B.5)}$$

---

[14] From Ross (1997).

Standard Deviation

The standard deviation of FR, denoted by σ_{FR}, is defined as

$$\sigma_{FR} = \sqrt{\text{Var}(FR)} \tag{A.2B.6}$$

It is estimated from a sample of n observations by

$$s_{FR} = \sqrt{\frac{\displaystyle\sum_{i=1}^{n}(FR_i - \overline{FR})^2}{n-1}} \tag{A.2B.7}$$

Appendix 2–C Continuous Random Variables: Expected Value, Variance, Standard Deviation, and Multivariate Random Variables

Suppose we have measured a large number of samples of front-to-rear leg span of the van seat discussed in Example 2.6 and that the histogram of the sample measurements is shown in Figure A.2C.1. Clearly there can be only a finite set of observations in any sample, but because leg span is a measurement, which can be any value within a given range, it is a continuous random variable. Because the histogram appears to be roughly symmetric, it can probably be fit reasonably well by a normal distribution.

If an FR is a continuous random variable, the probability P of a value of FR between any two given values is described by a probability density function $g(FR)$ as

$$P(a < FR < b) = \int_{a}^{b} g(FR)\, dFR \tag{A.2C.1}$$

Again the total probability is 1.0.

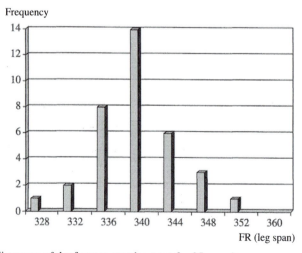

Figure A.2C.1 Histogram of the front-to-rear leg span for 35 samples.

$$\int_{-\infty}^{\infty} g(\text{FR}) \, d\text{FR} = 1 \qquad (A.2C.2)$$

Expected Value

The expected value of a continuous random variable FR is given by

$$E[\text{FR}] = \int_{-\infty}^{\infty} \text{FR} \, g(\text{FR}) \, d\text{FR} \qquad (A.2C.3)$$

Variance

$$\text{Var}(\text{FR}) = \sigma_{\text{FR}}^2 = E[(\text{FR} - \mu_{\text{FR}})^2] = E(\text{FR})^2 - \mu_{\text{FR}}^2 \qquad (A.2C.4)$$

Standard Deviation

$$\sigma_{\text{FR}} = \sqrt{\text{Var}(\text{FR})} \qquad (A.2C.5)$$

The mean, variance, and standard deviation of a continuous random variable are estimated exactly the same way as for a discrete random variable [Equations (A.2B.3), (A.2B.5), and (A.2B.7), respectively].

Normal Random Variables

The normal (bell-shaped) distribution is the most important and most prevalent of all continuous distributions. If the FR is normally distributed with mean μ and variance σ^2, the pdf is given by

$$g(x) = \frac{1}{\sqrt{2\pi}\sigma} e^{-(x-\mu)^2/2\sigma^2} \qquad -\infty < x < \infty \qquad (A.2C.6)$$

Multivariate Random Variables

If $\text{FR}_1, \text{FR}_2, \ldots, \text{FR}_m$ is a set of continuous random variables with joint probability density function $g(\text{FR}_1, \text{FR}_2, \ldots, \text{FR}_m)$, the probability P that all FR_i fall within their respective design ranges (e.g., dr_1^l–dr_1^u, $\ldots$, dr_m^l–dr_m^u) is described by

$$P = \text{Pr}(\text{dr}_1^l \leq \text{FR}_1 \leq \text{dr}_1^u, \ldots, \text{dr}_m^l \leq \text{FR}_m \leq \text{dr}_m^u)$$

$$= \int_{\text{dr}_1^l}^{\text{dr}_1^u} \cdots \int_{\text{dr}_m^l}^{\text{dr}_m^u} g(\text{FR}_1, \ldots, \text{FR}_m) d\text{FR}_1 \ldots d\text{FR}_m \qquad (A.2C.7)$$

If all FR_i are statistically independent, this joint probability can be expressed by the product of the separate (marginal) probabilities of each FR_i, i.e.,

$$g(\text{FR}_1, \text{FR}_2, \ldots, \text{FR}_m) = \Pi f_i(\text{FR}_i) \qquad (A.2C.8)$$

However, if some of the m FRs are not statistically independent of the others, then the joint probability is expressed by

$$g(\text{FR}_1, \text{FR}_2, \ldots, \text{FR}_m) = \Pi f_i(\text{FR}_i|\text{FR}_j) \; j = 1, \ldots, i-1 \qquad \text{(A.2C.9)}$$

where the appropriate conditional probabilities are indicated by the design matrix.

HOMEWORK

2.1 The windshield-wiper design discussed is a "one-FR" design problem. Because there is only one FR, the Independence Axiom is always satisfied. The only remaining task is to minimize information and satisfy the functional requirement within the specified design range. Formulate the design problem and show how you can develop a solution that always satisfies the FR with zero information content.

2.2 Derive the condition for ensuring that the specified design range is always greater than the random variation in DPs.

2.3 Robust design may be defined as a design that can satisfy functional requirements even when there are large variations in DPs and PVs. Develop criteria for robust design.

2.4 To improve the productivity of automobile assembly, the interior headliner of a car roof is to be attached to the steel exterior panel by means of a "snap fit," molded by injection molding of nylon. Design the "snap fit." The gap between the steel exterior panel and the interior panel must be between 0.435 to 0.500 inch. The weight of the interior panel is 30 lb. The part must be usable for 20 years.

2.5 Design a mechanism that can be used in an advanced lithography machine to control the linear motion of a stage to an accuracy of 0.01 μm.

2.6 A toggle mechanism is to be used to control the motion of a slider as shown in Figure H.2.6. Determine the dimensions of the linkages for the highest accuracy of the slider when the central joint of the toggle mechanism is controlled by a piezoelectric actuator that has an accuracy of 1 μm. How accurately can you control the motion of the slider?

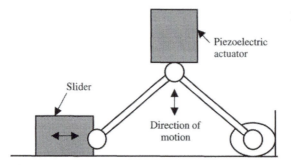

Figure H.2.6 Linear motion controller.

2.7 Automobile engines have fan belts and pulleys to drive accessory equipment such as air-conditioning pump, cooling water pump, and alternators. If the belt and the pulleys are not properly designed, the belt slips and makes undesirable noise. Design a fan belt/pulleys system that will satisfy the functional requirement of driving the accessory equipment without making any noise. Assume that the diameter of the pulley attached to the crankshaft of the engine is 8 inches and all other pulleys are 5 inches in diameter. Clearly state your FRs and DPs. Write your design equation(s).

2.8 Revisit Example 2.6. Based on the final solution given by Equation (g), which link is the best link to set last after all other linkages are fixed?

2.9 Propose a new design for the van seat assembly that can satisfy the goal of manufacturing interchangeable seats that will fit all vans at all positions.

3 Multi-FR Design

3.1 INTRODUCTION

This chapter deals with design that involves many functional requirements at a given level of the design hierarchy—multi-FR design. What distinguishes multi-FR design from one-FR design is the need to satisfy the Independence Axiom, i.e., to maintain the independence of FRs. A design that violates the Independence Axiom produces a coupled design, which either cannot satisfy the FRs or is not robust enough to fulfill the FRs at all times. These fundamental concepts are illustrated using many examples. Many of them are related to hardware design because hardware is easier to visualize and facilitates the understanding of the basic principles. However, the theory applies equally well to other multi-FR designs, i.e., software, systems, organizations, and processes.

A coupled design can be redesigned to be an uncoupled or decoupled design based on these axioms. The design procedure for developing uncoupled and decoupled designs is illustrated by many examples. The principles behind robust design are discussed and illustrated through examples. Nonlinear multi-FR design is also discussed.

Just like the one-FR design case, multi-FR designs must also satisfy the Information Axiom, which demands that we minimize the need for additional information in satisfying the FRs. The information content of a multi-FR design can be reduced when the design satisfies the Independence Axiom. Based on the information content, the best design among those that satisfy the Independence Axiom can be selected when many FRs must be satisfied simultaneously.

The *complexity* of a design is related to its information content and functional independence. Therefore, to reduce complexity, both the Independence Axiom and the Information Axiom must be satisfied. Examples are given to illustrate how complexity can be reduced through redesign.

In this chapter, we will consider multi-FR design—the design task that must satisfy several FRs at the highest level of the design hierarchy more or less simultaneously. Many designs are of this type. Even a design task with a single FR can become a multi-FR design when the FR must be decomposed. The functional independence of FRs is the most important issue in a multi-FR design in contrast to the one-FR design, in which the Independence Axiom is always satisfied and only the Information Axiom is left to be satisfied.

A simple multi-FR design example. The simplest case of a multi-FR design has only two functional requirements. As an example, suppose we want to design a knob that can grasp the end of a shaft and turn it. It must satisfy two FRs:

$FR_1 =$ Grasp the end of the shaft tightly.
$FR_2 =$ Turn the shaft.

One of the designs proposed for the knob is shown in Figure 3.1. A flat end (Surface B) has been machined at the end of the shaft (Surface A) so that the knob can turn the shaft. The knob has an axial slot and its inside diameter is slightly smaller than the shaft diameter. When the knob is pushed onto the shaft, the slotted section of the knob opens elastically to grasp the shaft by interference fit. Is the proposed design a good design? Why?

The design shown in Figure 3.1 is not a good design. When the knob is turned to rotate the shaft, it loses its grip of the shaft because the slotted section of the knob opens up. In other words, the two FRs are coupled by the proposed design. How should we improve this design? It will be shown later that a new design—a slight modification of the design shown in Figure 3.1—will lower the cost of the product and improve the performance.

This knob is the simplest case of a multi-FR design involving hardware. It has two FRs and two DPs:

$DP_1 =$ Interference fit and the slot
$DP_2 =$ The flat surface

The theory that will explain why it is not a good knob design is equally applicable to the design of other things, such as software, systems, and materials. The fundamental concepts are illustrated using many examples.

What is the status of multi-FR design in industry? In the past, designers used their experience to come up with clever and creative designs by trying out many different ideas—the "trial-and-error" method. A typical approach would be to develop a design, often empirically and intuitively, that satisfies some loosely defined design goals without

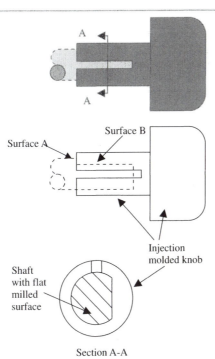

Figure 3.1 Knob design for a shaft.

specifying FRs. The next step would be an attempt to improve the design through extensive testing, debugging, modifying, and optimizing prototype hardware. Because the designers had no absolute reference frame for making correct design decisions, poor designs ensued. A great deal of time was wasted testing the defective design in attempts to correct it. It is difficult to improve poorly designed systems—they cannot be made robust, built readily at low cost, and function reliably or as originally intended.

In many organizations, the prevailing culture accepts and justifies the time-consuming process described above as being the norm. In other words, they do not have time to design things correctly to begin with, but have lots of time to fix past mistakes. Often it is done under the rubric of optimization. Optimization of design typically involves coming up with the best compromise—a design solution that compromises one of the original goals to satisfy another goal better. This practice is a result of not satisfying the Independence Axiom. If the design is a linear design that satisfies the Independence Axiom, each FR can be satisfied independently of the other FRs. If the design is a nonlinear design that satisfies the Independence Axiom, we may try to choose the best design window where the design is most robust. This is one of the subjects discussed in this chapter.

One of the goals of axiomatic design is to help designers make the correct decision first and do the job right. The goal of axiomatic design is to design it right to begin with, so that we do not spend endless time testing, debugging, and modifying the system at the end of the development cycle. For those uninitiated in axiomatic design, it may appear to take a little bit longer to design according to axiomatic principles—to develop FRs, DPs, and PVs, and to decompose them to create a complete design hierarchy. However, in the end we can easily justify the time spent, as it results in reduction of overall development time by minimizing the need to debug, test, and modify. The adage that "it takes longer to correct a mistake than to do it right in the first place" applies to engineering and design as well as to our daily experience.

The multi-FR design is presented using the design of hardware as examples, because many engineers find it easier to comprehend the basic principles when the design ideas can be visualized through hardware. However, all the theories presented in this chapter apply equally well to all other designs: software, systems, organizations, manufacturing processes, and materials. Chapters 4 through 8 discuss the design of these items.

3.2 BRIEF REVIEW OF AXIOMATIC THEORY FOR MULTI-FR DESIGNS

In Chapter 1, the basics of axiomatic design are presented. In this section, several key points related to multi-FR design will be reviewed.

1. **Recognize the existence of domains.** The world of axiomatic design consists of domains: the customer domain, the functional domain, the physical domain, and the process domain. Designers must know which domain they are in as they proceed with the design.

2. **Design involves mapping between the domains.** In the case of discrete mechanical products, design of products involves the mapping from the functional domain to the physical domain, whereas process design involves the mapping from the physical domain to the process domain.

3. **Define functional requirements (FRs) and constraints (Cs).** We must define the FRs of the design task and the associated Cs, based on the inputs from the ultimate user (sometimes called the customer) as well as from other relevant resources. This is not a trivial task! Sometimes it takes weeks and months to agree on a common set of FRs and Cs.

4. **Design equation and the independence of FRs.** The Independence Axiom is applied as we map from domain to domain. To satisfy the Independence Axiom, the design equation $\{FR\} = [DM]\{DP\}$ must be satisfied so that the FRs are independent of one another when the DPs are changed. That is, when any one of the FRs is changed by its DP, other FRs should not be affected. This condition is satisfied when the design matrix [DM] is either diagonal or triangular, regardless of whether the design is linear or nonlinear.

How do we determine the independence of FRs? To determine the nature of the design matrix, we have to arrange the design equation in such a manner that it can be put into a diagonal or triangular form. This can be done readily when the matrix is diagonal. When the design matrix is large and not diagonal, the rearrangement process can be time consuming. However, to make design decisions correctly, the design matrix must be rearranged correctly. This can be done with the help of computers (e.g., Axiomatic Design Software) or manually.

The procedure for rearranging the design matrix is as follows (from Appendix 10B of *The Principles of Design*, Suh, 1990):

a. Find the row that contains one nonzero element. Rearrange the order of FRs and DPs by putting the row and the column that contain the nonzero element first.
b. Excluding the first row and column, find the row that contains one nonzero element. Rearrange the components of FRs and DPs by putting the row and column that contains the nonzero element second.

c. Repeat the procedure until there are no more submatrices to analyze.

At any time during the execution of the above procedure, if all of the remaining rows contain more than one nonzero element, the design is coupled.

In the case of some nonlinear designs, the magnitude of the elements of the design matrix changes, depending on the specific values of DPs. Therefore, the above procedure can be applied only for a given set of DPs. This issue of nonlinearity is discussed further in Section 3.8.

Can we use coordinate transformation to make the [DM] a diagonal matrix? It should be noted that [DM] is a second-order tensor and therefore should be amenable to coordinate transformation so as to be converted to a diagonal matrix. However, it does not serve a useful purpose in design, as the resulting diagonal matrix will be a meaningless juxtaposition of physical elements that does not fulfill the FRs in the case of product design. Similarly, in the case of other designs, coordinate transformation may not yield useful results.

5. **Decomposition.** At any given level of the design hierarchy, a DP chosen to satisfy a given FR may not be implemented because the DP does not have sufficient details for implementation. For example, in dealing with the refrigerator design (Example 1.5), we had to decompose the highest level FRs (e.g., FR_1 = freeze food for long-term preservation) as the chosen DPs (e.g., DP_1 = the freezer section) lacked details for implementation. If DP_1 is an available commercial product, we do not have to decompose.

When a given set of FRs and DPs is decomposed, the next-level FRs are the functional requirements of the parent DP, which must be consistent with the parent FR. That is, the set of the child FRs must be able to yield the parent FR when they are integrated according to the design matrix at the child level. We cannot arbitrarily introduce a child-level FR if it has no bearing to the parent FR. The lower level FR must have a clear lineage by being derivable from the parent FR. If the child-level FR is absolutely necessary to proceed with the design and yet there is no obvious parent FR, introduce the missing FR at a higher level.

6. **Constraints.** It should be noted that all the Cs introduced at the beginning of the design process apply throughout the design process. Furthermore, all the decisions made at higher levels act as constraints at the lower levels.

The more constraints there are, the smaller the design window is, as the choice of DPs becomes limited. The available set of DPs that does not violate the constraints decreases with increases in the number of Cs.

7. **The Information Axiom.** There can be many different ways of decomposing FRs and DPs, all of which satisfy the same set of the highest level FRs and Cs. Different decomposition will lead to different designs in that the designs will have different components and different lower level functionality. However, all of these designs are *equivalent* if all of them satisfy the same set of highest level FRs and Cs.

What distinguishes these different designs is their information content. Some of the designs may satisfy the FRs all the time with 100% certainty, i.e., the system range is always inside the design range. These designs are superior to other designs with a lower probability of satisfying the FRs.

8. **Physical integration of DPs.** Sometimes, in the case of product design, the integration of DPs into fewer physical pieces can reduce the information content if and only if the physical integration does not lead to functional coupling. A good example of physical integration is the beverage can design discussed in Example 1.2.

9. **Nonhardware design.** The above argument for hardware design is equally valid for all other design. In the case of organizational design, for example, people satisfy the FRs by providing DPs. In some cases, one person may be able to satisfy many different functions equally well if the person can switch on different "DPs" in his or her mind as the FRs change. Unfortunately, it depends on the person—some people can deal with many FRs at the same time and some cannot. Many people try to deal with a variety of different FRs using the same DP, creating disastrous results. The one-dimensional person—who can deal with only one FR at a time—should not be in a leadership position because leaders must be able to deal with multi-FR design situations. Unfortunately, given the Peter Principle, many organizations have a one-dimensional person in the driver's seat for many years before his or her errors are discovered.

3.3 THE INDEPENDENCE AXIOM AND THE INFORMATION AXIOM: THEIR IMPLICATIONS FOR A MULTI-FR DESIGN TASK

To understand the significance of the Independence Axiom in multi-FR design, we must recall the definition of the FRs. FRs are defined as a minimum set of independent requirements that completely characterizes the desired functions of the design. FRs constitute "what we want to achieve." To achieve these FRs, we have to go to the physical domain and conceive a design that can be characterized by DPs. The chosen DPs must be able to satisfy the FRs without violating the Cs imposed on the design. The Independence Axiom states that the functional requirements, not the physical elements of the design, must remain independent of each other when we make design decisions.

The Information Axiom provides the basis for decision making when there are many choices. The Information Axiom states that of all the designs that satisfy the Independence Axiom, the best design is the one with the minimum information content. Because the information content is a measure of the probability of achieving the FRs, the axiom states that the design with the highest probability of achieving the stated FRs is the best one to choose, although other designs may also satisfy the Independence Axiom. Therefore, the Information Axiom can deal with the selection of the "best" design when the design task involves many FRs.[1]

The fact that a given design has a finite amount of information content means that to achieve the stated FRs successfully, additional information must be supplied to the design. The additional information is often supplied by human operators, such as when a machinist has to repeatedly measure the dimensions of a part being machined and gradually approach the desired dimensions of the part, because the machine is not accurate enough to produce the desired dimensions by itself.

The first and second axioms produce simple designs that can be more easily implemented than designs that violate them. Corollary 3 states that to reduce the information

---

[1] The Information Axiom is the best means of selecting the optimum solution when many FRs must be satisfied at the same time.

content, parts should be integrated, provided that the functional independence can be maintained after the physical integration. This integration is best illustrated by the beverage can in Example 1.3. The typical aluminum can must satisfy as many as 12 FRs, although the can consists of only three physical pieces: the body, the lid, and the opener. Some of the FRs of a beverage can are the following: contain the pressure of a carbonated beverage, allow stacking of cans, support the stacking load, resist impact loading, open the can, minimize the use of aluminum, etc. The DPs of the can are the thickness, curvatures at the lower section and the bottom of the can, shape of the can near the top, opening tab, etc. When the number of parts is reduced through physical integration according to Corollary 3, the information content can be reduced or at least remains the same. When there are many physical parts, the probability that the information content of some of the parts may not be zero increases with the number of parts.

These axioms have other implications:

1. If someone claims that a coupled design has less information content than a particular uncoupled (or decoupled) design, there is another uncoupled or decoupled design that has less information content than the coupled design (Theorem 18).
2. If someone has developed a design that violates the Independence Axiom (i.e., a coupled design), the design is inferior to a design that satisfies this axiom in terms of satisfying the specified FRs. Therefore, strive to find an uncoupled or decoupled design (Theorem 19).
3. Information content is always measured relative to the design range given for the FRs; that is, the information content is measured by determining if the system range is inside the design range. This means that if the design range is tightened, an uncoupled design may become a coupled design and vice versa (Theorem 20).

Are these two axioms independent from each other? The question of whether the Independence Axiom and the Information Axiom are independent of one another has been raised from time to time. The argument goes as follows:

Argument A—Because an uncoupled design is superior to and has less information content than a coupled design, the Information Axiom is identically satisfied. Therefore, the Information Axiom is not needed, i.e., these two axioms are not independent of each other.

Argument B—When we try to satisfy the Information Axiom, we will end up developing an uncoupled or decoupled design. Therefore, the Independence Axiom is a subset of the Information Axiom.

There are at least two important reasons why we need both the Independence Axiom and the Information Axiom as two independent axioms. First, there can be many designs that satisfy the Independence Axiom, but some may be superior to others in terms of the probability of achieving the FRs. Second, in some cases, you may have an uncoupled design (that satisfies the Independence Axiom) and yet have a larger information content than a coupled design, indicating that the designer should seek another uncoupled design that has less information content than the coupled design. Therefore, we need two independent axioms that must be satisfied by all designs. El-Haik and Yang (1999) come to the same conclusion based on an analysis of the information content; there can be coupled designs that have less information content than an uncoupled design, and therefore both of the axioms are required to come up with a rational design (see Appendix 3A).

■ 3.4 ON IDEAL MULTI-FR DESIGN

Many designers sometimes try to satisfy a given set of FRs using a random set of DPs. However, according to the Independence Axiom, the number of DPs must be matched to the number of FRs. There cannot be any arbitrary number of DPs for a given set of FRs.

Consider a design that is represented by the following design equation:

$$\{FR\}_m = [A]_{m \times n}\{DP\}_n \tag{3.1}$$

where $\{FR\}_m$ = a vector of m FRs
$\{DP\}_n$ = a vector of n DPs
$[A]_{m \times n}$ = an $m \times n$ design matrix

What is an ideal design? When the number of FRs equals the number of DPs (i.e., $m = n$) and the design is uncoupled, the design is an ideal design if the information content is zero because the system range is inside the design range (Theorem 4). Equation (3.2) shows an ideal design, provided that the information content is zero.

$$\begin{Bmatrix} FR_1 \\ FR_2 \\ FR_3 \end{Bmatrix} = \begin{bmatrix} X & 0 & 0 \\ 0 & X & 0 \\ 0 & 0 & X \end{bmatrix} \begin{Bmatrix} DP_1 \\ DP_2 \\ DP_3 \end{Bmatrix} \tag{3.2}$$

When $n < m$, the design is coupled (Homework 3.1 and Theorem 1). When $n > m$, the design is redundant, which may be coupled, uncoupled, or decoupled (Theorem 3).

We will consider three examples in this section. First, a redundant design will be examined to show how it can be improved by proper choice of DPs. The second example shows how a coupled design can be made uncoupled by proper choice of DPs. The third example illustrates a systematic means of developing an uncoupled design based on the Independence Axiom and also the means of reducing information content through the integration of DPs in a single physical part.

EXAMPLE 3.1 Redundant Design

Consider a redundant design given by the following design equation:

$$\begin{Bmatrix} FR_1 \\ FR_2 \\ FR_3 \end{Bmatrix} = \begin{bmatrix} X & 0 & X & 0 & 0 & X & X \\ X & X & 0 & X & 0 & 0 & X \\ X & X & 0 & 0 & X & X & 0 \end{bmatrix} \begin{Bmatrix} DP_1 \\ DP_2 \\ DP_3 \\ DP_4 \\ DP_5 \\ DP_6 \\ DP_7 \end{Bmatrix} \tag{a}$$

Is this a good design? What should we do with this design to make it robust and minimize random variation of FRs?

SOLUTION

When confronted with this kind of design, the temptation is to use a statistical technique to figure out how these DPs should be manipulated, but that is not the right thing to do.

Instead we should apply the Independence Axiom and see how the design can be made to work.

If we let all seven DPs be active design parameters, the design will behave like a coupled design. Furthermore, it will invite many sources of noise associated with random variation of each of the DPs and complicate the design unnecessarily. However, if we "freeze" some of the unnecessary DPs, we will produce a design that satisfies the Independence Axiom. We may obtain the best design if we select DP_3, DP_4, and DP_5 and eliminate or freeze all other DPs as shown by Equation (b). Then, the design will be an uncoupled design.

$$\begin{Bmatrix} FR_1 \\ FR_2 \\ FR_3 \end{Bmatrix} = \begin{bmatrix} X & 0 & 0 \\ 0 & X & 0 \\ 0 & 0 & X \end{bmatrix} \begin{Bmatrix} DP_3 \\ DP_4 \\ DP_5 \end{Bmatrix} \quad\quad (b)$$

If we select the set $\{DP_1, DP_2, DP_6\}$, the design will be coupled. On the other hand, if we select the set $\{DP_7, DP_2, DP_5\}$ or the set $\{DP_1, DP_2, DP_5\}$, we will have a decoupled design as shown in Equations (c) and (d).

$$\begin{Bmatrix} FR_1 \\ FR_2 \\ FR_3 \end{Bmatrix} = \begin{bmatrix} X & 0 & 0 \\ X & X & 0 \\ 0 & X & X \end{bmatrix} \begin{Bmatrix} DP_7 \\ DP_2 \\ DP_5 \end{Bmatrix} \quad\quad (c)$$

$$\begin{Bmatrix} FR_1 \\ FR_2 \\ FR_3 \end{Bmatrix} = \begin{bmatrix} X & 0 & 0 \\ X & X & 0 \\ X & X & X \end{bmatrix} \begin{Bmatrix} DP_1 \\ DP_2 \\ DP_5 \end{Bmatrix} \quad\quad (d)$$

Question: Both of the above designs are decoupled designs and thus are acceptable. How would you determine which is the better of the above two designs if you were given the actual magnitudes of the elements shown in Equations (c) and (d)?

When designers try to make design decisions without writing design equations, it is easy to make wrong decisions. In fact, if the number of FRs is greater than five, the number of permutations is so large that it cannot be done without the aid of a systematic approach as discussed in Section 3.2. When the number of FRs is greater than seven, it is advisable to use a heuristic approach as there will be too many combinations to try. These rules are incorporated in the Axiomatic Design Software[2] presented in Chapter 5.

How do we decouple a coupled design? Corollary 1 states:

Decouple or separate parts or aspects of a solution if FRs are coupled or become interdependent in the designs proposed.

There are many ways that some—not all—coupled designs may be decoupled or uncoupled. A simplest example of uncoupling a coupled design is given in Example 3.2.

---

[2] A simplified version of the Axiomatic Design Software developed by Axiomatic Design Software, Inc. can be found on the Internet. To download the demo, go to http://www.AxiomaticDesign.com/demo and enter: username—acclaro, password—demo, domain—(leave blank). The demo version is identical to the commercial version with three exceptions: (1) Windows only (the commercial version, written in Java, runs on any platform supporting Java 1.1.8, (2) single-user only (the commercial version supports multiuser), and (3) maximum of 15 FRs (there is no software limit in the commercial version).

EXAMPLE 3.2 Decoupling of Two Positioning Functions of a Beam[3]

The vertical position of the two end points of the beam is critical for the overall function of a beam. The two FRs are

FR_1 = Position the left end of the beam.
FR_2 = Position the right end of the beam.

It was proposed that FR_1 and FR_2 be satisfied by the two supports, DP_1 and DP_2, which are shown in Figure E3.2.

Initially, the supports were positioned such that $l_1 = l_2 = l_3$, resulting in the design equation:

$$\begin{Bmatrix} FR_1 \\ FR_2 \end{Bmatrix} = \begin{bmatrix} 2 & -1 \\ -1 & 2 \end{bmatrix} \begin{Bmatrix} DP_1 \\ DP_2 \end{Bmatrix} \tag{a}$$

This is a coupled design as the off-diagonal elements are $\neq 0$. It can also be noted that because the diagonal elements are > 1, this solution amplifies the input variation, which is characteristic of a sensitive (nonrobust) design. Show how the design can be decoupled.

SOLUTION

FR_1 and FR_2 can be satisfied by varying DP_1 and DP_2. The relation between the input parameters, DP_1 and DP_2, and the output parameters, FR_1 and FR_2, is given by the design equation as

$$\begin{Bmatrix} FR_1 \\ FR_2 \end{Bmatrix} = \begin{bmatrix} A_{11} & A_{12} \\ A_{21} & A_{22} \end{bmatrix} \begin{Bmatrix} DP_1 \\ DP_2 \end{Bmatrix} \tag{b}$$

For this case, the matrix elements A_{ij} representing the partial derivatives, $\partial FR_i / \partial DP_j$, may be written as

$$A_{11} = \frac{\partial FR_1}{\partial DP_1} = \frac{l_1 + l_2}{l_2} \tag{c}$$

$$A_{12} = \frac{\partial FR_1}{\partial DP_2} = -\frac{l_1}{l_2} \tag{d}$$

$$A_{21} = \frac{\partial FR_2}{\partial DP_1} = -\frac{l_3}{l_2} \tag{e}$$

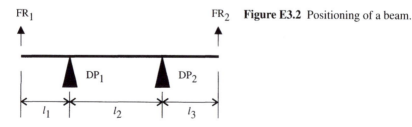

FR_1 $\qquad$ FR_2 **Figure E3.2** Positioning of a beam.

DP_1 $\qquad$ DP_2

l_1 $\qquad$ l_2 $\qquad$ l_3

---

[3] From Söderberg and Lindkvist (1999).

$$A_{22} = \frac{\partial FR_2}{\partial DP_2} = \frac{l_2 + l_3}{l_2} \tag{f}$$

To increase the robustness of the design and to make it uncoupled, a diagonal matrix with the diagonal elements ≤ 1 is preferable. In this case, the design is improved by choosing $l_1 = l_3 = 0$, i.e., moving the supports to the ends of the beam. This results in a fully diagonal matrix with the diagonal elements equal to 1, which is the best possible solution in this case.

$$\begin{Bmatrix} FR_1 \\ FR_2 \end{Bmatrix} = \begin{bmatrix} 1 & 0 \\ 0 & 1 \end{bmatrix} \begin{Bmatrix} DP_1 \\ DP_2 \end{Bmatrix} \tag{g}$$

If a given design is a fully coupled design with the number of FRs being equal to or smaller than the number of DPs, it may not be possible to decouple or uncouple the design. In this case, we have to start the design process all over again by finding a new set of DPs. Conversely, if the number of FRs is larger than the number of DPs and if the design matrix can be made diagonal, then a coupled design can be made decoupled by adding new DPs (Homework 3.2).

The purpose of Example 3.3 is to illustrate that an ideal uncoupled design can be created if we systematically develop the design, starting out with desired FRs and finding a right set of DPs.

EXAMPLE 3.3 Hot and Cold Water Faucet[4]

Suppose your job is to design a water faucet that has the following two functional requirements:

FR$_1$ = Control the water flow rate Q without affecting the water temperature.
FR$_2$ = Control the temperature T of the water without affecting flow rate.

This faucet will enable the user to set the temperature of the water without affecting the flow rate, and also to change the flow rate of the water without affecting its temperature. Most users would prefer this faucet rather than the faucet that has two knobs, one for control of the flow rate of hot water and the other for cold water. According to the Independence Axiom and Theorem 4, the user should be able to do this with two controls, one for each requirement. However, the designer's task is complicated by the fact that water comes in two pipes (hot and cold); thus the two things that are easiest to design controls for (flow rate of hot water and flow rate of cold water) are different from the specified functional requirements.

There are several faucet types that satisfy FR$_1$ and FR$_2$ available on the market today. The case study will show how a faucet designer could employ axiomatic design to approach the problem of developing a system that enables the user to easily and accurately control both the temperature and the flow rate of the water that comes out of the faucet.

---

[4] From "Axiomatic Design of Water Faucet," unpublished report by Anders Swenson and Mats Nordlund of Saab AB, Linkoping, Sweden, 1996.

Solution[5]

The water faucet designer's job is to develop a system that allows the user to easily and exactly control the temperature and flow rate of the water coming out of the faucet.

Figure E3.3.a shows a faucet with two valves, one that can control the flow rate of hot water and the other that can control the flow rate of cold water. In this two-valve faucet, there are two dials that can be turned an angle of ϕ_1 and ϕ_2. These angles are the DPs: $DP_1 = \phi_1$ and $DP_2 = \phi_2$. The faucet shown in Figure E3.3a does not satisfy the Independence Axiom because DP_1 and DP_2 each affects both FR_1 and FR_2. We derive the following design equation for this system:

$$\begin{Bmatrix} Q \\ T \end{Bmatrix} = \begin{bmatrix} X & X \\ X & X \end{bmatrix} \begin{Bmatrix} \phi_1 \\ \phi_2 \end{Bmatrix} \tag{a}$$

This two-valve faucet is a coupled design because the two FRs, Q and T, cannot be independently changed as shown by the full design matrix.

A good design would have a design matrix in which each of the diagonal elements is X and each of the off-diagonal elements is 0. Such a design would be uncoupled. In between the coupled design and the uncoupled design is the decoupled design, which has a triangular design matrix; decoupled and uncoupled designs are acceptable, whereas coupled designs are unacceptable.

The FRs are always independent by definition. According to the Independence Axiom, DPs must be so chosen that we can maintain the independence of these FRs. The faucet designer's job is to come up with a design that satisfies the FRs—ideally one that is uncoupled.

Design of the Next-Generation Faucet

We want to create a faucet system that can provide independent control of the flow rate Q and the temperature T to satisfy the Independence Axiom. One possible design for such a system may be represented by the design equation:

$$\begin{Bmatrix} Q \\ T \end{Bmatrix} = \begin{bmatrix} X & 0 \\ 0 & X \end{bmatrix} \begin{Bmatrix} \text{Knob A} \\ \text{Knob B} \end{Bmatrix} \tag{b}$$

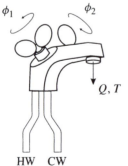

Figure E3.3.a Hot water (HW) and cold water (CW) faucet.

---

[5] The analysis of the water faucet used to explain the implications of the Independence Axiom was first introduced by Professor Gunnar Sohlenius of the Royal Institute of Technology (KTH) in Stockholm, Sweden. A comprehensive study of water-faucet design is presented by D. A. Norman, *Design of Everyday Things*, Currency Doubleday, New York, 1990.

Equation (b) states that we will somehow design a faucet system with two knobs that can control the flow rate and the temperature independently. One possible embodiment is shown in Figure E3.3.b. This design shows Knob A, which is a regular faucet valve that controls the overall flow rate. Knob A is located downstream where the hot and cold water have already been mixed. Knob B must now be designed to control the ratio of the hot and the cold water so as to control temperature.

Figure E3.3.c shows a proposed design that gives the details of Knob B. In this proposed design, Knob A with ϕ_1 as the DP_1 controls the flow Q. In addition, Knob B consists of the hot-water and cold-water valves, which have been connected in such a way that a turn ϕ_2 of Knob B causes one valve to close as the other opens. This means that ϕ_2 controls the temperature T.

The design equation for this design is

$$\begin{Bmatrix} Q \\ T \end{Bmatrix} = \begin{bmatrix} X & 0 \\ 0 & X \end{bmatrix} \begin{Bmatrix} \phi_1 \\ \phi_2 \end{Bmatrix} \tag{c}$$

This is an uncoupled design and thus satisfies the Independence Axiom. This design is better than the coupled design represented by Equation (a).

Design That Reduces Information through Physical Integration

Although the design shown in Figure E3.3.b is functionally uncoupled, the faucet designer should still try to integrate the design features into a single physical part, provided that FRs can be independently satisfied in the integrated design (Corollary 3). The faucet designer's aim is to identify a physical (not functional) integration of the DPs that would require only two valves. This physical integration is most likely to reduce the information content.

The FRs are, just to remind us, independent control of the flow rate Q and the temperature T of the water. From fluid mechanics, we learned that when the pressure drop is the same and the fluid is the same, the flow rate is proportional to the cross-sectional area. Therefore, we may choose the following as the design matrix:

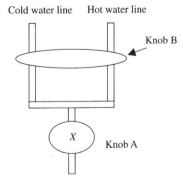

Figure E3.3.b Conceptual design of a system that controls the flow rate and the temperature of the water independently.

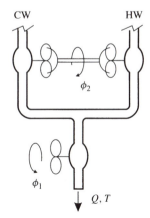

Figure E3.3.c An uncoupled hot water (HW) and cold water (CW) faucet.

$Q = f$ [Total cross-sectional area $(A_c + A_h)$] $= f(A_c + A_h)$
$T = f$ [Ratio of the cross-sectional area A_c/A_h] $= f(A_c/A_h)$

Then the design equation can be written as

$$\begin{Bmatrix} Q \\ T \end{Bmatrix} = \begin{Bmatrix} f(A_c + A_h) & 0 \\ 0 & g\left(\frac{A_c}{A_h}\right) \end{Bmatrix} \begin{Bmatrix} \text{Knob A} \\ \text{Knob B} \end{Bmatrix} \qquad \text{(d)}$$

One of the physical embodiments that can be consistent with the above design equation is given in Figure E3.3.d. It consists of a connecting rod whose length can be changed by turning the two threaded ends of the connecting rod in opposite directions, ϕ. This is achieved by turning the threaded connector of the valves, which are mechanically connected (the thread of both valves must be the same, either right handed or left handed). The temperature can be adjusted by moving the connecting rod that connects the two valves in the system horizontally. The DP for the temperature is position of the rod Y.

For the design shown in Figure E3.3.d, Equation (d) may be rewritten as

$$\begin{Bmatrix} Q \\ T \end{Bmatrix} = \begin{bmatrix} X & 0 \\ 0 & X \end{bmatrix} \begin{Bmatrix} \phi \\ Y \end{Bmatrix} \qquad \text{(e)}$$

The flow rate Q is controlled by turning the rod by an angle ϕ. The temperature T is controlled by the position of the rod Y.

Figure E3.3.e shows another design that is simpler to implement. To control flow rate and temperature, the faucet designer uses two plates that can be moved in a plane. The flow rate is a function of the total area, and the temperature is a function of the ratio of the openings (i.e., areas) for hot and cold water.

To control the flow rate, the top plate is moved along the X direction; to control temperature, the other plate is moved along the Y direction. The resulting design equation is

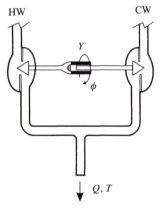

Figure E3.3.d Another version of an uncoupled hot water (HW) and cold water (CW) faucet.

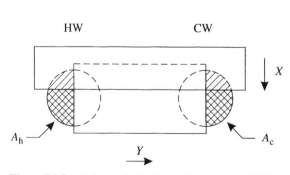

Figure E3.3.e A decoupled design with two plates. HW, hot water; CW, cold water.

$$\begin{Bmatrix} Q \\ T \end{Bmatrix} = \begin{bmatrix} X & 0 \\ X & X \end{bmatrix} \begin{Bmatrix} X \\ Y \end{Bmatrix} \tag{f}$$

This is now starting to look like a solution that would fit in a faucet for a bathroom basin.

Implications of the Information Axiom

The Independence Axiom has taken us to this point. This axiom prescribes that the DPs should be chosen such that the independence of the FRs is maintained. But what is the implication of the Information Axiom for this design task?

The Information Axiom prescribes that the best design is the one that has the lowest information content. Ideally, the information content should be zero.

Applied to the faucet design, a system with two moving parts has a lower probability of success than a system with one moving part (assuming that the mean time between failures—MTBF—is the same for any moving part). If, for example, the probability of success for a moving part is 0.99, then a system consisting of one moving part would have an information content $= \log_2(1/0.99)$. Similarly, an uncoupled system of two statistically independent moving parts would have twice the information content $= \log_2(1/0.99) + \log_2(1/0.99) = 2 \log_2(1/0.99) > \log_2(1/0.99)$.

Based on these assumptions about the probabilities of success, the faucet designer should try to realize the design with one movable part that has two (or more) degrees of freedom of movement that can be used to satisfy the two FRs.

Axiomatic design does not provide any methodological support to integrate DPs in a way that maintains functional independence while minimizing information content. The faucet designer has to rely on experience and analogies to do this part of the work.

Figure E3.3.f shows the principle for a design integrating DPs in one physical part. Both FRs can be satisfied with the movable plate. The plate has one triangular hole that affects A_h and A_c. Turning the plate ϕ controls the temperature, while moving the entire plate along the X direction controls the flow rate. Unfortunately, this is true only when the circular holes are symmetrically located with respect to the triangular opening.

The resulting design equation for the symmetric position is

$$\begin{Bmatrix} Q \\ T \end{Bmatrix} = \begin{bmatrix} X & 0 \\ 0 & X \end{bmatrix} \begin{Bmatrix} X \\ \phi \end{Bmatrix} \tag{g}$$

When the valve is rotated from this symmetric position and moved along the X direction, the temperature will change, in addition to the flow rate. Depending on the tolerance of

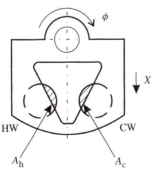

Figure E3.3.f An uncoupled design that integrates the part to reduce the information content. HW, hot water; CW, cold water.

FRs, this design may not be acceptable, except in a narrow design window. How would you design an uncoupled faucet with one moving part that satisfies the Independence Axiom?

The faucet designer has now reached the goal, having designed several valves in which flow and temperature can be controlled independently of each other. Which design is the best among those proposed?

3.5 UNCOUPLED AND DECOUPLED MULTI-FR DESIGNS

As stated in the preceding section, an ideal design is an uncoupled design with a diagonal design matrix with zero information content. In this case, a multi-FR design is almost identical to a one-FR design problem. For each FR, we can write a design equation, relating the FR to a single DP. If there are m FRs, there are m design equations, each of which can be solved independently. Modeling of the design also becomes simple because the modeling can be limited to relating one FR to one DP. The element of the design matrix can be expressed quantitatively or analytically. Furthermore, the design can be made robust using the techniques discussed in Section 2.4.

Decoupled designs can also be modeled similarly, although this involves additional consideration of the off-diagonal elements and the sequence of the operation. However, there is a substantial difference between the uncoupled design and the decoupled design in the allowable DP and PV tolerances.

3.5.1 Propagation of Tolerances in Uncoupled, Decoupled, and Coupled Designs and Its Implication for Design Robustness

How does the tolerance propagate from domain to domain in the case of an uncoupled design? Tolerance specification is simple in the case of an uncoupled design. If the specified design range for FR_i is ΔFR_i, then the tolerance for DP_i is simply

$$\Delta DP_i = \frac{\Delta FR_i}{A_{ii}} \tag{3.3}$$

Because the goal of a robust design is to make ΔDP_i as large as possible, A_{ii} should be made small. Similarly, the tolerance for PV_i is

$$\Delta PV_i = \frac{\Delta DP_i}{B_{ii}} \tag{3.4}$$

The *design range* is defined by ΔFR. The actual variation of FR, which is determined by the variation of DPs and PVs as well as by the magnitude of the design matrix elements, defines the system range. If the system range determined by the random variation of FR is less than the specified design range ΔFR_i and if the target in the design range and the peak of the system pdf coincide, then the information content is equal to zero as the system range is always inside the design range.

Is the propagation of tolerance for a decoupled design different from the case of an uncoupled design discussed so far? Can a decoupled design be as robust as an uncoupled design? Why? How are they different? Consider the decoupled design shown below.

$$\begin{Bmatrix} FR_1 \\ FR_2 \\ FR_3 \end{Bmatrix} = \begin{bmatrix} A_{11} & 0 & 0 \\ A_{21} & A_{22} & 0 \\ A_{31} & A_{32} & A_{33} \end{bmatrix} \begin{Bmatrix} DP_1 \\ DP_2 \\ DP_3 \end{Bmatrix} \tag{3.5}$$

The Independence Axiom can be satisfied if we change the DPs in the order shown. However, to have a robust design, we must be sure that the off-diagonal elements are much smaller than the diagonal elements, i.e., $A_{ii} >> A_{ij}$.

If the specified design ranges for FRs are ΔFR_1, ΔFR_2, and ΔFR_3, the maximum allowable tolerances for DPs may be expressed as

$$\Delta DP_1 = \frac{\Delta FR_1}{A_{11}}$$

$$\Delta DP_2 = \frac{\Delta FR_2 - |A_{21} \Delta DP_1|}{A_{22}} \tag{3.6}$$

$$\Delta DP_3 = \frac{\Delta FR_3 - |A_{31} \Delta DP_1| - |A_{32} \Delta DP_2|}{A_{33}}$$

For ΔDP_2 of Equation (3.6), the fluctuation of ΔDP_2 due to the term $A_{21} \Delta DP_1$ can make the term ΔDP_2 larger or smaller depending on its sign. However, the maximum allowable ΔDP_2 corresponds to the worst possible case, i.e., when ΔDP_2 is made smaller by the term ($A_{21} \Delta DP_1$). A similar argument holds true for ΔDP_3. Therefore, the absolute value represented by $|x|$ is used to represent the worst possible case.

According to Equation (3.6), the maximum tolerances for DPs of the decoupled design are less than the corresponding tolerances for DPs of an uncoupled design. This means that the decoupled design is inherently less robust than the uncoupled design. This may be stated as Theorem 22.

Theorem 22 (Comparative Robustness of a Decoupled Design)

Given the maximum tolerances for a given set of FRs, decoupled designs cannot be as robust as uncoupled designs in that the allowable tolerances for DPs of a decoupled design are less than those of an uncoupled design.

Equation (3.6) was for a decoupled design with three FRs and three DPs. Extending the argument given above to the case of m FRs and m DPs, it becomes obvious that as m increases, the allowable tolerance for the last DP of the triangular matrix becomes increasingly smaller. This means that the robustness of a decoupled design diminishes as the number of FRs increases.

Theorem 23 (Decreasing Robustness of a Decoupled Design)

The allowable tolerance and thus the robustness of a decoupled design with a full triangular matrix diminish with an increase in the number of functional requirements.

Variance of uncoupled, decoupled, and coupled designs. If the estimated variance of DP_i is designated by $s_{DP_i}^2$ and the DPs are all statistically independent, then for the decoupled design given by Equation (3.5), the variance of the FRs is given by

$$s_{FR_1}^2 = A_{11}^2 \, s_{DP_1}^2$$

$$s_{FR_2}^2 = A_{21}^2 \, s_{DP_1}^2 + A_{22}^2 \, s_{DP_2}^2 \qquad (3.7)$$

$$s_{FR_3}^2 = A_{31}^2 \, s_{DP_1}^2 + A_{32}^2 \, s_{DP_2}^2 + A_{33}^2 \, s_{DP_3}^2$$

Equation (3.7) indicates that the variance of a decoupled design is larger than that of an uncoupled design, other things equal. This additional system variation for a decoupled design is what makes it more difficult for the information content to be zero.

How does the tolerance propagate in the case of coupled designs? Can a coupled design be robust? In the case of a coupled design, the maximum allowable tolerance is even smaller than was the case for a decoupled design. Consider the following coupled design represented by the design equation:

$$\begin{Bmatrix} FR_1 \\ FR_2 \\ FR_3 \end{Bmatrix} = \begin{bmatrix} A_{11} & A_{12} & A_{13} \\ A_{21} & A_{22} & A_{23} \\ A_{31} & A_{32} & A_{33} \end{bmatrix} \begin{Bmatrix} DP_1 \\ DP_2 \\ DP_3 \end{Bmatrix} \qquad (3.8)$$

The above equation may be solved for DPs if the determinant of the design matrix |DM| is not equal to zero, which is likely to be the case. The solution for DP_1 is

$$DP_1 = \frac{1}{|DM|} \{\alpha \, FR_1 - \beta \, FR_2 - \gamma \, FR_3\} \qquad (3.9)$$

where

$$\alpha = A_{22} \, A_{33} - A_{23} \, A_{32}$$

$$\beta = A_{12} \, A_{33} - A_{32} \, A_{13}$$

$$\gamma = A_{22} \, A_{13} - A_{12} \, A_{23}$$

The expressions for DP_2 and DP_3 are of a similar form.

For a given set of design ranges of FRs, the maximum allowable tolerances for DPs may be expressed as

$$\Delta DP_1 = \frac{1}{|DM|} \{\alpha \Delta FR_1 - |\beta \Delta FR_2| - |\gamma \Delta FR_3|\} \qquad (3.10)$$

As argued before, although the magnitudes of $\beta \Delta FR_2$ and $\gamma \Delta FR_3$ can be either positive or negative, the maximum allowable ΔDP_1 is given by Equation (3.10). Therefore, the allowable tolerances of DPs for coupled designs are smaller than those for uncoupled or decoupled designs.

The main point of this section is to emphasize the importance of making a design more robust by uncoupling it or if it cannot be uncoupled, by creating a decoupled design. And the more robust the design is, the more likely it is to have low information content.

3.5.2 Examples of Multi-FR Design

Three case studies of multi-FR designs from industrial firms are presented here. Engineers who took the axiomatic design course taught in their companies solved these case studies.

EXAMPLE 3.4 Liquid Crystal Display Holder[6]

The liquid crystal display (LCD) is a projection display system that contains three LCD panels that project the red, green, and blue images of a TV signal. The configuration of an LCD projector is shown in Figure E3.4.a. To display an exact color image by an LCD projection system, the pixels of the three panels should be aligned with respect to the blue image within a tolerance that is a fraction of the pixel dimension.

To align the pixels, the LCD projector uses adjusting mechanisms that enable the translation and rotation of each LCD panel. By using the pixels of one of the three panels as a reference, the pixels of the remaining two panels can be properly aligned.

Each LCD panel is attached to an adjusting mechanism called an LCD holder, which enables pixel alignment. To obtain pixel alignment, at least two LCD holders should have three degrees of freedom—translation along the X-axis and Y-axis and rotation with respect to the Z-axis.

The Daewoo team is developing an actuated mirror array (AMA) projector shown in Figures E3.4.b and E3.4.c. This projector provides a much brighter (i.e., higher lumens) color picture as intense light is reflected from a mirror surface. The AMA projector uses three AMA panels and needs to align all of the pixels of the three AMA panels to provide an exact color display. This AMA system is unique with respect to the LCD system because the AMA system is a reflection type whereas the LCD system is a transmissive type. In the AMA projector, the light from the lamp passes through a lens system, reflects off the AMA mirrors, and travels back through the lens system. The AMA projector needs more precise alignment than the LCD projector because the light passes through the lens system twice for the AMA projector compared with only once for the LCD projector. Therefore, the tolerance of the AMA holder will be less than that of the LCD holder.

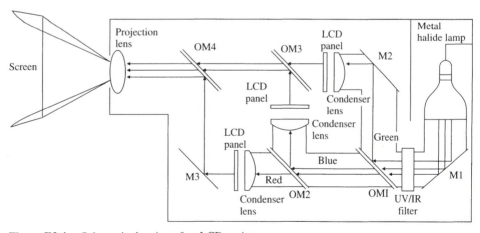

Figure E3.4.a Schematic drawing of an LCD projector.

[6] This case study was conducted by Dong Seon Yoon of Daewoo Electronics Co., as part of the axiomatic design course offered at the Institute for Advanced Engineering (IAE), Korea. It was edited by Vigain Harutunian of MIT.

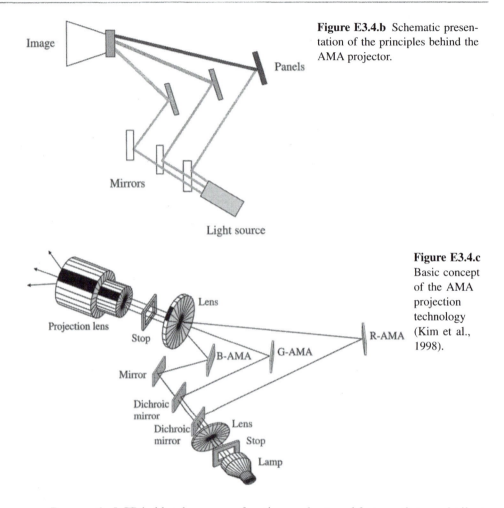

Figure E3.4.b Schematic presentation of the principles behind the AMA projector.

Figure E3.4.c Basic concept of the AMA projection technology (Kim et al., 1998).

Because the LCD holders have many functions and external features that are similar to those of the AMA holder, they were investigated prior to the design of the AMA projector. The two LCD holders that were chosen are the ones made by the Sanyo Corporation and the Sharp Corporation.

Your job is to determine the superior design that can align the pixels of three LCD panels based on the Independence Axiom. When this question was asked at a conference in Washington attended mostly by vice presidents of engineering of major corporations, more than half of them came up with a wrong answer. (They were given only 10 minutes to come up with an answer.)

SOLUTION

Determination of FRs

The goal of this project is to develop a mechanism that can align the pixels of three LCD panels. Thus, the highest-level FR and DP can be stated as follows:

FR = Align the pixels of three LCD panels.
DP = LCD holder that can align the pixels of three LCD panels

To align the pixels of all three LCD panels (see Figure E3.4.d), the FR can be decomposed as

FR_1 = Translate along X-axis = $T(X)$.
FR_2 = Translate along Y-axis = $T(Y)$.
FR_3 = Rotate with respect to Z-axis = $R(Z)$.

The LCD projector uses three LCD panels, of which one can serve as a reference. Therefore, the holder with three degrees of freedom is used for alignment.

We now have to evaluate the design of two commercially available LCD projectors that were chosen for analysis.

CASE 1 Sanyo Holder

Figure E3.4.e shows a sketch of the Sanyo holder. The Sanyo holder has three mechanisms, all of which are lead-screw structures. If a screw is rotated, its corresponding plate moves. Thus, the rotation of a screw can be transduced into the translation or rotation of the plate.

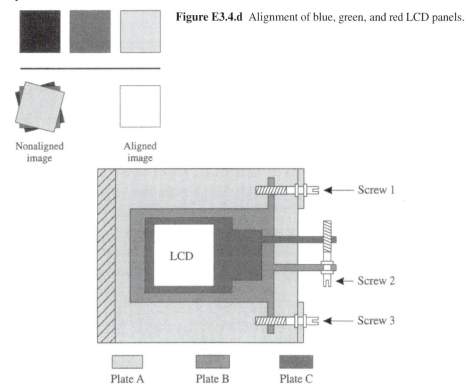

Figure E3.4.d Alignment of blue, green, and red LCD panels.

Nonaligned image

Aligned image

Figure E3.4.e Sanyo LCD plate alignment mechanism.

Plate A is fixed to the side frame and the other plates move on Plate A. The LCD panel is attached to Plate C and thus moves simultaneously with it. These three mechanisms are regarded as DPs and can be stated as

DP_1 = Lead-screw composed of Plate B and Screw 1
DP_2 = Lead-screw composed of Plate C and Screw 2
DP_3 = Lead-screw composed of Plate B and Screw 3

Because the center of the LCD panel is distant from the rotating center, the rotation of Plate B can be projected into the $T(X)$- and $R(Z)$-axes. In this case, Plate C and the LCD panel move with Plate B. Thus, the movement of Plate B can be regarded as that of the LCD panel. Therefore, rotating Screw 1 moves the LCD panel in the $T(X)$ and $R(Z)$ directions. Accordingly, DP_1 affects FR_1 and FR_3.

DP_2 consists of Screw 2 and Plate C. Rotating Screw 2 moves Plate C and the LCD panel in the Y direction toward Plate B. Therefore, rotating Screw 2 moves the LCD panel in the $T(Y)$ direction and DP_2 affects only FR_2.

DP_3 has the same structure as DP_1. Therefore, rotating Screw 3 moves the LCD panel in the $T(X)$ and $R(Z)$ directions. Thus DP_3 affects both FR_1 and FR_3.

The design equation for the Sanyo design may be written as

$$\left\{ \begin{array}{c} T(X) \\ T(Y) \\ R(Z) \end{array} \right\} = \left[\begin{array}{ccc} X & 0 & X \\ 0 & X & 0 \\ X & 0 & X \end{array} \right] \left\{ \begin{array}{c} DP_1 \\ DP_2 \\ DP_3 \end{array} \right\} \qquad (a)$$

The design matrix is neither diagonal nor triangular. Therefore, it is a coupled design and violates the Independence Axiom.

This design requires several iterations in order to align the pixels. For example, if there is an angular misalignment, DP_1 may be used for realignment. However, DP_1 also changes the X position of the LCD panel and thus creates another misalignment. Because there is no DP that can affect only FR_1, several iterations are required. This kind of mechanism is subject to error if there is a machining error of the parts and thus would require custom calibration.

CASE 2 Sharp Holder

The Sharp holder also has three mechanisms. One is a simple lead-screw structure and the other two are more robust lead-screw structures, each with a guide way and guide boss. Figure E3.4.f shows a sketch of the Sharp holder.

Plate A is fixed on the side frame and the remaining plates move on Plate A. The LCD panel is mounted to Plate C and thus moves simultaneously with it. These three mechanisms are regarded as DPs and are defined as

DP_1 = Lead-screw composed of Plate B and Screw 1
DP_2 = Lead-screw composed of Plate C, Plate D, Boss 1, and Screw 2
DP_3 = Lead-screw composed of Plate B, Plate E, Boss 2, and Screw 3

DP_1 consists of Screw 1 and Plate B. Rotating Screw 1 moves Plate B in the X direction toward Plate A (fixed on the side frame). Because the LCD panel is mounted on Plate C, which moves with Plate B, the LCD panel moves along the X-axis. Thus, rotating Screw 1 moves the LCD panel in the X direction and DP_1 affects only FR_1.

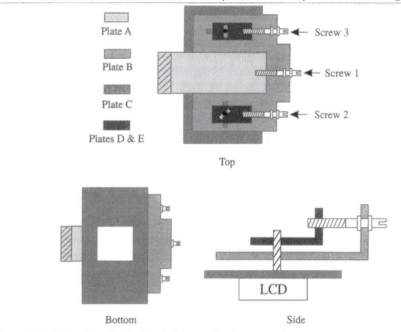

Figure E3.4.f The Sharp LCD plate holder mechanism.

DP$_2$ consists of Screw 2, Plate C, Plate D, and Boss 1. Rotating Screw 2 moves Plate D in the X direction, and the wall of the guide way in Plate D pushes Boss 1 (riveted on Plate C). As a result, Plate C moves along the Y-axis because the vertical groove in Plate B guides the movement of Boss 1 in the Y direction.* Thus, rotating Screw 2 moves the LCD panel in the Y direction, so DP$_2$ affects only FR$_2$.

DP$_3$ consists of Screw 3, Plate C, Plate E, and Boss 2. Rotating Screw 3 moves Plate E in the X direction, and the wall of the guide way in Plate E pushes Boss 2 (riveted on Plate C). Because there are no directional constraints to Boss 1, Plate C rotates with respect to Boss 1 (inside Plate D guide way). Because the center of the LCD panel is distant from the rotating center of Plate C, the rotation of the LCD panel can be projected into the X- and Y-axes. Therefore, rotating Screw 3 moves the LCD panel in the X direction and rotates about the Z-axis. DP$_3$ affects both FR$_3$ and FR$_1$.

The design equation may be written as

$$\begin{Bmatrix} T(X) \\ T(Y) \\ R(Z) \end{Bmatrix} = \begin{bmatrix} A_{11} & 0 & A_{13} \\ 0 & A_{22} & 0 \\ 0 & 0 & A_{33} \end{bmatrix} \begin{Bmatrix} DP_1 \\ DP_2 \\ DP_3 \end{Bmatrix} \qquad (b)$$

The design matrix is triangular. Thus, the design is decoupled and satisfies the Independence Axiom, provided that the DPs are changed in the correct order. Alignment of the LCD in the $R(Z)$ direction by DP$_3$ should precede the other alignments by DP$_1$ or DP$_2$.

To have a robust design, the magnitude of the diagonal elements should be small but much larger than the magnitude of the off-diagonal element A_{13}. Then small errors in the DPs will not be magnified in terms of the random error in the FRs.

*The width of the top slot in Plate B is slightly larger than the boss diameter.

The Sanyo holder is a coupled design. The DPs cannot independently change the FRs. Therefore, the Sanyo holder is not a good design because it is difficult to align the pixels for the three panels. Even if the pixels are aligned successfully, the time required to complete this alignment is unpredictable and even small errors in part tolerances can create misalignment problems.

The Sharp holder is a decoupled design. The DPs maintain the independence of the FRs if the holder is aligned in the order specified by the design matrix. This design allows the exact alignment of the pixels.

Some of the people who chose the Sanyo design stated that the Sanyo design is superior because it has fewer parts and appears to be simple to manufacture. However, the fact that there are fewer parts does not have much merit if the FRs cannot be satisfied easily and reliably. Any small error during the operation will make the Sanyo design more prone to poor pixel alignment. The Sanyo design will take longer to assemble because every set of parts must be individually adjusted to fit within the specified tolerance. Although there are fewer parts, the overall cost, including the loss of customer good will, may be much greater for the Sanyo design.

An interesting question is: Can we come up with an uncoupled or decoupled design that is simpler than the Sharp design?

EXAMPLE 3.5 Parking Mode of Automatic Transmission

The automatic transmission of an automobile has a parking mode (i.e., position), that is designed to prevent an unattended vehicle from moving on its own by locking the transmission mechanism.

An automobile company has been investigating a new design concept to overcome the shortcomings of current products. Their customers have complained that it is difficult to disengage the gear from the parked position when the vehicle is parked on a steep hill. Also there can be excessive vibration when the gear is removed from the parking position on a steep hill. Their current design is shown in Figure E3.5.a.

Study how the current parking mechanism works and develop a better mechanism. Justify your design process and solution based on axiomatic design. Develop design equations for your design. Also show how you can optimize your design through modeling.

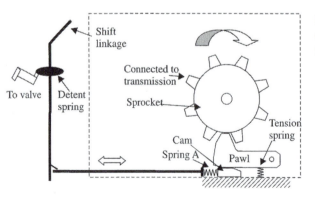

Figure E3.5.a Schematic drawing of parking mechanism for automatic transmission.

SOLUTION

Analysis of the Current Design

Before we can come up with the solution to the poor design, we need to understand how the current mechanism works, unless, of course, we want to abandon the current design completely and start with a clean sheet. In this example, our task is to improve the current design. The figure shows the parked position where the pawl is locked into the "sprocket" wheel by means of a shift-linkage mechanism. The sprocket wheel is attached to the transmission.

When the shift linkage is moved to put the car in the parked position, it activates the detent spring and turns the notched cam disk that activates the hydraulic system, etc. The shifting action also pushes Spring A that is attached to the cam that supports the load transmitted by the pawl. The surface of the cam is so profiled that when the spring load due to the shifting action is greater than the force resisting the engagement of the pawl with the sprocket wheel, the cam pushes the pawl into the engagement position, locking the sprocket wheel in a stationary position. Now the car is in parked position.

The pawl cannot accidentally engage with the sprocket wheel while the car is in motion because of the impact load between the sprocket and the pawl when the car is moving at more than 3 miles an hour. The tooth profile of the sprocket wheel and the profile of the pawl are so shaped that they cannot be engaged when the impact load between them is larger than the load exerted by Spring A. When the impact load is greater than the spring force, the pawl is kicked back, thus preventing the pawl from moving into the locked position.

When the car is parked on a hill, the entire vehicle weight exerts a torque on the sprocket wheel, which is transmitted through the teeth and the pawl to the cam. Thus, when the automatic transmission is to be disengaged from the parked position by moving the shifting linkage, the frictional force between the cam and the pawl must be overcome to slide the cam out of the engagement position. The pawl is attached to a tension spring so that as the cam is pulled out of its parked position, the pawl is pulled out of its engaged position with the sprocket wheel.

Figure E3.5.b is a free-body diagram that shows the forces acting on the pawl. F_R is the reaction force between the pawl and the sprocket, F_C is the reaction force between the pawl and the cam, F_S is the spring force, F_P is the force exerted on the pawl by the pin, and μ is the coefficient of friction between the pawl and the cam. F_R increases as the slope of the hill increases, which in turn increases the reaction force F_C and the friction force μF_C. The spring force F_S remains constant until the cam is removed and is set by the extension of the spring.

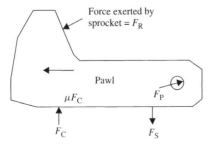

Force exerted by sprocket = F_R

Pawl

μF_C

F_P

F_C

F_S

Figure E3.5.b Free-body diagram showing the forces acting on the pawl.

It should be noted that the force F_R exerted on the pawl by the sprocket acts on the opposite side of the pawl surface when the van is parked in the opposite direction on the hill.

The reaction force F_C can be expressed as a function of the applied load F_R by taking a moment about the pin and applying the equilibrium condition.

The FRs of the cam/pawl/sprocket wheel assembly may be stated as

FR_1 = Engage the pawl in the locked position.
FR_2 = Disengage the pawl from the locked position.
FR_3 = Prevent accidental engagement.
FR_4 = Keep the pawl in the engaged position.
FR_5 = Carry the load transmitted by the vehicle.

The current design has the following DPs to satisfy these FRs:

DP_1 = The tapered section of the cam profile
DP_2 = Tension spring
DP_3 = The tooth profile of the sprocket wheel and the profile of the pawl "teeth"/Spring A/Linkage/Tension spring (Spring B)
DP_4 = The flat surface of the cam profile
DP_5 = The flat surfaces of pawl/cam

The design equation is

$$
\begin{Bmatrix} FR_1 \\ FR_2 \\ FR_3 \\ FR_4 \\ FR_5 \end{Bmatrix} = \begin{bmatrix} X & 0 & 0 & X & X \\ X & X & 0 & 0 & X \\ 0 & X & X & 0 & 0 \\ 0 & X & 0 & X & X \\ X & X & 0 & X & X \end{bmatrix} \begin{Bmatrix} DP_1 \\ DP_2 \\ DP_3 \\ DP_4 \\ DP_5 \end{Bmatrix} \tag{a}
$$

The current design is a coupled design because the vehicle load transmitted by the transmission is carried by the pawl and the cam. As the slope of the hill increases, the normal load and friction force on the cam increase, making it difficult to disengage from the park mode on a steep hill.

New Proposed Design

Figure E3.5.c shows the new proposed design. The new design has a sprocket with a different tooth shape from that of the previous design. The tapered section near the outer edge of the tooth is to prevent accidental engagement of the pawl, and the flat section of the pawl teeth is to transmit the vehicle load. The leading edge of the pawl is also tapered to prevent accidental engagement when the vehicle is moving at 3 miles per hour. The pin holding the pawl is now located at the same height as the flat section of the pawl. The vehicle load acting on the pawl is now colinear with the center of the pin, and therefore the vehicle load is carried by the pin, not by the cam (see Figure E3.5.c). In this new design, the cam does not carry the vehicle load. F_P is now approximately equal to F_R.

The FRs are the same as before. The DPs are shown in Figure E3.5.c. They are

DP_1 = The tapered section of the cam profile
DP_2 = Tension spring
DP_3 = The tooth profile of the sprocket wheel and the tapered section of the pawl "teeth"/Spring A/Linkage

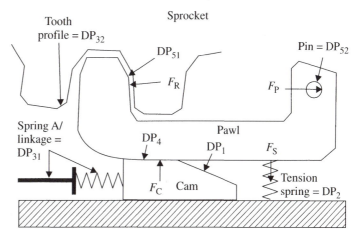

Figure E3.5.c Schematic diagram of the cam/pawl/sprocket assembly in the engaged position showing the physical parts and DPs. The forces acting on the pawl are shown by thick arrows.

$DP_4 =$ The flat surface of the cam profile
$DP_5 =$ The flat surfaces of pawl/sprocket and pin

The design equation may now be written as

$$\begin{Bmatrix} FR_4 \\ FR_2 \\ FR_3 \\ FR_1 \\ FR_5 \end{Bmatrix} = \begin{bmatrix} X & 0 & 0 & 0 & 0 \\ X & X & 0 & 0 & 0 \\ 0 & X & X & 0 & 0 \\ X & 0 & 0 & X & 0 \\ 0 & 0 & 0 & 0 & X \end{bmatrix} \begin{Bmatrix} DP_4 \\ DP_2 \\ DP_3 \\ DP_1 \\ DP_5 \end{Bmatrix} \qquad \text{(b)}$$

This is a decoupled design.

FR$_3$ and FR$_5$ should be decomposed, as many ideas are embedded in the description of DP$_3$ and DP$_5$. FR$_3$ may be decomposed as

$FR_{31} =$ Control the force that pushes the pawl into sprockets.
$FR_{32} =$ Create reaction load if the sprocket is turning.

The corresponding DPs are

$DP_{31} =$ Spring A that connects the linkage to the cam and the displacement of the linkage
$DP_{32} =$ Sprocket tooth profile

The design matrix is a triangular matrix.

FR$_5$ and DP$_5$ may be decomposed as

$FR_{51} =$ Transmit the force from the sprocket to the pawl.
$FR_{52} =$ Carry the load transmitted.

$DP_{51} =$ "Nearly vertical" surface of the pawl and the sprocket tooth profile
$DP_{52} =$ Pin located colinearly with the force vector acting on the vertical surface of the pawl

DP_{51} is a nearly vertical surface in order to make the reaction force between the sprocket and pawl nearly horizontal so as to minimize the reaction force between the cam and the pawl. A small slope of the pawl and the sprocket profile may be useful to make the retraction of the pawl small.

The design matrix is a diagonal matrix.

A Potential Shortcoming of the Proposed Design

In Figure E3.5.c, it was shown that the load-carrying surface of the pawl (DP_{51}) was straight. (In a real situation, it cannot be straight because of the fact that the sprocket is circular.) The consequence of having an almost vertical DP_{51} is that the tension spring DP_2 may have to exert a large force to remove the pawl from the engaged position. This problem can be partially alleviated by introducing a taper to the surface DP_1 to allow a vertical component of the force F_R.

Modeling the Details of the Proposed Design

To be certain of the proposed design, we need to model the FR–DP relationships based on the application of first principles. This will lead to the replacement of each X in the design matrix in Equation (b) with physical variables. This step is left to the reader.

Example 3.5 illustrated the importance of defining FRs correctly, as well as determining the cause of coupling, coming up with an uncoupled design, the importance of the basic principles of engineering science in developing a design, and the role of decomposition. However, a complete modeling was not done to replace each X in the design matrix with equations for determination of the final dimensions and configurations. In the following example, after a decoupled design is created, modeling is done to represent the design equations more exactly so that the final dimensions of the design can be given. It also shows how to choose a correct DP so as to eliminate coupling of FRs.

EXAMPLE 3.6 Design and Assembly of the Injection-Molded Vacuum Cleaner Wheel

To produce a low-cost vacuum cleaner, the wheels of a vacuum cleaner are to be injection molded, as shown in Figures E3.6.a and E3.6.b. The stem of the wheel is designed to be snap-fit onto the body frame of the vacuum cleaner. The wheel must rotate freely and should withstand a pulling force of 500 N. It must be easily assembled during the manufacturing operation with an axial force of less than 50 N.

CleanVac Corporation has developed the design shown in Figures E3.6.a and E3.6.b. They discovered that this design causes problems during assembly, requiring the use of a hammer to insert the stem of the wheel into the main body of the vacuum cleaner. The use of such an assembly technique yielded many broken parts. CleanVac is also concerned about long-term durability.

Your job as a designer is to review and improve this design. For this purpose, you are asked to provide the following:

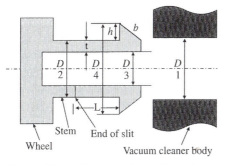

Figure E3.6.a Cross-sectional view of the wheel/shank for the CleanVac vacuum cleaner. The shank has three equally spaced slots that extend a distance L from the end of the shank. The shank bends to allow a snap-fit of the wheel/shank in the vacuum cleaner body.

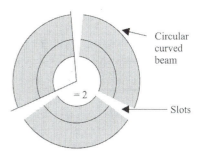

Figure E3.6.b End view of the shank for the CleanVac design.

a. Define the FRs
b. Develop DPs
c. For your chosen DPs, determine the design matrix
d. Model the relationship between FRs and DPs
e. Optimize the design based on the design matrix and the model

SOLUTION

Determination of FRs and Cs

We may state the functional requirements at the highest level as follows:

FR_1 = Make the wheel rotate easily by maintaining low friction between the wheel and the vacuum cleaner body and by making the torque exerted by the wheel/floor contact larger than the friction of the contact of plastic components.

FR_2 = Retain the wheel in the vacuum cleaner body under 500 N of pulling force.

FR_3 = Provide a means of easy assembly with an axial force of less than 50 N.

FR_4 = Carry the weight of the vacuum cleaner and accidental load applied when people step on the vacuum cleaner (200 lb).

The constraints for this design are:

C_1 = No fracture

C_2 = No fatigue failure

C_3 = No plastic yielding of the wheel

C_4 = Torque due to the traction at the wheel and the floor is greater than torque at the shaft surface due to the friction between plastic components

C_5 = Manufacturing considerations, e.g., injection-molded part should have approximately a constant thickness to prevent secondary flow caused by nonuniform cooling

C_6 = Minimize the manufacturing cost

Please note that C_4 may also be stated as a functional requirement.

Selection of DPs

The design process consists of selecting the appropriate DPs in order to satisfy the above FRs. The DPs selected for the initial design evaluation are

$DP_1 = \delta = (D_1 - D_2)/2$ (i.e., the clearance between the diameters of the wheel and the vacuum cleaner body)

$DP_2 = t = (D_2 - D_3)/2$

$DP_3 = \theta$

$DP_4 = \pi D_2 t$ (i.e., the area of the tubular stem without the axial cut)

The friction that the wheel is subjected to depends on the clearance between the two diameters, $D_1 - D_2$. Once D_2 is determined, D_1 can be determined to allow a minimum clearance to prevent seizure of the two parts. When the clearance is bigger than the minimum value, Coulomb friction may be assumed, i.e., the friction force $F = \mu N$, where μ is the coefficient of friction and N is the normal load.

When the wheel is subjected to a pulling force of 500 N, the failure is most likely to occur at the neck of thickness t. The stress that is generated due to this axial load will tend to rip the wheel off the vacuum cleaner body and will be a result of tensile failure at the cross section. Note that we must always round the sharp corners to reduce the stress concentration.

The assembly process consists of applying an axial force to bend the circular spring sections. The force acting on the inclined surface of the interlocking key is the resultant force consisting of the normal load and the tangential load due to interaction with the hole of the vacuum cleaner body. The maximum force may be assumed to be normal to the surface (i.e., no frictional component as it occurs when the contact occurs at the extreme outer rim of the interlock key).

The force P generates a bending moment that bends the circular cantilever beam, which permits the assembly of the wheel to the body. It is obvious that the deflection at the end would be a function of the total deflection of the curved beam required to clear the hole (Figure E3.6.c). The maximum tensile stress is experienced by the outer surface of the circular beam. One of the design constraints is that this maximum tensile bending stress must be less than the yield stress of the material.

The weight of the vacuum cleaner must be carried by the unslitted tubular section of the shaft. It must be thick enough to carry the load. Therefore, DP_4 is chosen to be D_2 as

F = Assembly force
P = Reaction force
F_s = Shear force
M = Moment

Figure E3.6.c Free-body diagram of one of the curved beams (approximate).

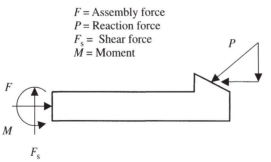

this solid section must carry the shear stress that must be less than the shear strength of the material, i.e., approximately $(1/2)\sigma_y$.

Design Equation

$$\begin{Bmatrix} FR_1 \\ FR_2 \\ FR_3 \\ FR_4 \end{Bmatrix} = \begin{bmatrix} A_{11} & 0 & 0 & 0 \\ 0 & A_{22} & A_{23} & A_{24} \\ 0 & A_{32} & A_{33} & 0 \\ 0 & A_{42} & 0 & A_{44} \end{bmatrix} \begin{Bmatrix} DP_1 \\ DP_2 \\ DP_3 \\ DP_4 \end{Bmatrix} \qquad (a)$$

All off-diagonal elements are zeros except A_{32}, A_{42}, A_{23}, and A_{24}. A_{32} is a nonzero element as the bending stiffness is affected by the thickness t. A_{42} is a nonzero element because the thickness $(DP_2 = t)$ determines the inner-hole size, which is to be the same throughout the stem section to make injection molding easier. A_{23} also is a nonzero element as the width of the curved beam affects the cross-sectional area and, thus, the axial stress. A_{24} is nonzero because the area over which tensile failure will occur is likely to be the slitted section. To make the design a decoupled design we must make the off-diagonal elements A_{23} and A_{24} zero.

A_{23} is nonzero because we have chosen the width of the circular beam θ as DP_3. As the angle θ gets smaller, the gap between the circular beam sections increases and the cross-sectional area of the beam decreases, affecting the axial load-carrying capacity. A_{23} can be made to be zero if we make the total circular length of all the curved beams remain constant by adding more sections of the curved beams, but this may increase the manufacturing cost.

Another way of decoupling the design is to choose either the height h of the interlock key or the length L of the beam as DP_3. The greater the length of the beam, the lower the bending stress. However, the longer beam has a higher manufacturing cost. If the beam is too short, there is an increased likelihood of binding the bearing surface.

A_{24} can be made zero by choosing the area of the unslitted section A as DP_4. This might mean that the thickness of the unslitted section would be different from the thickness of the slitted section, which would seem to violate C_5. However, if we find that the thickness of the unslitted section needs to be smaller than t in order to carry the vacuum cleaner weight, then we can make this thickness equal to t in order to avoid violating C_5. In this case, we shall be overdesigning the solid section. However, if we find from the calculations that the thickness of the unslitted section has to be greater than t, then we can increase the thickness of the slitted section t to this value. In this case, we shall be overdesigning the slitted section. Thus, we shall now settle on an appropriate value of D_2 to start the design process.

To have a good bearing surface, the length of the bearing surface must be larger than the diameter. We will set the length $L = 2D_2$. We will then select the height h of the interlock key as DP_3. Then the design matrix becomes triangular.

$$\begin{Bmatrix} FR_1 \\ FR_2 \\ FR_3 \\ FR_4 \end{Bmatrix} = \begin{bmatrix} A_{11} & 0 & 0 & 0 \\ 0 & A_{22} & 0 & 0 \\ 0 & A_{32} & A_{33} & 0 \\ 0 & 0 & 0 & A_{44} \end{bmatrix} \begin{Bmatrix} DP_1 = \delta \\ DP_2 = t \\ DP_3 = h \\ DP_4 = A \end{Bmatrix} \qquad (b)$$

Modeling the Relationship between the FRs and DPs

Evaluating A_{11}

The friction between the wheel and the vacuum cleaner body is a function of the clearance between the diameters D_1 and D_2. If the clearance is too small, the shaft and the hole can lock up by seizure when dirt particles or wear particles lock in at the interface. However, too large a clearance will make the wheels wobble. We will choose the clearance to be 0.010 inch (0.25 mm) on each side. It will also be a good idea to make a surface with shallow slots (about 50 μm deep and 100 μm wide) to keep the friction low by trapping particles.[7]

The friction force will then be a function of the coefficient of friction μ between the vacuum cleaner wheels and the body. The friction force will also depend on the weight W of the vacuum cleaner because this is the weight that is being supported. However, the friction force is not affected by the nominal diameter of the wheel as the Coulomb friction is independent of the contact area.

The friction force F is

$$F = \mu W \tag{c}$$

Then A_{11} is given by

$$A_{11} = \frac{\mu W}{\delta} = \text{constant} \tag{d}$$

For the wheel to turn properly, we must satisfy C_4:

$C_4 = $ Torque due to the traction at the wheel/floor interface is greater than torque at the shaft surface due to the friction between plastic components

If the friction coefficient is the same, this constraint is satisfied as long as the diameter of the wheel is larger than the shaft diameter, which is easily satisfied.

Evaluating A_{22}

This is a matter of modeling the failure due to a tensile load. We assume that there are three circular sections and that θ is the included angle.

$$F_{\text{pull}} = 3 \left(\frac{\theta}{8} \right) (D_2^2 - D_3^2)\sigma = \left(\frac{3\theta}{2} \right) (D_2 - t)t\sigma \tag{e}$$

The stress σ should not exceed the yield stress σ_y. Furthermore, if we want to avoid fatigue, then a good rule is that σ should not exceed $\sigma_y/2$. We can obtain A_{22} by differentiating the above expression with respect to t as

$$A_{22} = \frac{\partial \text{FR}_2}{\partial \text{DP}_2} = \left(\frac{3\theta}{4} \right) (D_2 - 2t)\sigma_y \tag{f}$$

It must be noted that

$$A_{42} = \partial \text{FR}_2/\partial \text{DP}_4 = (\pi/2)t\,\sigma_y$$

which is not zero. However, because this is much smaller than the rest of the elements (because t is small) we may set it equal to zero.

Evaluating A_{33}

FR$_3$ is the force necessary to push the wheel into the body, i.e.,

---

[7] See Suh (1986), Chapter 3.

$$FR_3 = P \tan \beta$$

$$DP_3 = h \tag{g}$$

The deflection at the end of the cantilever is given by

$$h = \frac{PL^3}{3EI} \tag{h}$$

To determine the maximum bending stress, we have to determine the moment of inertia of the section and the neutral axis of the circular beam. The height of the centroid relative to the centerline of the wheel is given by

$$\bar{y} = \frac{\iint y \, dA}{\iint dA} = \frac{\int_{-\phi}^{\phi} \int_{r_3}^{r_2} (r \cos \phi) \, r \, dr \, d\phi}{\int_{-\phi}^{\phi} \int_{r_3}^{r_2} r \, dr \, d\phi}$$

$$= \frac{\dfrac{2[(r_2)^3 - (r_3)^3]}{3} \sin \phi}{[(r_2)^2 - (r_3)^2]\phi} = \frac{\dfrac{2[(r_2)^3 - (r_3)^3]}{3} \sin \dfrac{\theta}{2}}{[(r_2)^2 - (r_3)^2]\dfrac{\theta}{2}}$$

The moment of inertia of the beam is given by

$$I_{xx} = \int_{-\phi}^{\phi} \int_{r_3}^{r_2} y^2 \, r \, dr \, d\phi = \int_{-\phi}^{\phi} \int_{r_3}^{r_2} (r \cos \phi)^2 \, r \, dr \, d\phi$$

$$= \frac{(r_2)^4 - (r_3)^4}{4} \left(\phi + \frac{\sin 2\phi}{2} \right) = \frac{(r_2)^4 - (r_3)^4}{4} \left(\frac{\theta}{2} + \frac{\sin \theta}{2} \right)$$

The above equation may be expressed as

$$I_{xx} = \left(\frac{\theta}{32} \right) [D_2^2 - (D_2 - 2t)^2]$$

$$\left\{ \frac{[D_2^2 + (D_2 - 2t)]^2}{4} + \left(\frac{4}{3\theta} \right)^2 \left[\sin \frac{\theta}{2} \frac{D_2^3 - (D_2 - 2t)^3}{D_2^2 - (D_2 - 2t)^2} \right]^2 \right\}$$

Therefore, A_{33} is given by

$$A_{33} = \left(\frac{3EI}{L^3} \right) \tan \beta \tag{i}$$

Evaluating A_{44}

$$FR_4 = W = A\tau = A \frac{\sigma_y}{2}$$

$$DP_4 = A \tag{j}$$

$$A_{44} = \frac{\sigma_y}{2}$$

Evaluating A_{32}

$$A_{32} = \frac{\partial FR_3}{\partial DP_2} = \frac{\partial \left[(3EIh/L^3) \tan \beta \right]}{\partial t} = \frac{3Eh}{L^3} \tan \beta \frac{\partial I}{\partial t} \qquad (k)$$

Thus we have all the elements of the design matrix. It is obvious that the equations are nonlinear. We can numerically determine t, h, diameters, and the length of the circular cantilever beam after setting the value of θ, which was set to be $100°$ after trying several possibilities. Then according to the design equation, we must first determine the thickness t (i.e., DP_2) and then h (i.e., DP_3).

The material properties for nylon[9] are

$$\text{Coefficient of friction} = 0.4$$

$$E = 362,590 \text{ psi}$$

$$\sigma_y = 7,250 \text{ psi}$$

The solution of the design equations for the dimensions are approximated as

$$D_1 = 0.425 \text{ inch}$$

$$D_2 = 0.405 \text{ inch}$$

$$D_3 = 0.315 \text{ inch}$$

$$D_4 = 0.443 \text{ inch}$$

$$t = 0.045 \text{ inch}$$

$$h = 0.019 \text{ inch}$$

3.6 INFORMATION CONTENT, COMPLEXITY, AND NOISE OF MULTI-FR DESIGN

So far in this chapter, the use of the Independence Axiom was discussed in developing rational multi-FR designs. The Information Axiom also has many implications for multi-FR designs. When there are many designs that satisfy the Independence Axiom, the one with the minimum information content is the best design, as discussed in Chapter 1.

3.6.1 The Relationship between Complexity and Information Content

The Information Axiom states that the information content must be minimized in design. Information was defined in Chapter 1 as a logarithmic function of the probability of achieving a given FR. In the case of multi-FR design, the sum of the information associated with all of the FRs must be minimized. In Chapter 1, the example of buying a house

---

[9] See *Encyclopaedia of Modern Plastics*, McGraw-Hill, New York. Published annually.

(Example 1.8) was used to illustrate how the Information Axiom can be used in choosing the best town for purchasing a house.

Complexity and information content are related—the design that requires more information content is more complex. A coupled design is more complex than an uncoupled design in nearly all cases. If there is a coupled design that requires less information than an uncoupled design, we can always find a better uncoupled design with less information content. The purpose of this section is to illustrate the relationship between complexity and information content using the simple example of the knob presented in Section 3.1. This is done in Example 3.7.

EXAMPLE 3.7 Knob Design

Let us consider again the design of the knob for grasping and turning a shaft shown in Figure 3.1. It was stated there that this was not a good design, but the explanation was not given. The knob was designed to satisfy the following two FRs:

FR_1 = Grasp the end of the shaft tightly with axial force of 30 N.
FR_2 = Turn the shaft by applying 15 N-m of torque.

The DPs for the design shown in Figure 3.1 are

DP_1 = Interference fit between the shaft and the inside diameter of the knob
DP_2 = The flat surface

The design equation may be written as

$$\begin{Bmatrix} FR_1 \\ FR_2 \end{Bmatrix} = \begin{bmatrix} X & X \\ x & X \end{bmatrix} \begin{Bmatrix} DP_1 \\ DP_2 \end{Bmatrix} \tag{a}$$

Equation (a) states that the interference fit between the shaft and the inside diameter of the knob provides the gripping force on the shaft and also affects the ability to turn the shaft. Similarly, DP_2 (the flat surface on the shaft and the knob) satisfies not only FR_2 (turn the shaft) but also affects the gripping force because the turning action opens the slot. The lower-case x is used to signify the fact that the effect of DP_1 on FR_2 is much less than the other effects indicated by upper-case X. Given the exact magnitude of forces and material properties, we should be able to come up with a unique set of DP_1 and DP_2. However, it is a coupled design or, at best, a decoupled design (depending on the design ranges of the FRs) and thus cannot be the best design.

For instance, when the torque is applied by the knob, the grip force on the shaft decreases with increases of the slot opening as a result of the normal load acting on the flat surface. Eventually, when the grip force is less than the force required to keep the knob on the shaft, the knob will slide off the shaft. How do we solve this problem?

Some will suggest that the solution to this coupled design problem is to make the outer diameter of the knob shaft (i.e., the thickness of the cylindrical section) thicker, which will make the slot open up less and thus minimize the reduction of the gripping force. However, this solution has its cost; not only does it require more materials but also higher information content, which ultimately means a higher manufacturing cost.

The process of increasing the thickness of the cylindrical section is equivalent to making the information content larger. This is illustrated in Figure E3.7.a. The design range for the force is given by ΔFR, which is specified in the functional domain. The manufacturing capability is given by the bell-shaped distribution of DPs in the physical domain (i.e., along the DP axis). The figure shows two bell-shaped distribution curves in the physical domain, which are assumed to be the same because the same manufacturing process is used. Depending on the stiffness, the same bell-shaped distribution along the DP axis translates into very different system distributions in the functional domain as shown along the FR axis. When the stiffness is lower, the system pdf (i.e., the solid bell-shaped curve along the FR axis) fits in the design range, but when the stiffness increases, the system pdf (i.e., the dotted bell-shaped curve) is outside the design range.

When the thickness of the cylinder wall increases, the stiffness increases, and thus the interference between the shaft and the inner diameter of the cylinder (DP_1) must be manufactured to a tighter tolerance to be within the design range. Therefore, the allowable tolerance between the shaft and the inner diameter of the knob must be made smaller as small changes in the tolerance will result in a large difference in the gripping force. The overlap between the system range and the design range will become smaller with an increase in wall thickness, thus increasing the information content. The probability of success will decrease as the tolerance becomes tighter.

To overcome this problem, the manufacturing operations must be made increasingly accurate. This probably requires that the machining of the shaft diameter be done accurately and the tolerance of the injection-molded part (if the knob is made by injection molding) be kept tight. The latter may require special molds and molding cycles or measuring all the parts to select the parts that meet the specification. Thus, the manufacturing process becomes more complex when the design is coupled.

An alternate design is shown in Figure E3.7.b. In this design, the slot terminates where the flat part of the knob begins. Because the flat surface is completely away from the slot, the turning action does not force the slot to open and, therefore, the axial grip is not affected. This is a completely uncoupled design. The information content is zero as the system range for each FR can be made to lie inside the design range. It is much *less complex* to make than the original design because the thickness of the cylinder wall can be made to provide the

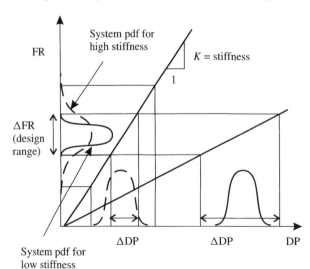

Figure E3.7.a Dependence of system pdf on stiffness. The design range of the gripping force is given by ΔFR. The manufacturing capability is given by the bell-shaped pdf in the physical domain (i.e., the DP axis). Depending on the stiffness, the same DP distribution translates into very different system distributions in the functional domain.

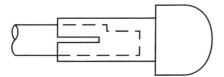

Figure E3.7.b A new uncoupled design. This design has the flat surface used for turning away from the slot. The bottom corners of the slot should be rounded to reduce the stress concentration.

right gripping force at the knob/shaft interface and thus the manufacturing tolerance can be made as large as possible.

How do we actually determine the wall thickness and the desired interference? The gripping force can be controlled either by changing the stiffness of the slitted section of the knob by changing the thickness or by adjusting the interference fit between the shaft and the knob. What we want to do is to make the thickness as thin as possible to reduce cost and to make the interference as large as possible. The thickness must be determined by considering two limiting factors: manufacturability and failure of the knob under stress. Can it be manufactured by injection molding? Does the maximum stress at the bottom corner of the slit, which is the stress concentration point, cause either fracture or plastic deformation? (Obviously we should put fillet to round the sharp inside corners.)

What we want to do is to maximize the grip force and, at the same time, minimize the stress. This can be done if we make the shaft engage the knob near the end of the knob by reducing the diameter of the shaft slightly as shown in Figure E3.7.c. Then the concentrated force will act near the end of the knob. Ultimately, a finite element analysis is needed to determine the best geometry of the knob and shaft.

Although a more exact solution for the curved beam should be analyzed before the design is finalized, insight can be gained about the rough relationship between the stress and deflection by examining the case of a cantilever beam shown in Figure E3.7.d.

The maximum deflection is given by

$$y_{\max} = -\frac{FL^3}{3EI} \tag{b}$$

where

$$I = \frac{bh^3}{12}$$

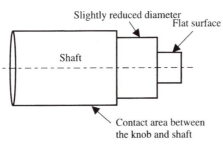

Figure E3.7.c Modified shaft. The diameter near the end is slightly reduced to concentrate the gripping force near the end of the knob.

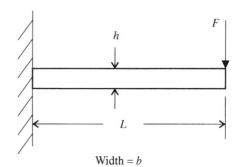

Figure E3.7.d A cantilever beam loaded at the end by a concentrated load.

The maximum stress is given by

$$\sigma_{\max} = \frac{FLh}{2I} = \frac{6FL}{bh^2} \tag{c}$$

The stiffness $K = |\sigma_{\max}/y_{\max}|$ is given by

$$K = \frac{3Eh}{2L^2} \tag{d}$$

To minimize K for robustness, h should be made as small as possible. The limit is reached when $\sigma_{\max}$ reaches σ_y. Then the smallest h is obtained as

$$h_{\min} = \left(\frac{\sigma_y b}{6FL}\right)^{1/2} \tag{e}$$

At this point, we must determine if the minimum thickness $h_{\min}$ is feasible and advisable from the injection-molding point of view. If the answer is affirmative, the maximum interference fit is obtained by substituting $h_{\min}$ into Equation (b). If the thickness must be larger than the minimum value of h obtained, a larger thickness must be used that will allow injection molding, which may then be used to determine the maximum interference fit.

This example illustrates how complexity and information content are related in a product design. Although it is less obvious, the same reasoning should apply to designs of other kinds, i.e., software, systems, organizations, and materials.

How does the above example illustrate the relationship between complexity and the design axioms? The foregoing example illustrates the following aspects of the Information Axiom and the Independence Axiom:

1. Complexity is related to the probability of achieving the functional requirement. The coupled design made it much more difficult to make the knob.
2. Even the same design can have a very different information content and complexity, depending on the stiffness of the system.
3. The greater the information content, the more complex is the task of achieving the FR because the probability of success decreases. Therefore, information content is a measure of design complexity.
4. A design that violates the Independence Axiom, i.e., a coupled design, is more complex and requires more information content than a design that satisfies the Independence Axiom.

3.6.2 Determination of Information Content of Uncoupled, Decoupled, and Coupled Designs

The determination of the information content of uncoupled designs is simple and straight-forward, as all FRs are maintained independent of one another. The total information content is the sum of the information content for each FR, which was expressed in Chapter 1 as

$$I = -\sum_{i=1}^{m} \log_2 P_i \tag{3.11}$$

This is stated in Theorem 13 as

Theorem 13 (Information Content of the Total System)

If each DP is probabilistically independent of other DPs and affects only its corresponding FR, the information content of the total system is the sum of the information of all individual events associated with the set of FRs that must be satisfied.

It should be noted that in axiomatic design, FRs are independent of each other in the functional domain by *definition* and thus each FR is orthogonal to other FRs. Only in the case of uncoupled designs are the DPs orthogonal to each other and parallel to the FR axes. Thus only in the case of uncoupled designs are the FRs *maintained* independent of one another in the design realization that provides the system pdf. This is why the probability that all *m* FRs are satisfied by uncoupled designs can be computed by the product of the probabilities for each FR.

How do we compute the information content of a decoupled design? In the case of decoupled designs, the DPs chosen compromise the orthogonality of FRs and one DP may affect more than its corresponding FR. Therefore, the design realization of the FRs may result in a correlated system.

The probability that all *m* FRs will be satisfied by a decoupled design can be computed by the product of the probabilities for each FR, provided that appropriate conditional probabilities are used where necessary. Then the information for the entire set of FRs may be obtained by the negative sum of the log of these conditional probabilities. This is stated as Theorem 12.

Theorem 12 (Sum of Information)

The sum of information for a set of events is also information, provided that proper conditional probabilities are used when the events are not statistically independent.

The information content for decoupled designs must include the effect of the off-diagonal elements of the triangular design matrix. Consider the two-dimensional case represented by the following design equation:

$$\begin{Bmatrix} FR_1 \\ FR_2 \end{Bmatrix} = \begin{bmatrix} A_{11} & 0 \\ A_{21} & A_{22} \end{bmatrix} \begin{Bmatrix} DP_1 \\ DP_2 \end{Bmatrix} \tag{3.12}$$

The information content of FR_1 can be determined by computing the area of the system pdf in the common range just as for an uncoupled design, as DP_2 does not affect FR_1. However, to compute the information content associated with FR_2, we have to include the change in the information content due to the off-diagonal element as illustrated in Figure 3.2.

The solid curve of Figure 3.2 is the system pdf of FR_2 when the off-diagonal element A_{21} is equal to zero, i.e., for an uncoupled design. However, the system pdf of FR_2 for a decoupled design may be shifted to the right or left by the off-diagonal element A_{21}, which changes the mean of the system pdf for FR_2 when DP_1 changes because FR_2 is affected by DP_1. The figure illustrates the fact that when DP_1 changes, the system pdf shifts left horizontally to the dotted pdf. For the particular case shown in Figure 3.2, the information content decreases and the system performance actually improves when the mean decreases,

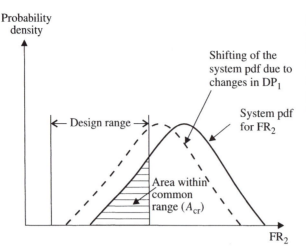

Figure 3.2 Shift of the FR_2 system pdf due to change in DP_1. In the decoupled design represented by Equation (3.12), the solid curve for the system pdf may move horizontally to the dotted curve.

thus shifting the system pdf left to the dotted pdf. Conversely, when the mean increases, the curve shifts to the right and the information content increases.

In some cases, the variance of the system pdf can change as well as the mean. This is shown in Figure 3.3. This occurs when the random variation of other DPs changes the variance of the system pdf.

When the system pdf is symmetrical with its mean in the middle of the design range, the effect of the off-diagonal element is to change the spread of the system pdf, as shown in Example 3.8.

The foregoing argument has several implications. First, the effect of the off-diagonal elements on the information content is specific to a given design. It can either decrease or increase the information content. Second, the information content of a decoupled design can be determined by summing up the information content of each FR if and only if the DPs are varied in a proper sequence. Third, the best setting for a decoupled design can be determined by evaluating the minimum information point for each FR when the following two conditions are satisfied:

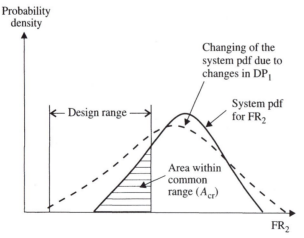

Figure 3.3 Change in variance of system pdf of FR_2 due to change in DP_1.

For FRi:

$$\sum_{j=1}^{i} \frac{\partial I_i}{\partial DP_j} = 0$$

$$\sum_{j=1}^{i} \frac{\partial^2 I_i}{\partial DP_j^2} > 0 \tag{3.13}$$

Equation (3.13) must be evaluated in the sequence given by the design equation, as the design matrix is triangular for a decoupled design.

Equation (3.6)—the relationship that defines the allowable tolerances for coupled designs—showed that for a given set of design ranges for FRs, the allowable tolerances for DPs and PVs are less for decoupled designs than for uncoupled designs. Therefore, in most cases, the information content of a decoupled design is expected to be larger than that of an uncoupled design, as a decoupled design cannot be as robust as an uncoupled design.

Consider the decoupled design in Equation (3.12), which can be expressed as

$$\begin{Bmatrix} FR_1 \pm \Delta FR_1 \\ FR_2 \pm \Delta FR_2 \end{Bmatrix} = \begin{bmatrix} a & 0 \\ b & c \end{bmatrix} \begin{Bmatrix} DP_1 \pm \Delta DP_1 \\ DP_2 \pm \Delta DP_2 \end{Bmatrix} \tag{3.14}$$

where ΔFR_1 and ΔFR_2 are the design ranges for FR_1 and FR_2, respectively, and ΔDP_1 and ΔDP_2 are the specified tolerances of the DPs needed to satisfy the FRs within their design ranges.

Then from Equation (3.6),

$$DP_1 = \frac{FR_1}{a}$$

$$DP_2 = \frac{FR_2 - b\,DP_1}{c}$$

and

$$\Delta DP_1 = \frac{\Delta FR_1}{a}$$

$$\Delta DP_2 = \frac{\Delta FR_2 - b\,\Delta DP_1}{c}$$

We can design for FR_1 without being concerned about DP_2 because FR_1 is affected only by DP_1. Once FR_1 is set by choosing the right DP_1, we can design for FR_2 by considering only DP_2. This means that if we follow the triangular matrix in the proper sequence, each FR can be treated sequentially, which makes the design process simple.

For FR_s and DP_s, Equation (3.14) can be expressed as

$$\begin{Bmatrix} FR_1 \\ FR_2 \end{Bmatrix} = \begin{bmatrix} a & 0 \\ b & c \end{bmatrix} \begin{Bmatrix} DP_1 \\ DP_2 \end{Bmatrix} \tag{3.15}$$

Equation (3.15) states that the variation of FR_1 is affected only by the variation of DP_1. However, the variation of FR_2 is affected by the variation of both DP_1 and DP_2 as

$$FR_2 = b\,DP_1 + c\,DP_2$$

Because DP_1 affects both FR_1 and FR_2, the FRs are not statistically independent. This is illustrated in Example 3.8.

EXAMPLE 3.8 Information Content of a Decoupled Design

Consider a decoupled design given by Equation (3.12). Suppose FR_1 is "Turn the shaft" and FR_2 is "Grip the shaft." The design is done in a manner very similar to the design shown in Figure 3.1. DP_1 is the "flat surface" and DP_2 is the "interference fit." The design ranges for FR_1 and FR_2 are $\pm 5\%$, i.e.,

$$FR_1 = 1 \pm 0.05$$

$$FR_2 = 1 \pm 0.05$$

The design equation and the design matrix are, after normalization, given by Equation (a).

$$\begin{Bmatrix} FR_1 \\ FR_2 \end{Bmatrix} = \begin{bmatrix} 1 & 0 \\ 0.2 & 1 \end{bmatrix} \begin{Bmatrix} DP_1 \\ DP_2 \end{Bmatrix} \tag{a}$$

The system ranges for both FR_1 and FR_2 have uniform distributions.

There are two suppliers for this part. Supplier A has a process that can manufacture within the required $\pm 5\%$ for the DP_1 tolerance, but Supplier B's process can manufacture within $\pm 10\%$. Both suppliers use the same process for DP_2 and therefore both can meet the same tolerance of 6%.

Determine the information content of this design.

SOLUTION

The DPs are determined as

$$DP_1 = FR_1/A_{11} = 1.0$$

$$DP_2 = (FR_2 - A_{21}DP_1)/A_{22} = [1 - 0.2(1)]/1 = 0.8$$

To satisfy the FRs within the design range, the DPs must be set to the following tolerances:

$$\Delta DP_1 = \Delta FR_1/A_{11} = 0.05$$

$$\Delta DP_2 = (\Delta FR_2 - A_{21} \Delta DP_1)/1 = [0.05 - 0.2(0.05)]/1 = 0.04$$

It should be noted that the design range for FR_2 is conditional on the design range of FR_1. If the manufacturing process cannot hold the DP_1 tolerance to within ΔDP_1, two things will happen for those parts that are outside the DP_1 tolerance: FR_1 will not be satisfied and meeting the DP_2 tolerance will not ensure satisfaction of FR_2.

Therefore, to calculate the probability that FR_2 is satisfied, we must first determine the probability that FR_1 is satisfied. The probabilities that FR_1 is satisfied by Supplier A and Supplier B are shown in Figures E3.8.a and E3.8.b, respectively. So although Supplier A can always meet the DP_1 tolerance and thus the design range for FR_1, only 50% of Supplier B's parts are within the DP_1 tolerance and thus within the design range for FR_1.

Now we can determine the probability that FR_2 is satisfied. The required tolerance for DP_2 is 5% (i.e., $.8 \pm .04$). Both suppliers use the same process for DP_2, which can meet the tolerance only within 6% (i.e., $.8 \pm .048$). Therefore, only 83.3% of the parts meet the DP_2 tolerance for both suppliers.

The computation of the information content for Supplier A is simple. Because all of the parts satisfy FR_1, the conditional probability of satisfying FR_2 given that FR_1 satisfied is

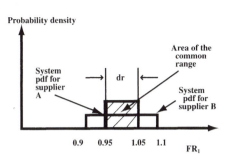

Figure E3.8.a System pdf and common range of FR_1 for Supplier A and Supplier B.

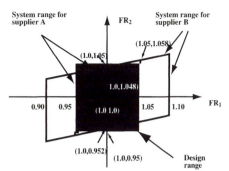

Figure E3.8.b System pdf and common range of FR_1 for Supplier B.

just the marginal probability of satisfying FR_2. Therefore, the joint probability of satisfying both FR_1 and FR_2 is $1 \times .833$. So for Supplier A, the information content is

$$I = -\log_2[\Pr(\text{satisfy both } FR_1 \text{ and } FR_2)]$$

$$= -\log_2[\Pr(\text{satisfy } FR_1)] - \log_2[\Pr(\text{satisfy } FR_2|\text{satisfy } FR_1)]$$

$$I = -\log_2(1) - \log_2(0.833) = 0.264$$

Even though Supplier B has the same probability of meeting the DP_2 tolerance, its conditional probability of satisfying FR_2 is not the same as that for Supplier A. Because Supplier B's parts have only a .5 probability of satisfying the DP_1 tolerance, the probability of satisfying FR_2 for these parts is not assured by being within the DP_2 tolerance. Unfortunately, the conditional probability that FR_2 is satisfied given FR_1 cannot be computed based on the information given in the problem statement. We need to make either statistical measurements—the best way—or make some assumptions to estimate the information content.

In fact, a study of Supplier B's parts revealed that of the parts that are within the DP_1 tolerance, only 90% are also within the DP_2 tolerance. So the joint probability of satisfying both FR_1 and FR_2 is $.5 \times .9 = .45$. Thus, the information content for Supplier B is

$$I = -\log_2(0.45) = -\log_2(0.5) - \log_2(0.9) = 1.152$$

Because Supplier A always satisfies FR_1, 83.3% of the parts will satisfy both FRs. But only 45% of Supplier B's parts satisfy both FRs because only half satisfy FR_1.

If we assume a uniform density function for the system range, the information content for the fixed value of ΔDP_2 is given as

$$I = -\log_2(0.5) - \log_2(0.833) = 1.264$$

The tolerances given ΔDP_1 and ΔDP_2 are fixed tolerances. But in an actual production situation, the value ΔDP_2 can be larger than the fixed tolerance if the actual variation is smaller than ΔDP_1. This may be called *moving tolerance*. Then, the actual system range can be computed and plotted as shown in Fig. E3.8b. This is done by plotting the actual value of FR_2 for a given value of FR_1 (this is equivalent to making ΔDP_2 a moving tolerance—not a practical thing to do). Under the assumption of uniform density function, the information content for this case can be computed as

$$I = -\log_2(0.5) - \log_2(0.967) = 1.0484$$

The difficulty with the moving tolerance for DP_2 is that it depends on the value of DP_1. Even if we have a pretty good idea of the DP_1 pdf and the DP_2 pdf from past experience with the manufacturing processes, we still do not know the joint pdf. Because the DPs are independent, their values can combine with one another in any random manner, and there is no way to know which combinations exist in what relative frequency until the product is actually produced.

With the fixed tolerance approach, any combination of DP_1 and DP_2 within their respective tolerances will satisfy both FRs. Because the same fixed tolerance applies to all candidates for design choice, there is no bias in evaluating their information content relative to one another. Thus, we can use instead the conditional probability of satisfying FR_2 given that FR_1 is satisfied, eliminating the need to know the joint density. For these reasons, the concept of fixed tolerance is used in axiomatic design.

3.6.3 Accommodating Noise in the Design Process

During manufacturing and use of a product, random variation from various sources affects the performance of a machine or system. The variation so introduced is given the generic name "noise." There are five generic noise sources: manufacturing variation, customer usage, environmental variation, degradation/wear-out, and system-to-system interaction.

In Example 2.7 (Joining of Aluminum Tube to Steel Shaft), noise was introduced by the random variation of the machining processes and by the temperature fluctuation in service, all of which contributed to the random variation of the interference fit of the shaft and the cylinder. In this example, the random noises were simply aggregated as the random variation of the chosen DP (i.e., the interference fit Δr). Based on the introduction of noise δDP, the element of the design matrix A_{11} was adjusted so as to satisfy the specified FR within the acceptable design range.

Multi-FR designs must also accommodate noise by adjusting the stiffness, i.e., the elements of the design matrix, to develop a design that satisfies all FRs within the design range.

Suppose that the aggregated random noise from all sources can be lumped into the variation of the DP_i. Then the effect of the random variation δDP_i on the variation of the FR_i can be written as

$$\begin{Bmatrix} \delta FR_1 \\ \delta FR_2 \\ \cdots \\ \cdots \\ \delta FR_n \end{Bmatrix} = [A] \begin{Bmatrix} \delta DP_1 \\ \delta DP_2 \\ \cdots \\ \cdots \\ \delta DP_n \end{Bmatrix} \tag{3.16}$$

where the design matrix $[A]$ is made up of elements A_{ij}.

When the design is uncoupled with a diagonal design matrix, the variation of each FR_i is affected by variation of only the corresponding δDP_i. Because the system range of the FR_i needs to be smaller than the design range ΔFR_i, the variation of DP_i can be accommodated if the corresponding element of the design matrix A_{ii} can be changed to lower its magnitude, i.e., stiffness.

When the design is a decoupled design with a triangular design matrix $[A]$, the variation of FR_i is caused by the random variations of many DPs, which may be expressed as

$$\delta\text{FR}_i = \sum_{j=1}^{i} A_{ij}\,\delta\text{DP}_j \tag{3.17}$$

To satisfy FR_i, the elements A_{ij} that correspond to large values of the δDP_j must be made smaller. Equation (3.17) may be expressed as

$$\delta\text{FR}_i = M_i\,\delta\text{DP}_i \tag{3.18}$$

where M_i is defined as a module that is equal to

$$M_i = \sum_{j=1}^{j=i} A_{ij}\frac{\text{DP}_j}{\text{DP}_i} \tag{3.19}$$

Therefore, in a decoupled design, to minimize the effect of random noise, the magnitude of the corresponding module must be decreased if the random variation in FR is larger than the specified design range of FR.

3.7 INTEGRATION OF DPs TO MINIMIZE THE INFORMATION CONTENT

In the case of hardware design, it is generally desirable to combine DPs in a physical piece if the FRs can be maintained independent from each other, as stated in Corollary 3. For example, in Chapter 1, the integration of 12 DPs for a beverage can in three physical pieces was discussed. In the case of the faucet, the valve arrangement was designed so that the DPs were integrated. Similarly, a can and bottle opener made by punching a single steel plate also integrates two different DPs in a single physical piece. These are examples of physical integration of several different DPs in a single physical body without violating the Independence Axiom.

In general, physical integration reduces the information content by removing the uncertainty associated with assembling several physical pieces. Boothroyd and Dewhurst (1989) developed a rule-based physical integration scheme that has been used by many industrial firms.

The general guidelines for physical integration of DPs that may reduce the information content are based on general observations. Provided that the Independence Axiom is not violated, DPs may be integrated in a single physical part under the following circumstances:

1. DPs do not undergo relative motion.
2. DPs can be made of the same material.
3. Integration does not create a problem such as excess stress and fracture.
4. Integration does not violate a cost constraint.
5. The integrated parts can be manufactured.

In addition to the integration of DPs to reduce the number of physical parts, the physical parts must be integrated to create a system. The system must perform the desired functions within the specified constraints (e.g., footprint of a machine, weight, or cost). The integration of the physical parts must be consistent with the DP hierarchy in the physical domain, where all leaf-level DPs are related to other leaf-level DPs according to the specified relationship.

3.8 NONLINEAR MULTI-FR DESIGN

When the elements of the design matrix are not constants, but instead are functions of DPs, the design is a nonlinear design. Even in the case of nonlinear design, a multi-FR design must be designed so that the design matrix is either diagonal or triangular. There are three kinds of situations in nonlinear design:

1. The design matrix is always either diagonal or triangular, regardless of how DPs change.
2. The elements of the design matrix may vary, depending on the specific values of DPs, so that the design behaves as a coupled, uncoupled, or decoupled design in different parts of the design window.
3. The design is always coupled regardless of the specific values of DPs.

The difference between the linear and the nonlinear design of the second kind is that in the case of nonlinear design, we may strive to find a better design window, because the elements of the design matrix change as functions of DPs. This can be illustrated graphically.

A graphic representation of the relationship between the FRs and the DPs is obtained by superimposing the physical domain onto the functional domain as shown in Figure 3.4A, B, and C. In these figures, the FR axes are plotted orthogonal to each other, as FRs are independent from each other by definition. When each DP axis is parallel to one of the

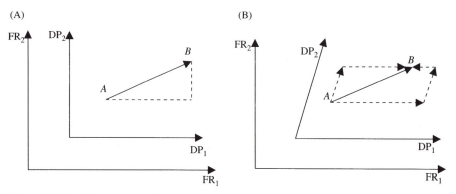

Figure 3.4 Graphic representation of two-dimensional design by superposition of the functional and the physical domains. (A) A completely uncoupled two-FR design. Both of the DP axes are parallel to the FR axes. Therefore, any one of the DPs can be changed in any order to go from A to B. (B) A decoupled two-FR design. One of the DP axes is parallel to one of the FR axes. Therefore, the order of the DP changes can affect the FR changes, i.e., a decoupled design is path dependent. (C) A coupled design. In this case, neither of the DP axes is parallel to the FR axes. It is very complicated to go from A to B. In some cases, they may never converge. (D) A case of nonlinear design. The curved lines are lines of constant DP_1 and DP_2. When the DP_1 line is parallel to the FR_1 axis and the DP_2 line is parallel to the FR_2 axis (Region A), the design is uncoupled. When one of the DP lines is parallel to an FR axis (Region B), the design is decoupled. When neither of the DP lines is parallel to an FR axis (Region C), the design is coupled.

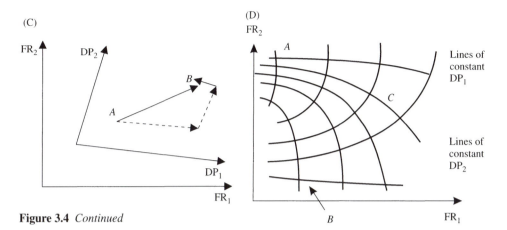

Figure 3.4 *Continued*

FR axes, as shown in Figure 3.4A, the design is an uncoupled design. When only one of the DP axes is parallel to one of the FR axes, as shown in Figure 3.4B, the design is a decoupled design. Otherwise, the design is a coupled design, as shown in Figure 3.4C. [For further discussion of graphic representation and the use of reangularity and semangularity, see Chapter 4 of *The Principles of Design* (Suh, 1990).]

In nonlinear design, the lines of constant DPs are curved as the elements of the design matrix are functions of DPs (rather than constants) as shown in Figure 3.4D.

The design that lies in Region A of Figure 3.4D—where the magnitude of FR_1 is small and that of FR_2 is large—is nearly uncoupled. Therefore, in this region, the design matrix will be diagonal with perhaps very small off-diagonal elements. Similarly, the design is decoupled in Region B. The design behaves as a coupled design in Region C.

The graphic representation given in Figure 3.4 explains the independence of FRs. However, when there are more than two FRs, it is difficult to use a graphic means of determining the nature of the design. As an alternate means of measuring the independence of FRs, we define two scalar metrics—reangularity R and semangularity S.

Reangularity R and semangularity S are defined as (Rinderle and Suh, 1982; Suh, 1990)

$$R = \prod_{\substack{i=1,n-1 \\ j=1+i,n}} \left[1 - \frac{\left(\sum_{k=1}^{n} A_{ki} A_{kj} \right)^2}{\left(\sum_{k=1}^{n} A_{ki}^2 \right) \left(\sum_{k=1}^{n} A_{kj}^2 \right)} \right]^{1/2} \tag{3.20}$$

$$S = \prod_{j=1}^{n} \left[\frac{|A_{jj}|}{\left(\sum_{k=1}^{n} A_{kj}^2 \right)^{1/2}} \right] \tag{3.21}$$

These measures are useful when there are many FRs and DPs.[9] Reangularity R measures the angular relationship between the DP axes. Semangularity S measures the magnitude of the diagonal elements of a normalized design matrix.

R and S are defined so that when they are both equal to one, the design is an uncoupled design, and when $R=S$ but not equal to 1, the design approaches a decoupled design. (When there are only two FRs and two DPs, $R = S$ represents a decoupled design.) In all other cases, the design is a coupled design.

Because in Region A of Figure 3.4C the design is uncoupled, both reangularity R and semangularity S should be nearly equal to 1. For small values of FR_1 and FR_2, the design is a decoupled design (Region B) because only one of the DPs is parallel to an FR axis, i.e., the FR_2 axis. In this region, R is nearly equal to S. In Region C, neither of the DP lines is parallel to either of the FR axes, and therefore the design is coupled. In this region, both R and S are less than 1.

Reangularity R and semangularity S are reasonably good—but not perfect—metrics for measuring the independence of FRs when there are many FRs. R measures the angle between the DP axes and S measures the magnitude of the diagonal elements. However, they do not measure the actual alignment of DP axes with FR axes, which more accurately measures the independence of FRs. In general, as R and S approach zero, the degree of coupling increases. Although the significant difference in coupling between the case ($R \rightarrow 1$ and $S \rightarrow 1$) and the case ($R << 1$ and small S) can be measured by R and S, one cannot judge the difference in the quality of two coupled designs when there is only a small difference between these values. R and S are convenient measures, but not unique or absolute. After trying many different metrics, R and S were adopted because they provide reasonable means of identifying uncoupled and decoupled designs.

◼ 3.9 DESIGN OF DISPATCHING RULES AND SCHEDULES: AVOIDING TRAFFIC CONGESTION

When a large number of parts must be processed by a system (e.g., an automated machine system with many process modules, job shops, factories, and airlines), what is the rational way of supplying (dispatching) parts and scheduling machines? Dispatching and scheduling are important tasks in many situations, such as production of mechanical parts in job shops, scheduling of robot tasks in automated manufacturing systems, and scheduling of airline flights. Even the traffic-congestion problem involving commercial airplanes belongs to this class of problems. These are important problems because to maximize the productivity of systems, we must design correct dispatching and scheduling algorithms or rules for these systems.

These scheduling problems have been solved using the mathematical tools of operations research or through simulations of the actual situations. However, the results of these efforts have not always been successful. If the system is poorly designed, these techniques do not always yield sufficiently improved viable results, because a design that violates the Independence Axiom cannot be improved through optimization. They must be designed right (to satisfy the Independence Axiom) before the parameters can be adjusted to obtain the correct FRs, and the Information Axiom should be applied to minimize the information content.

---

[9] We could use other definitions, but after trying many other expressions, we adopted this one.

There are two kinds of basic problems that can be solved only through rational design. First, we encounter a situation in which two (or more) parts are finished at two different machines at the same time, competing for the attention of the same robot at the same time. However, both of them cannot be handled for the next event, because there is only one robot. This is a coupled design. When there are many machines and many parts being processed, this kind of conflict can occur frequently. To maximize productivity, we must deal with this kind of problem quickly and rationally.

Some have tried to solve these coupled scheduling problems using "if–then" rules, but there are so many unexpected events that these rules cannot respond quickly enough, delaying the throughput rate of a production system. Furthermore, if there is no appropriate rule built into the system for an unexpected event, the entire system may crash. Also, the process can quickly result in chaos because decisions made earlier affect all subsequent events, making it difficult to predict the future outcomes. Design axioms provide a powerful insight to the scheduling and dispatching problem.

All dispatching and scheduling algorithms must satisfy the Independence Axiom and the Information Axiom to be able to come up with a rational strategy.[10] The strategy of satisfying these axioms may be different, depending on the specific situation. When an identical set of parts is processed through a variety of different machines but the same set of processes, rational scheduling and dispatching algorithms can be developed based on the Independence Axiom so that the scheduling and transport of the part will be uncoupled from the manufacturing processes. In this case, we can come up with a "push" type process that can maximize the productivity, which is discussed in this section. When a random set of parts is processed through a variety of different processes, a "push" system can no longer maximize the throughput rate. In this case, the independence of FRs can be satisfied by designing a cellular manufacturing system—a "pull" system. This "pull" system will control the production rate based on the demand rate, an approach that satisfies the Independence Axiom. This is discussed in Chapter 6.

A solution to scheduling problems can be obtained through rational design of the system. The transport system must be designed to eliminate coupling among the tasks (such as manufacturing processes). The system must be designed correctly by decoupling the tasks from each other using "decouplers."[11] Decouplers play the role of buffers that take care of differences in process times. The role of the decoupler is to eliminate coupling when more than one part requires the attention of the same robot (or person) at the same time, or when a machine is not ready to accept the next part that has just been completed by a preceding machine. Decouplers have been used to resolve the conflict that arises when one machine finishes its operation before the next machine in line is ready to accept the part in a cellular manufacturing system.

---

[10] In the past, many dispatching rules have been used. Some are really ad hoc, such as "first in, first served," "parts with shortest processing time served first," "parts with earliest due dates served first," and "minimum slack time." Some of these rules were compared with the method based on minimizing the information content according to the Information Axiom (see Chapter 8, Section 3 of Suh, 1990). Some other ad hoc scheduling methods have been advanced, such as "mean flow time," "mean tardiness," and "mean slack time." These were also compared with the predictions made by the information minimization method in Chapter 8, Section 5 of Suh (1990).

[11] J. T. Black (1991) used the term "decoupler" in the context of manufacturing systems.

3.9.1 Dispatching Rules and the Independence Axiom

There are several special cases of dispatching situations:

1. *Frequency of dispatches for identical parts.* Consider a manufacturing system in which a certain part must be processed by N machines in a sequential arrangement. Each machine processes the part sequentially and the processing time at each machine is different, which will be denoted as τ_i. The transport time between the machines is τ_t. The longest processing time is τ_m, which is the time taken by Machine m. Then the parts should be dispatched for processing at an interval τ_d given by

$$\tau_d = \tau_m + \tau_t \tag{3.22}$$

If the dispatching rate must be increased to a higher rate than that given by Equation (3.22), the number n of the slowest machines must be increased to

$$n = \text{Int}\left(\frac{\tau_m}{\tau_d}\right) \tag{3.23}$$

where $\text{Int}(x)$ is an integer rounded to the next whole number for any x.

2. *Dispatch rate when random parts are processed.* When random parts are processed, it is difficult to create a dispatching and scheduling algorithm a priori. If the parts are "pushed" into the manufacturing system, it is difficult to satisfy the Independence Axiom and thus to avoid coupling of tasks. In this case, the only way we can satisfy the Independence Axiom is to "pull" the part through the system as the finished part leaves the last machine. The system can be designed to meet the given demand rate. The demand rate cannot exceed the "dispatching" rate given by Equation (3.22).[12] When the demand rate is greater than that given by Equation (3.22), more machines must be added according to Equation (3.23). This is discussed further in Chapter 6.

3.9.2 Scheduling

Scheduling depends on whether an identical set of parts or different random parts are being processed by the system (i.e., manufacturing system, automated machines, robot motions, or airlines). Consider the following scheduling and traffic-congestion problem involving identical parts, which is solved by proper design of the system.

EXAMPLE 3.9 Scheduling of Robots in the Pharmaceutical Industry

A mixture of chemicals (hereinafter called Product) must be subjected to various heating and cooling cycles at various temperatures for different durations before the medicine can be shipped to a packaging section of the factory. There are four process steps involving four different kinds of modules. A robot must place the container with the chemicals into one of these modules, take it out of the module on completion of the process step, and transport it to the next process module according to a preset sequence. We want to maximize the

---

[12] The term "dispatching" is a misnomer in this case as the parts are "pulled" from the last machine as the last manufacturing process is completed.

throughput rate by using the robot so as to utilize the modules most effectively to maximize productivity. The desired throughput rate is 60 units an hour (i.e., one every 60 seconds). A constraint is that we must use a minimum number of modules. It takes the robot 7.5 seconds to put the Product into a module and 7.5 seconds to take the Product out of a module, for a total transport time of 15 seconds.

The chemicals are processed through the following sequence:

Process Steps	Modules	Temperature (°C)	Duration ± Tolerance (seconds)
1	A	35	$40 + 15$ and -5
2	B	80	$20 + 0$
3	C	10	$60 + 10$ and -5
4	D	50	$50 + 10$ and -5

The process times in Modules B must be precise because of the critical nature of the process as indicated by the tight tolerance.

The robot must pick up the chemical in a container from a supply bin (sometimes called load-lock) and deliver it to Module A. When the process is finished, it must pick up the container from Module D and place it on a conveyor belt. These transport operations take 15 seconds each.

SOLUTION

The TAKT time (which is defined as the available time for processing at each station = total cycle time/number of parts required) is 60 seconds. Because the desired throughput rate is 60 parts an hour, the Product must be fed into the machine every 60 seconds.

It takes 40 seconds to process in Step 1 (Module A), 20 seconds in Step 2 (Module B), 60 seconds in Step 3 (Module C), and 50 seconds in Step 4 (Module D). The process times include the overhead time of Product handling.

Because the TAKT time is 60 seconds, we must have two each of Module C and Module D.

The design range is anything greater than 60 parts an hour as shown in Figure E3.9.a. For the information content to be zero, the system must be able to produce more than 60 Products an hour.

One way of solving this problem is to develop an AI-based software program based on "if–then" rules of sending robots to different modules. That is, at each stage of the completion of individual processes, the robot will have to make a decision about which Product to pick up at that instant. However, this becomes an impossible task for a reasonably complicated case, as there are too many combinations to deal with. In fact, there can be an infinite number of combinations. If one of the "if–then" rules is missing, the machine will eventually crash. Furthermore, if we follow this strategy, we will end up affecting the throughput rate because the part will not be fed on a regular interval from the supply bin. This becomes a coupled design because

$$\begin{Bmatrix} \text{Process} \\ \text{Transport} \end{Bmatrix} = \begin{bmatrix} X & X \\ X & X \end{bmatrix} \begin{Bmatrix} \text{Modules} \\ \text{Robot} \end{Bmatrix} \tag{a}$$

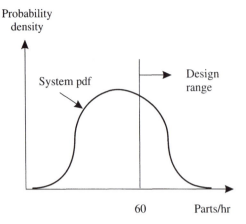

Figure E3.9.a Design range and system pdf for the production rate.

One of the major shortcomings of the above "if–then" scheduling method is that it is ad hoc and cannot be generalized because we would have to anticipate all possible situations.

To create a system in which the robot does not affect the process but accommodates the changes in the process, we must make the off-diagonal element of the upper right-hand corner equal to zero. This can be done if we can make the function of process totally independent of the transport by the robot. In this case, the design equation becomes

$$\left\{ \begin{array}{c} \text{Process} \\ \text{Transport} \end{array} \right\} = \left[\begin{array}{cc} X & 0 \\ X & X \end{array} \right] \left\{ \begin{array}{c} \text{Modules} \\ \text{Robot} \end{array} \right\} \qquad (b)$$

This decoupling of Equation (a) to obtain Equation (b) can be done using "decouplers" that eliminate coupling between tasks the robot must perform. The role of decouplers is to delay the removal of Products or store Products until the robot is ready to pick up the Product. The modules themselves can play the role of decouplers.

To make the off-diagonal element of the upper right-hand corner of Equation (a) zero, we can design the schedule of the robot as follows.

a. Simulate the feeding of the Product once every 60 seconds by completely separating the transport issues from the process issues. Then in 1 hour, we will produce the 60 parts required by the system.

b. Then, 40 seconds after the first Product was placed in Module A, it must be taken out of Module A and placed in Module B. Because Module A has already been vacated, the second Product will be placed in Module A. After 20 seconds in Module B, the first Product will be placed in Module C, and 20 seconds later, the second Product will be placed in Module B.

c. The robot transport time is 15 seconds. While the first Product is in Module C and the second part is in Module B, another Product must be fed into the system in Module A. Through this process, the location of various Products can be determined through the entire cycle, which is defined as the time taken by each Product to undergo various production processes before emerging from the machine as the final product. For this process, the total cycle time is 230 seconds. Note that the cycle time within which the robot must make all the necessary moves is the sending rate of 60 seconds. This is shown in Figure E3.9.b.

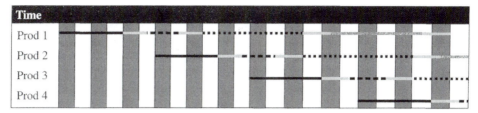

Figure E3.9.b Time-line chart showing the process sequence of the first four products. Each box is 10 seconds long. The black line is Module A (Process 1), the gray line is the overhead time, the long dotted line is Module B (Process 2), the short dotted line is Module C (Process 3), and the darker gray line is Module D (Process 4). There are two each of Modules C and D. Note that at $t = 170$ seconds, there is a coupling for transport requirement for Products 1 and 3. Also at $t = 230$ seconds, there is another scheduling conflict.

d. This simulation of the entire cycle will show that there are spots of conflict in timing. That is, some of the Products will be done nearly at the same time (within the time required for the robot to transport parts between the modules). In some cases, the next module may not be ready to receive the Product from the preceding module because it still has the Product that came in earlier. Figure E3.9.b shows that at $t = 170$ seconds, there is a coupling for transport requirement for Products 1 and 3. Also at $t = 230$ seconds, there is another scheduling conflict.

Within a given sending time period of 60 seconds, the robot can make only four moves. Figure E3.9.b shows that Processes 2, 3, and 4 are in conflict. This is illustrated in Figure E3.9.c, which shows when the robot has to pick up the Product within a given 60-second cycle. The horizontal axis is the normalized time, i.e., 1 is equal to 60 seconds. Within this one cycle (i.e., 60 seconds), the robot can make four moves.

e. When such scheduling conflicts arise, introduce a "decoupler" that will keep the Product in a given module for a fixed duration longer until the congestion is eliminated. The decoupler may be either physically separate hardware or simply the extra time the Product is allowed to stay in the preceding module if the longer stay does not do any harm to the Product.

f. Priority should be given to the Products coming out of Module B as Process Step 2 must be completed in exactly 20 seconds (i.e., zero tolerance).

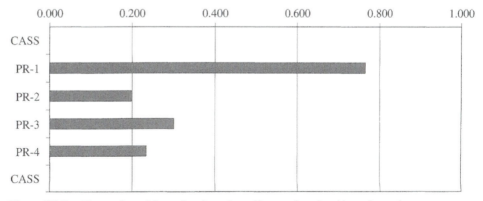

Figure E3.9.c The product pick-up time in a given 60-second cycle without decouplers.

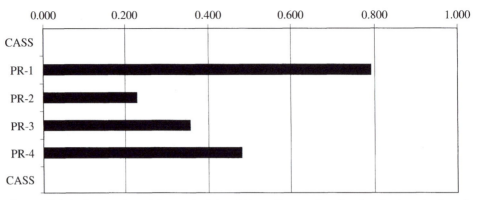

Figure E3.9.d New product pick-up schedule within a 60-second cycle after the introduction of decouplers.

g. The conflict can be eliminated if the decoupler for Process 1 is given a queue time of 1.1 seconds, 0 seconds for Process 2, 0.5 second for Process 3, and 10.4 seconds for Process 4. The robot schedule after these decouplers are introduced is given in Figure E3.9.d.

If we follow this method, the robot schedule will be completely independent of the process and vice versa. Furthermore, the robot schedule is deterministic and also optimum. This observation may be stated as Theorem 24.

Theorem 24 (Optimum Scheduling)

Before a schedule for robot motion or factory scheduling can be optimized, the tasks must be so designed to satisfy the Independence Axiom by eliminating coupling by adding decouplers, which may be in a form of queue or separate hardware or buffer.

A closed-form analytical solution can be obtained to the above task of designing scheduling based on the Independence Axiom (Homework 3.10). The design of a schedule is discussed further in Chapter 6.

3.10 AXIOMATIC DESIGN BASIS FOR ROBUST DESIGN

All designs—hardware, software, organizations, and systems—must be robust for two reasons. First, the design should be robust so that the original design goals can be achieved easily and faithfully. Second, the product must be reliable and durable. One of the goals of axiomatic design is to develop robust designs in all fields of synthesis. The necessary condition for robust design is the fulfillment of FRs within the bounds established by constraints under all operating conditions, including situations where variables not under the engineer's control—noise—affect the performance of the function.

3.10.1 One-FR Design

The one-FR robust design was discussed in Chapter 2. The basic concepts presented for one-FR design are equally applicable to multi-FR design. Therefore, the robust design concepts for one-FR design will be reviewed first before discussing multi-FR robust design.

Consider the following redundant design with one FR and a large number of DPs:

$$FR = f[DP^a, DP^b, \ldots, DP^n] \tag{3.24}$$

Since this is one-FR design, we need only one DP. However, this design has many extra DPs, which may be sources of random variation in performance of the function. The task is to make this design robust under all conditions, if possible. Some of the DPs of Equation (3.24) cannot be controlled by the engineer.

Under the robust design method practiced today in industry, noise cannot be a DP because noise, by definition, is a generic variable that cannot be controlled by the engineer but that causes variation in the performance of the desired function. Clearly some of these factors that contribute to noise cannot be fixed and the design must be done to accommodate the noise. According to this definition of noise, only variables that cannot be fixed or controlled in any way (e.g., ambient temperature and customer usage) can be considered noise. The consequence of noise can be treated either as a constraint or a functional requirement. If it is treated as an FR, we would take an active means of controlling the noise, whereas if it is treated as a constraint, we must be certain that the design does not violate it. However, for the purpose of the discussion presented in this section, we will assume that some of the DPs shown in Equation (3.24) are under the control of the designer and some are not.

FR may be expressed as

$$\Delta FR = \frac{\partial f}{\partial DP^a} \Delta DP^a + \frac{\partial f}{\partial DP^b} \Delta DP^b + \cdots + \frac{\partial f}{\partial DP^n} DP^n \tag{3.25}$$

In an ideal one-FR design, only one DP is needed (Theorem 4, Ideal Design). All other DPs are possible sources of random variation.

One way of making the design insensitive to such variation is to make the coefficient $(\partial f / \partial DP)$ associated with the extra DPs to be zero. This method of making a design *immune* to random variation is one of the basic concepts of robust design practiced in industry today. This method was effectively exploited when the one-FR design was nonlinear, as illustrated in Example 2.3 (Measuring the Height of the Washington Monument) and Example 2.4 (Windshield Wiper—Robust Mounting Design).

Another way of dealing with random variation for the one-FR design given by Equation (3.24) is to compensate for it by "fixing" the values of all DPs except one DP chosen so that it can vary the value of FR. Some DPs can be fixed during manufacturing operations (see Example 2.6 for Van Seat Assembly).

In the one-FR design represented by Equation (3.24), only one DP should be chosen as the primary DP and the rest of the DPs should be fixed to constant values, eliminating them as possible sources of random variation. Then, a change in FR can be achieved by varying the chosen DP. This may be expressed as

$$\begin{aligned} FR &= \frac{\partial f}{\partial DP^c} DP^c + \sum_{\substack{i=a \\ i \neq c}}^{i=n} \frac{\partial f}{\partial DP^i} DP^i \\ &= \frac{\partial f}{\partial DP^c} DP^c + [\text{constant}] \\ &= [\text{Module}]DP^c + [\text{constant}] \\ &= [\text{Stiffness}]DP^c + [\text{constant}] \end{aligned} \tag{3.26}$$

where DP^c is the DP chosen to satisfy the FR. All other DPs are fixed and thus eliminated as potential sources for variation. For the one-FR case, $\partial f / \partial DP^c$ is equal to both module and stiffness. The first term of the right-hand side of Equation (3.26) compensates for the variation generated by all other extra DPs and satisfies FR.[13]

The chosen DP^c of Equation (3.26) must have the following two characteristics:

1. The magnitude of the term $(\partial f / \partial DP)\,\Delta DP$ of the chosen DP should be larger than the sum of the constant terms so that the accumulated errors can be compensated.
2. If the magnitudes of two or more terms of the right-hand side of Equation (3.26) are approximately the same, the one with smaller $(\partial f / \partial DP)$ should be chosen to minimize the sensitivity of FR to the variation of DP.

If the desired (target) value of FR is equal to FR_0, then DP^c is given by

$$DP^c = \frac{1}{(\partial f / \partial DP^c)}[FR_0 - (\text{constant})]$$

$$= \frac{[(FR)_0 - (\text{constant})]}{\text{module}}$$

$$(3.27)$$

If the design range of FR is given as ΔFR, the maximum tolerance of DP^c, ΔDP^c, is given by

$$(\Delta DP^c)_{\max} = \frac{1}{(\partial f / \partial DP^c)}[\Delta FR - (\text{constant})]$$

$$= \frac{[\Delta FR - (\text{constant})]}{\text{module}}$$

$$(3.28)$$

To make the design robust by making ΔDP^c as large as possible, the module[14] (i.e., stiffness) must be small. If the variation caused by the fixed DPs is larger than ΔFR, it cannot be compensated. In this case, the design cannot be accepted because the FR is not in the design range, thus requiring the development of a new design concept.

The information content of a design with uniform pdf, symmetric with respect to FR_0, and with actual FR range = $(FR_{\max} - FR_{\min})$ is given by

$$I = -\log_2\left[\frac{\Delta FR}{FR_{\max} - FR_{\min}}\right] \quad \text{if } FR_{\max} - FR_{\min} > \Delta FR$$

otherwise

$$(3.29)$$

$$I = 0$$

The information content can be computed for any distribution of pdf.

---

[13] In Example 2.6 (Van Seat Assembly), the lengths of all linkages randomly varied from part to part. To make the seat assembly process robust, the lengths of all the linkages except one were fixed by either welding or using lock nuts, and then the seat was assembled. Through this process, the second term of the right-hand side of Equation (3.26) was fixed. Then, using the chosen DP, the accumulated random errors introduced by the extra DPs were compensated using a fixture.

[14] Module is defined in Chapter 4, Equation (4.13).

Some variables cannot be fixed, such as when random variation (noise) is introduced by the environment or by product usage, e.g., random temperature variation during use. In this case, we must design the product (or software, organization, etc.) correctly by specifying the appropriate FRs and Cs in anticipation of unpredictable random variation.

3.10.2 Multi-FR Design

Even when there are many FRs to be satisfied at the same time, the robust design concepts discussed with respect to one-FR design do apply if the design satisfies the Independence Axiom.

Consider a multi-FR design where the number of DPs is larger than the number of FRs—a redundant design. In such a design, each FR may be affected by many DPs. The design equation for such a design may be expressed as

$$\{FR\} = [\text{square } DM]\{DP\} + [\text{extra matrix}]\{DP\}^{\text{extra}} \tag{3.30}$$

where the vector $\{DP\}$—the first term of the right-hand side of Equation (3.30)—represents the DPs chosen to satisfy the vector $\{FR\}$. The number of DPs in $\{DP\}$ is the same as the number of FRs in $\{FR\}$. On the other hand, $\{DP\}^{\text{extra}}$ of the second term of right-hand side of Equation (3.30) is the vector of the redundant DPs, i.e., the extra DPs that are left over after $\{DP\}$ is chosen to satisfy $\{FR\}$. In some cases, the extra DPs may not be known explicitly. [Square DM] is the square design matrix that relates $\{FR\}$ to $\{DP\}$. [Extra matrix] gives the effect of the extra DPs on the FRs. To satisfy the Independence Axiom, [square DM] must be either diagonal or triangular. However, [extra matrix] can be any matrix, including a full matrix, and the Independence Axiom can still be satisfied if [square DM] is either diagonal or triangular.

Consider a special case of Equation (3.30)—a multi-FR design with three FRs and many DPs represented by Equation (3.31).

$$\begin{Bmatrix} FR_1 \\ FR_2 \\ FR_3 \end{Bmatrix} = \begin{bmatrix} X & 0 & 0 \\ 0 & X & 0 \\ 0 & 0 & X \end{bmatrix} \begin{Bmatrix} DP_1 \\ DP_2 \\ DP_3 \end{Bmatrix} + \begin{bmatrix} X & X & X & X & \cdots & X \\ X & X & X & X & \cdots & X \\ X & X & X & X & \cdots & X \end{bmatrix} \begin{Bmatrix} DP_4 \\ DP_5 \\ DP_6 \\ DP_7 \\ \cdots \\ DP_n \end{Bmatrix} \tag{3.31}$$

The first term of the right-hand side of Equation (3.31) is an uncoupled design, but the second term represents a fully coupled design. In spite of the second term, this design can be treated as an uncoupled design, if we fix the values of DP_4 through DP_n. Then DP_1, DP_2, and DP_3 can be used to satisfy the independence of FR_1, FR_2, and FR_3. In this case, the design range of each FR can be represented as

$$\Delta FR_i = \frac{\partial FR_i}{\partial DP_i} \Delta DP_i + \sum_{j=4}^{n} \frac{\partial FR_i}{\partial DP_j} \delta DP_j$$

$$= \frac{\partial FR_i}{\partial DP_i} \Delta DP_i + \sum_{j=4}^{n} [\text{extra terms}] \tag{3.32}$$

Equation (3.32) is similar to Equation (3.26) with similar implications for compensation by fixing the values of the extra DPs.

If the diagonal design matrix of the first term of the right-hand side of Equation (3.31) is replaced with a triangular matrix shown in Equation (3.33), the design may be treated as a decoupled design.

$$
\begin{Bmatrix} FR_1 \\ FR_2 \\ FR_3 \end{Bmatrix} = \begin{bmatrix} X & 0 & 0 \\ X & X & 0 \\ X & X & X \end{bmatrix} \begin{Bmatrix} DP_1 \\ DP_2 \\ DP_3 \end{Bmatrix} + \begin{bmatrix} X & X & X & X & \cdots & X \\ X & X & X & X & \cdots & X \\ X & X & X & X & \cdots & X \end{bmatrix} \begin{Bmatrix} DP_4 \\ DP_5 \\ DP_6 \\ DP_7 \\ \cdots \\ DP_n \end{Bmatrix} \tag{3.33}
$$

In this case, the effect of any random variation of DP_4 through DP_n can be compensated by means of DP_1, DP_2, and DP_3, if the extra terms are fixed first and then compensated. Then the design range ΔFR_i can be achieved with DP_i, which may be expressed as

$$
\Delta FR_i = \left(\frac{\partial FR_i}{\partial DP_i} \right) \Delta DP_i + \sum_{\substack{j=1 \\ j \neq i}}^{3} \left(\frac{\partial FR_i}{\partial DP_j} \right) \Delta DP_j + \sum_{k=4}^{n} \left(\frac{\partial FR_i}{\partial DP_k} \right) \delta DP_k
$$

$$
\tag{3.34}
$$

$$
= \left(\frac{\partial FR_i}{\partial DP_i} \right) \Delta DP_i + \sum_{\substack{j=1 \\ j \neq i}}^{3} \left(\frac{\partial FR_j}{\partial DP_j} \right) \Delta DP_j + (\text{extra terms})
$$

In Equation (3.34), DP_i is the primary DP chosen to control FR_i. DP_j is the DPs chosen to satisfy FR_j but with a secondary effect on FR_i because of the off-diagonal elements. DP_ks are not the primary DPs and thus the source for causing random variation δFR. In compensating for this, ΔDP_j must be set first—according to the sequence defined by the triangular matrix—to control the corresponding FR_j. The variation caused by the extra terms can be fixed if δDP_k is eliminated by fixing DP_k. Finally, the desired ΔFR_i can be controlled by means of ΔDP_i.

If the square design matrix of the first term of the right-hand side of Equation (3.30) is a full matrix, we have a coupled design, which violates the Independence Axiom. In this case, there is no way we can satisfy each FR independently when unknown DPs introduce random variation.

What are the implications of Equations (3.33) and (3.34) for robust design? The foregoing discussion has profound implications for robust design when there are many FRs that must be satisfied at the same time. For example, when data are taken to determine the best operating parameters of an existing machine, we may find that there are more DPs than FRs—a redundant design. In this case, we should select DPs that will be in the form of Equations (3.32) and (3.34) by making sure that the number of selected DPs is equal to the number of FRs and the selected DPs yield either a diagonal or a triangular matrix. Once this is done, the robust performance of the machine is ensured.

How do we select the primary DPs? The selection of DP_1, DP_2, and DP_3 in a multi-FR design must satisfy the same set of conditions as those discussed for one-FR design—robustness and sensitivity. However, in the case of a decoupled multi-FR design, we need to satisfy the following additional condition for the elements of the triangular design matrix:

$$\left(\frac{\partial f}{\partial DP} \right)_{\text{diagonal}} >> \left(\frac{\partial f}{\partial DP} \right)_{\text{off-diagonal}} \tag{3.35}$$

If the above condition is not satisfied, we have an unacceptable design due to wrong choice of DPs. For example, consider the following situation in which the magnitudes of diagonal elements are smaller than those of the off-diagonal elements. That means DP_2 and DP_3 chosen to control FR_2 and FR_3, respectively, have larger effects on FR_1 than DP_1 has on FR_1. In the limit, when one of the diagonal elements becomes much smaller than the off-diagonal elements, the design becomes a coupled design, since one DP can affect more than two FRs. When the number of DPs becomes less than the number of FRs—because one or more diagonal elements become zero—the design becomes a coupled design as per Theorem 1 (Coupling Due to Insufficient Number of DPs).

3.10.3 Information Content of Multi-FR Design

In most cases, the information content of a design that satisfies the Independence Axiom is expected to be less than the information content of a coupled design, because uncoupled and decoupled designs allow the elimination of bias and the reduction of variance by choosing low stiffness designs. However, it is conceivable that certain coupled designs—with small random variation and with a special relationship between DPs—can have lower information content than uncoupled designs with large random variation and large bias (Homework 3.21 and 3.22). That is, designs that violate the Independence Axiom may have smaller information content than an uncoupled design under a special situation. Furthermore, even among the designs that satisfy the Independence Axiom equally well, the information content will be different. Therefore, both the Independence Axiom and the Information Axiom are required. They are also independent from each other (see Appendix 3A).

A consequence of Axiom 1 and Axiom 2 is *Theorem 18 (Existence of an Uncoupled or a Decoupled Design)*, which states, "There always exists an uncoupled or a decoupled design that has less information content than a coupled design." According to this theorem, although we may not be able to readily advance such uncoupled designs for a variety of reasons—lack of suitable technology or lack of knowledge—there must be an uncoupled design with less information content than the coupled design.

The foregoing discussion may become clearer if we consider a decoupled design given by the design matrix [A] as

$$[A] = \begin{bmatrix} A_{11} & A_{12} \\ 0 & A_{22} \end{bmatrix} \tag{3.36}$$

The design represented by Equation (3.36) is always decoupled because $A_{21} = 0$. When the magnitude of A_{12} approaches zero, the design approaches that of an uncoupled design. The degree of coupling between FR_1 and FR_2 increases with the magnitude of A_{12}. Eventually when the magnitudes of A_{11} and A_{12} are about the same, DP_2 has as much influence on FR_1 as DP_1 does on FR_1. What this means is that when DP_2 is changed to satisfy FR_2, DP_1 must be adjusted as much to keep FR_1 where it was. In this case, the reangularity increases. If A_{12} becomes much larger than A_{11}, it almost behaves like a coupled design.

Therefore, in a multi-FR design, one must be sure that the DPs are chosen so that the magnitudes of the diagonal elements—after the design equations are normalized with respect to both FR and DP—remain much larger than the magnitude of the off-diagonal elements.

EXAMPLE 3.10 Robust Design of a Microgyroscope[15]

An electronics company in Korea designed a microgyroscope made of microelectrico-mechanical systems (MEMS), which is shown in Figure E3.10.a. It measures the motion of the object to which it is attached by detecting resonant vibratory responses of MEMS in response to external motion. The gyroscope is made from silicon wafer by means of photolithography and etching. A finite element model is shown in Figure E3.10.b.

The gyroscope functions by sensing the relative motion of the stationary members with respect to the moving part. It induces the translational motion along the X direction by applying electric current to the driving spring at a predetermined frequency. When an angular motion is present, the gimbal (the center plate) rotates because of Coriolis acceleration. The gyroscope measures the translational motion (the X directions) and the rotational motion about the X axis, as shown in Figure E3.10.c. When three of these gyroscopes are mounted along the three orthogonal directions, they can measure motions in six directions—three translational and three rotational motions—and thus define the motion of the body to which the system is attached. (Apparently, the actual device has four of these gyroscopes—perhaps one for redundancy.) The driving force generates the translational motion, deforming the four bending springs, which in turn induces the rotational motion of the central plate (gimbal) that is attached to the translational plate by the sensing spring.

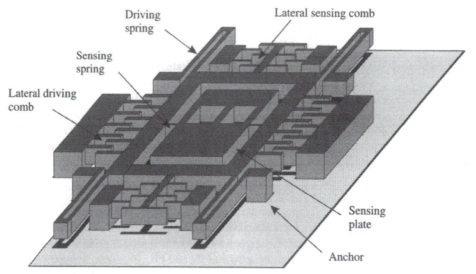

Figure E3.10.a Design of the original resonant vibratory gyroscope.

---

[15] Developed based on materials presented by Lee, Hwang, and Park (2000).

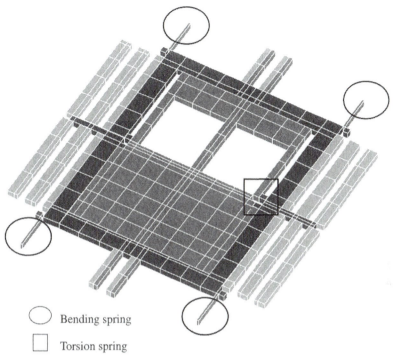

Bending spring

Torsion spring

Figure E3.10.b Finite element model of the gyroscope.

The gyroscope has to have correct resonant natural frequencies to be able to respond to the external motion in a manner for which it was originally designed and calibrated. One of the two important resonant frequencies is associated with the translational motion (along the X direction) of the entire moving plate assembly, which is connected the stationary part by four bending springs. The other important natural frequency is associated with the torsional resonance of the gimbal (the central plate), which is connected to the rest of the moving plate by the sensing spring. The actual measurement of the relative motion is done by means of the capacitance change between a series of capacitor plates between the stationary part and the moving part.

The gyroscope is designed to have two specific natural frequencies, $f_1 = \beta$ and $f_2 = \gamma$. f_1 is the driving mode for the translational motion and f_2 is the sensing mode corresponding to the rotational motion of the gimbal. These two frequencies must be exactly the same to have the best response and provide the most accurate measurement. The driving mode is tuned with a dc voltage on the sensing plate. The bias voltage on the sensing plate preloads the four bending springs, which affect the translational motion along the X direction, and thus vary the bending resonant frequency so as to match the rotational frequency. However, because the accuracy of the manufacturing processes is only 10%, there is a mismatch between the frequencies of the two modes and tuning is extremely difficult.

The current design has such a large random variation in dimensions that the yield of manufacturing operations is extremely low. Therefore, the company asked Professor Park of Hanyang University to develop a robust design so that the gyroscope could be manufactured more easily and reliably and the product could be commercialized.

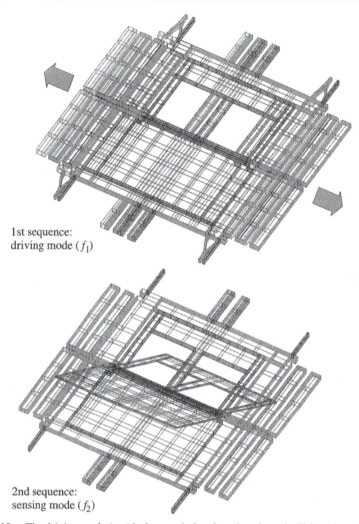

1st sequence:
driving mode (f_1)

2nd sequence:
sensing mode (f_2)

Figure E3.10.c The driving mode (top) is the translational motion along the X direction with a natural frequency of f_1 and the sensing mode (bottom) of the gyroscope is the rotational motion of the gimbal (central plate), which has a natural frequency of f_2.

SOLUTION

The functional requirements (FRs) of the gyroscope are the following:

$FR_1 = $ Set the frequency of the driving mode—the translational motion of the moving plate—at β.

FR_2 = Set the frequency of the sensing mode—the torsional motion of the central plate (gimbal) at γ.

The design parameters (DPs) are the following physical variables:

DP_1 = Stiffness of the four bending springs
DP_2 = Stiffness of the two torsional springs
DP_1^* = Bias voltage on the sensing plate.

DP_1^* is a redundant DP for FR_1 for the purpose of fine tuning β.
The design equation may be written as

$$\begin{Bmatrix} FR_1 \\ FR_2 \end{Bmatrix} = \begin{bmatrix} A_{11} & 0 & A_{11}^* \\ 0 & A_{22} & 0 \end{bmatrix} \begin{Bmatrix} DP_1 \\ DP_2 \\ DP_1^* \end{Bmatrix} \tag{a}$$

It is an uncoupled design. The random error of δFR_1 and δFR_2 must be within the design ranges ΔFR_1 and ΔFR_1. The random errors δFR_1 and δFR_2 are given by

$$\delta FR_1 = \delta A_{11} \, DP_1 + A_{11} \, \delta DP_1 + \delta A_{11}^* \, DP_1^*$$
$$\delta FR_2 = \delta A_{22} \, DP_2 \tag{b}$$

The design task is to make the frequency β the same as γ. The problem Samsung had was that the manufacturing variations were so large that the random error terms—the first two terms of the right-hand side of Equation (b)—were larger than that which can be compensated by the third term. Therefore, Professor Park and his team set out to make the random error terms as small as possible by making the design robust by making ($\delta FR_1 - \delta FR_2$) small. This can be done by lowering the "stiffness" terms A_{11} and A_{22}.

The natural frequencies of both f_1 and f_2 are affected by the distribution of mass and thus the moments of inertia of the gimbal (the central plate). Therefore, they chose the inertia of the gimbal as a means of minimizing the variance of ($\delta FR_1 - \delta FR_2$). Then, they conducted an orthogonal array experiment to determine the best values of the gimbal dimensions that yielded the least variance.

The gimbal plate has three dimensions (a, b, c), all of which affect FR_1, FR_2, and FR_3. The gimbal and the variable dimensions are shown in Figure E3.10.d. Depending on the specific values of these three dimensions, the moments of inertia along the translational and the rotational direction will be affected. By choosing these dimensions correctly, the effect of manufacturing variation on the random variation of FR_1 and FR_2 can be minimized.

The function that needs to be minimized may be expressed as

$$\delta FR_1 = \frac{\partial (f_2 - f_1)}{\partial a} \, \delta a + \frac{\partial (f_2 - f_1)}{\partial b} \, \delta b + \frac{\partial (f_2 - f_1)}{\partial c} \, \delta c \tag{c}$$

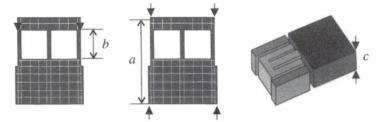

Figure E3.10.d Gimbal plate with three dimensions that can affect the moment of inertia along the translational direction and the rotational direction.

One way of minimizing Equation (c)—the best being $\delta FR_1 = 0$—is to conduct an orthogonal array experiment (see Section 5.8 of Suh, 1990) to determine the best combinations of a, b, and c that satisfy the equation when a, b, and c vary by the manufacturing tolerance δ. That is, a varies by $\delta a = \delta$, b by $\delta b = \delta$, and c by $\delta c = \delta$. Once Equation (c) is satisfied, then β can be varied to match γ by varying $DP_1{}^*$.

Professor Park and his students set up an orthogonal array experiment, and the variance of $(\delta FR_1 - \delta FR_2)$ was evaluated for 27 combinations of a, b, and c at three levels each as shown in Table E3.10.1, using the finite element method. Once Equation (h) is satisfied, DP_1 is changed to satisfy $FR_1 = f_1$ and DP_2 to satisfy $FR_2 = f_2$. They tried several different tolerances that can be achieved by the manufacturing processes to determine the best dimensions for the gyroscope. Their final recommended values are given in Table E3.10.2.

Then they determined the probability of success of their proposed design by determining the area of the system range (assuming a normal distribution) that is within the design range. The results show a significant increase in probability of making an acceptable gyroscope as shown in Table E3.10.3.

When the magnitude of the undercut (i.e., the thickness of the supporting beam) is assumed to be 50% less, the probability of success increased to 94.7%. The information content was reduced to 0.08.

It should be noted that if the design was done using axiomatic design from the beginning, a correct design of the gyroscope could have been achieved, which would have shortened the development process. Samsung found out about their problem after they made the gyroscope in large numbers. This is a very common mistake made by many companies. Most companies do not have a formal design methodology adopted throughout the company and depend on experienced designers who do not know how to make correct design decisions a priori and thus depend on trial and error processes. Such a practice lengthens the product development time and increases the R&D cost.

Many companies are willing to spend time and money to correct the mistakes made but are reluctant to provide resources at the beginning to develop a correct design for a given product. Then, they make many compromises to meet the deadline and produce

Table E3.10.1 Range of Orthogonal Array Experiments

Level	DP$_3$a: a	DP$_3$b: b	DP$_3$c: c
1	Lower bound	Lower bound	Lower bound
2	Current	Current	Current
3	Upper bound	Upper bound	Upper bound

Table E3.10.2 Final Values

Design	f_1	f_2	$f_2 - f_1$	$\mu(f_2 - f_1)$	$\sigma(f_2 - f_1)$	Improvement
Original	a_1	a_2	0.287	334.33	84.84	
Recommended	b_1	b_2	0.119	127.15	51.81	38.9%

Table E3.10.3 Comparison of the Probability of Success of the Original Design and the Recommended Design

Design	Probability of Success Pr	Information Content I
Original	5.7%	4.14
Recommended	86.0%	2.18

inferior or less reliable products. As a consequence, many products fail in the market place.

Universities do not help this situation either. Some of the universities do not teach design theories and methodologies in a rigorous way. Also some teach robust design concept from a statistical point of view rather than from the engineering and axiomatic design point of view. Consequently, students do not write design equations even when they have a multi-FR design task. These practices are beginning to change for better as more schools begin to teach more formal design methods.

Once students and engineers learn to think systematically based on axiomatic design, many seemingly complicated problems can be solved in a rational way. However, it takes a fair amount of practice. In this respect, the need to learn the principles taught in axiomatic design by solving problems cannot be over emphasized. To those who are not initiated in axiomatic design, axiomatic design appears to be so simple that it is almost trivial. To those who just learned the subject but who have not yet mastered the subject, axiomatic design may appear to be very difficult. However, once the axiomatic design theory is fully understood, it becomes fairly easy to use and practice.

■ 3.11 SUMMARY

In this chapter, multi-FR designs are considered. The implications of the Independence Axiom and the Information Axiom are presented with relevant theories that govern multi-FR designs. Many examples are given to illustrate the basic concepts. The robust-design concept is given for the multi-FR case as well as discussing the relationship between the complexity and the information content of a design.

REFERENCES

Black, J. T. *The Design of a Factory with a Future*, McGraw-Hill, New York, 1991.

Boothroyd, G., and Dewhurst, P. *Product Design for Assembly*, Boothroyd & Dewhurst Inc., Wakefield, RI, 1989.

El-Haik, B., and Yang, K. "The Components of Complexity in Engineering Design," *IIE Transactions*, Vol. 31, Issue 10, pp. 925–934, August 1999.

Kim, S.-G., Hwang, K.-H., Hwang, J., Koo, M.-K., and Lee, G.-W. "Actuated Mirror Array—A New Chip-Based Display Device for the Large Screen Display," SID Asia Display 98, 1998.

Lee, K.-H., Hwang, K.-H., and Park, G.-J. "Robust Design of a Micro Gyroscope Using the Axiomatic Approach," presented at UKC-2000, Chicago, IL, September 2000.

Rinderle, J. R., and Suh, N. P. "Measures of Functional Coupling in Design," *Transactions of A.S.M.E., Journal of Engineering for Industry,* Vol. 104, No. 4, pp. 383–388, 1982.

Simon, H. A. *The Science of the Artificial*, 2nd ed., MIT Press, Cambridge, MA, 1981.

Söderberg, R., and Lindkvist, L. "Computer Aided Assembly Robustness Evaluation," *Journal of Engineering Design*, Vol. 10, No. 2, pp. 165–181, 1999.

Suh, N. P. *Tribophysics*, Prentice Hall, Englewood Cliffs, NJ, 1986.

Suh, N. P. *The Principles of Design,* Oxford University Press, New York, 1990.

APPENDIX 3–A Independence of the Two Design Axioms[16]

In engineering design, the array $\{FR\}_m$ is the minimum set of m independent functional requirements that completely characterizes the design objectives (Suh, 1990). The array $\{DP\}_n$ is the set of n design parameters needed to fulfill the FRs, and the array $\{PV\}_p$ is the set of p process variables needed to produce the product specified.

In this section, we assume that $[A]_{m \times n}$ is a square (i.e., $n = m$) nonsingular matrix with constant entries. Also we assume that the DPs are normally distributed random variables

---

[16] Contributed by El-Haik and Yang (1999).

where $f_{DP}(\{DP\}_n)$ is the joint probability density function characterizing the variation of the DPs.

The entropy of the $\{FR\}_m$ array, $h(f\{FR\}_m)$, is a good measure of complexity for the overall design solution. Therefore, since

$$\{FR\}_m = [A]_{m \times n}\{DP\}_n$$

then

$$h(f\{FR\}_m) = h([A]_{m \times n}\{DP\}_n)$$

Theorem 3A1

If the design matrix $[A]_{m \times n}$ is a square (i.e., $n = m$) nonsingular constant matrix, then the complexity of a design can be expressed by the entropy of the FRs, $h(f\{FR\}_m)$.

$$h(f\{FR\}_m) = h(f_{DP}\{DP\}_n) + \ln|A| \qquad (A.3A.1)$$

where $|A|$ is the determinant of the design matrix $[A]_{m \times n}$.

Proof

Let $f_{DP}(\{DP\}_n)$ be the joint probability density function characterizing the variation of the DPs. If $[A]_{m \times n}$ is a square nonsingular matrix (i.e., $n = m$) and the DPs are normally distributed, then $f_{FR}\{FR\}_m) = (1/|A|)f_{DP}([A]^{-1}\{FR\}_m)$. By using Boltzmann measure, we have

$$h[f(FR)] = -\int f_{FR}\{FR\} \log f_{FR}\{FR\}d\text{FR}$$

$$= -\int \left(\frac{1}{|A|}\right) f_{DP}([A]^{-1}FR) \ln[(1/|A|)f_{DP}([A]^{-1}FR)]d\text{DP}$$

$$= -\int f_{DP}\{DP\} \ln f_{DP}\{DP\}d\text{DP} + \ln|A|$$

$$= h[f_{DP}\{DP\}] + \ln|A|$$

There are two components of complexity in the FR array. The first component, *variability*, is expressed by $h(f_{DP}\{DP\}_n)$, whereas the second component is due to the determinant of the design matrix $\ln|A|$. It would be logical to call this the *sensitivity* component as the determinant is a function of the sensitivity coefficients, the elements of $[A]_{m \times n}$. However, the term *vulnerability* is used instead because $\ln|A|$ has broader meaning than the numerical values of the sensitivity coefficients. There are three ingredients that collectively make the complexity component due to vulnerability: *mapping*, *sensitivity*, and *dimension*.

The mapping ingredient refers to the topological structure of the design matrix that corresponds to the position of the nonzero sensitivity coefficients in $[A]_{m \times n}$. The sensitivity ingredient refers to the magnitude and sign of nonzero coefficients $A_{ij} = \partial FR_i/\partial DP_j$, whereas the dimension ingredient refers to the size of the design problem itself, i.e., the number m of FRs. We view our interpretation of this complexity component as the mathematical form of Simon's (1981) definition of complexity.

For an uncoupled design, the value of $|A|$ is the product of the diagonal elements:

$$|A| = \prod_{i=1}^{n} A_{ii}$$

and the complexity component due to vulnerability is

$$\sum_{i=1}^{n} \ln |A_{ii}|$$

The total independent design complexity (assuming all DPs are normal information sources) is

$$\sum_{i=1}^{n} \ln \left(\sqrt{2\pi e} \sigma_i \cdot A_{ii} \right) \qquad \text{nats}$$

Corollary 1

For linear process mapping, we have

$$h(\{DP\}_n) = h(\{PV\}_p) + \ln |B|$$

Then by substitution in Equation (A.3A.1), the total design complexity is given by

$$h(\{FR\}_m) = h(\{DP\}_n) + \ln |A|$$
$$= h(\{PV\}_p) + \ln |B| + \ln |A|$$
$$= h(\{PV\}_p) + \ln |C|$$

where $[C]_{m \times p} = [A]_{m \times n} [B]_{n \times p}$, the overall design matrix.

From Corollary 1, we conclude that the PVs are the primary sources of the variability component, whereas the vulnerability component has a multiplicative coupling effect from the physical and process mappings. That is, the term $\ln |C| = ln|[A][B]|$ is the aggregated complexity component due to vulnerability.

Theorem 1 suggests another interesting point on the relationship between the Independence Axiom and the Information Axiom. The vulnerability component of complexity is denoted by $\ln |A|$ per Theorem 1. The fact is that a totally uncoupled design may not be the simplest design, based on $\ln |A|$. Assume that there are two solutions for an engineering design problem: an independent (uncoupled) design and a coupled design. For the independent design,

$$|A| = \prod_{i=1}^{n} A_{ii}$$

and the complexity component due to vulnerability is

$$\sum_{i=1}^{n} \ln |A_{ii}|$$

For the coupled design, it is possible that

$$\ln |A| < \sum_{i=1}^{n} \ln |A_{ii}|$$

even if both designs have the same dimensionality and the same diagonals A_{ii}. So it is always possible to conceive a coupled design with less overall complexity than an uncoupled design. That is, Axiom 1 is not a consequence of Axiom 2; both are mutually exclusive design principles because each addresses a particular conceptual weakness. Always seek to find an uncoupled design with lower information content (less complexity) than a coupled design (Corollary 7, Suh, 1990).

Appendix 3–B Corollaries and Theorems Related to Information and Complexity

There are many corollaries and theorems given in Appendix 1A that guide the designer in reducing information content. The most relevant ones for complexity and information content are the following:

1. COROLLARIES

Corollary 1 (Decoupling of Coupled Designs)   Decouple or separate parts or aspects of a solution if FRs are coupled or become interdependent in the designs proposed.

Corollary 2 (Minimization of FRs)   Minimize the number of FRs and constraints.

Corollary 3 (Integration of Physical Parts)   Integrate design features in a single physical part if FRs can be independently satisfied in the proposed solution.

Corollary 4 (Use of Standardization)   Use standardized or interchangeable parts if the use of these parts is consistent with FRs and constraints.

Corollary 5 (Use of Symmetry)   Use symmetrical shapes and/or components if they are consistent with the FRs and constraints.

Corollary 6 (Largest Design Range)   Specify the largest allowable design range in stating FRs.

Corollary 7 (Uncoupled Design with Less Information)   Seek an uncoupled design that requires less information than coupled designs in satisfying a set of FRs.

2. THEOREMS OF GENERAL DESIGN

Theorem 1 (Coupling Due to Insufficient Number of DPs)   When the number of DPs is less than the number of FRs, either a coupled design results or the FRs cannot be satisfied.

Theorem 2 (Decoupling of Coupled Design)   When a design is coupled because of a greater number of FRs than DPs (i.e., m > n), it may be decoupled by the addition of new DPs to make the number of FRs and DPs equal to each other if a subset of the design matrix containing m × m elements constitutes a triangular matrix.

Theorem 3 (Redundant Design)   When there are more DPs than FRs, the design is either a redundant design or a coupled design.

Theorem 4 (Ideal Design)   In an ideal design, the number of DPs is equal to the number of FRs.

Theorem 6 (Path Independence of Uncoupled Design)   The information content of an uncoupled design is independent of the sequence by which the DPs are changed to satisfy the given set of FRs.

Theorem 7 (Path Dependency of Coupled and Decoupled Design)   The information content of coupled and decoupled designs depends on the sequence by which the DPs are changed to satisfy the given set of FRs.

Theorem 8 (Independence and Tolerance)   A design is an uncoupled design when the design range is greater than

$$\sum_{\substack{j=1 \\ i \neq j}}^{n} (\partial FR_i / \partial DP_j) \Delta DP_j$$

in which case, the nondiagonal elements of the design matrix can be neglected from design consideration.

Theorem 12 (Sum of Information)   The sum of information for a set of events is also information, provided that proper conditional probabilities are used when the events are not statistically independent.

Theorem 13 (Information Content of the Total System)   If each DP is probabilistically independent of other DPs and affects only its corresponding FR, the information content of the total system is the sum of the information of all individual events associated with the set of FRs that must be satisfied.

Theorem 14 (Information Content of Coupled versus Uncoupled Designs)   When the state of FRs is changed from one state to another in the functional domain, the information required for the change is greater for a coupled process than for an uncoupled process.

Theorem 16 (Equality of Information Content)   All information contents that are relevant to the design task are equally important regardless of their physical origin, and no weighting factor should be applied to them.

Theorem 17 (Design in the Absence of Complete Information)   Design can proceed even in the absence of complete information only in the case of decoupled design if the missing information is related to the off-diagonal elements.

Theorem 18 (Existence of an Uncoupled Design)   There always exists an uncoupled design that has less information than a coupled design.

Theorem 19 (Robustness of Design)   An uncoupled design and a decoupled design are more robust than a coupled design in the sense that it is easier to reduce the information content of designs that satisfy the Independence Axiom.

Theorem 20 (Design Range and Coupling)   If the design ranges of uncoupled or decoupled designs are tightened, the designs may become coupled designs. Conversely, if the design ranges of some coupled designs are relaxed, the designs may become either uncoupled or decoupled designs.

Theorem 22 (Comparative Robustness of a Decoupled Design)   Given the maximum design ranges for a given set of FRs, decoupled designs cannot be as robust as uncoupled designs in that the allowable tolerances for DPs of a decoupled design are less than those of an uncoupled design.

3. THEOREMS RELATED TO DESIGN AND DECOMPOSITION OF LARGE SYSTEMS

Theorem S8 (Complexity of a Large Flexible System)   A large system is not necessarily complex if it has a high probability of satisfying the FRs specified for the system.

APPENDIX 3–C Probability of Success of Decoupled and Uncoupled Designs When There Is No Bias[17]

Notation

δFR_i = Random variation in FR_i
ΔFR_i = Design range of FR_i
δDP_j = Random variation in DP_j
ΔDP_j = Tolerance of DP_j
A_{ij} = Elements of the design matrix
u, d = Superscripts used to denote quantities associated with uncoupled or decoupled design, respectively.

Problem Statement

Consider two alternative designs used to achieve two values of FR_i^*s. One is an uncoupled design,

$$\text{FR}_1 = f_1^u(\text{DP}_1^u)$$

$$\text{FR}_2 = f_2^u(\text{DP}_2^u)$$

(A.3C.1)

---

[17] Contributed by Hilario L. "Larry" Oh, March 2000 (personal communication).

The other design is a decoupled design:

$$FR_1 = f_1^d(DP_1^d)$$

$$FR_2 = f_2^d(DP_1^d, DP_2^d)$$

(A.3C.2)

The random variation in FR_i due to the random variation in DP_j can be derived from Equations (A.3C.1) and (A.3C.2). For the uncoupled design[18]:

$$\begin{bmatrix} \delta FR_1 \\ \delta FR_2 \end{bmatrix} = \begin{bmatrix} \dfrac{\partial FR_1}{\partial DP_1^u} & \dfrac{\partial FR_1}{\partial DP_2^u} \\ \dfrac{\partial FR_2}{\partial DP_1^u} & \dfrac{\partial FR_2}{\partial DP_2^u} \end{bmatrix} \begin{bmatrix} \delta DP_1^u \\ \delta DP_2^u \end{bmatrix} = \begin{bmatrix} A_{11}^u & 0 \\ 0 & A_{22}^u \end{bmatrix} \begin{bmatrix} \delta DP_1^u \\ \delta DP_2^u \end{bmatrix}$$

(A.3C.3)

For the decoupled design:

$$\begin{bmatrix} \delta FR_1 \\ \delta FR_2 \end{bmatrix} = \begin{bmatrix} \dfrac{\partial FR_1}{\partial DP_1^d} & \dfrac{\partial FR_1}{\partial DP_2^d} \\ \dfrac{\partial FR_2}{\partial DP_1^d} & \dfrac{\partial FR_2}{\partial DP_2^d} \end{bmatrix} \begin{bmatrix} \delta DP_1^d \\ \delta DP_2^d \end{bmatrix} = \begin{bmatrix} A_{11}^d & 0 \\ A_{21}^d & A_{22}^d \end{bmatrix} \begin{bmatrix} \delta DP_1^d \\ \delta DP_2^d \end{bmatrix}$$

(A.3C.4)

For a design to be successful, the random variation δFR_i must stay within a range $\pm\Delta FR_i$ of targeted values FR_i^*. Namely,

$$Pr(\text{Successful design}) \equiv Pr(-\Delta FR_1 \le \delta FR_1 \le \Delta FR_1, -\Delta FR_2 \le \delta FR_2 \le \Delta FR_2)$$

We assumed that the A_{ij} are constant and that the δDP_j are statistically independent. For the uncoupled design, δFR_2 is statistically independent of δFR_1 per Equation (A.3C.3). For the decoupled design, δFR_2 is statistically dependent on δFR_1 because of the coupling per Equation (A.3C.4). Thus the probability of success for the uncoupled design is

$$Pr(-\Delta FR_1 \le \delta FR_1 \le \Delta FR_1, -\Delta FR_2 \le \delta FR_2 \le \Delta FR_2)$$

$$= Pr(-\Delta FR_1 \le \delta FR_1 \le \Delta FR_1) \cdot Pr(-\Delta FR_2 \le \delta FR_2 \le \Delta FR_2)$$

(A.3C.5)

$$= Pr(-\Delta FR_1 \le \left| A_{11}^u \delta DP_1^u \right| \le \Delta FR_1) \cdot Pr(-\Delta FR_2 \le \left| A_{22}^u \delta DP_2^u \right| \le \Delta FR_2)$$

The probability of success for the decoupled design is

$$Pr(-\Delta FR_1 \le \delta FR_1 \le \Delta FR_1, -\Delta FR_2 \le \delta FR_2 \le \Delta FR_2)$$

$$= Pr(-\Delta FR_1 \le \delta FR_1 \le \Delta FR_1) \cdot Pr(-\Delta FR_2 \le \delta FR_2 \le \Delta FR_2|$$

$$- \Delta FR_1 \le \delta FR_1 \le \Delta FR_1)$$

(A.3C.6)

$$= Pr(-\Delta FR_1 \le A_{11}^d \delta DP_1^d \le \Delta FR_1) \cdot Pr(-\Delta FR_2 \le \left| A_{21}^d \delta DP_1^d + A_{22}^d \delta DP_2^d \right|$$

$$\le \Delta FR_2| - \Delta FR_1 \le \left| A_{11}^d \delta DP_1^d \right| \le \Delta FR_1)$$

We show that an uncoupled design has a higher probability of success than a decoupled

---

[18] We need to be careful about the notation. δ is used to imply a random variable. We will use the symbol $|x|$ as in $|A_{ij} \delta DP_j|$ to indicate the extreme value of a random variable x, which, in this case, is equal to the system range of FR_i. More generally, the system range of FR_i is given by $\sum A_{ij} \delta DP_j$.

design because a combination of $(A_{11}^u, A_{22}^u, \delta DP_1^u, \delta DP_2^u, A_{11}^d, A_{21}^d, A_{22}^d, \delta DP_1^d, \delta DP_2^d)$ exists such that the probability in Equation (A.3C.5) is higher than that in Equation (A.3C.6).

Analysis

Assuming a uniform probability distribution for $Pr(DP_j)$, probabilities as stated in Equations (A.3C.5) and (A.3C.6) may be expressed in terms of the system range, the design range, and their intersection called the common range as shown in Figures A.3C.1A, A.3C.1B, A.3C.2A, and A.3C.2B. Figures A.3C.1A and A.3C.2A depict situations in which the design range is inside the system range. Figures A.3C.1B and A.3C.2B depict situations in which a portion of the design range is outside the system range. The latter situations are the more realistic. Based on these figures, the probability of a successful design is given by

$$Pr(Success) = \frac{common\ range}{system\ range}$$

Note that the figures are symmetric about the (FR_1, FR_2) axes. Therefore, analysis can be carried out for the first quadrant using only absolute value of quantities involving δDP_j. For the uncoupled design, the system range is

$$system\ range = 4\left|A_{11}^u \delta DP_1^u\right| \cdot \left|A_{22}^u \delta DP_2^u\right|$$

The common range is

$$common\ range = 4 \cdot \min\left(\left|A_{11}^u \delta DP_1^u\right|, \Delta FR_1\right) \cdot \min\left(\left|A_{22}^u \delta DP_2^u\right|, \Delta FR_2\right)$$

which for the situation shown in Figure A.3C.1A is

$$common\ range = 4\Delta FR_1 \Delta FR_2$$

$$Pr(Success) = \frac{\Delta FR_1 \Delta FR_2}{\left|A_{11}^u \delta DP_1^u\right| \cdot \left|A_{22}^u \delta DP_2^u\right|} \tag{A.3C.7}$$

And for the situation shown in Figure A.3C.1B,

$$common\ range = 4\Delta FR_1 \cdot \left|A_{22}^u \delta DP_2^u\right|$$

$$Pr(Success) = \frac{\Delta FR_1}{\left|A_{11}^u \cdot \delta DP_1^u\right|} \tag{A.3C.8}$$

For the decoupled design as shown in Figures A.3C.2A and A.3C.2B, the system range is the area of the parallelogram:

$$system\ range = 4\left|A_{11}^d \delta DP_1^d\right| \cdot \left|A_{22}^d \delta DP_2^d\right|$$

For the situation in Figure A.3C.2A,

$$common\ range = 4 \cdot \min\left(\left|A_{11}^d \delta DP_1^d\right|, \Delta FR_1\right) \cdot \min\left(\left|A_{22}^d \delta DP_2^d\right|, \Delta FR_2\right)$$

$$= 4\Delta FR_1 \Delta FR_2 \tag{A.3C.9}$$

$$Pr(Success) = \frac{\Delta FR_1 \Delta FR_2}{\left|A_{11}^d \delta DP_1^d\right| \cdot \left|A_{22}^d \delta DP_2^d\right|} \tag{A.3C.10}$$

(A)

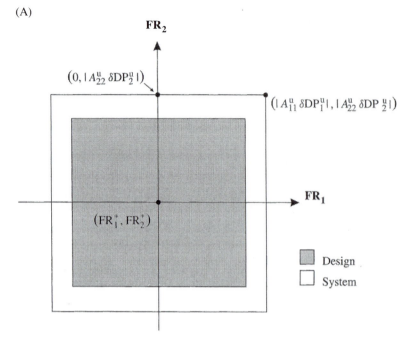

Figure A.3C.1 (A) Design range inside system range for an uncoupled design.

(B)

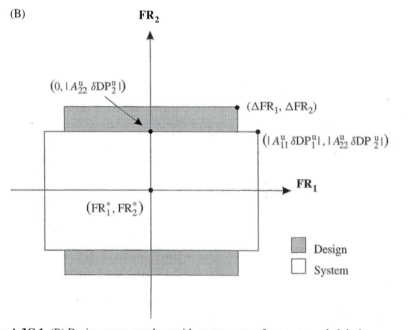

Figure A.3C.1 (B) Design range overlaps with system range for an uncoupled design.

(A)

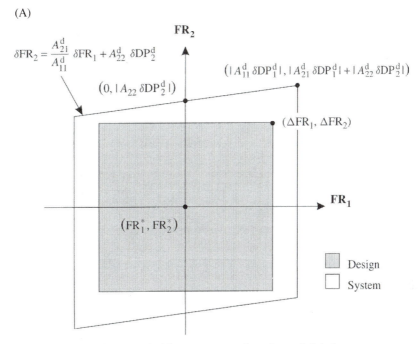

Figure A.3C.2 (A) Design range inside system range for a decoupled design.

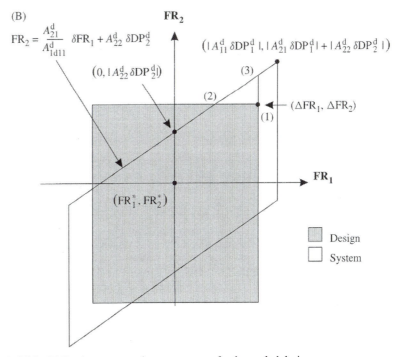

Figure A.3C.2 (B) Design range and system range of a decoupled design.

For the situation in Figures A.3C.2B, the common range is that given in Equation (A.3C.9) reduced by twice the area of the triangle 123 as denoted by vertices 1, 2, and 3 in Figure A.3C.2B. The coordinates of these vertices are solutions to intersections of lines depicted in Figure A.3C.3.

vertex (1): $\left(\min\left(\left|A_{11}^{d}\delta DP_{1}^{d}\right|,\Delta FR_{1}\right),\Delta FR_{2}\right)$

$= (\Delta FR_{1},\Delta FR_{2})$

vertex (2): $\left[\left|\dfrac{A_{11}^{d}}{A_{21}^{d}}\right|\left(\Delta FR_{2}-\left|A_{22}^{d}\delta DP_{2}^{d}\right|\right),\Delta FR_{2}\right]$

vertex (3): $\left[\min\left(\left|A_{11}^{d}\delta DP_{1}^{d}\right|,\Delta FR_{1}\right),\left|\dfrac{A_{21}^{d}}{A_{11}^{d}}\right|\min\left(\left|A_{11}^{d}\delta DP_{1}^{d}\right|,\Delta FR_{1}\right)+\left|A_{22}^{d}\delta DP_{2}^{d}\right|\right]$

$= \left(\Delta FR_{1},\left|\dfrac{A_{21}^{d}}{A_{11}^{d}}\right|\Delta FR_{1}+\left|A_{22}^{d}\delta DP_{2}^{d}\right|\right)$

Twice the area of triangle 123

$$= \left[\Delta FR_{1}-\left|\frac{A_{11}^{d}}{A_{21}^{d}}\right|\left(\Delta FR_{2}-\left|A_{22}^{d}\delta DP_{2}^{d}\right|\right)\right]\left(\left|\frac{A_{21}^{d}}{A_{11}^{d}}\right|\Delta FR_{1}+\left|A_{22}^{d}\delta DP_{2}^{d}\right|-\Delta FR_{2}\right)$$

$$= \left|\frac{A_{11}^{d}}{A_{21}^{d}}\right|\left(\left|\frac{A_{21}^{d}}{A_{11}^{d}}\right|\Delta FR_{1}+\left|A_{22}^{d}\delta DP_{2}^{d}\right|-\Delta FR_{2}\right)^{2}$$

$$\text{common range} = 4\Delta FR_{1}\left|A_{22}^{d}\delta DP_{2}^{d}\right|-\left|\frac{A_{11}^{d}}{A_{21}^{d}}\right|\left(\left|\frac{A_{21}^{d}}{A_{11}^{d}}\right|\Delta FR_{1}+\left|A_{22}^{d}\delta DP_{2}^{d}\right|-\Delta FR_{2}\right)^{2}$$

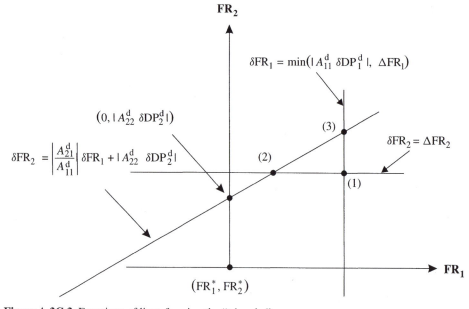

Figure A.3C.3 Equations of lines forming the "triangle."

$$\text{Pr(Success)} = \frac{\Delta\text{FR}_1}{\left|A_{11}^d \delta\text{DP}_1^d\right|} - \frac{\left|\frac{A_{11}^d}{A_{21}^d}\right|\left(\left|\frac{A_{21}^d}{A_{11}^d}\right|\Delta\text{FR}_1 + \left|A_{22}^d\delta\text{DP}_2^d\right| - \Delta\text{FR}_2\right)^2}{4\left|A_{11}^d\delta\text{DP}_1^d\right| \cdot \left|A_{22}^d\delta\text{DP}_2^d\right|} \quad\quad\text{(A.3C.11)}$$

Comparison of the Probability of Success of the Two Designs

We first compare the two situations in Figures A.3C.1A and A.3C.2A. The probability of success as expressed in Equations (A.3C.7) and (A.3C.10) is similar in form. However, their values will be different depending on the tolerances specified for δDP_j^u and δDP_j^d. As a basis for comparison, we set the tolerances such that their system range would be the same:

$$A_{11}^u \delta\text{DP}_1^{u^*} = A_{11}^d \delta\text{DP}_1^{d^*}$$

$$A_{22}^u \delta\text{DP}_2^{u^*} = A_{22}^d \delta\text{DP}_2^{d^*}$$

The probability of success for both designs will therefore be the same. However as we proceed to compare on the same basis the two situations in Figures A.3C.1B and A.3C.2B, we find that the uncoupled design always has a higher probability of success than the decoupled design because the second term in Equation (A.3C.11) is always positive.

In conclusion, a decoupled design used to achieve a set of "target" FRs can never have a higher probability of success than an uncoupled design used to achieve the same set of "target" FRs, if there is no bias. If there is bias, the information content of an uncoupled design can be larger than that of decoupled designs. However, if the design satisfies the Independence Axiom, bias can be eliminated and thus the information content of an uncoupled design can be made smaller than that of a decoupled design.

APPENDIX 3–D Why Coupling in Design Should Be Avoided[19]

Given a design that exhibits a system range shown in Figure A.3D.1, the measure of success used was the information content, which is based on the probability of success given by the area of the system pdf within the design range. The probability of success was expressed as

$$P = \int_{-a}^{a} p(\text{FR})\,d\text{FR} \quad\quad\text{(A.3D.1)}$$

where the integration limits are the bound of the design range. On the other hand, if we choose as a metric for success the closeness to the target FR_0, then the idea for the quadratic loss function L can be used, which may be expressed as

$$L = \int_{-\infty}^{\infty} (\text{FR})^2\, p(\text{FR})\, d\text{FR} \qu\quad\text{(A.3D.2)}$$

---

[19] H. L. Oh, personal communication, 2000.

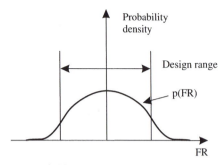

Figure A.3D.1 pdf of FR, the system range, and the design range.

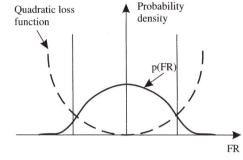

Figure A.3D.2 pdf of the FR and quadratic loss function.

This is shown in Figure A.3D.2. The metric L measures the variability of FR around its target value FR_0, smaller values being better. For small deviations from FR_0, the variance of FR can be expressed in terms of the variances of DPs. Consider the two-FR case given by the following matrix, which represents a decoupled design:

$$[A] = \begin{bmatrix} A_{11} & A_{12} \\ 0 & A_{22} \end{bmatrix} \tag{A.3D.3}$$

For this decoupled design, the magnitudes of reangularity R and semangularity S are equal. R and S may be expressed as

$$R = \left[1 - \frac{(A_{11}A_{22} + A_{21}A_{22})^2}{(A_{11}^2 + A_{21}^2)(A_{12}^2 + A_{22}^2)} \right]^{1/2} \tag{A.3D.4}$$

$$R^2 = \left[1 + \left(\frac{A_{12}^2}{A_{22}^2} \right) \right]^{-1}$$

$$S = \left[\frac{|A_{11}|}{\sqrt{A_{11}^2 + A_{21}^2}} \right] \left[\frac{|A_{22}|}{\sqrt{A_{12}^2 + A_{22}^2}} \right] \tag{A.3D.5}$$

$$S^2 = \left[1 + \left(\frac{A_{12}}{A_{22}} \right)^2 \right]^{-1} = R^2$$

The degree of coupling increases as R and S approach zero.

We now express the variance in FR in terms of the variance in the DPs and their degree of coupling. Let Σ denote the variance-covariance matrix. All letters in bold are vector or matrix quantities. Then the random variation for the design can expressed as

$$\delta\mathbf{FR} = [\mathbf{A}]\delta\mathbf{DP}$$

$$\delta\mathbf{FR} \cdot \delta\mathbf{FR}^T = [\mathbf{A}]\delta\mathbf{DP} \cdot \{[\mathbf{A}]\delta\mathbf{DP}\}^T = [\mathbf{A}]\delta\mathbf{DP} \cdot \delta\mathbf{DP}^T[\mathbf{A}]^T \tag{A.3D.6}$$

$$E\{\delta\mathbf{FR} \cdot \delta\mathbf{FR}^T\} = [\mathbf{A}] \cdot E\{\delta\mathbf{DP} \cdot \delta\mathbf{DP}^T\} \cdot [\mathbf{A}]^T$$

$$\sum_{\mathrm{FR}} = [\mathbf{A}] \sum_{\mathrm{DP}} [\mathbf{A}]^T \tag{A.3D.7}$$

If we assume that the DPs are probabilistically independent, then

$$\sum_{\mathrm{DP}} = \begin{bmatrix} \sigma_{11}^2 & 0 \\ 0 & \sigma_{22}^2 \end{bmatrix}$$

$$\sum_{\mathrm{FR}} = \begin{bmatrix} \sigma_{11}^2\,A_{11}^2 + \sigma_{22}^2\,A_{12}^2 & \sigma_{22}^2\,A_{12}\,A_{22} \\ \sigma_{11}^2\,A_{11}\,A_{22} + \sigma_{22}^2\,A_{12}\,A_{22} & \sigma_{22}^2\,A_{22}^2 \end{bmatrix} \tag{A.3D.8}$$

Note that because $A_{12} \neq 0$, covariance of FRs is not zero, i.e., the FRs are not probabilistically independent. They are correlated. The total system variance is the sum of the diagonal terms of the variance-covariance matrix Σ_{FR}, which is given as

$$\text{System variance} = E\{\delta\mathbf{FR} \cdot \delta\mathbf{FR}^T\} = \sigma_{11}^2\,A_{11}^2 + \sigma_{22}^2\,A_{12}^2 + \sigma_{22}^2\,A_{22}^2$$

$$= \sigma_{11}^2\,A_{11}^2 + \sigma_{22}^2\,A_{22}^2\left(1 + \frac{A_{12}^2}{A_{22}}\right) \tag{A.3D.9}$$

Substituting the expression for square of reangularity and semangularity, the system variance may be expressed as

$$\text{System variance} = \sigma_{11}^2\,A_{11}^2 + \frac{\sigma_{22}^2\,A_{22}^2}{R^2} \tag{A.3D.10}$$

As the coupling increases with an increase in the magnitude of A_{12}, R decreases. As R approaches zero, the system variance increases for given variance of DPs. It is best to avoid coupling.

HOMEWORK

3.1 Consider a design that has m FRs and n DPs, which is represented by the design equation: $\{FR\}_m = [DM]_{m \times n}\{DP\}_n$. Show that when $m > n$, the design is a coupled design.

3.2 Prove the following statement. If a given design is a fully coupled design with $m \leq n$, then the design cannot be decoupled. In this case, we have to start the design process all over again by finding new DPs. Conversely, if $m > n$ and if the design matrix can be made diagonal or triangular, then a coupled design can be made uncoupled or decoupled by adding new DPs.

3.3 Engineer Kim is in charge of a manual transmission design for DDD Automobile Co. He wanted to design a manual transmission with five speed reduction ratios using several sets of gears so as to utilize the engine power and torque in an optimum manner. The desired speed ratios would be different, depending on the specific requirements of a vehicle and the specific engines used in the vehicle. During the course of his design, Engineer Kim found that he could eliminate a gear (and thus reduce the weight of the transmission, which is desirable) and still get five speed reductions. However, in this case, the fourth speed reduction ratio cannot be independently set from the third speed reduction ratio, i.e., they are not independent of one another. Therefore, Engineer Kim claims that a coupled design is not a bad design, contrary to the Independence Axiom. How would you explain to Engineer Kim the meaning of the Independence Axiom, using his design as an example?

3.4 In some design situations, we may find that we have to make design decisions in the absence of sufficient information. In terms of the Independence Axiom and the Information Axiom, explain when and how we can make design decisions even when we do not have sufficient information. What kinds of information can we do without and what kinds of information must we have in design? Illustrate your argument using a design task with three FRs as an example (used in Chapter 1).

3.5   Figure H.3.5 is a drawing of a cylinder with an internal piston used in a control valve. One side of the cylinder is filled with hydraulic liquid and the other side is filled with a liquid lubricant, which is at 3 bars at all times. When the hydraulic fluid is pressurized to 30 bars, the piston moves to the right, and when the pressure of the hydraulic fluid is decreased to 0 bar, the piston moves to the left. The fluids are separated from each other by piston rings, which are made of nitrile butadiene rubber (NBR). It has been found that the hydraulic fluid leaks into the lubricant side after a month of testing, which consisted of displacing the piston cyclically a distance of 1 cm. How would you solve this design problem? Develop a new design of either the seals or the entire mechanism.

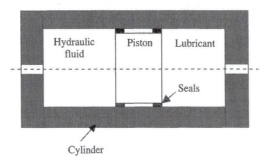

Figure H.3.5 Floating piston in a cylinder with seals.

3.6   An electric switch is activated with an external magnet as shown in Figure H.3.6. When the fluid in the pipe flows up along the pipe, the magnet rotates and lines up against the pipe wall. Then the electric switch closes. When the fluid stops flowing in the pipe, the magnet returns to the original position, and the electric switch also returns to its open position. The electric switch consists of two "leaf-type" electric contacts that are gold-plated to prevent the formation of oxidation and thus to guarantee a constant electric contact resistance at all times. One of the contacts is stationary and the other contact bends under the magnetic field to make the electric contact. When the magnetic field is removed, the deflected electric contact returns to its original position because of its own spring force. It has been found that some of the electric switches

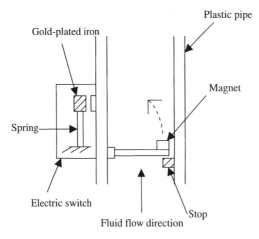

Figure H.3.6 Flow-activated electric switch.

are returned to the manufacturer after a few months because they do not function properly in the field. Your design task is to state the functional requirements and the design parameters and to suggest the reasons for failure. Prescribe a solution.

3.7. A precision electric furnace must heat 100 sheets of large round ceramic plates uniformly at the same time. The 1-mm-thick plates are stacked vertically—one plate on top of another—with a spacing of 2 cm between the plates. Because the environment must be kept extremely clean during heating, an electric coil is used. The diameter of the heater is 1 m and the coil diameter is 2 cm. The wire diameter is 2 mm. Because the heating coil is heated to 1000°C, it must be supported and expansion of the coil must be accommodated. One hundred coils are used in the furnace. Your design task is to come up with a design that can support the heating coil by stating clearly what the functional requirements and design parameters are.

3.8 The plates must be placed in the support rack in the furnace described in Problem 3.7. This is done by a robot arm that transports the plate to the correct position and moves it horizontally to place the plate on the supports in the support bracket as shown in Figure H.3.8. Each plate weighs 2 lb. A decision was made to narrow the gap between the ceramic plates to 4 mm from 2 cm. To put the plates in the narrow space, the robot arm must be very thin and accurate. Your design task is to design the robot arm so that it will be able to place the plates without ever touching other plates. State your FRs and DPs clearly and develop the design matrix. Model your solution to give the quantitative dimensions of your robot arm after optimization of the design.

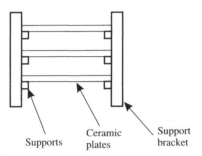

Figure H.3.8 Support bracket for ceramic plates.

Supports   Ceramic plates   Support bracket

3.9 A hydraulic cylinder with a piston must be mounted so that when the piston exerts 2000 lb of load against a structure, the cylinder can deflect 3 mm and return to the original position on unloading. Moreover, the piston must be perfectly aligned at all times. The mounting device must last 10 million cycles. Design the mounting device, which is sometimes called flexure. State the FRs and design the device showing all DPs of the physical part. Show how the information has been used to reduce the information content.

3.10 Consider the injection-molded part for a vacuum cleaner discussed in Example 3.6. An additional constraint is added that calls for minimum weight of the injection-molded part. Develop a better set of dimensions for the wheel assembly.

3.11 Determine the time the part has to stay in the modules by obtaining a mathematical solution to the scheduling problem in Example 3.7.

3.12 Determine the ratio of the weights of the knob shown in Figure 3.1 with that shown in Figure E3.6.b, assuming that the FRs of the knobs are

$FR_1 = $ The knob gripping force will vary from 0.5 to 25 lb.
$FR_2 = $ The knob must be able to carry a torque of 10 to 40 ft-lb.

The diameter of the steel shaft is 1 inch. We may wish to make three different size knobs to meet these FRs. The knob will be made of injection-molded nylon. You may obtain the material properties of plastics from the *Modern Plastics Encyclopaedia*, which is published annually by McGraw-Hill Publishing Co. Please specify the tolerances.

3.13 A gasoline fuel tank for trucks and passenger cars is to be designed. The FRs are

$FR_1 = $ Provide in-flow of gasoline into the tank.
$FR_2 = $ Provide a means of stopping the pump when the tank is full.
$FR_3 = $ Prevent gasoline from surging back out through the inlet tube as a result of the vapor pressure of the gasoline when the gasoline level is higher than the end of the pipe.
$FR_4 = $ Control vapor pressure of the gasoline.

Design a fuel tank that can satisfy the above four FRs. The cost cannot exceed the current cost of making fuel tanks.

3.14 Earthmoving equipment for coal mining, etc. uses many pin joints, which are subject to heavy loads, abrasive environment, friction, and wear. When these pins fail, entire equipment cannot perform its functions. Design a pin that can carry 2000 lb of cyclic load at a frequency of 0.01 Hz, withstand the abrasive environment, last 5000 hours, have a friction coefficient of 0.15, and withstand a corrosive environment. Give the quantitative values of DPs.

3.15 A furnace is being designed to rapidly heat an optical disk to a uniform temperature. One proposed design requires that the entire surface of the optical disk be brought to the heater surface within 0.2 mm ± 0.001 mm. The optical disk is brought to a lifting mechanism, which is designed to raise the optical disk 25 cm to reach the heater surface. The diameter of the optical disk is 30 cm. The disk weighs 10 lb. The optical disk must be located concentric with the axis of the lifting device within 0.5 mm. Design the lifting device.

3.16 A conventional design of a submarine rudder consists of two orthogonal plates arranged in a cross "+" shape. The horizontal surface controls elevation and the vertical surface controls lateral motion. The two surfaces are controlled by a stick at the fore of the vessel. An acoustic expert suggests that if the two control surfaces are rotated by 45° (from a "+" to a "×"), turbulence would be minimized. Perform top-level design of the original mechanism and then redesign for the suggested improvement. What difficulty does the improvement introduce and how can it be overcome?

3.17 Compare elements of axiomatic design theory to those of other design methodologies, specifically quality function deployment (QFD), robust design (Taguchi methods), and Pugh concept selection. Where do they agree and where do they differ?

3.18 Given below is an equation of a decoupled design consisting of two DPs and two FRs. One of the stiffness coefficients of the design matrix is denoted by x, which can be set to 0.5, 1, or 2.

$$\begin{Bmatrix} 0 < FR_1 < 1 \\ 0 < FR_2 < 1 \end{Bmatrix} = \begin{bmatrix} 1 & 0 \\ 1 & x \end{bmatrix} \begin{Bmatrix} 0 < DP_1 < 1 \\ 0 < DP_2 < 1 \end{Bmatrix}$$

Consider the effect of each of these values on both the following aspects of the design:

a. The probability of success
b. The amount of adjustment in DP_2 required to overcome a system perturbation of magnitude δ.

Illustrate the effect graphically for all cases. What can you conclude as to the effects of magnitude of elements of the design matrix?

3.19 The drag coefficient of various paint coatings is measured in a wind tunnel using the experimental setup shown in Figure H.3.19. The coating is applied to a sphere of radius R held on a rod of length L and diameter d. Air at velocity V flows through the tunnel of diameter D. The drag force applied to the sphere causes deflection of the rod. The deflection of the rod is measured using a strain gauge attached to the surface of the rod at its root. The DPs are R, L, and d. The functional requirements of this design are specified in terms of

a. Sensitivity of measurement
b. Auxiliary drag introduced by the measuring apparatus itself
c. Efficient use of air velocity profile in tunnel

Derive the design matrix for this instrument including explicit equations, and determine the order in which the parameters should be set.

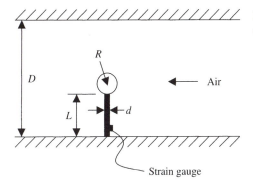

Figure H.3.19 Drag coefficient measuring device.

3.20 In Chapter 2 (Example 2.6), the design of the van seat was treated as a one-FR design task. The automobile company found the optimum linkage lengths by minimizing an objective function numerically. A simpler and more elegant design may result if this design task is treated as a two-FR design task with the following two specific FRs:

FR_1 = Make the latch mechanism work at $F \pm \Delta F = 340 \pm 2$ mm.
FR_2 = Accommodate the random variations in y, Δy, over the range of 336–344 mm.

Design a modified van seat locking mechanism that can satisfy the above two FRs without adding to the manufacturing cost and with improved reliability.

3.21 Using Figure 3.4, show when the information content of a coupled design can have less information content than an uncoupled design, if the random variations in DP_1 and DP_2 are the same in both cases. Derive a mathematical relationship that can generalize your explanations when there are more than two FRs.

3.22 Consider Homework 3.21 again. What is the consequence of having the same variations in DP_1 and DP_2 on FR_1 and FR_2 in uncoupled design and coupled designs?

3.23 Most people are surprised to learn that there are more microprocessors used in passenger vehicles than in most industrial equipment. For example, the engine and the transmission—often called the "power train" of automobiles—are managed by microprocessors to satisfy the stringent requirements for vehicle performance and to meet the emission standards of local and federal government. Unfortunately, integrated chips (microprocessors) cannot function when their temperature exceeds a critical value. Therefore, any automotive component that employs microprocessors must be kept at a temperature below 50°C. Suppose that you are given the job of designing the engine compartment of an automobile. You have to place the engine, starting motor, alternator, compressor for air-conditioning, battery, etc. inside the compartment. Please state FRs and Cs for packaging of various components in the engine compartment. Design the engine compartment, specifying the design parameters.

4

Design of Systems

4.1 INTRODUCTION

This chapter presents a general theory for system design based on axiomatic design. The theory is applicable to many different kinds of systems, including machines, large systems, software systems, organizations, and systems consisting of a combination of hardware and software. Systems are designed to satisfy a set of highest functional requirements (FRs) and constraints (Cs). This is done by choosing the right set of design parameters (DPs), which are in turn satisfied by selecting the right set of process variables (PVs). These FRs, DPs, and PVs must be decomposed until the leaves of FRs, DPs, and PVs are established.

Systems are represented by means of a system architecture, which takes the form of the FR/DP/PV hierarchies, a "junction-module" diagram, and the "flow chart" (also called the "flow diagram"). The flow chart for system architecture concisely represents the system design, the relationship among modules, and the control sequence in operating systems. The flow diagram of the system architecture can be used for many different tasks: design, construction, operation, modification, and maintenance of the system. It is also useful for distributed design and operation of systems, diagnosis of system failures, and archival documentation.

Why are we concerned about system design? Recently newspaper articles featured stories on how a major airplane manufacturer had been losing money despite the fact that it had a large number of orders for new airplanes, apparently because it could not manufacture at quoted prices and deliver the airplanes on time. Similarly, in the health-care sector, major hospitals had to consolidate their operations because they could not operate them with efficiency and at a profit in the new environment of managed care. In the transportation area, railroad companies do not know exactly where their freight cars are located at a given instant and thus do not utilize their resources effectively. Also, it is well known that two automobile manufacturers can have a major difference in productivity, although both are making essentially the same kinds of vehicles.

It is also a well-known fact that software development is so ad hoc that it often requires extensive debugging, and it is difficult to trace mistakes. Also machines with many parts, sometimes integrated by software, are developed through prototyping, testing, and redesign of many parts, all of which are expensive and add to the lead time. All of these problems and issues are caused by poor design of systems. System design and operation is an important topic for which the engineering profession has not had adequate tools and knowledge. Axiomatic design can be applied to systematize the design and operation of systems. This chapter deals with the general design theory of systems based on axiomatic design. In later chapters, the design of several specific systems (i.e., software, manufacturing, products, and materials) is presented in detail.

What is a system? A system may be defined as an assemblage of subsystems, hardware and software components, and people designed to perform a set of tasks to satisfy specified functional requirements and constraints.

Most engineering and societal issues exist in a systems context. Engineers build systems. Machines, airplanes, software systems, and automobile assembly plants are all systems—albeit systems of different kinds—and each has subsystems and components. Systems often consist of hardware, software, and people. Systems are hierarchical, with many layers of subcomponents and subsubcomponents. Such human-made systems must be designed, fabricated, and operated to achieve their intended functions. Each of these systems performs many functions. Some, like the automobile, perform a large number of different but dedicated functions. Others, such as a job shops, perform a variety of different functions during their lifetimes and as such may be categorized as large flexible systems (Suh, 1995b).

Why do we need a theory for system design? The design of effective systems is the ultimate goal of many fields, including engineering, business, and government. Yet system design has lacked a formal theoretical framework and thus has been done heuristically or empirically (Rechtin, 1991). Heuristic approaches emphasize qualitative guidelines, exemplified by use of the phrases "Murphy's laws," "make it simple," and "ask five why's." After systems are designed, they are sometimes modeled and simulated. In many cases, they have to be constructed and tested. All these very expensive and unpredictable processes are done to debug and improve the design after heuristic design solutions are implemented in hardware and software. Such an approach to systems design entails both technical and business risks because of the uncertainties associated with the performance and quality of a system that is created by means of empirical decisions.

Some people use dimensional analysis, decision theory, and other techniques to check or optimize a system that has already been designed. There are three issues with these approaches. First, they do not provide tools for coming up with a rational system design beginning from the definition of the design goals. Second, some of these methods simply confirm the result to determine whether systems are correctly designed. For example, any physical system that is properly configured should satisfy the π-theorem. However, the fact that a design is dimensionally consistent does not make it a good design. Third, they are not general principles for system design because they cannot be applied to nonphysical systems, such as software systems and organizations.

Systems with many FRs, physical components, and many lines of computer code can be complex in the sense that the probability of satisfying the highest FRs decreases with an increase in the number of FRs and DPs. One of the goals of the axiomatic approach to design is to reduce this complexity by being able to make the right design decisions at all levels. From the axiomatic design point of view, the design of systems is not fundamentally different from the design of simple mechanical products and software. All of them—systems, machines, and software—must satisfy FRs, Cs, the Independence Axiom, and the Information Axiom. However, the specific relationships among FRs, DPs, and PVs are different depending on what the system must do. Also, the governing physical laws may be different depending on the FRs the system must satisfy, the nature of the DPs chosen, and the specific PVs used to achieve the design objectives.

Notwithstanding its obvious importance, no effective way of representing system architecture has been advanced until the introduction of axiomatic design theory. Dealing with a system design without a system architecture is equivalent to trying to build an electronic instrument without a circuit diagram. Just as an electrical system is described and defined by an electric circuit diagram, a system—either physical, software, or a combination thereof—must be represented by a diagram that shows how the "modules" must be put together and how the system should be operated.

In this chapter, the concept of "module" is defined, and the "system architecture" is represented in several forms: FR/DP/PV hierarchies, junction-module diagrams, and the flow chart (flow diagram). The "system control command" (SCC) provides the operational and procedural instruction for operation of the system (Suh, 1997).

The design of the system must be documented to capture all of its relevant information. The document must show the complete structure of the system and the relationships among the decomposed FRs, DPs, and PVs. The relevant information about the system includes the specific functional requirements for which the system is designed, the rationale for choosing specific design parameters, and the causality relationship among the FRs, DPs, and PVs. Also, the system architecture must be clearly known in order to construct the system, distribute the design responsibilities, track the effect of engineering change orders, create a maintenance procedure, and organize and coordinate the complex tasks of managing a large project.

Maintenance of complex systems is expensive—the annual cost being roughly 10% of the system cost. Documentation of the system is needed to provide a means of determining the causes of system failures or detecting impending failures. It should provide ready instruction for identifying the components to be checked when the system does not perform its specified functions. Effective documentation is also useful for other purposes: tracking design history, improving system performance through modification, and creating a knowledge-based approach to the system.

What is this chapter about? In this chapter, the general issues involved in the design of systems are discussed and several new concepts are introduced: the module, the module-junction diagram, the flow chart, and the system command and control.

The concepts introduced in this chapter are applied to the design of specific systems in later chapters. In Chapter 5, the design of software systems is presented, followed by Chapter 6, which deals with the design of manufacturing systems. Chapter 7 presents the design of materials and materials processing, and Chapter 8 deals with product design.

4.2 ISSUES RELATED TO SYSTEM DESIGN

There are several issues related to design that should be considered in dealing with systems.

1. *How should a complex system be designed?* A system must be designed; it cannot evolve by adding on subsystems and components without designing the entire system. A system can be designed within the basic framework of axiomatic design theory and methodology. It is a top-down approach. The final design of the system can be constructed from the modules (defined in this chapter) associated with the lowest level leaves of the FR and DP hierarchies and design matrices.

2. *How should the complex relationships between various components of a system be coordinated and managed?* When the system design is based on axiomatic design theory, a flow chart of the system architecture can be generated. This can be the basis for distributed project management, engineering change orders, collaborative design among many designers located over a wide geographic area, and the maintenance of the system, in addition to providing the basic documentation in system implementation.

3. *How can the stability and controllability of a system be guaranteed?* A system must be stable and controllable. If the system is not stable and changes randomly, it will not be reliable and controllable. If the system is not controllable, it cannot satisfy its functional requirements at all times. In this sense, it is argued qualitatively that when a system satisfies the Independence and Information Axioms, the system is stable and controllable within its design range, because the DPs are chosen to satisfy each one of the FRs independently.

4. *What is the role of human operators in a system?* People are an integral part of many systems. Human–machine interface is an important consideration in these systems because the performance of human beings is a determining factor in the outcome of the system.

4.3 CLASSIFICATION OF SYSTEMS

Why does it make more sense to classify systems based on functions instead of physical size? Notwithstanding the fact that all systems can be designed using the same set of basic principles, it is nevertheless useful to classify systems. A proper classification of systems will shift the thinking and focus away from the conventional identification based on physical size to a functionally based identification of systems.

The conventional way of characterizing systems is based on the physical size or the number of components of the system. However, when making design decisions, physical size is of less significance than the number of functional requirements that the system must satisfy at the highest level and the number of levels of decomposition required to arrive at a complete design solution. Furthermore, in many design situations, physical size does not have much significance, e.g., software design, organizational design, design of policies, and even the design of integrated circuit chips and microdevices. Therefore, a better classification scheme is to define the systems according to the number and nature of FRs the system must satisfy.

Functional classification of systems can be done using a number of different features or characteristics: large systems vs. small systems, static systems vs. dynamic systems, fixed systems vs. flexible systems, passive systems vs. active systems, and automated systems vs. manual systems. Some systems cannot be classified in a simple way, e.g., large, flexible, dynamic systems. Some systems are open systems—systems whose constituents change throughout their lifetimes—in contrast to closed systems that are made up of the same components at all times. Examples of open systems are factories, universities, and many machine tools; closed systems include inertial guidance systems and television sets.

Axiomatic design theory can be applied to all of these systems with slight variations. The differences among these systems are typically the functional requirements, constraints, and accompanying database. The same theory can be applied to design software, products, processes, manufacturing systems, organizations, and others.

▨ 4.4 AXIOMATIC DESIGN THEORY FOR FIXED SYSTEMS

One of the most basic engineering systems is a fixed system. We will define a *fixed* system as a system that has to satisfy the same set of functional requirements at all times and whose components do not change as a function of time. Many machines and robots may be classified as closed fixed systems because they are designed to satisfy a fixed set of functions at all times, in contrast to a flexible system, whose functional requirements change as a function of time. Closed fixed systems can be complex or simple, depending on the difficulty or ease in achieving the functional requirements.

Is the design of systems different from the design of other things? The steps involved in designing a system are not any different from those involved in designing a product or a process (Suh, 1990, 1995b). Although the basic steps involved in designing systems using axiomatic design theory are the same as those discussed in Chapters 1 through 3, they will be repeated here for clarification when they are applied to systems.

1. *The first step in designing a system: Define FRs of the system.* The first step in designing a system is to determine the customer needs/attributes (CAs) in the customer domain that the system must satisfy. Then the FRs and Cs of the system in the functional domain are determined to satisfy the CAs. The FRs must be determined in a *solution-neutral environment*—defining FRs without thinking about the solution in order to come up with creative ideas. FRs must satisfy the CAs with fidelity. It is important to remember that the FRs are defined as the minimum set of independent functional requirements that the system must satisfy.

2. *Mapping between domains: A step in creating system architecture.* The next step in axiomatic design is to map these FRs of the functional domain into the physical domain—conceiving a design embodiment and identifying the DPs. At the highest level of the system design, DPs may be conceptual entities, which must be decomposed to complete the design. DPs must be so chosen that there is no conflict with the constraints. In the case of products, DPs may be physical parameters or parts or assemblies, whereas in the case of software, DPs may be inputs to software modules or program codes.

Once the DPs are chosen, designers must go to the process domain and identify the PVs, based on the creation of a new process or the use of an existing process. When existing machines are to be used, the PVs are given and therefore act as Cs since we may not be free to create new processes and choose new PVs. In the case of organizations, the PVs are typically resources—both human and financial. In the case of software, PVs may be subroutines or machine codes or compilers.

During the mapping process, the design must satisfy the Independence Axiom, which requires that functional independence be satisfied through the development of an uncoupled or decoupled design. In an ideal design, the number of FRs must be equal to the number of DPs, a consequence of the Independence Axiom and the Information Axiom. In choosing DPs, we must also be mindful of the Cs and the information content. The Independence Axiom does not require that the DPs must be independent or that each DP must correspond to a separate physical piece. For example, as discussed in Chapter 1, a beverage can may have 12 FRs and 12 DPs, but only three physical pieces.

The mapping process between the domains can be expressed mathematically in terms of the characteristic vectors that define the design goals and design solutions. At a given level of the design hierarchy, the set of FRs that defines the specific design goals constitutes a vector in the functional domain. Similarly, the set of DPs in the physical domain that is the "how" for the FRs also constitutes a vector. The relationship between these two vectors can be written as

$$\{FR\} = [A]\{DP\} \tag{4.1}$$

where $[A]$ is the design matrix that characterizes the design. The design matrix is of the following form for a design with three FRs and three DPs:

$$[A] = \begin{bmatrix} A_{11} & A_{12} & A_{13} \\ A_{21} & A_{22} & A_{23} \\ A_{31} & A_{32} & A_{33} \end{bmatrix} \tag{4.2}$$

Equation (4.1) is a *design equation* and may be written in terms of its elements as

$$FR_i = \sum_j A_{ij} DP_j$$

When the change in FRs is related to changes in the DPs, the elements of the design matrix are given by

$$A_{ij} = \frac{\partial FR_i}{\partial DP_j} = Constant \qquad \text{for linear design}$$

$$\tag{4.3}$$

$$A_{ij} + \frac{\partial A_{ij}}{\partial DP_j} DP_j = \frac{\partial FR_i}{\partial DP_j} \qquad \text{for nonlinear design}$$

For a linear design, the A_{ij} are constants, whereas for a nonlinear design, the A_{ij} are functions of DPs.

For the design of processes, which involves mapping from the physical domain to the process domain, the design equation may be written as

$$\{DP\} = [B]\{PV\} \tag{4.4}$$

$[B]$ is the design matrix that defines the characteristics of the process design and is similar in form to $[A]$.

3. *The independence of system functions.* To satisfy the Independence Axiom, the design matrix must be either diagonal or triangular. When $[A]$ is diagonal, each of the FRs can be satisfied independently by means of its corresponding DP. Such a design is called an uncoupled design. When the matrix is triangular, the independence of FRs can be guaranteed if and only if the DPs are changed in the proper sequence. Such a design is called a decoupled design. All other designs violate the Independence Axiom; they are called coupled designs. Therefore, when several FRs must be satisfied, designers must develop designs that have either a diagonal or a triangular design matrix.

A large number of corollaries and theorems can be derived from these axioms. They are listed in Appendix 1A of Chapter 1.

4. *Information content for systems: The best design.* The Information Axiom states that the design that has the least information content is the best design. The relative merit of "different but equally acceptable" designs that satisfy the Independence Axiom can be compared by means of their information contents. The Information Axiom is a powerful tool in identifying the best design when a design must satisfy more than one functional requirement or when we must choose a DP from among many possible DPs. Through the use of the Independence Axiom and the Information Axiom, the original design goals of a multi-FR system can easily be satisfied.

Among all the designs that satisfy the Independence Axiom, the design that has the least information content is the best design according to the Information Axiom. Information in axiomatic design is defined in terms of the logarithmic probability of satisfying the FRs. Information content is a relative concept; it is a function of the area under the system pdf that is inside the specified design range for the FR, also called the area over the common range A_{cr}. Thus, the information associated with a given FR is obtained by computing the probability of achieving the FR.

How do we measure the information content of a system that has many decomposed layers in its hierarchy? Even in the case of a system design with many layers of FRs and with many FRs at all levels of the design hierarchy, the information content I of the system is still the information needed to satisfy the highest level FRs. Therefore, given the information content for each of the highest level FRs, the information content for the system can be determined by

$$I_{sys} = \sum_i \log \left[Pr(FR_i | \{FR_j\}_{j=1,\dots,i-1}) \right] \tag{4.5}$$

where $Pr(FR_i | \{FR_j\}_{j=1,\dots,i-1})$ is the conditional probability that FR_i is satisfied given that the other relevant (correlated) FRs in the total set are satisfied, i.e., $Pr(FR_i | FR_1, FR_2, \dots, FR_{i-1})$.

For uncoupled designs, where all FRs are statistically independent, Equation (4.5) can be written as

$$I_{sys} = -\sum_i \log \{Pr(FR_i)\} = -\sum_i \log (A_{cr})_{\text{highest level } FR_i} \tag{4.6}$$

How do we determine I_{system} when $I_{\text{highest level } FR_i}$ is not known? In some cases, the probability of success at the highest level may not be known a priori, but it may be possible to determine the information content associated with the leaves. Then to compute the information content of the system, we need to determine the system probability distribution and the A_{cr} of each leaf.

The probability of satisfying the highest level FRs is related to the probability of satisfying the lowest level FRs (i.e., leaves) because the lowest level FRs yield the highest level FRs when they are combined according to the instruction given by the design matrices. Therefore, the probability of satisfying a given highest level FR is the product of the probabilities associated with all of its decomposed lowest level FRs in the system architecture, provided that they are all statistically independent. Then the information content of the total system is the sum of the information contents associated with all lowest level FRs, which may be expressed as

$$I_{\text{system}} = -\sum \log(p_{\text{leaf}}) = -\sum \log(A_{cr})_{\text{leaf}} \tag{4.7}$$

where $(A_{cr})_{\text{leaf}}$ is the area over the common range associated with each leaf.

Equating Equations (4.6) and (4.7), we obtain

$$\sum \log (A_{cr})_{\text{leaf}} = \sum \log(A_{cr})_{\text{highest level FR}_i} \tag{4.8}$$

Equation (4.8) is valid only for designs where the FRs are statistically independent and where the integration of the lowest level modules does not introduce a new element of uncertainty. In the case of a coupled design, in most cases it is expected that

$$\sum \log (A_{cr})_{\text{leaf}} < \sum \log(A_{cr})_{\text{highest level FR}_i} \tag{4.9}$$

as the A_{cr} for a given FR will be affected by a change in any other FR at the same hierarchical level.

When the integration of the modules introduces a new element of uncertainty, Equations (4.7), (4.8), and (4.9) must be modified to account for this additional uncertainty associated with the assembly of modules. Equation (4.7) should be modified as

$$I_{\text{system}} = -\sum \log (p_{\text{leaf}}) + I_a = -\sum \log (A_{cr})_{\text{leaf}} + I_a \tag{4.10}$$

where I_a is the information associated with the assembly of modules.

The ultimate goal of design is to minimize the additional information required to make the system function as designed by making all p_{leaf} equal to 1.0. To achieve this goal, the design must satisfy the Independence Axiom. When the design satisfies functional independence, the bias can be eliminated and the variance of the system pdf may be reduced so that the system range lies inside the design range, thus reducing the information content to zero.

A design that can accommodate large variation in DPs and PVs and yet satisfy the FRs is called a robust design. The Information Axiom provides a theoretical foundation for robust design.

Based on the Information Axiom, it was shown in Chapters 1 and 2 that there are four different ways of reducing the bias and the variance of a design to develop a robust design, provided that the design satisfies the Independence Axiom. If robust design is practiced at each level of the hierarchy and with each FR, then the information content will be minimal and thus the design and operation of the system will be done efficiently and reliably.

5. *Definition of modules.* The concept of "module" is important in system design. To avoid confusion, it must be defined carefully based on basic principles. For example, a module is not a piece of hardware, although in some cases, it may coincidentally correspond

to a hardware piece. In axiomatic design, module is defined in terms of the (FR/DP) or the (DP/PV) relationship.

A *module* is defined as the row of the design matrix that yields an FR when it is provided with (or multiplied by) the input of its corresponding DP. For example, consider the following design equation:

$$\begin{Bmatrix} FR_1 \\ FR_2 \end{Bmatrix} = \begin{bmatrix} a & 0 \\ b & c \end{bmatrix} \begin{Bmatrix} DP_1 \\ DP_2 \end{Bmatrix} \tag{4.11}$$

M_1 is the module that corresponds to the combination of the first row of the design matrix and DP_1, i.e., when DP_1 is supplied as input to M_1, FR_1 results as the output. Similarly, FR_2 is obtained when DP_2 is provided as an input to the module M_2.

M_1 and M_2 are given by

$$FR_1 = aDP_1 + 0DP_2 = M_1 \cdot DP_1 \qquad \text{where } M_1 = a$$

$$FR_2 = bDP_1 + cDP_2 = M_2 \cdot DP_2 \qquad \text{where } M_2 = b(DP_1/DP_2) + c \tag{4.12}$$

The definition of modules can be generalized as

$$M_i = \sum_{j=1}^{j=i} \frac{\partial FR_i}{\partial DP_j} \frac{DP_j}{DP_i} \tag{4.13}$$

The above definition of module has an advantage in that it simplifies the representation of the system architecture as shown later in this chapter.

6. *Decomposition of FRs, DPs, and PVs. How does the decomposition process affect the outcome of the design process?* The FRs, DPs, and PVs must be decomposed by zigzagging between the domains until the design can be implemented without further decomposition. For example, if the DPs are an electric motor and telephone, it will not be necessary to decompose them further, as we do not wish to invent a different kind of electric motor or telephone. The resulting hierarchies of FRs, DPs, and PVs and the corresponding matrices represent the system architecture.

The decomposition of these vectors cannot be done by remaining in a single domain, but can be done only by zigzagging between the domains. Each designer may decompose the highest level FRs and all other FRs in a unique way because of his particular experience and knowledge base. Therefore, the final design will be different in terms of its components, physical arrangement, etc. However, no matter how they have been decomposed, all of them should perform the same set of highest level FRs. From the functional point of view, all of these designs are "equivalent," although their physical components and detailed physical layout, in the case of a product design, may be substantially different.

Based on the foregoing reasoning, the following definition of "equivalent" and "identical" design will be adopted in this book:

Definition S1 (Equivalent Designs): Two designs are defined to be *equivalent* if they satisfy the same set of highest level FRs within the bounds established by the same set of constraints, even though the mapping and decomposition process may have yielded designs that have substantially different lower level FRs and DPs.

Definition S2 (Identical Designs): Designs that fulfill the same set of highest level FRs and satisfy the Independence Axiom with zero information content are defined to be *identical* if their lower level FRs and all DPs are also the same.

The following theorems may be stated based on the foregoing reasoning and definitions:

Theorem S1 (Decomposition and System Performance)

The decomposition process does not affect the overall performance of the design if the highest level FRs and Cs are satisfied and if the information content is zero, irrespective of the specific decomposition process.

Theorem S2: (Cost of Equivalent Systems)

Two "equivalent" designs can have a substantially different cost structure, although they perform the same set of functions and may even have the same information content.

7. *System Architecture* The hierarchical structure of FRs, DPs, and PVs, together with the corresponding design matrices created by the decomposition process, represents the system architecture. The system architecture can also be represented by the module-junction diagram and the flow diagram, which are discussed in Section 4.6.

4.5 DESIGN AND OPERATION OF LARGE SYSTEMS

4.5.1 Introduction to Large-System Issues

In recent years, the design and operation of large systems have attracted the interest of many researchers and industrialists. This is in part due to the ever-increasing size of technical systems, industrial enterprises, and government organizations. This trend is likely to continue in the future, because the rapid access and management of information make efficient operation of such large systems possible. Yet the design and operation of large systems are still based on empiricism and heuristics. In the future, however, to be competitive, industrial firms should have a rational strategy for design and management of large products, their enterprises, and manufacturing operations. Also, to be efficient in handling societal programs such as health care, governments need administrative systems that are rationally designed. Cost-efficient and responsive designs of such systems can be created reliably only through a firm scientific base for large-system design.

4.5.2 What Is a Large System?

Intuitively most people, including researchers and engineers, seem to know what we mean by large systems. The telephone system for Boston, the government bureaucracy, an assembly plant for automobiles, a software system that controls nuclear power plants, and the Boeing 747 airplanes seem to fit the term "large system" because they are physically large, having many components or many lines of computer code. It turns out that the examples cited above are also large systems from the functional point of view because they satisfy many different sets of FRs throughout their lifetime.

The conventional definition of large systems is based on the number of physical parts or the physical size of the system. However, this definition has some shortcomings. The number of components in a piece of hardware or the number of lines of code in a software program will depend on the number of highest level FRs and the number of layers of decomposition necessary to generate a complete design solution. A good designer will minimize both the number of upper-level FRs and the number of layers of hierarchical decomposition. Therefore, the definition of a large system based on the number of physical components or the physical size of the system will make it difficult to provide a true measure (or metric) of a large system. We need a definition that separates the size effects due to the inherent complex nature of the design task from those that result from poor design decisions. In other words, a poor designer can make anything into a large system.

From the axiomatic design point of view, which focuses on functions and functional requirements, the definition of a large system based on physical size or number of components is not sufficient to capture the thought process involved in designing systems. In this section, a different definition based on the number of functions the system must satisfy will be given.

4.5.3 Definition of a Large Flexible System

What kinds of large systems are there? A large system has a large number of FRs at the highest level of specification or at the problem-definition stage. Some large systems satisfy only a fixed set of FRs at all times. This type of large system may be classified as a *large fixed system*. The design methodology for a large fixed system is the same as that for a small system, which was discussed in Section 4.4.

There is another class of large systems. A system may have a large number of FRs at the highest level, but only a subset of these FRs must be satisfied at a given time, and different subsets must be satisfied at different points in time. Such a system must be able to reconfigure itself to satisfy different subsets of FRs throughout its life. This type of system may be classified as a *large flexible system*. For example, a large telephone company serves many customers, but the particular set of customers that must be linked together changes as a function of time. Therefore, the system is dynamic and must be reconfigurable on demand. Similarly, the Boeing 747 airplane must be able to deal with different requirements throughout its life. On the other hand, simple machine tools, such as a lathe, satisfy the same set of FRs at all times, i.e., they are designed to satisfy a *fixed* set of FRs at the highest level, which does not change as a function of time.

Regardless of the number of FRs and their temporal variation, the system must be designed so that the Independence Axiom is always satisfied. When this is done incorrectly, the system is difficult to operate because of the coupling of FRs by the DPs chosen. Such a coupled system often cannot satisfy all of its required set of FRs, and the information content is relatively large. The information content of a coupled system is also large relative to that of an ideal uncoupled system because coupling introduces conditional probabilities. These poorly designed systems appear to be complex because much information must be supplied to increase the probability of satisfying the FRs.

From the functional point of view, a system is defined as a *large flexible system* if the total number of FRs that the system must satisfy during its lifetime is large, and if at different times, the system is required to satisfy different subsets of FRs.

According to this definition, a large building or even a large castle is not necessarily a large flexible system, although they may have many physical components and many rooms, if their set of highest level FRs is small and remains fixed throughout their lifetime. However, an intelligent skyscraper or a hotel with a thousand beds is a large flexible system according to the above definition, as the set of FRs that each of these must satisfy is large and changes continuously and unpredictably over its lifetime because of changes in customer requirements.

In this section, only the design of *large flexible systems* will be considered.

4.5.4 Axiomatic Design of a Large Flexible System

How do we design a large flexible system? Suppose that we have to design a system that has to satisfy a set of m FRs. To achieve the design goal, we have to conceive of a design that has a set of m DPs. Designers must find these DPs from their knowledge or with the help of a *knowledge base* or *database*.

The knowledge base can be structured as follows:

$$FR_1 \$ \left(DP_1^a, DP_1^b, \ldots, DP_1^r\right)$$

$$FR_2 \$ \left(DP_2^a, DP_2^b, \ldots, DP_2^q\right)$$

$$FR_3 \$ \left(DP_3^a, DP_3^b, \ldots, DP_3^w\right)$$

$$\ldots \tag{4.14}$$

$$\ldots$$

$$\ldots$$

$$FR_m \$ \left(DP_m^a, DP_m^b, \ldots, DP_m^s\right)$$

Equation (4.14) simply states that FR_1 can be satisfied (indicated by \$) by DP_1^a or DP_1^b or DP_1^r, etc. Similarly, FR_m is satisfied by DP_m^a or DP_m^s, etc. Equation (4.14) represents the *knowledge base* or the *database* for a large flexible system. Adding additional DPs to these equations is equivalent to expanding the knowledge base or the database. The richness of the system is defined by the size and quality of the database, as the larger the number of available DPs, the greater is the likelihood of better designs.

As we search for DPs in the physical domain that will enable us to satisfy the FRs, we may find that there is more than one DP_i that can satisfy a given FR_i. Equation (4.14) does not state which of the candidates for DP_3 (i.e., DP_3^a, DP_3^b, ..., DP_3^w), for example, is the best choice for FR_3. Furthermore, because all or a subset of the FRs must be satisfied at any given instant, one cannot say a priori which DP_3 is the best solution without considering the other FRs that must be satisfied at the same time. That is, the choice of DP_3 may be different depending on the chosen subset of FRs.

As defined earlier, a large flexible system is often required to satisfy only a subset of its FRs at any given time. Suppose the subsets of FRs change as a function of time as follows:

$$@t = 0 \qquad \{FR\}_0 = \{FR_1, FR_5, FR_7, FR_m\}$$

$$@t = T_1 \qquad \{FR\}_1 = \{FR_3, FR_5, FR_8, FR_z\} \qquad (4.15)$$

$$@t = T_2 \qquad \{FR\}_2 = \{FR_3, FR_9, FR_{10}, FR_m\}$$

Equation (4.15) is written in vector notation form. For example, the equation $\{FR\}_0 = \{FR_1, FR_5, FR_7, FR_m\}$ should be read as "the vector of FRs at time $t = 0$ is composed of the components $FR_1, FR_5, FR_7,$ and FR_m."

To satisfy $\{FR\}_0$, the chosen $\{DP\}_0 = \{DP_1, DP_5, DP_7, DP_m\}$ must ensure the independence of $FR_1, FR_5, FR_7,$ and FR_m, as required by the Independence Axiom. Once these DPs are chosen, a large system operates just like any other system. However, at $t = T_1$, a different subset of FRs must be satisfied. This means that the system must *reconfigure* or *switch* to satisfy $\{FR_3, FR_5, FR_8,$ and $FR_z\}$ independently. Similarly, at $t = T_2$, the system must again reconfigure in order to satisfy $\{FR_3, FR_9, FR_{10}, FR_m\}$.

The search process for DPs is straightforward, in principle. Consider the case of choosing $\{DP\}$ to satisfy the specified $\{FR\}_2$ at $t = T_2$. When we are searching for an uncoupled design, we can choose DP_3 that affects only FR_3 and has no effect on any of the other FRs in the subset. Then we choose DP_9 that affects only FR_9. This process continues until the entire subset of FRs is satisfied.

If an uncoupled design cannot be developed because of the lack of proper DPs in the database, then we must seek a decoupled design. In this case, we can choose any DP_3 first to satisfy FR_3. Then we can look for a DP_9 that does not affect FR_3, yet DP_9 may affect all other FRs. Similarly, we can choose DP_{10} that does not affect FR_3 and FR_9. Finally, when we are looking for a DP_m that can satisfy FR_m, DP_m cannot affect FR_3, FR_9, and FR_{10}. Obviously, there are many combinations of DPs that may yield a decoupled solution. It would be difficult to try all these combinations by a purely trial-and-error process when the number of FRs and the database for DPs are large.

The switching mechanism to go from a given subset of DPs to another must operate at an acceptable speed. In the case of the telephone system, the switching rate has become ever faster over the years. In the case of the U.S. government, the time constant to change and satisfy the subset of FRs is at least 6 months, as indicated by the time it takes for a new administration to appoint subcabinet-level government officials. In the case of automotive companies, it used to take at least 6 months to change over the manufacturing line when new car models were introduced. Now, because of global competition, the change over time has been shortened considerably. In this sense, competition is a good thing!

For a given subset of FRs, there may be many different subsets of DPs that are acceptable from the functional point of view. The best solution can be chosen by measuring the information content of each of the proposed solutions.

An ideal large flexible system is an uncoupled design with infinite adaptability or flexibility. Infinite adaptability means that an acceptable set of DPs can always be selected to satisfy the given subset of FRs. In this case, the database [i.e., the knowledge base given by Equation (4.14)] must be expandable because the system must be configured to adapt to a new situation by acquiring new data or knowledge. Otherwise, the system will have limited usefulness and flexibility. Furthermore, a large system can function most effectively if any user can modify the database, and, therefore, an ideal large system is designed so that users can adapt it to a variety of applications.

To map from the physical domain to the process domain, we must establish a database similar to Equation (4.14) for the relationship between DPs and PVs. Following the format and the system of notation used in Equation (4.14), the database may be expressed as

$$\mathrm{DP}_i \$ \left(\mathrm{PV}_i^a, \mathrm{PV}_i^b, \ldots, \mathrm{PV}_i^s \right) \tag{4.16}$$

The nature of the databases of plausible DPs and PVs will be different, depending on the hierarchical level of the FRs or DPs. For example, the word "vehicle" is at a higher level of abstraction than "bus" or "car." Similarly, the word "extruder" is at a higher level of abstraction than the specification of a particular combination of "screw" and "barrel" that comprise the extruder. Thus, the lower level database in the design hierarchy contains more specific and detailed information, whereas the highest level database tends to be more conceptual or abstract.

4.5.5 System Synthesis through Physical Integration of DPs

After the DPs are chosen to satisfy the FRs, the Information Axiom demands that the disparate DPs be integrated physically, if possible, to minimize the information content. This is easier to see in the case of hardware. The beverage can presented in Chapter 1 is one of the best examples where many DPs are integrated in a few physical components. The beverage can has 12 functional requirements that are satisfied by 12 design parameters, but has only three physical parts. However, physical integration is not always possible. For example, the engine of a car and its steering system are a part of the same system, but they are physically separate.

For assembly of physical parts, rules for integration of parts have been developed by a number of companies and researchers (Boothroyd and Dewhurst, 1987). These rules have been effective at a component level. However, some of these rule-based algorithms can cause a major problem if the physical integration of DPs is done at the expense of the independence of FRs. Remember that physical integration is a good thing as long as the functional independence of all FRs is maintained.

The design parameters of a system can sometimes be integrated through the use of software. For example, the automatic transmission of automobiles is linked to the engine by a software engine management system (EMS) through the use of sensors so that the FRs of the engine and the transmission are satisfied independently. They constitute a decoupled design.

4.5.6 On Designing the Best Large Flexible System

The quality of the designs that satisfies the Independence Axiom depends on the richness of the database—represented by Equations (4.14) and (4.16)—available to the designer. The database can be expanded by an accumulation of experience and knowledge. Therefore, there is no "unique" or "best" design solution. Available knowledge and technology determine the best design we can develop at a given point in time. When new knowledge or technology emerges rapidly, system design also undergoes rapid change. This is happening today in the telecommunications industry.

When there are several equally acceptable system designs that satisfy the Independence Axiom, we may be able to choose the best design from among these solutions if all the subsets of FRs that the system must satisfy during its life are known. For example, if there are four possible designs for satisfying the three subsets of FRs specified by Equation (4.15), we can compute the information content for each design. The design with the least information content is the best of the four proposed alternatives.

The difficulty arises in choosing the best design for a system when the subsets of FRs the system must satisfy vary unpredictably over the system's lifetime. In this case, the best design cannot be determined a priori. That is, the one design that requires the least amount of information for a given subset of FRs may not be the best for another subset of FRs.

What then should be the strategy for designing a large flexible system when the specific subsets of FRs the system will be required to satisfy in the future are not known? There may be at least the following three different possibilities:

1. The first option is to include as many FR/DP relationships in the database as possible for unknown but likely contingencies.
2. If it is possible to identify at least some of the subsets of FRs that the large flexible system will encounter in its lifetime, then we may choose the design that is the best at least for these known subsets. Obviously, the probability of choosing the best design among all those possible increases as the number of subsets identified during the design stage approaches the complete set to be encountered during the actual operation.
3. The third possibility is to endow the system with enough intelligence so that it can determine the set of FRs it must satisfy on the fly when the need arises and then choose the right set of DPs consistent with the Independence Axiom.

4.5.7 Theorems Related to the Design of Large Systems

Based on the foregoing discussions and reasoning, we may state the following theorems that are related to the design of large systems:

Theorem S3 (Importance of High-Level Decisions)

The quality of design depends on the selection of FRs and the mapping from domain to domain. Wrong selection of FRs made at the highest levels of design hierarchy cannot be rectified through the lower level design decisions.

Theorem S4 (The Best Design for Large Systems)

The best design for a large flexible system that satisfies m FRs can be chosen among the proposed designs that satisfy the Independence Axiom if the complete set of the subsets of FRs that the large flexible system must satisfy over its life is known a priori.

Theorem S5 (The Need for a Better Design)

When the complete set of the subsets of FRs that a given large flexible system must satisfy over its life is not known a priori, there is no guarantee that a specific design will always have the minimum information content for all possible subsets and thus there is no guarantee that the same design is the best at *all times*.

Theorem S6 (Improving the Probability of Success)

The probability of choosing the best design for a large flexible system increases as the known subsets of FRs that the system must satisfy approach the complete set that the system is likely to encounter during its life.

Theorem S7 (Infinite Adaptability versus Completeness)

A large flexible system with infinite adaptability (or flexibility) may not represent the best design when the large system is used in a situation in which the complete set of the subsets of FRs that the system must satisfy is known a priori.

Theorem S8 (Complexity of a Large Flexible System)

A large system is not necessarily complex if it has a high probability of satisfying the FRs specified for the system.

Theorem S9 (Quality of Design)

The quality of design of a large flexible system is determined by the quality of the database, the proper selection of FRs, and the mapping process.

4.6 REPRESENTATION OF THE SYSTEM ARCHITECTURE OF FIXED SYSTEMS

Is there a need to represent the system architecture? How do we present it in a concise manner? A formal way of representing a system is important for many reasons. First, when people located in many different parts of the world work together to create a system, they must be able to communicate among themselves without allowing the possibility of misunderstanding. Second, when the system is large with many physical components or lines of code, an accepted means of representing their relationship will minimize the likelihood of improperly configuring the system during actual operation. Third, it is often the case that a system requires many weeks of debugging after it is completed because the system has been put together without the proper level of thoroughness, which results in long lead time, low reliability, and high cost. Fourth, engineering change orders and after-sale service cannot be performed with confidence in the absence of precise understanding of the system configuration. Fifth, computer simulation and automated diagnosis cannot be done without a means of clearly representing the system.

There are three different but equivalent ways of representing a system: (1) FR/DP/PV hierarchies with corresponding design matrices, (2) a module-junction diagram, and (3) the flow diagram or flow chart. Although all these different representations of the system architecture are equivalent, they emphasize different aspects of the system.

The hierarchical diagram gives the entire decomposition steps and all FRs and DPs. The module-junction diagram is created to show the hierarchical structure of modules and their interrelationships. The flow diagram illustrates the design relationships of all modules at the leaf level and the precedence of implementation based on design matrices of each level of design decomposition. The flow diagram is a concise and powerful tool that provides a comprehensive view of the system design and a road map for implementation of the system design.

The purpose of this section is to describe these three means of representing the system architecture of a fixed system.

4.6.1 Hierarchies in Design Domains through Decomposition of FRs, DPs, and PVs: A Representation of the System Architecture

What is a design hierarchy and how does it represent the system architecture? In Chapter 1 and in Section 4.4, the creation of the FR/DP/PV hierarchies through zigzagging between the domains was discussed. To create the hierarchies, the FRs, DPs, and PVs must be decomposed until the design can be implemented without further decomposition. These hierarchies of FRs, DPs, and PVs and the corresponding matrices are one way of representing the system architecture. The decomposition of these vectors cannot be done by remaining in a single domain, but can be done only through zigzagging between the domains as discussed in Chapter 1. As shown by Example 1.7 of Chapter 1, which dealt with the refrigerator design, the decomposition must be done until the design task is completed.

Suppose that we have completed a system design such that the FR and the DP hierarchies are as shown in Equation (4.17) and Figure 4.1. This design required four layers of decomposition to satisfy the highest level FRs.

$$\begin{Bmatrix} FR_1 \\ FR_2 \end{Bmatrix} = \begin{bmatrix} X & 0 \\ 0 & X \end{bmatrix} \begin{Bmatrix} DP_1 \\ DP_2 \end{Bmatrix}$$

$$\begin{Bmatrix} FR_{11} \\ FR_{12} \end{Bmatrix} = \begin{bmatrix} X & 0 \\ X & X \end{bmatrix} \begin{Bmatrix} DP_{11} \\ DP_{12} \end{Bmatrix}$$

$$\begin{Bmatrix} FR_{21} \\ FR_{22} \\ FR_{23} \end{Bmatrix} = \begin{bmatrix} X & 0 & 0 \\ X & X & 0 \\ 0 & 0 & X \end{bmatrix} \begin{Bmatrix} DP_{21} \\ DP_{22} \\ DP_{23} \end{Bmatrix} \tag{4.17}$$

$$\begin{Bmatrix} FR_{121} \\ FR_{122} \\ FR_{123} \end{Bmatrix} = \begin{bmatrix} X & 0 & 0 \\ X & X & 0 \\ X & 0 & X \end{bmatrix} \begin{Bmatrix} DP_{121} \\ DP_{122} \\ DP_{123} \end{Bmatrix}$$

$$\begin{Bmatrix} FR_{1231} \\ FR_{1232} \end{Bmatrix} = \begin{bmatrix} X & 0 \\ X & X \end{bmatrix} \begin{Bmatrix} DP_{1231} \\ DP_{1232} \end{Bmatrix}$$

The nonzero elements of design matrices in Equation (4.17) are indicated by X for simplicity.

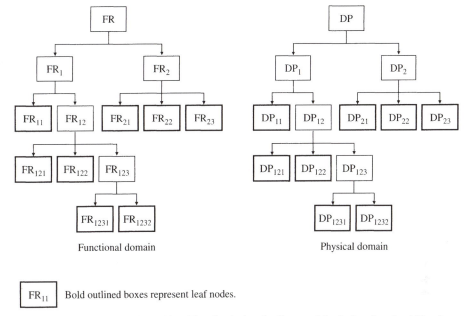

Functional domain

Physical domain

FR$_{11}$ Bold outlined boxes represent leaf nodes.

Figure 4.1 The FR and DP hierarchies. The physical embodiment of the design that should be shown at each level of decomposition is not shown here.

The first layer is an uncoupled design, but the second, third, and fourth layers are decoupled designs. There is no coupling in this design, and thus it represents an acceptable design. The decoupled designs must be controlled by following the sequence dictated by the design matrices. FR$_1$ and FR$_2$ can be simply combined to obtain the highest level FR because they are uncoupled with respect to each other. If the design matrix is a triangular matrix, the lower level FRs must be satisfied in the sequence given by the design matrix to satisfy the higher level FR. Each design matrix represents a junction and shows how the lower level modules must be assembled to yield the desired FR.

Each FR of the lowest layer of each branch (i.e., each FR that does not need further decomposition) is defined as a "leaf." The "leaves" of the system shown in Figure 4.1 are FR$_{11}$, FR$_{21}$, FR$_{22}$, FR$_{23}$, FR$_{121}$, FR$_{122}$, FR$_{1231}$, and FR$_{1232}$. The higher level FRs are satisfied by combining the leaves according to the information contained in the design matrix. For example, the leaves FR$_{1231}$ and FR$_{1232}$ yield FR$_{123}$ when they are combined following the sequence given by the design equation because it is a decoupled design.

The FR/DP hierarchies shown in Figure 4.1 and the design matrices given in Equation (4.17) represent the *system architecture* of a small fixed system. This is one of the three ways of representing system architecture.

4.6.2 Design Matrix and Module-Junction Diagrams: Another Means of System Representation

What is a module? How is it defined? In Section 4.4, a *module* is defined as the row of the design matrix that yields an FR when it is provided with (or multiplied by) the input of its corresponding DP. For instance, in this example, M$_{123}$ is a module that corresponds to the third row of the design matrix that yields FR_{123} when it is multiplied by (or provided with the input of) DP_{123}.

M_{123} can be obtained by combining M_{1231} and M_{1232} as per the design matrix shown in Equation (4.17). M_{123} does not appear in the design explicitly, as it is not the leaf or the terminal module. Only those DPs that are leaves need to be brought in to execute the system architecture. Lower level modules are combined to obtain higher level modules that yield the higher level FRs.

At each level, each FR is controlled by its corresponding DP through a module. The module is given by the elements in each row of the design matrix. The DP can be a simple input or data. For example, M_{1231}, which is also a leaf, is the module that enables the determination of FR_{1231} when M_{1231} is "multiplied" by DP_{1231}. In many situations, DP_{1231} is simply an input to M_{1231}.

To obtain the higher level FR, FR_{123}, the design matrix states that FR_{1231} should be determined first and then FR_{1232} because the design of FR_{123} is a decoupled design. Therefore, the lower level modules M_{1231} and M_{1232} can be combined to obtain FR_{123} in accordance with Equations (4.17).

It is important to note that a module in a hardware system does not correspond to a physical component of the hardware. In a hardware component, there can be many DPs and, therefore, there can be many modules associated with a given piece of the hardware system. This confusion has generated unnecessary debates among researchers in design.

Because there can be many modules distributed throughout a system, how can we represent the interrelationship among modules in a system design? To represent the properties of junctions at each level of decomposition, a "module-junction structure diagram" is defined (Kim, Suh, and Kim, 1991). The module-junction structure diagram constitutes another means of representing the system.

One of the key points of this section is that in software design as well as in hardware design, some designers regard a design that consists of modules—so-called "modular design"—as being a good design. However, a modular design may not necessarily be a good design. These modules must be properly arranged and must satisfy the Independence Axiom, which is the subject of this section.

To represent the relationship between modules, circled symbols are used at the junctions of modules to represent the operations needed to combine the modules. The following circled symbols will be used. For an uncoupled design, a circled S is used to represent a simple summation of FRs. A circled C indicates that DPs and Ms must be controlled in a sequence as suggested by the design matrix for a decoupled design. A circled F is used for a coupled design, indicating that it requires feedback and violates the Independence Axiom.

The module-junction structure diagram is shown in Figure 4.2 for the design given by Equation (4.17). It should be noted that the higher level modules are obtained by combining lower level modules. For example, by combining modules M_{1231} and M_{1232} with a control junction (C), we obtain the next higher module M_{123}. Then by combining modules M_{121}, M_{122}, and M_{123} with a control junction (C), we obtain the next higher level module M_{12}. Then, M_{11} and M_{12} are combined to obtain M_1 using a control junction (C) again. Also, M_{21}, M_{22}, and M_{23} are combined with a control junction (C) to get M_2. Finally, M_1 and M_2, although not shown as they are not leaves, are simply summed as indicated by symbol S.

A main module is defined as a module that contains all the junctions of all levels (Figure 4.3). A system has one main module and n modules corresponding to n FR leaves. Starting

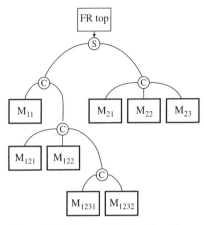

Figure 4.2 The module-junction diagram for the design described by Equation (4.17).

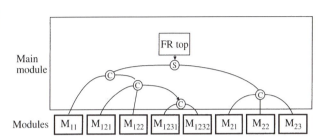

Figure 4.3 Modules and main module for the design represented by Equation (4.17).

from the leaves, modules can be combined to obtain higher level FRs and ultimately the highest level FRs in accordance with the module-junction structure diagram shown in Figure 4.3. These modules, which correspond to leaves, must be multiplied by their corresponding DPs to obtain the corresponding FRs.

4.6.3 Flow Diagram: A Representation of System Architecture

Is there a better way of representing system architecture? After defining the modules for the system, a flow diagram or flow chart may be used as another means of representing the system architecture. To construct the flow diagram, we will adopt a specific means of representing the relationship between modules as shown in Figure 4.4 (Kim, Suh, and Kim, 1991).

In an uncoupled design, because the child FRs are independent of each other, their parent FR is satisfied by combining all the outputs of its child modules in any random sequence. This is represented by the summation junction with the modules being parallel to each other as shown in Figure 4.4a. When the design is decoupled, the parent FR is determined by combining the child modules in a given sequence indicated by the design matrix, which is shown as a serial arrangement of modules. In this case, the output of the left-side module is controlled first and then the right-side module is executed next, as

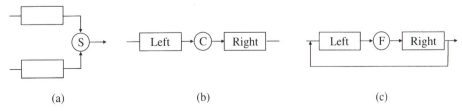

(a)   (b)   (c)

Figure 4.4 Representation of the design at each junction of a flow chart. (a) Summing junction (uncoupled design). (b) Control junction (decoupled design). (c) Feedback junction (coupled design).

shown in Figure 4.4b. For a coupled design, which violates the Independence Axiom, the feedback junction requires that the output of the right-side module be fed back to the left-side module, requiring a number of iterations until the solution converges, if it converges, which is shown in Figure 4.4c. Some feedback junctions may never converge. When there are feedback junctions, especially at the higher level of the system hierarchy, we should stop and seek an uncoupled or decoupled design.

Again consider the design given by Equation (4.17). FR_1 has to be decomposed further to FR_{11} and FR_{12} with corresponding DPs and modules (Ms). This design at this level consists of two parallel boxes as shown in Figure 4.5.

However, these modules are not the leaves, requiring that M_1 and M_2 be decomposed as shown in the design equations [Equation (4.17)]. Figure 4.6 shows the second-level flow diagram. It should be noted that the lower level modules are inside the higher level boxes. This process can continue until the complete flow diagram of the system architecture is obtained.

Figure 4.7 shows the final flow diagram for the system design represented by Equation (4.17), indicating how the system should be structured in terms of its operational sequence and modules. It consists of modules that correspond to the leaves. It is constructed by following the FR and DP hierarchies and design matrices.

The flow chart for the system given by Equation (4.17) shown in Figure 4.7 is an acceptable system design as the design satisfies the Independence Axiom at all levels of the design hierarchy. Note that Figure 4.7 does not have any feedback junctions with feedback loops. During the operation of the system (which may consist of hardware and software as shown in Figure 4.7), the lowest level DPs are the input variables to the system that ultimately control the final output of the system.

The system architecture shown in Figure 4.7 can be generated automatically by the Axiomatic Design Software[1] (Do, 1998). The design of this program is discussed in Chapter 5.

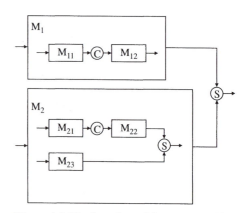

Figure 4.6 The flow chart of the system architecture given by Equation (4.17) at the second-level decomposition.

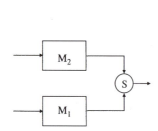

Figure 4.5 The highest level flow diagram for the system architecture given by Equation (4.17).

---

[1] The trade name is Acclaro, developed by Axiomatic Design Software, Inc. of Boston, MA.

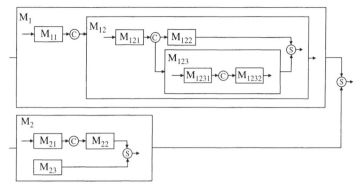

Figure 4.7 Flow chart of the system in Equation (4.17). This flow chart represents the system architecture, which consists only of summation and control junctions because the Independence Axiom is satisfied.

4.6.4 System Control Command (SCC)

How do we operate a system? Once the system is designed, it can be operated using the flow diagram of the system architecture represented by Figure 4.7. The flow diagram shows the sequence of computation/operation of the system. The goal of the system operation is to achieve the highest level FRs by means of the leaves.

To achieve the highest level FRs of the system, the implementation must always follow the sequence given in the flow diagram for the system architecture. When the modules are parallel, they can be operated on simultaneously and if they are in a series, the module on the left must be operated on first. The outer modules can be determined only from the input of the innermost boxes that represent the leaves in each major branch. Then we can go to the next higher level modules following the sequence indicated by the system architecture. Therefore, the operation is always from the innermost boxes to the outermost boxes in the sequence given by the flow diagram.

All the innermost boxes can be simultaneously executed followed by the next higher level boxes. When the design is decoupled, the execution of operation must be sequential as given by the flow diagram.

The rule for execution of the system architecture, which is called the system control command (SCC), may be stated as follows:

1. Establish the order of execution following the flow diagram of the system architecture, starting with the highest level modules.
2. The highest level modules can be determined by combining the lower level modules. Ultimately, all high-level modules must be constructed from the input data for the lowest level DPs (i.e., leaves) and the corresponding lowest level modules, following the sequence given by the flow diagram. All higher level FRs and modules are made of the lowest level modules and DPs (i.e., leaves).
3. Execute this procedure simultaneously at all branches at which such an action can be achieved independently so as to minimize the operational time.

Therefore, the SCC for the system architecture shown in Figure 4.7 may be programmed to perform in the following sequence:

1. M_1 and M_2 can be executed at the same time because, at this level, the design is uncoupled as indicated by parallel boxes.
2. M_1 can be obtained by combining M_{11} (leaf) with the results of the decoupled design output of M_{12}. However, because M_{12} is not a leaf, it must be determined by combining the leaves M_{121} and M_{122}, and M_{123} that consists of leaves M_{1231} and M_{1232}, as shown in the flow diagram.
3. Similarly, M_2 can be established from the leaves shown.
4. Combine M_1 and M_2 as indicated by the system architecture to obtain the final desired output.

EXAMPLE 4.1 Design of Wafer Processing Equipment[2]

A track machine is designed to coat silicon wafers with a thin layer of photoresist, which is then exposed to ultraviolet light to make an electric circuit (transistors, etc.) in a lithography machine (called a "stepper"). After the exposure, the wafer is returned to the track to develop the exposed circuit.

The goal of this exercise is to develop a new machine that can handle shorter wavelength light for manufacturing circuits and transistors with smaller dimensions in order to increase the speed of microprocessors and the storage density of DRAMs. This example is an oversimplified description of the actual process because of lack of space and to protect the proprietary interests of SVG, Inc. This exercise is done to document a machine that has been developed in the past.

The customer requests certain functionality for the track machine to deal with the DUV (deep ultraviolet) process technology. There are also several constraints (Cs) imposed on the track design. The Cs are

- Cost
- Footprint
- Reliability
- Safety
- Serviceability
- Manufacturability
- Contamination
- Minimization of variation in wafer temperature

1. The Highest Level FRs, DPs, Design Matrix, and Constraints

The system may be developed from the highest level by defining FRs, DPs, and Cs.

Description of FRs and DPs

This is the highest level FR/DP specification, and thus it defines the main functions of the track machine.

---

[2] This example is provided as a courtesy of Silicon Valley Group, Inc., San Jose, CA, a leading manufacturer of semiconductor processing equipment. This project was executed by MIT graduate student T.-S. Lee and Dr. H. L. Oh, Vice President of SVG, Inc. The details of the example are modified to protect the proprietary interest of SVG, Inc.

Index: 0.#			Highest Level
	Functional Requirements (FRs)		Design Parameters (DPs)
	Name	Description	Description
		Fabricate wafer surface during photoresist process	Track system
1	Process	Coat wafers with desired resist film	Coating process modules
2	Process	Develop the exposed film	Developing process modules
3	Transport/support	Process wafers at the desired rate	System configuration
4	Control	Control the system functions	Command and control algorithm (CCA)

- **FR₁**: As a part of the photolithography system, the track machine *coats the wafer surface* with photoresist before sending the wafer to a stepper. The specifications for FR_1 are listed below.

Coating thickness = 0.8–1.2 μm
Coating uniformity
    within wafer = 20 Å(3σ)
    wafer-to-wafer = 25 Å(3σ)
    cassette-to-cassette (24 hours) = 25 Å

- **DP₁**: This design parameter is primarily responsible for performing the coating processes. It indicates that the overall coating process, FR_1, is achieved through the use of separate subprocess modules.
- **FR₂**: As a part of a photolithography system, the track machine *develops images* that are replicated onto the resist film during the exposure process. The specification for FR_2 is listed below.

Critical dimensions (CD)
    within wafer = 0.015 μm
    wafer-to-wafer = 0.015 μm
    cassette-to-cassette (24 hours) = 0.015 μm

- **DP₂**: This design parameter is primarily responsible for performing the developing processes. Like DP_1, it indicates that the overall developing process is achieved through the use of separate subprocess modules.
- **FR₃**: It is important for a track system to process wafers at the desired rate, which is called *throughput*. FR_1 and FR_2 must be performed while satisfying the system throughput rate. The throughput of the clustered stepper is the throughput requirement for a track system. A stepper is among the most expensive tools in the FAB system, and its maximum utilization must be guaranteed. The track system should not limit throughput of the track-stepper cluster for that reason. Various process flows typically required by customers are

Standard (simple) coating/developing flow
TARC (top antireflective coating)-included flow
BARC (bottom antireflective coating)-included flow
Stand-alone coating
Stand-alone developing
Throughput rate—the range of throughput requirement varies significantly, depending on process type and time. It varies typically from 50 to 100 WPH.

- **DP$_3$**: The system configuration is primarily responsible for meeting the process flow and throughput requirements. Given DP$_1$ and DP$_2$, the system must be configured such that the required flow can be processed at the desired throughput rate. DP$_3$ necessarily incorporates the physical layout and the transport issue, which are discussed in the next-level decomposition.
- **FR$_4$**: Because the track system is a complex, *automated* machine, it has to control the functions within the system to make it operate smoothly.
- **DP$_4$**: A command and control algorithm (CCA) coordinates the process tasks at this level. This includes all the system-level activities, such as administration and communication.

The DPs chosen at this level are abstract and conceptual, providing limited information. Some of the information available here from DPs is as follows:

- The system will have a modular structure.
- DP$_3$, the system configuration, covers system capacity and transport requirements.
- DP$_4$, the CCA, will have all the information for system operation.

With the above information, the design matrix needs to be determined. Determining the high-level design matrix has a large impact on the further decompositions because they must be consistent with the earlier design decisions.

Design Matrix

$$\begin{Bmatrix} FR_1 \\ FR_2 \\ FR_3 \\ FR_4 \end{Bmatrix} = \begin{bmatrix} X & 0 & 0 & 0 \\ 0 & X & 0 & 0 \\ X & X & X & 0 \\ X & X & X & X \end{bmatrix} \begin{Bmatrix} DP_1 \\ DP_2 \\ DP_3 \\ DP_4 \end{Bmatrix}$$

Elements of Design Matrix

A_{11}, A_{22}, A_{33}, A_{44}: The diagonal elements are X because each DP has been chosen in such a way that it affects the corresponding FR.

A_{12}, A_{21}: These elements are very clearly 0; DP$_1$ has no effect on FR$_2$ and vice versa because the track system is based on modular structure, and these two sets of modules are completely separated functionally and physically.

The remaining elements in the design matrix indicate the direction for the later design process. The decisions regarding those elements need careful discussion.

A_{13}, A_{23}: These elements are evaluated by asking whether the choice of DP_3 affects FR_1 or FR_2. The answer is that it should not. Because both FR_1 and FR_2 are directly responsible for the on-wafer result, there should be no adverse effect on those functions from any other sources within the system.

A_{31}, A_{32}: These elements are evaluated by asking whether the choice of DP_1 or DP_2 affects FR_3. As indicated in the FR/DP description section, FR_3 includes process flow requirements and the throughput requirement. By the nature of FR_3, there is a strong relationship between DP_3 and both DP_1 and DP_2. For example, the accessibility to each module will affect DP_3, and consequently sub-$FR_{3,x}$. These off-diagonal nonzero elements can be viewed as a design decision that the interface between system configuration, DP_3, and process modules, DP_1 and DP_2, belongs to the DP_3 side. It is evident that it will be better if we can make these elements 0. The conditions will be flexibility and standard interface.

A_{41}, A_{42}, A_{43}: The CCA, which is responsible for system control, must be designed with the information regarding the objects to be controlled. The objects for this case are DP_1, DP_2, and DP_3.

A_{14}, A_{24}, A_{34}: Change of DP_4 should not affect FR_1, FR_2, or FR_3. The implication of these zeros is that FR_1, FR_2, and FR_3 must be immune to any change of system control DP_4.

The design matrix $[A]$ is a lower triangular matrix, which corresponds to a decoupled design. It implies that for further design of DP_3 and DP_4, information should be transferred from other DP design. For example, design of DP_3 requires some of the information from DP_1 and DP_2.

Constraints

Once the design matrix is determined, the constraints need to be clarified. These constraints are the guidelines to make the subsequent decomposition consistent with high-level design decisions.

At this high level, the constraints are stated straightforwardly, and they do not contain details.

- Decomposition of DP_1 must not affect FR_2.
- Decomposition of DP_2 must not affect FR_1.
- Decomposition of DP_3 must not affect FR_1 or FR_2.
- Decomposition of DP_4 should not affect FR_1, FR_2, or FR_3.

More explicit statements will be available in later decompositions.

2. The Second-Level Decomposition for Coating Process Modules

Description of FRs and DPs

The inputs for this decomposition activity are as follows:

- The parent FR is FR_1 (Coat wafers with desired resist film), and the parent DP is DP_1 (coating process modules). Constraints are all of the constraints imposed on the system and those identified in the previous-level design.

Index: 1.#			Coating Process Modules Branch
	Functional Requirements (FRs)		Design Parameters (DPs)
	Name	Description	Description
		Coat wafers with desired resist film	*Coating process modules*
1	*Process*	Improve resist adhesion of a wafer surface	Vapor-prime module
2	*Process*	Bring wafer temperature down to coat-process temperature	Chiller module
3	*Process*	Coat wafer with resist	Spin coater
4	*Process*	Cure the resist layer	Soft-bake module
5	*Process*	Bring wafer temperature down to expose—process temperature	Chiller module

• To coat a resist film onto a wafer surface, it needs a precoating thermal treatment and a postcoating thermal treatment in addition to the coating process itself.

Those FRs and corresponding DPs are stated more precisely.

• **FR_{11}**: To promote the wafer-surface adhesion to the photoresist material, the wafer surface should be free from any contamination and moisture. A clean silicon surface still does not provide a good bonding site for novolac-based photoresists. The surface must be chemically compatible with the resist. A couple of substeps are required to achieve sufficient adhesion, which will be discussed in the next-level decomposition.

Temperature range: 100°C–140°C
Temperature control: ±0.5°C

• **DP_{11}**: Vapor prime is conventionally used to achieve the wafer-surface adhesion. The vapor-prime module is basically a heating module, which is a kind of small oven, and it first gets rid of the moisture on the wafer surface. The module also performs a surface modification task using chemical adhesion promoters. The most common adhesion promoter for silicon photoresist is hexamethylenedisiloxane (HMDS), which is a volatile liquid at room temperature. A vapor-priming method for coating is used for reproducible results.

• **FR_{12}**: To be processed in subsequent steps, a heated wafer needs to cool to the preset temperature. This is a critical FR because it determines the temperature distribution of a wafer before the coating process.

Temperature range: 18°C–24°C
Temperature control: ± 0.2°C

• **DP_{12}**: The chiller module is primarily responsible for cooling the wafer temperature. Tight temperature control is required to achieve FR_{12} satisfactorily. Design of a chiller module—decomposition of DP_{12}—has to reflect the tight temperature control.

• **FR_{13}**: Liquid photoresist material has to be deposited and coated onto the wafer surface. Successful achievement of this FR is primarily responsible for the coating quality.

It must produce uniform and repeatable resist film conforming to the specifications shown in the FR_1 description. As a subrequirement, it should be able to carry various photoresist coatings.

• **DP_{13}**: Spin coating has long been accepted as the best coating method for satisfying this FR. The spin-coater module is responsible for the process of flooding the substrate with resist solution and drying the surface by spinning.

• **FR_{14}**: After the spin coating and drying, the polymer film contains up to 15% residual solvent and may contain built-in stresses due to the shear forces encountered during the spinning process. The wafer surface has to be thermally treated to get the final photoresist film on it. Because the dominant mechanism in this thermal process is solvent evaporation, whose rate is extremely temperature dependent, tight temperature control is required.

Temperature range: 85°C–100°C
Temperature control: ± 0.2°C

• **DP_{14}**: Bake module—so called soft-bake module—is primarily responsible for removing residual solvent and annealing any stress in the film. Design of the soft-bake module should reflect the tight temperature control.

• **FR_{15}**: Resist thickness continues to change until the temperature of the wafer is chilled. The wafer temperature must be brought down to a set point.

Temperature range: 18°C–24°C
Temperature control: ± 0.2°C

• **DP_{15}**: Chiller module is primarily responsible for cooling the wafer temperature. FR_{15} is the same as FR_{12}, and DP_{15} will have practically the same features as DP_{12}.

As discussed in Level 1, we must determine the design matrix only for the given information at this level.

• DP_{11}, DP_{12}, DP_{14}, and DP_{15} are the DPs of thermal processes—heating and cooling.
• Each DP will be separated physically.
• There is no direct interaction among DPs.

Based on the above information, the evaluating questions can be answered.

Design Matrix

$$
\begin{Bmatrix} FR_{11} \\ FR_{12} \\ FR_{13} \\ FR_{14} \\ FR_{15} \end{Bmatrix} = \begin{bmatrix} X & 0 & 0 & 0 & 0 \\ 0 & X & 0 & 0 & 0 \\ 0 & 0 & X & 0 & 0 \\ 0 & 0 & 0 & X & 0 \\ 0 & 0 & 0 & 0 & X \end{bmatrix} \begin{Bmatrix} DP_{11} \\ DP_{12} \\ DP_{13} \\ DP_{14} \\ DP_{15} \end{Bmatrix}
$$

Elements of Design Matrix

All off-diagonal terms are zero. The design of DP_{11}–DP_{15} is modular, and no interface is required between the modules.

The resulting design matrix is diagonal, which represents the best design at this level. This design decision implies that each DP must satisfy its corresponding FR exactly.

Otherwise, the design matrix would be lower triangular because of the process sequence: the wafer is processed in DP_{11} first, then sent to DP_{12} and processed in DP_{13}, etc. Any failure of one of these FRs will propagate along the process.

Constraints

In order to ensure that the design matrix is diagonal, there are some constraints for the off-diagonal zero elements.[3]

$B_{21}, B_{31}, B_{41}, B_{51}$: $DP_{12}, DP_{13}, DP_{14}$, and DP_{15} have to be protected from any thermal effects caused by DP_{11}.

$B_{14}, B_{24}, B_{34}, B_{54}$: $DP_{11}, DP_{12}, DP_{13}$, and DP_{15} have to be protected from any thermal effects caused by DP_{14}.

$B_{13}, B_{23}, B_{43}, A_{53}$: Neither mechanical vibration nor particle generation from DP_{13} may be transferred to $DP_{11}, DP_{12}, DP_{14}$, or DP_{15}.

Because $DP_{11}, DP_{12}, DP_{13}, DP_{14}$, and DP_{15} are physically separate modules, there are no other constraints over them.

3. The Second-Level Decomposition for Developing Process Modules

Index: 2.#			Developing Process Modules Branch	
	Functional Requirements (FRs)		**Design Parameters (DPs)**	
	Name	**Description**	**Description**	
		Develop exposed film	*Developing process modules*	
1	*Process*	Activate photoacid created by the exposure process	Postexposure bake (PEB) module	
2	*Process*	Bring wafer temperature to coat temperature	Chiller module	
3	*Process*	Develop the exposed film	Spray/Spin developer	
4	*Process*	Dehydrate the wafer to harden the film	Hard-bake module	
5	*Process*	Bring wafer temperature to a setpoint	Chiller module	

Description of FRs and DPs

In order to produce the final three-dimensional (3-D) image, we need a series of developing processes.

 • **FR$_{21}$**: For the DUV process, the exposure process creates a photoacid that acts as a catalyst to mediate a cascade of reactions. The photoacid is used to generate a latent

---

[3] In this section, *A* refers to the elements of design matrix. To simplify the notation for subscripts, the subscripts referring to the high levels are not used. For example, B_{21} is the design element that relates FR_{x2} to DP_{x1}.

image, and a heating process is required for the reaction. With the photoacid as a catalyst, a chain reaction is initiated such that it modifies the resist chemistry to allow latent image generation.

Temperature range: 90°C–110°C
Temperature control: ± 0.2°C

- **DP_{21}**: The postexposure bake (PEB) module is responsible for latent image generation. Because chemical reactions after the photoacid generation are all thermally activated, temperature uniformity control is important. It is also important to control the extent of the chemical reactions by controlling process time.
- **FR_{22}**: The baked wafer needs to be chilled until its temperature reaches a set point. Until the chill process starts, the chain reaction for DUV resist continues. FR_{22} is basically the same as FR_{15}.
- **DP_{22}**: The PEB chiller module is primarily responsible for bringing the wafer temperature down. DP_{22} will have the same features as DP_{15}.
- **FR_{23}**: Producing the final 3-D image of the replicated mask pattern from the latent image. This is known to be the most complex of the processing steps, and has the greatest influence on pattern quality. Typical criteria for a developing FR are as follows:

Uniform CD control: stated in FR_2 specification
Maximum defect count: 0.02 defects/cm^2 ≥ 0.2 μm size
Minimum contamination and minimum reduction in film thickness

- **DP_{23}**: The spray-type development process module basically sprays fresh developer and dries the substrate of all residual developers and rinses after the completion of the developing cycle.
- **FR_{24}**: Increase the etch resistance of the developed resist film. A wafer needs to be dehydrated to improve resist adhesion, and residual solvent needs to be evaporated to harden the film. Fundamental considerations for FR_{24} are the same as those discussed for FR_{14}, soft bake, except that it has less limitation and thus does not require strict control.

Temperature range: 90°C–110°C

- **DP_{24}**: The hard-bake module is primarily responsible for satisfying FR_{24} and is essentially the same as the soft-bake module (DP_{14}).
- **FR_{25}**: The wafer temperature must be brought down to a set point. This is a noncritical process requirement as compared to other chilling FRs.
- **DP_{25}**: The chiller module is primarily responsible for cooling the wafer temperature. DP_{25} is essentially the same as DP_{15}.

The information given in the FR/DP statements is

- DP_{21}, DP_{22}, DP_{24}, and DP_{25} are the modules of thermal processes—heating and cooling.
- Each DP will be separated physically.
- There is no direct interaction among DPs.

Based on the above information, the evaluating questions can be answered.

Design Matrix

$$
\begin{Bmatrix}
FR_{21} \\
FR_{22} \\
FR_{23} \\
FR_{24} \\
FR_{25}
\end{Bmatrix}
=
\begin{bmatrix}
X & 0 & 0 & 0 & 0 \\
0 & X & 0 & 0 & 0 \\
0 & 0 & X & 0 & 0 \\
0 & 0 & 0 & X & 0 \\
0 & 0 & 0 & 0 & X
\end{bmatrix}
\begin{Bmatrix}
DP_{21} \\
DP_{22} \\
DP_{23} \\
DP_{24} \\
DP_{25}
\end{Bmatrix}
$$

Elements of Design Matrix

All of the off-diagonal terms are zero. The design of DP_{21}–DP_{25} is modular, and no interface is required between them.

The resulting design matrix is diagonal, which represents the best design at this level. Because of the similar nature of modular design, the same argument used in the decomposition of the coating process modules is applied at this level.

Constraints

The three constraints for the coating process modules will be applied to the developing process modules equivalently.

$A_{21}, A_{31}, A_{41}, A_{51}$: $DP_{22}, DP_{23}, DP_{24}$, and DP_{25} have to be protected from any thermal effects caused by DP_{21}.

$A_{14}, A_{24}, A_{34}, A_{54}$: $DP_{21}, DP_{22}, DP_{23}$, and DP_{25} have to be protected from any thermal effects caused by DP_{24}.

$A_{13}, A_{23}, A_{43}, A_{53}$: Neither mechanical vibration nor particle generation from DP_{23} must be transferred to $DP_{21}, DP_{22}, DP_{24}$, and DP_{25}.

Because $DP_{21}, DP_{22}, DP_{23}, DP_{24}$, and DP_{25} are physically separate modules, there are no other constraints over them.

The DUV process will impose some constraints on DP_3, system configuration; the wafer has to be moved from the stepper side to the PEB module with no delay. This constraint will be discussed in the consistency-checking section.

4. The Second-Level Decomposition for System Configuration

Index: 3.#			**System Configuration Branch**	
	Functional Requirements (FRs)		**Design Parameters (DPs)**	
	Name	Description	Description	
		Process wafers at desired rate	*System configuration*	
1	*Support*	Support the system physically	System layout	
2	*Process (Transport)*	Move wafer when process is finished	Transport robot system	

Description of FRs and DPs

To satisfy the process flows and throughput requirements, the system has to be designed for two aspects: capacity and transport. Both are captured by the FRs at this level.

-   **FR_{31}**: As the decomposition for FR_1/DP_1 and FR_2/DP_2 shows, the track system is composed of a number of process modules. Sufficient capacity to meet the flow and throughput requirements must be ensured, and the modules need to be physically supported (allocated within a frame).
-   **DP_{31}**: System layout includes capacity planning and module allocation. In the next-level decomposition, those two issues are discussed in detail.
-   **FR_{32}**: Because a wafer has to go through a series of different process modules, wafer transport is one of the essential FRs. Basically, FR_{32} states that every wafer must be transported when its specified process is completed.
-   **DP_{32}**: Transport is achieved by a transport robot system. As the throughput requirement becomes higher, the robot is currently the most effective means of transport. The number (or kinds) of robots is determined with more information, which is obtained from further decomposition of FR_{1x}/DP_{1x}, FR_{2x}/DP_{2x}, and FR_{31}/DP_{31}.

Design Matrix

$$\begin{Bmatrix} FR_{31} \\ FR_{32} \end{Bmatrix} = \begin{bmatrix} X & X \\ X & X \end{bmatrix} \begin{Bmatrix} DP_{31} \\ DP_{32} \end{Bmatrix}$$

Elements of the Design Matrix

A_{21}: The design of DP_{32}, or sub-DP_{32}, must definitely be in accordance with the choice of DP_{31}, the system layout. For example, the layout drives the requirements for the robot system, such as moving range of certain joint in the robot, low bound of robot speed, etc. B_{21} indicates the transfer of that type of information.

A_{12}: Preferably, the transport robot system (DP_{32}) should not affect FR_{31}. However, in the system analyzed in this case study, it turns out that DP_{32} has an effect on FR_{31}, and thus neither DP_{31} nor DP_{32} can be chosen—designed—without iteration. In fact, the manufacturer of this track system had to spend a lot of time to find the optimum operating point for this coupled design, and ended up with an unsatisfactory solution for both internal and external customers.

Constraints

There is no specific constraint within the design matrix of this level. However, more important constraints for both DPs are imposed by other branches, such as FR_{2x}/DP_{2x}. They are discussed at the end of the second-level decomposition.

5. The Second-Level Decomposition for Command and Control Algorithm

Description of FRs and DPs

As mentioned in the FR_4 description, the parent FR (operate the system functions) is derived from the fact that the track machine is a complex automated machine. The parent DP (system

Index: 4.#		**Command and Control Algorithm Branch**	
Functional Requirements (FRs)			*Design Parameters (DPs)*
Name	**Description**		**Description**
	Operate the system functions		*Command and Control Algorithm (CCA)*
1 CCA	Administer system		System administrator
2 CCA	Command/control system operation		Operation controller
3 CCA	Build internal communication		Communication manager

control command) will have access to all information required to control the system, such as the process recipe, status of modules, and error. Based on the parent FR/DP, three sub-FRs/sub-DPs are identified. At this level, they are defined in general terms simply to clarify the main features of system CCA in a structured way.

- **FR_{41}**: Need for administrating the system. Detailed FRs are presented in next-level decomposition.
- **DP_{41}**: System administrator is responsible for the various tasks regarding administration, which are clarified as sub-FR_{41}. A group of software modules constitutes DP_{41}.
- **FR_{42}**: Need for generating actual command to control system operation.
- **DP_{42}**: Operational controller is primarily responsible for generating control commands. A group of software modules constitutes DP_{42}.
- **FR_{43}**: All information, essentially electric signals, needs to be transferred accurately. There is a strong requirement for robust communication.
- **DP_{43}**: Communication manager is responsible for building robust internal communication. It comprises communication protocol, data-transfer management, etc.

FR_{41} concerns the process recipe to operate the system. The recipe is defined by the user and contains information such as process flow, process time, setting temperature, spin speed, priority of each process, and type of chemicals. The system must be designed to have enough flexibility to support this variation. (This is represented by the fourth row Xs in the first-level design matrix.)

To control the system, we need to know the states of wafers and modules in real time. FR_{42} states this requirement and DP_{42} will be constructed to perform this function through further decomposition.

FR_{43} is the key concept of DP_4 (system command control) for the system. Based on the recipe, this document will capture process flow knowledge and design information, which will be resources of DP_{43} (command module). It will do all the tasks for system operation: sequencing operations, determining operating parameters, etc

Each of the DPs is more logical than physical; therefore the design matrix will be determined mainly on the information flow.

Design Matrix

$$\begin{Bmatrix} FR_{41} \\ FR_{42} \\ FR_{43} \end{Bmatrix} = \begin{bmatrix} X & 0 & 0 \\ X & X & 0 \\ X & X & X \end{bmatrix} \begin{Bmatrix} DP_{41} \\ DP_{42} \\ DP_{43} \end{Bmatrix}$$

Elements of the Design Matrix

A_{21}: DP_{41}, system administrator, has basic information necessary for the operation and distributes it to other portions of CCA. DP_{41} also has the authority of high-priority interrupt.

A_{31}, B_{32}: The information to be transferred from DP_{41} to DP_{42} and vice versa affects FR43 and consequently the design of DP_{43}.

The design matrix is decoupled. Further decomposition reveals specific contents of the off-diagonal terms, and those off-diagonal terms are formulated in the form of software modules or data transfer.

Constraints

As mentioned above, the decomposition for FR_4/DP_4 focuses mainly on information flow. There are no physical constraints as in the other decompositions. The only constraint derived from this design matrix is to follow the information flow sequence in design.

6. Check Design Decision Consistency (Level 1 and Level 2)

Because the two levels of FRs/DPs and every design matrix are determined, the consistency of the design decisions needs to be checked. For example, the first-level design matrix has $A_{21} = 0$. Does any one of DP_{1x} have a strong effect on FR_{2x}? If that is the case, either proper constraints need be imposed or the first-level design matrix needs to be reevaluated.

The following table shows the consistency-checking full matrix for the first-level and second-level designs. The particular example given below is related to the first two levels, where the concrete information is not yet available. Therefore, an effort is made to impose the proper constraints in order to maintain the higher level design decisions. The higher level design matrix is a 4 × 4 lower triangular matrix. The shaded cells indicate X elements in the first-level design matrix. The elements with superscripts indicate that a certain constraint is specified to ensure the design decision, and the constraint is described.

		$DP_1.\#$					$DP_2.\#$					$DP_3.\#$		$DP_4.\#$		
		1	2	3	4	5	1	2	3	4	5	1	2	1	2	3
$FR_1.\#$	1	X	0	0	0	0	0^1	0	0	0^1	0	0^8	0	0	0	0
	2	0	X	0	0	0	0^1	0	0	0^1	0	0^8	0	0	0	0
	3	0	0	X	0	0	0^1	0	0^2	0^1	0	0^8	0^4	0	0	0
	4	0	0	0	X	0	0^1	0	0	0^1	0	0^8	0^5	0	0	0
	5	0	0	0	0	X	0^1	0	0	0^1	0	0^8	0	0	0	0
$FR_2.\#$	1	0^1	0	0	0^1	0	X	0	0	0	0	0^8	0^6	0	0	0
	2	0^1	0	0	0^1	0	0	X	0	0	0	0^8	0	0	0	0
	3	0^1	0	0^2	0^1	0	0	0	X	0	0	0^8	0^7	0	0	0
	4	0^1	0	0	0^1	0	0	0	0	X	0	0^8	0	0	0	0
	5	0^1	0	0	0^1	0	0	0	0	0	X	0^8	0	0	0	0
$FR_3.\#$	1	X	X	X	X	X	X	X	X	X	X	X	X	0	0	0
	2	X^3	X^3	X^3	X^3	X^3	X^3	X^3	X^3	X^3	X^3	X	X	0	0	0
$FR_4.\#$	1	X	X	X	X	X	X	X	X	X	X	0	X	X	0	0
	2	X	X	X	X	X	X	X	X	X	X	0	X	X	X	0
	3	X	X	X	X	X	X	X	X	X	X	0	0	X	X	X

Constraints

1. Thermal effect from the modules of bake processes to other modules in the system must be prevented; the system must either use thermal shields or have appropriate layout.
2. Spin module must not affect others in terms of vibration, particle generation, etc.
3. Standard mechanism for wafer hand-off at the different process modules will minimize the effect of these elements.
4. Having no delay in wafer transport from a coating module to a soft-bake module is preferred. If delay is inevitable, it should be repeatable for consistent on-wafer result.
5. Having no delay in wafer transport from a soft-bake module to a chiller module is strongly preferred. The delay, if any, should be minimized and repeatable.
6. Having no delay in wafer transport from a PEB module to a chiller module is one of the most important requirements because it affects the critical chemical reaction within the photoresist film.
7. If unavoidable, repeatable transport delay after developing process is desirable.
8. The design of the system layout must be done in such a way that the cross-effects among $DP_{1.x}$ and $DP_{2.x}$ are minimized.

The constraints listed above are derived from examining the above multilevel design matrix. Those constraints are considered in the subsequent decomposition process for each branch of hierarchy.

7. The Third-Level Decomposition for the Spin Coater

The third-level decomposition is done for each of the 15 branches from the second level and involves a fair amount of detail. In this section, a few of the branches are selected for illustrative purposes. As an example of process module design, decomposition of the spin coater (FR_{13}/DP_{13}) is shown. Decomposition of system layout (FR_{31}/DP_{31}) is presented on behalf of system configuration design. Finally, operation controller (FR_{42}/DP_{42}) is decomposed to demonstrate system software design.

Index: 1.3.#			Spin Coater Branch	
	Functional Requirements (FRs)		*Design Parameters (DPs)*	
	Name	**Description**	**Description**	
		Coat wafer with photoresist	*Spin Coater*	
1	*Process*	Send/receive wafer from transport device	Wafer I/O mechanism, which changes	
2	*Process*	Deposit resist onto wafer surface	Resist delivery/deposit system	
3	*Process*	Spin wafer with desired velocity profile	Spin system	
4	*Process*	Clean wafer edge with solvent material	Solvent delivery/deposit system for top and bottom sides	
5	*Process*	Dispose of wasted material	Material disposal system	
6	*Process*	Control local environment	Environmental factor controller	
7	*Control*	Control coating process functions	Local process controller	

Description of FRs and DPs

At the highest level, the parent FR_1 is to coat wafers with desired resist film and the parent DP_1 is the coating-process module. At the second level of decomposition of the coating-process module, FR_{13} is to coat wafer with resist and DP_{13} is the spin coater. For the third-level decomposition of the spin coater, seven sub-FRs/sub-DPs are identified. These FRs capture key functions that a spin coater has to perform, and consequently their DPs identify the essential features of a spin coater.

- **FR_{131}**: The spin-coating process starts with accepting a wafer from a certain type of transport robot of DP_{32x}, and ends with sending the wafer out. It should provide the robot with as easy access as possible.
- **DP_{131}**: The wafer in/out (I/O) mechanism is responsible for receiving/sending a wafer from/to a transporter. Basically, it moves a wafer between process position and I/O position. Alternatively, it could be a mechanism that makes the spin coater accessible by the robot without changing the wafer position. In our design, the former concept is selected and designed in detail with further decomposition.
- **FR_{132}**: Once a wafer is in process position, the resist solution is to be flooded over the substrate. The initial resist application must be accomplished with precise position control in order to obtain a liquid film of uniform thickness. The resist material should be under tight temperature control before/during application because resist temperature has a huge effect on uniformity and a significant effect on mean thickness.
- **DP_{132}**: The resist delivery/deposit system is primarily responsible for FR_{132}. It is in charge of the delivery of the resist solution from reservoir to deposit device and of the application of the resist over the substrate. To satisfy the FR, further decomposition should focus on precise position control and resist-solution temperature control.
- **FR_{133}**: With the resist solution applied to the substrate, a wafer has to spin to flood the solution over the entire surface uniformly and subsequently to dry it at a constant rate to yield a uniform, solid polymeric film. The initial coating can be accomplished either by performing FR_{132} first prior to spinning (static dispense) or by performing FR_{132} and FR_{133} simultaneously (dynamic dispense). For both cases, the spinning-speed profile should be precisely determined. It is known that spinning speed has a strong dominant effect on film thickness and a significant effect on the uniformity of the coated film.

Spinning-speed range: 0–7,500 rpm
Spinning speed control: $\pm$ 1 rpm in the range of 10–5000 rpm

- **DP_{133}**: The spin system must satisfy a high-accuracy speed requirement. From the constraints stated at the second-level decomposition, minimum vibration is required for the design of DP_{133}.
- **FR_{134}**:[4] During the spinning process, excess resist is deposited on the edges and backside of the wafer, which can become a source of particle contamination in subsequent process steps. The excess resist needs to be removed for better quality image.
- **DP_{134}**: A bead-removal technique is required to achieve FR_{134}, which is done by applying solvent to the edges in a precisely controlled manner. DP_{134} is basically similar

---

[4] If imaging is needed on the outer edges, removal of the bead is not acceptable.

to DP_{132}; it delivers solvent instead of resist solution and deposits solvent onto unwanted polymer.

• **FR_{135}**: This is a "non-value-added" FR, but still indispensable. The spin coater needs to handle the splashed material, such as the resist solution and solvent.

• **DP_{135}**: A material-disposal system is responsible for collecting the splashed material and disposing it out of the system. The design of DP_{135} must guarantee minimum splashback of the splashed material. Other considerations are discussed in the constraints section.

• **FR_{136}**: Because the resist coating is dominated by evaporation, and the resist solution is initially applied in a sticky liquid state, strict environmental control is required. The factors to be controlled are temperature, humidity, airflow, and cleanliness.

Ambient air temperature range: 18°C–30°C
Ambient air temperature control: ± 0.1°C
Air humidity: 35–45%
Air humidity control: ± 0.5%
Environment: class 0.1

• **DP_{136}**: The environmental-factor controller is primarily responsible for controlling temperature, humidity, airflow, and particle contamination.

• **FR_{137}**: Each of the coating-process functions should be coordinated and controlled. This involves sequencing/synchronizing the functions and setting local process parameters.

• **DP_{137}**: The local process controller is responsible for coordinating FR_{13x} and their sub-FRs in conjunction with the top-level controller, DP_4 and sub-DP_4, and also with the lower level controller, which is included in the decomposition of FR_{13x}/DP_{13x}.

The information given at this level is summarized as follows:

• DP_{132}, resist delivery/deposit system, and DP_{134}, solvent delivery/deposit system, are placed over a wafer during its deposit process.
• DP_{133}, spin system, generates a certain amount of heat and particles.
• DP_{135}, material disposal system, has to enclose the main body of the spin coater to block splashing resist solvent.
• DP_{136}, environmental-factor controller, is primarily in charge of temperature, humidity, airflow, and particle contamination control throughout the process.

Design Matrix

$$\begin{Bmatrix} FR_{131} \\ FR_{132} \\ FR_{133} \\ FR_{134} \\ FR_{135} \\ FR_{136} \\ FR_{137} \end{Bmatrix} = \begin{bmatrix} X & 0 & 0 & 0 & 0 & 0 & 0 \\ 0 & X & 0 & 0 & 0 & 0 & 0 \\ 0 & 0 & X & 0 & 0 & 0 & 0 \\ 0 & 0 & 0 & X & 0 & 0 & 0 \\ 0 & 0 & 0 & 0 & X & 0 & 0 \\ 0 & X & X & X & X & X & 0 \\ X & X & X & X & 0 & 0 & X \end{bmatrix} \begin{Bmatrix} DP_{131} \\ DP_{132} \\ DP_{133} \\ DP_{134} \\ DP_{135} \\ DP_{136} \\ DP_{137} \end{Bmatrix}$$

The design matrix is triangular, indicating a decoupled design. The foregoing decomposition needs to follow the sequence indicated in the design matrix. It should be noted that the coating process is highly sensitive to the temperature and humidity of the environment and the temperature of the resist solution. All of the sources affecting them need to be minimized. In terms of the design matrix, the off-diagonal B_{62}, B_{63}, B_{64}, B_{65} must be minimized.

Elements of the Design Matrix

A_{31}: The wafer I/O mechanism should not affect the performance of the spinning function.

A_{62}: The resist delivery/deposit system, specifically the deposit device, has an effect on airflow control when it is located on top of the wafer surface. The resist-solution dispensing device should have minimum effect on airflow over the wafer surface. A constraint can be imposed on the shape and size of the dispensing device.

A_{63}: The spin system generates a certain amount of heat while spinning, which affects temperature control. Particle generation during its spinning process affects particle contamination control. Heat generation in the spin system, either from actuator or from friction, must be minimized or the heat should not be transferred to the wafer-processing area. The spin system design should prevent any particle from coming out to the environment.

A_{64}: The solvent delivery/deposit system, specifically the deposit device, has an effect on airflow control when it is located on top of the wafer surface. The effect of DP_{134}, however, is smaller than that of DP_{132}, because DP_{134} operates after the resist-coating process is done. Airflow at that point may not be as significant as that in the midst of resist coating. The solvent-dispensing device should have minimum effect on airflow over the wafer surface. A constraint can be imposed on the shape and size of the dispensing device.

A_{65}: The enclosure of the material-disposal system around the spinning system affects the pattern of airflow. Recirculating flow may also cause a particle contamination problem. The enclosure of the disposal system should be designed such that it generates minimum recirculation of airflow.

A_{71}, A_{72}, A_{73}, A_{74}: The local process controller is designed to control the operation of each DP. Change of any of those DPs potentially requires redesign of certain features of DP_{137}.

Constraints

At the second-level design, the constraint related to the coater module is that neither mechanical vibration nor particle generation from DP_{13} must be transferred to DP_{11}, DP_{12}, DP_{14}, or DP_{15}.

The above constraint is not specifically addressed in the third-level design, and needs to be considered for the next-level decomposition.

8. The Third-Level Decomposition for the System Layout

Description of FRs and DPs

- **FR$_{311}$**: We define TAKT time as the time interval (in seconds) between two consecutive wafers coming out of the system. Thus, the TAKT time of the system is directly driven from the required throughput.

Index: 3.1.#		System Layout Branch	
	Functional Requirements (FRs)		*Design Parameters (DPs)*
	Name	**Description**	**Description**
		Support the system physically	*System layout*
1	*Support*	Keep TAKT$_{\text{process } i}$ below TAKT$_{\text{sys}}$	Number of each process module
2	*Support*	Maintain number of movements by main robot not to degrade target throughput	Number of IBTA
3	*Support*	Locate process modules into 200-APS frame	Layout (module arrangement)

$$\text{TAKT}_{\text{sys}}(\text{seconds}) = \frac{1}{\text{Throughput (WPH)}} \times 3600$$

For example, if the required throughput is 60 WPH, then the TAKT time of the system is 60 seconds. To meet the throughput requirement, the TAKT time for every process step must be less than or equal to the system TAKT time.

- **DP$_{311}$**: TAKT time is computed for a given process time, number of modules, and transport time. It is termed a fundamental period (FP).[5] For process i, FP$_i$ is defined as

$$\text{FP}_i = \frac{\text{PT}_i + 4T}{N_i}$$

where PT$_i$ =process time, N_i =number of process modules, and T =transport time. T is more precisely defined as the sum of translation time and the time for wafer handling before being ready to move to another module.

The condition to satisfy system throughput is

$$\text{FP}_i \leq \text{TAKT}_{\text{sys}}$$

Combining the above two equations,

$$N_i \geq \frac{\text{PT}_i + 4T}{\text{TAKT}_{\text{sys}}}$$

Therefore, N is selected as DP$_{311}$ to satisfy FR$_{311}$. However, as is clear from the equation, N involves transport time, which is determined from DP$_{32x}$. This is the source of the coupling indicated in the second-level decomposition for system configuration.

- **FR$_{312}$**: The transport module is simply a process module with a certain process time, from the perspective of system throughput. Yet, TAKT time for a transport module is slightly different from that of a process module.

Within a period of TAKT$_{\text{system}}$, a transport module must complete all required wafer moves. TAKT time of a transport module is defined as the sum of transport time for all moves. For example, suppose that there are two process steps. Then a total of three moves must be performed in one cycle, which is TAKT$_{\text{system}}$. The TAKT time

---

[5] The concept of fundamental period is presented in Perkinson, Gyurcsik, and McLarty (1996).

of a transport module is $\tau_1 + \tau_2 + \tau_3$ or, for simplicity, $3 \times \tau$. If $\text{TAKT}_{\text{system}}$ is 30 seconds, the necessary condition to meet the throughput requirement is $\tau \leq 10$ seconds.

Now that the transport time is mostly limited by transport module design, although the time is also a function of location (DP_{313}), there are some instances when $\text{TAKT}_{\text{transport}}$ exceeds $\text{TAKT}_{\text{system}}$. In that case, $\text{TAKT}_{\text{transport}}$ must be reduced until it is less than system TAKT time.

- **DP_{312}**: The definition of $\text{TAKT}_{\text{transport}}$ is

$$\text{TAKT}_{\text{transport}} = N \times \tau$$

where N is number of moves and τ is $2T$.

To reduce $\text{TAKT}_{\text{transport}}$, we have to decrease either τ or N. Decreasing τ is related to DP_{312} performance improvement and is not easily achieved. Decreasing N is decreasing the number of moves that one (main) transport module has to perform by assigning a small and cheap transporter for some of the moves. It should be noted that $\text{TAKT}_{\text{transport}}$ involves transport time, and both of the options to reduce it are dependent on the transport time—another source of the coupling indicated in the second-level decomposition.

The company's design selected decreasing N because of cost of the main transport device, layout design (DP_{313}), etc. They added a number of subsidiary transport devices to the system. The device is named the Inter-Bay Transfer Arm (IBTA). IBTA is a relatively simple transport device that handles wafer moves between two consecutive thermal process modules. Yet it turns out that the introduction of IBTA causes another coupling, resulting in system configuration inflexibility.

- **FR_{313}**: A number of process modules (DP_{311}) need to be allocated into proper geometric position within the system.
- **DP_{313}**: The system adopts a "bay-and-spin station" frame structure. It is divided into two portions: a thermal stack on one side and a spin station on the other. The previous-level constraint, minimization of thermal cross-talk, must be considered in the design of the frame. The arrangement of modules in the frame is also considered.

Design Matrix

$$\begin{Bmatrix} \text{FR}_{311} \\ \text{FR}_{312} \\ \text{FR}_{313} \end{Bmatrix} = \begin{bmatrix} X & 0 & 0 \\ 0 & X & X \\ X & X & X \end{bmatrix} \begin{Bmatrix} \text{DP}_{311} \\ \text{DP}_{312} \\ \text{DP}_{313} \end{Bmatrix}$$

The design matrix is coupled at this level. It turns out that the coupling results from the adoption of IBTA and the inflexible nature of the bay-and-spin station structure. Note that the previous-level design matrix is also coupled, and causes of the coupling are mentioned in the description of DP_{311} and DP_{312} above.

Elements of the Design Matrix

A_{21}, A_{12}: The number of modules is dependent only on the throughput requirement and process time (plus transport time), and the number of moves a transport device must make is dependent on process flow. Mathematically,

$$A_{21} = \frac{\partial \mathrm{FR}_{312}}{\partial \mathrm{DP}_{311}} = \frac{\partial N\tau}{\partial N} = 0$$

$$A_{12} = \frac{\partial \mathrm{FR}_{311}}{\partial \mathrm{DP}_{312}} = \frac{\partial \{(\mathrm{PT}_i + 4T)/N_i\}}{\partial (\text{Number of IBTA})} = 0$$

A_{31}: This element indicates how many modules are in the system. DP_{311} is the direct object addressed by FR_{312}.

A_{32}: DP_{312}, IBTA, is a simple device performing transport between two thermal modules. Because of the limited capability, two consecutive modules must be located side by side once IBTA is selected as a means of wafer transfer. This limitation affects FR_{313} and decreases the freedom in allocating the modules in the frame.

A_{23}: Because the space for juxtaposition of consecutive thermal modules is limited in the 'bay-and-spin station' structure, determining the number of IBTA is not free from DP_{313}.

9. The Third-Level Decomposition for the Operation Controller

Index: 4.2.#		Operation Controller Branch	
	Functional Requirements (FRs)		*Design Parameters (DPs)*
Name	**Description**		**Description**
	Command/control system operation		*Operation controller*
1 CCA	Initialize the system		Initiate-mode algorithm
2 CCA	Acquire operational data		Read/refer to the recipe database
3 CCA	Compile module-level information		Process data collector
4 CCA	Command the "action"		Event handler/scheduler
5 CCA	React to malfunction		Error handler

Description of FRs and DPs

Operation of the system involves three major activities: initialization, routine operation, and error handling. The FRs identified here address those three activities.

- **FR$_{421}$**: Before starting the operation, the system must be initialized, including both hardware and software initialization.
- **DP$_{421}$**: The inititate-mode algorithm is responsible for coordinating the hardware/software during startup procedures.
- **FR$_{422}$**: To perform the operation, the operation controller must have data regarding all of the operations.
- **DP$_{422}$**: The software module of reading/referring is responsible for acquiring the data from the recipe database (DP$_{4111}$ and DP$_{4112}$).
- **FR$_{423}$**: All of the necessary information related to each process module state needs to be known by the operation controller.

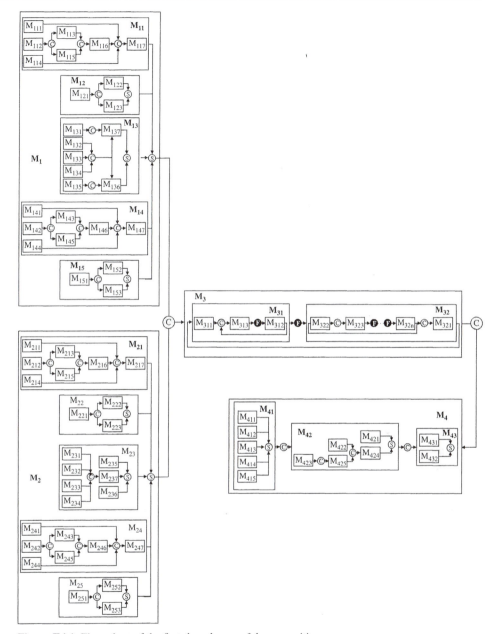

Figure E4.1 Flow chart of the first three layers of decomposition.

- **DP$_{423}$**: The software module is responsible for collecting the information.
- **FR$_{424}$**: Various commands for the operation are to be generated.
- **DP$_{424}$**: These software modules (algorithms) are responsible for generating system commands.

• **FR$_{425}$**: When there is malfunction within the system, the operation controller has to react to the malfunction.

• **DP$_{425}$**: The error handler is primarily responsible for coordinating the system during malfunctioning.

Design Matrix

$$\left\{ \begin{array}{c} FR_{421} \\ FR_{422} \\ FR_{423} \\ FR_{424} \\ FR_{425} \end{array} \right\} = \left[\begin{array}{ccccc} X & 0 & 0 & 0 & 0 \\ 0 & X & 0 & 0 & 0 \\ 0 & 0 & X & 0 & 0 \\ 0 & X & X & X & X \\ 0 & 0 & X & 0 & X \end{array} \right] \left\{ \begin{array}{c} DP_{421} \\ DP_{422} \\ DP_{423} \\ DP_{424} \\ DP_{425} \end{array} \right\}$$

Elements of Design matrix

A_{42}: DP$_{422}$ acquires recipe data and transfers it to DP$_{424}$ such that it can generate operation command based on the data.

A_{43}: Process data are transferred to DP$_{424}$, the event handler.

A_{45}: The event handler should acquire error information from DP$_{425}$, error handler, and deal with error situation.

A_{53}: Process data are transferred to DP$_{425}$, the error handler.

10. Graphic Representation of System Architecture

Part of the system architecture for the track system is represented using the flow chart shown in Figure E4.1, which shows the system architecture with the three decomposition levels. The simple format of the flow chart provides visual information about the structure of the design decisions without reexamining all of the design matrices.

4.7 MATHEMATICAL MODELING, SIMULATION, AND OPTIMIZATION OF SYSTEMS

Much of the discussion so far has been qualitative in the sense that the design equations were given in terms of X and 0. However, we must model the system to make the design equations that relate FRs to DPs quantitative in order to have confidence in the design decisions we made. The modeling effort will improve the original qualitative decisions made. Based on the model, each X in the design matrices can be replaced with an analytical expression. This should be done at the level of leaves. Because the higher level modules are obtained by combining the leaf-level modules after all the leaf-level modules are expressed analytically, a comprehensive quantitative model for the entire system can be developed. The system model thus created can be used to simulate the behavior of systems and evaluate the effectiveness of the system designed.

Optimization is often a process of obtaining the most acceptable values for all FRs through trade-offs or compromises. However, a rational design that satisfies the independence of functions will not require compromise because the FRs of such a system are independent of each other. This should also be the case for nonlinear designs. In a design that satisfies the independence of functions, the optimization of each FR can be attained without adversely affecting other FRs. If a design is coupled, compromising of FRs is necessary. Such compromise is normally achieved by optimizing some FRs at the expense of other FRs.

4.8 APPLICATION OF THE FLOW DIAGRAM OF THE SYSTEM ARCHITECTURE

What is the system architecture good for? The system architecture for the software represented by the flow chart can be used in many different applications and for many different purposes. Some of these applications are related to system design and operation:

1. Diagnosis of system failure—The failure of a complex system (be it hardware, software, hardware/software, or manufacturing) can be diagnosed based on the system architecture. Once the flow chart is available, it can be used to trace the probable cause of failure. The flow diagram should also reduce the service cost of maintaining software systems.

2. Engineering change orders (ECO)—During the development process, it is inevitable that some of the design decisions made in the past must be changed. The system architecture and the flow diagram can be used to identify all other modules that must also be changed when a change is made in one of the modules.

3. Job assignment and management of a system-development team—The flow diagram can be used to organize development projects and to assign tasks to various participants because it defines the interrelationships among various modules. This will aid in project execution and management.

4. Distributed systems—When a large development project is executed, the lower level tasks are performed by many people located at distant places, sometimes involving subcontractors and subsubcontractors. The flow diagram of the system architecture will provide the road map by which the tasks can be coordinated.

5. System design through assembly of modules—Ultimately, a system may be designed by assembling existing modules in a database. The system architecture will provide a guideline for system development.

6. System consisting of hardware and software—When a complex machine is being designed, some of the modules may be hardware and some software. The flow diagram shows how the software and hardware must be integrated to have the system function as an integrated unit.

4.9 ON HUMAN–MACHINE INTERFACE

Most systems involve human beings as part of a system. How do we deal with human–machine interface? The human–machine interface has been studied in relation to automotive vehicles, nuclear reactors, aeronautics, astronautics, and other fields (Sheridan, 1974). The studies have dealt with the interaction of human operators with specific machines. They have modeled the interactions, sometimes analytically and sometimes based on the results of simulations, to understand and create an ideal human–machine interface.

From the axiomatic design point of view, human–machine interactions can occur at all levels of a design hierarchy. In the case of modern aircraft, the human–machine interface is at the highest level—between the pilot and the aircraft cockpit. The rest of the system is made up of hardware and software.

In the case of large factories, human–machine interactions may occur at all levels—machinists running automated machines, who must interface with automatic guided vehicles (AGV), which in turn interface with many other hardware or software modules. In such a system, the human–machine interface can be embedded at all levels of the hierarchy. Therefore, in these systems, the system functions depend on hardware, machine-to-machine interactions, and human–machine interactions at many different levels.

Large systems involving many human–machine interfaces have both positive and negative factors. The positive aspects are related to the fact that human intelligence is a powerful module that can easily relate the DPs to FRs without the need to model the system and to depend on software systems. Human operators can be trained. They also generate operational knowledge from experience. Human operators are flexible and can acquire a database. Human operators provide both intelligence and motive power.

On the other hand, the shortcomings of the human–machine interface are that the human operator is not precise and sometimes may not be dependable. The probability of error by human operators is greater than that by well-proven hardware/software systems. Therefore, the system must be able to detect and compensate for human operators when the consequence of a system failure cannot be tolerated.

The system architecture, such as the one shown in Figure 4.7, can have many modules at many levels that are performed by human operators. In the extreme case, the entire system can be made of modules, all of which are performed by human operators, e.g., government agencies and consulting companies. Many management organizations represent systems with only human modules.

4.10 SUMMARY

In this chapter, axiomatic design of systems is considered. Systems can be products, software, processes, systems, and organizations. These systems are physically different, but functionally alike. Therefore, system design can be generalized for all of these physically different systems. Some systems perform only a fixed set of functions. Other systems must perform different sets of functions over their lifetime.

To design a system, we must construct a database or a knowledge base. This database can be used to construct a system to satisfy a given set of functional requirements. During this process, the Independence Axiom must be satisfied. In selecting the right set of DPs, we must be concerned about the Information Axiom to be certain that DPs with the highest probability of success are chosen.

Large systems are those with a large number of functional requirements at the highest level. Large systems are further subdivided into large fixed systems and large flexible systems. Large fixed systems can be designed through the same design process as that used for small systems. However, the design process for large flexible systems requires different design processes because different sets of functional requirements must be satisfied at different points in time. The design and operation of large flexible systems require a large database (or knowledge base). The selection of DPs and PVs must be consistent with the design axioms.

System representation is important for a variety of reasons. There are three different kinds of representation of systems:

1. FR/DP/PV hierarchies with appropriate design equations
2. Module-junction diagram
3. Flow diagram

Methods of creating these system architectures are given. The flow diagram for the system architecture of large systems is based on design hierarchies and design matrices.

The general methodology presented here for system design will be applied to various specific systems in later chapters.

REFERENCES

Boothroyd, G., and Dewhurst, P. *Product Design for Assembly*, Boothroyd & Dewhurst Inc., Wakefield, RI, 1987.

Do, S.-H., *Axiomatic Design Software*, MIT Axiomatic Design Group Report, unpublished, 1998.

Kim, S. J., Suh, N. P., and Kim, S.-K. "Design of Software Systems Based on Axiomatic Design," *Annals of the CIRP*, Vol. 40, No. 1, pp. 165–170, 1991 (also *Robotics & Computer-Integrated Manufacturing*, 3:149–162, 1992).

Perkinson, T. L., Gyurcsik, R. S., and McLarty, P. K. "Single-Wafer Cluster Tool Performance: An Analysis of the Effects of Redundant Chambers and Revisitation Sequences on Throughput," *IEEE Transactions on Semiconductor Manufacturing*, Vol. 9, pp. 384–400, 1996.

Rechtin, E. *Systems Architecting: Creating and Building Complex Systems,* Prentice-Hall, Englewood Cliffs, NJ, 1991.

Sheridan, T. *Man-Machine Systems: Information, Control, and Decision Models of Human Behavior*, MIT Press, Cambridge, MA, 1974.

Suh, N. P. *The Principles of Design*, Oxford University Press, New York, 1990.

Suh, N. P. "Axiomatic Design of Mechanical Systems," *Special 50th Anniversary Combined Issue of the Journal of Mechanical Design and the Journal of Vibration and Acoustics, Transactions of the ASME*, Vol. 117, pp. 1–10, 1995a.

Suh, N. P. "Design and Operation of Large Systems," *Journal of Manufacturing Systems*, Vol. 14, No. 3, pp. 203–213, 1995b.

Suh, N. P. "Design of Systems," *Annals of CIRP,* Vol. 46, No. 1, pp. 75–80, 1997.

HOMEWORK

4.1 Develop a system architecture for trading of stocks. First establish functional requirements and create the design hierarchy.

4.2 A major office equipment manufacturer is interested in developing a system that can remotely monitor the performance and the remaining life of their machines in use at their customer sites. This will do several things for the company. It will reduce the cost of sending service people, who cost about $170.00 an hour including wages, overhead, transportation, and other associated costs. It will also enhance the reputation of the company by increasing customer satisfaction because preventive maintenance can be performed on individual machines when the service is needed before the machine fails to function. They may also be able to check the performance and life of the machines before they are shipped. Your task is to design a system that can perform these desired functions for this company.

4.3 Design a system for a new Internet company that plans to market sporting goods (such as golf clubs and tennis rackets) over the world-wide web. Show the flow chart for your design.

4.4 In 1999, AT&T bought many cable television companies with the view of using the cable network wires to provide local telephone service, long-distance telephone service, television programs, and home shopping service. Design the first four layers of a system that AT&T needs to achieve their ambitious business plan. Show the flow chart for your design.

4.5 Design a university system that will teach students using only the Internet.

4.5 Design a paper copier based on xerography.

4.6 Prove Theorem S1.

4.7 Prove Theorem S2.

4.8 Prove Theorem S3.

4.9 Prove Theorem S4.

4.10 Prove Theorem S5.

4.11 Prove Theorem S6.

5 Axiomatic Design of Software

5.1 INTRODUCTION

This chapter presents the design of software systems based on axiomatic design (AD) theory, distinct from specific programming languages or computer algorithms. It provides a framework for software design for all systems, including those that involve only software and those that involve both hardware and software. The framework for software design is based on the same set of principles and methodologies of axiomatic design that were presented for products, processes, and various systems in the preceding chapters. This framework provides a general approach to designing software and software architecture regardless of the specific computer language used for programming.

Software design based on axiomatic design is self-consistent, provides proper interrelationship and arrangement among modules, and is easy to change, modify, and extend. This is a result of making correct decisions at each stage of mapping and decomposition.

The final design of software is represented by a flow chart that displays the entire system architecture of the software, which can aid software programmers. The flow chart can also be used as a management tool during the software development phase. It provides clear guidelines to software engineers engaged in a collaborative development effort and gives the order of execution of the resulting program. Extensionality and reusability at any level are guaranteed when the software design is based on axiomatic design theory. When a system consists of hardware and software, the system architecture defines the interface between the software modules and the hardware modules.

As a case study, the development of Acclaro[1]—a commercial software system to aid designers who use axiomatic design—is presented. This software design is based on axiomatic design and is implemented using a modified version of object-oriented techniques (OOT) and the Java programming language, a platform-independent language. The use of OOT was necessary because the use of Java requires OOT. When axiomatic design is used, OOT can be considerably simplified.

Many other case studies—software for ABS used in automobiles, design of television tubes—are presented to illustrate the application of axiomatic design in designing hardware/software systems.

What are the key issues in software programming? Software is now ubiquitous in our modern life-style. It has transformed the way people live, work, and think. Even the way consumers purchase a variety of goods is undergoing changes, as evidenced by the latest Wall Street excitement created by the Internet stocks and electronic commerce. Software has also changed the inner workings of organizations and the governance of nations. It has changed the industrial infrastructure, hotels, transportation, and education. It forms the core of many machines, systems, and organizations. Indeed when software crashes, the world comes almost to a standstill until the problems are fixed. All the frantic activity to deal with the Y2K bug is an indication of the importance of computers in all sectors of the economy.

Software performs important functions in many physical products. Software controls individual components of a machine as well as integrating and controlling all functions of a hardware system. In some large systems consisting of hardware, sensors, and software, the software is embedded among hardware components and also integrates the system functions. Smart weapons, automobiles, machines, airplanes, instruments, telecommunications, and space stations would not exist without software. Even the functions of ordinary products, such as toys, movies, and appliances, are controlled by software. Software has replaced many mechanical components to give lower cost products that are more flexible and intelligent. The software in engine management systems (EMS) regulates the performance of automobile engines and matches the impedance of engines with that of the automatic transmission.

In many companies, the hardware part of the system is developed first and then turned over to the software engineers to develop software code to make the system function. This process does not work well, causing delays in product introduction. This current practice is also costly because many debugging and field-service problems are associated with or caused by poor interface between hardware and software. These symptoms are the result of ad hoc, heuristic approaches to software design. In addition, more and more complex applications have become possible because computers have become faster, cheaper, and more powerful. As a consequence, computational efficiency has become less of an issue relative to the cost of developing software and the reliability of the software. This low productivity associated with software development can be attributed to the lack of a proper design procedure and basic decision-making criteria.

To meet the ever-increasing need for software systems for all of these diverse applications, the software industry has become the most rapidly growing industry in the United States. The demand for software engineers is far greater than the supply as we begin the

---

[1] Trademark of Axiomatic Design Software, Inc., Boston, MA.

twenty-first century. American companies routinely hire programmers from all over the world, especially India, China, Russia, and Korea. Indeed, software has become a global industry, in part driven by the need to reduce the labor cost of software programming. To increase the productivity of software development, a systematic means of creating and maintaining software through the use of a design theory must be implemented.

How can axiomatic design be used in software design? The axiomatic approach to software design is intended to overcome some of the known shortcomings of the current practice and create a scientific framework for software synthesis. In many respects, it differs significantly from various methodologies used in software engineering:

1. Its framework is based on the idea that the software must be designed first to satisfy a set of functional requirements and constraints before commencing coding activities. It also provides a methodology for design that includes the idea of mapping, decomposition, and the creation of leaves.

2. The most important aspect of axiomatic design is that it provides a rigorous means of decomposition and the criteria for acceptable software design. Therefore, it is explicit and thus simple. The flow chart—a consequence of axiomatic design—describes the precise relationship between all modules, and provides a complete description of the software architecture and the computational sequence.

3. The axiomatic design framework is equally applicable for all software designs, including (a) software that consists of only software modules, (b) software that controls both hardware and software, and (c) rapid prototyping software that allows modification based on current sensor input.

4. The axiomatic design framework is not field specific, i.e., it can be used in telecommunications, machine control, and others. The framework consists of all the elements of axiomatic design: the design domains, mapping, decomposition through zigzagging, creation of hierarchies, the design equation, the design axioms, and the design matrix. The additional elements introduced for system design in Chapter 4 are equally valid in software design: the creation of flow charts for software design, the concept of "leaf," and specific characteristics of each domain.

5. The flow chart for system architecture is generated as a result of the existence of the axiomatic design framework—in contrast to a heuristic approach that depends solely on the intuitive ideas and skills of individual software programmers. The use of the flow chart enables a large collaborative effort as the role of each programmer is explicitly identified by the flow diagram.

6. Axiomatic design creates software for real-time control of systems in an explicit manner. Thus the software can be efficient, minimizing the information content of the system.

7. Axiomatic design is a good framework for modeling systems that consist of hardware and software because the design matrix defines the FR/DP relationship that needs to be modeled.

8. Axiomatic design allows reusability and extensionality of software modules in an explicit manner.

What is software? Software may be defined as an information transformer that generates certain desired outputs given a set of inputs and an algorithm for execution within the bounds established by constraints.

Software development has typically consisted of the following stages:

1. Analysis of the need and gathering of requirements, functions, data, tools, and various information
2. Design—abstract design of the software
3. Coding—creating programs according to the design
4. Debugging—correcting errors
5. Testing—site experiments
6. Maintenance—customer service after release
7. Extension—major extension of the developed software to add new functions and features

The cost of correcting and changing software escalates as the software is coded at lower and lower levels. Therefore, significant effort should be expended to design the software correctly and completely before undertaking any coding activity. Axiomatic design is a particularly powerful tool for design of complex software systems.

A historical perspective on software technology development. Once a software system has been designed, it has to be coded in a suitable programming language. Programming language has evolved over the years. A brief history of programming language at this time may be appropriate.

Most popular programming languages of the late 1990s use the object-oriented techniques (OOT). However, even the use of OOT and modular design does not guarantee reliable software unless the coding is preceded by axiomatic design of the software. In fact, because OOT does not have a fundamental basis for design, it has created many complicated ways of describing the design with many definitions, some of which describe variants of the same thing, e.g., class, object, and behavior. Apparently this is done because OOT does not have an explicit means of dealing with decomposition of FRs and DPs as well as their relationships. In the following section, OOT will be described, followed by a slight modification of OOT based on axiomatic design to simplify the software structure.

OOT is the latest in the history of computer languages, which have evolved with changes in software design methodology. During the mid-1970s and 1980s, software designers used structured analysis and design techniques (SADT) to design software, and wrote code using BASIC, FORTRAN, and C programming languages (DeMarco, 1979; Ross, 1977, 1985). Commercial software products were generated using these languages, which were appropriate in representing the idea of SADT. However, it was difficult to reuse some of the modules created using SADT, as the attributes (i.e., programming variables) were used concurrently in various modules.[2] The global variables such as COMMON block in FORTRAN and Header file in C coupled the modules. Therefore, to reuse some of the modules that were developed previously, they had to be modified. Consequently, reusability became a key issue, as it affected the productivity of software development.

Object-oriented methodology was introduced to overcome this limited modularity and productivity of SADT (Cox, 1986). However, OOT had remained merely a concept until programming languages such as C++ and Java were developed. In these programming

---

[2] The term "module" used in software engineering may be different from the definition given in Chapter 4, which is restated later in this chapter. The definition used in axiomatic design enables the design of complex systems.

languages, the attributes and methods are combined together to act like an object in the real world. No global variables are used in C++ or Java.

C++ and Java are structured quite differently from older languages, such as FORTRAN and C. Knowing how to program in FORTRAN or C offers only minimal advantages when developing software with C++ or Java. In order to develop software products with C++ or Java, the programmer must know about *objects, classes*, and *relationships* as these languages are the only tools used to represent the object-oriented method. Objects, classes, and relationships structure the code itself. The Java language depends on object-oriented methodology.

Object-oriented techniques emphasize both the need to design software correctly during the early stages of software development and the importance of modularity. However, even with object-oriented techniques, there are many problems that software programmers face in developing and maintaining software over its life cycle. Although there are several reasons for these difficulties, the main reason is that current software design methodologies do not provide any criterion for good software design.

What are the current shortcomings? In software engineering, it is well known that software has to be created with independent modules (Pressman, 1997). However, an assembly of independent modules does not in itself constitute effective software. To be good, software must be designed with functional independence in mind to make the relationship between the modules effective and explicit without coupling. The axiomatic design framework ensures that the modules are in the right place in the right order.

Most software products require extensive debugging when they are first developed and also require extensive maintenance during the course of use—both expensive propositions. When software must be modified or changed after the original programmer has left the organization, it is extremely difficult to understand the intention of the previous designer and to implement the changes. In fact, many software programs developed, especially with government money, cannot be used at all. To overcome these problems, we must make software engineering more of a science than an art.

What is the relationship between axiomatic design and object-oriented programming? A direct comparison of axiomatic design of software with the object-oriented programming is difficult. There are both similarities and differences. Axiomatic design enables the design of complex systems from the top down, whereas OOT is a bottom-up process. Axiomatic design requires a clear definition of FRs, as it divides the design tasks into domains and requires mapping between the domains for decomposition. Furthermore, the design matrix explicitly defines the relationships between sets of FRs and DPs. OOT mixes FRs and DPs.

Although OOT emphasizes modularity, not every combination of the separate modules constitutes an acceptable design, as some combinations may violate the Independence Axiom. OOT does not provide criteria as to how these objects can be combined to deal with design tasks that have many highest level FRs and how the modules should be related to ensure acceptable performance of the overall system. Perhaps the greatest shortcoming with object-oriented techniques is that the decomposition of the software has to be done in the heads of the programmers based on their experience, as they do not map or zigzag or write down design equations. OOT can be very cumbersome when there are many objects that must be dealt with at the same time. It uses various words to describe variants of the same thing to deal with issues that arise from decomposition of the FR, DP, and PV hierarchies.

Notwithstanding these shortcomings, object-oriented techniques have been used extensively to develop software systems. OOT has been incorporated in existing commercial codes such as the Java language. Therefore, by combining axiomatic design theory with a modified version of OOT, we can develop practical, productive, and reliable software systems, using the OOT tools.[3] This idea is explored further in Section 5.3.

5.2 AXIOMATIC DESIGN THEORY FOR SOFTWARE DESIGN

Is the design of software any different from the design of systems in general? From the axiomatic design point of view, a software system is a subset of system design, and in this sense, software design is not much different from the design of other systems discussed in Chapter 4. Axiomatic design of software is based on the Independence and Information Axioms, the existence of domains, mapping, decomposition through zigzagging, design axioms, the design matrix, and the creation of hierarchy. Several new ideas have been introduced to deal with system design, such as a *leaf*, the *module-junction structure diagram*, and the creation and use of a *flow chart* that defines the sequence of execution of the system. All of these ideas are equally valid in software design (Kim, Suh, and Kim, 1991; Do, 1998).

There are four domains in the software design world as shown in Figure 5.1. As discussed in Chapter 1, the relationship between the two adjacent domains is "what we want to achieve" and "how to achieve what we want to achieve." The "what" is represented by the domain on the left and the "how" is represented by the domain on the right.

The specific characteristics of CAs, FRs, DPs, and PVs depend on the specific nature or goals of the software and the "what"/"how" relationship as shown in the following:

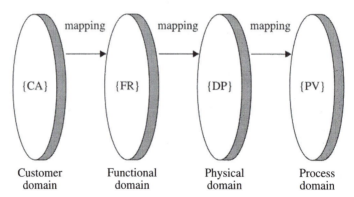

| | Customer domain | Functional domain | Physical domain | Process domain |

Figure 5.1 Concept of domains. The customer domain is where the user's needs or attributes are specified as CAs. The functional domain is where the desired software output or software specifications are given as FRs through mapping from the customer domain. FRs are equivalent to "objects" in OOT. In the physical domain, the DPs—equivalent to data in OOT—are obtained through mapping from the functional domain. DPs are typically inputs that control FRs through modules. In the process domain, the PVs are the actual codes, algorithms, etc. to accomplish the DPs.

---

[3] This was first done by Dr. Sung-Hee Do, while he was a postdoctoral researcher at MIT. He is now Vice President for Technology of Axiomatic Design Software, Inc.

CAs: The user's needs or customer attributes that the software program must satisfy.

FRs: The outputs, specifications, or requirements of a software system. FRs are "objects" in object-oriented programming.

DPs: 1. The inputs to modules in the case of pure algorithms. In the case of object-oriented programming, DPs can be treated as "data."

2. Signal from hardware sensors and application-specific integrated circuits (ASICs) in the case of systems that involve both software and hardware.

3. Program code that generates input to the modules.

PVs: 1. Subroutines

2. Machine codes

3. Compilers

Modules: The row of the design matrix (DM) is a "module" that corresponds to "relationship" when the DPs are provided.

5.2.1 Review of the Axiomatic Design Process for Software

Are the steps involved in designing software any different from those for other systems? The steps involved in designing software are identical to those used in designing other systems discussed in Section 4.4, but require a different knowledge base or database (Suh, 1995, 1997, 1998). For completeness, these steps are reviewed here.

STEP 1
Define FRs of the Software System

The first step in designing a software system is to determine the customer needs (CAs) or attributes in the customer domain that the software system must satisfy. Then the functional requirements (FRs) and constraints (Cs) of the software in the functional domain are determined to satisfy the customer needs. The FRs must be determined in a *solution-neutral environment*—defining FRs without thinking about DPs (i.e., the "how" or the solution). FRs must satisfy the customer needs with fidelity. The FRs are defined as the minimum set of independent requirements that the design must satisfy.

STEP 2
Mapping between the Domains and the Independence of Software Functions

The next step in axiomatic design is to map these FRs of the functional domain into the physical domain by identifying the design parameters (DPs). DPs are "how" we are going to satisfy specific FRs. DPs must be so chosen that they are consistent with the constraints. Once the DPs are chosen, designers must go to the process domain and identify the process variables (PVs), which are the "how's" with respect to DPs. PVs may be subroutines or machine codes or compilers.

Consider, for example, a decoupled software design at the leaf level that can be stated as:

FR_1 = Add FRs.
FR_2 = Delete FRs.

We are going to satisfy these FRs by developing two algorithms: an ADD algorithm and a DELETE algorithm. To achieve FR_1 (add FRs) by means of the software, we will choose

DP_1 to be an input that triggers the ADD algorithm. Similarly, we will choose DP_2 to be an input that triggers the DELETE algorithm. In this case, DPs are physical descriptors or data that trigger some action that enables the modules to produce FRs. Therefore, DPs may be written as

$DP_1 = $ Input that activates the ADD algorithm
$DP_2 = $ Input that activates the DELETE algorithm

Then the design equation may be written as

$$\left\{ \begin{array}{c} FR_1 \\ FR_2 \end{array} \right\} = \left[\begin{array}{cc} A_{11} & 0 \\ A_{21} & A_{22} \end{array} \right] \left\{ \begin{array}{c} DP_1 \\ DP_2 \end{array} \right\} \tag{5.1}$$

A_{21} is a nonzero element, because "Add FRs" must precede the need to "Delete FRs." Equation (5.1) may also be written as

$$FR_1 = M_1 \, DP_1$$

$$FR_2 = M_2 \, DP_2 \tag{5.2}$$

where M_1 and M_2 are modules[4] defined as

$$M_1 = A_{11}$$

$$M_2 = A_{21} \frac{DP_1}{DP_2} + A_{22} \tag{5.3}$$

M_1 is the ADD algorithm and M_2 is the DELETE algorithm.

During the mapping process, the design must satisfy the Independence Axiom, which requires that the independence of FRs be maintained. FRs are, by definition, a minimum set of independent FRs. The Independence Axiom requires that the functional independence be satisfied through the development of an uncoupled or decoupled design. In an ideal design, the number of FRs is equal to the number of DPs as per Theorem 4. This is a consequence of the Independence Axiom and the Information Axiom. In choosing DPs, we must also be mindful of the Cs and the information content.

STEP 3
Information Content for Software Systems to Select the Best Design

There can be many software designs that are equally acceptable from a functional point of view if they satisfy the Independence Axiom. However, one of these designs may be superior to others in terms of the Information Axiom, which states that the one with the highest probability of success (i.e., the lowest information content) is the best design.

The Information Axiom requires the minimization of the information content. According to the Information Axiom, the design that has the least information content is the best

---

[4] In Section 4.6.2, a module M_i is defined as the row of the design matrix that yields an FR when it is provided with the input of its corresponding DP, which may be expressed as

$$M_i = \sum_{j=1}^{j=i} \frac{\partial FR_i}{\partial DP_j} \frac{DP_j}{DP_i}$$

design. The information content is defined in terms of the probability of achieving the FRs. The relative merit of equally acceptable designs that satisfy functional independence is determined by measuring their information contents. The Information Axiom is a powerful tool in identifying the best design when there are many FRs that the software must satisfy.

How do we measure the information content of a software system? Conventionally, the complexity of software programs is measured in terms of the number of lines of software code or the central processing unit (CPU) time, as the number of lines is a function of the number of layers of decomposition and the number of the highest level FRs of the software. Unfortunately, these are not absolute measures of the inherent complexity of the software program, as less skilled software designers may use more lines of software code than highly skilled designers would to achieve the same objective. We need a more absolute measure for complexity of the software program than either the number of lines of code or the CPU time.

In Chapter 1, complexity is equated with the information content, which is defined in terms of the probability of achieving the specified FRs.[5] Thus, the design that has a lower probability of success is more complex and has a higher information content. This definition implies that two different software programs that satisfy the same highest level FRs equally well and have the same information content are equally complex, irrespective of the length of the software program. This defies common sense, as the shorter software program that uses less CPU time is likely to be faster and cheaper to use, although its information content is the same as that of a longer program. We may draw two conclusions from this. First, the longer software program is likely to have higher information content because the probability of making a mistake is greater as the length of the code increases. Second, the information content is not necessarily a measure of the efficiency or the cost of the computing, although there may be a weak inverse correlation between the efficiency and the information content.

Many software programs have "bugs," which means that the FRs are not fully satisfied some of the time. Programs with bugs have a finite information content, as they cannot satisfy the FRs under all conditions. Such a program runs some of the time, but once in a while, it crashes because of bugs. In this case, the information content may be computed by determining the probability of the "up-time" of the program.

In some cases, there is no design range of an FR for a software program—either it fulfills the FR or it does not. That means that the information content is either zero or infinity. When the design range can be specified for an FR, the information content can be determined by whether the actual output of the software meets the intended functional requirements.

STEP 4
Decomposition of {FR}, {DP}, {PV}, and Software Modules

As discussed in Chapter 4, FRs, DPs, and PVs must be decomposed to the leaf level, where the design can be implemented without further decomposition. The programming then consists of coding the rows of a design matrix, which are called modules. When these modules are combined according to the flow chart or the module-junction diagram, we have the complete software system. This is discussed further in Section 5.5.

---

[5] The issue of complexity is discussed further in Chapter 9.

STEP 5
Object-Oriented Programming and the Modules for Leaves

At the leaf level, any software programming language, e.g., C++ or Java, can be used to write the modules as long as they are internally consistent. Java requires the use of object-oriented techniques.

STEP 6
Design Matrix and Module-Junction Diagrams

In Chapter 4, a "module-junction structure diagram" is presented as a means of representing the relationship between the modules. The summation junction, S, is used to represent an uncoupled design; the control junction, C, is used for the decoupled junction; and the F junction is used for a coupled design. The same notation system will be used for software design.

STEP 7
Flow Chart Representation of Software System Architecture

In Chapter 4, the concept of flow chart (sometimes called flow diagram) is presented as a means of representing system architecture. The flow chart consists of only the modules of leaves. Of the three different ways discussed in Chapter 4 of representing the system—the hierarchical trees, module-junction diagram, and the flow chart—the flow chart is the most convenient means of representing a software system. Based on the flow diagram, individual modules can be developed and linked to create a software system that can yield the highest level FRs.

STEP 8
System Control Command (SCC)

The flow chart also provides the software implementation scheme, which is simple and graphic. The flow chart shows the sequence of computation/operation of the software to achieve the highest level FRs. To operate the software, the operation must always begin from the innermost modules that represent the lowest level leaves in each major branch. Then we move to the next higher level modules following the sequence indicated by the system architecture.

The execution of the system architecture may be done using the system control command (SCC), which is sometimes called the system command and control module. The sequence of operation may be stated as follows:

a. Construct the lowest level modules of all branches. Multiply these modules with appropriate DPs (or make these modules operational by providing DPs as inputs) to obtain lowest level FRs corresponding to "leaves."

b. Combining these lowest level FRs according to the system architecture, obtain the next higher level FR. Repeat this process until the highest level module is obtained.

c. Execute this procedure simultaneously at all branches where such an action can be achieved independently so as to minimize the operational time.

Sometimes the inverse process may be necessary. When the highest level FRs and their tolerances are specified, the allowable values for the lowest level DPs can be determined. However, the sequence of the programming should follow the process described in this section.

5.2.2 Application of the Flow Diagram

The system architecture for the software represented by the flow chart can be used in many different ways, some of which are listed below:

a. Diagnosis of software failure—The failure of a complex software system can be diagnosed based on the system architecture. The flow chart should reduce the cost of maintaining the software system.

b. Software change orders—During the development process, it is inevitable that changes will need to be made. The system architecture and the flow diagram can be used to identify all the related modules that must be changed when a single change in the system architecture is proposed.

c. Job assignment and management of the software development team—The flow diagram can be used to organize development projects and to assign tasks to various participants, as it defines the interrelationships between various modules. This will aid in project execution and management.

d. Distributed systems—When a large development project is executed, the lower level tasks are often performed by various people including employees, subcontractors, and subsubcontractors, many of whom are located at remote locations. The flow diagram of the software system provides the road map by which various tasks can be assigned and coordinated.

e. Software design through assembly of modules—Ultimately, software must be designed by assembling existing software modules in a software database or knowledge base. The systems architecture can guide the software development process.

f. Systems consisting of hardware and software—When a complex machine is being designed, some of the modules are hardware and some are software. The flow diagram shows how the software and hardware must be integrated to enable the system to function as an integrated unit.

g. Interactive software program—When human operators interact with software, some of the modules in the system architecture must be performed by human beings. The human–machine interface can be defined clearly using the system architecture.

▓ 5.3 SOFTWARE DESIGN PROCESS

In this section, the software design process discussed in Section 5.2.1 is illustrated, using two simple examples.

EXAMPLE 5.1 Design of a Software System for a Library[6]

Consider the design of a software system for use by libraries. The task of the library software system is to assign a call number to a new incoming book, update the keyword database, and process a search query without missing any book that is relevant to the query.

---

[6] Adapted from Kim, Suh, and Kim (1991).

Creation of the FR and DP Hierarchy

STEP 1
Establishment of the Highest Level FRs and Mapping in the Physical Domain

The following FRs of the library software system are established to meet the user's needs:

FR_1 = Generate the call number and keyword database for new incoming books.
FR_2 = Provide a list of books that corresponds to subject keywords of a search query.

We must search for a DP_1 that can satisfy FR_1. Then we must select a DP_2 that satisfies FR_2 but does not affect FR_1 in order to create an uncoupled or at least a decoupled design. An appropriate set of DPs that satisfies the FRs may be chosen as

DP_1 = A classification system based on the content of the book
DP_2 = A search system based on the set of subject keywords

The design matrix at this level is a triangular matrix that indicates that the design is a decoupled design, i.e.,

$$\begin{Bmatrix} FR_1 \\ FR_2 \end{Bmatrix} = \begin{bmatrix} X & 0 \\ X & X \end{bmatrix} \begin{Bmatrix} DP_1 \\ DP_2 \end{Bmatrix}$$ (a)

where X represents a nonzero element.

STEPS 2 and 3
Zigzagging and Decomposition

Because the DPs at the first level are not detailed enough to provide the desired output, these FRs must be decomposed. For the design defined by the selected set of DPs, FR_1 may be decomposed into FR_{11} and FR_{12} as

FR_{11} = Assign a call number (ID number) to a new book.
FR_{12} = Generate subject keywords for the new book.

FR_2 is similarly decomposed as

FR_{21} = Find relevant reference books.
FR_{22} = Generate a list of relevant references in response to a request.

STEP 4
Mapping FR₁₁ and FR₁₂ to Lower Level DPs

The second-level DPs, DP_{11} and DP_{12}, are now selected to satisfy the second-level FRs, FR_{11} and FR_{12}, as

DP_{11} = Information on the title information page of the book (field title, author, publisher, year, etc.)
DP_{12} = The table of contents of the book

It should be noted that these DPs are inputs that enable the module to generate FR_{11} and FR_{12}.

This library uses the Library of Congress Classification (LCC) system. The field, title, authors, and other headline information are the key inputs (DP_{11}) to the call number system that can classify and assign a number to a new incoming book (FR_{11}). Because the LCC system already exists as a subroutine in the designer's database, FR_{11} does not need to be decomposed any further.

FR_{12} requires that the new book be registered in the keyword database for future search purposes. A typical user does not have the exact information about the book being searched. Therefore, the user must generate a query with a series of most appropriate and commonly understood keywords that can properly represent the context of the book. Assuming the table of contents of the book has sufficient information about its contents, it is selected as DP_{12}. However, because DP_{12} is not a complete design entity that can be implemented to satisfy FR_{12} without further decomposition, Steps 6 and 7 must be taken.

Because the call number must be assigned before the keyword database is generated, DP_{11} also affects FR_{12}. Thus, the design is a decoupled design with a triangular design matrix given by

$$\begin{Bmatrix} FR_{11} \\ FR_{12} \end{Bmatrix} = \begin{bmatrix} X & 0 \\ X & X \end{bmatrix} \begin{Bmatrix} DP_{11} \\ DP_{12} \end{Bmatrix} \tag{b}$$

STEP 5
Mapping FR_{21} and FR_{22} to Lower Level DPs

Effective DPs that can control FR_{21} and FR_{22} may be selected as

DP_{21} = Number of keywords that describe the field of interest
DP_{22} = String of keywords to find references that contain all of the relevant books

To process a search/query without missing any relevant book that corresponds to the query (FR_{21}), the query should consist of as many subject keywords as possible. However, a large number of keywords will generate a longer list of references that must be screened— a time-consuming task. Therefore, a cross-indexing method of search using a string of keywords connected by the Boolean operator OR is adopted here. This search selects the call numbers of books only in the common domain of the keywords when the keywords are connected by the OR operator. The cross-indexing method effectively reduces the size of the reference list, but may increase the probability of missing relevant references.

This mapping for FR_{21} and FR_{22} has resulted in a coupled design, which violates the Independence Axiom. Under a normal design situation, it is necessary to go back to the "drawing board" and search for an uncoupled design. However, to illustrate the use of the flow chart, the coupled solution will be used as part of this design solution. The design equation for this coupled design may be written as

$$\begin{Bmatrix} FR_{21} \\ FR_{22} \end{Bmatrix} = \begin{bmatrix} X & X \\ X & X \end{bmatrix} \begin{Bmatrix} DP_{21} \\ DP_{22} \end{Bmatrix} \tag{c}$$

STEPS 6 and 7
Decomposition of FR_{12}

FR_{12} (generate keywords for the new book) is decomposed for the selected DP_{12} (the table of the contents of the book) as

FR_{121} = Extract nouns in the table of contents to generate keywords for the new book.
FR_{122} = Generate a cross-indexing keyword database.

The keyword database created from the table of contents should be rich enough not to miss any search query. This can be accomplished by putting every noun in the table of contents of the book into the keyword database. To reduce the size of the keyword list, we want to create a cross-indexing system. A synonym dictionary may be used to reduce the size of the keyword database to a small set by replacing the number of nouns extracted from the table of contents with a small set of key nouns. This synonym-based keyword database can be used later to retrieve the desired book by generating a new set of keywords before beginning the search. Then the selected set of DPs for FR_{121} and FR_{122} may be stated as

$DP_{121} =$ Nouns extracted from the table of contents
$DP_{122} =$ Synonym dictionary

The design matrix is triangular, indicating that we created a decoupled design.

$$\begin{Bmatrix} FR_{121} \\ FR_{122} \end{Bmatrix} = \begin{bmatrix} X & 0 \\ X & X \end{bmatrix} \begin{Bmatrix} DP_{121} \\ DP_{122} \end{Bmatrix} \tag{d}$$

The resulting design hierarchy is shown in Figure E5.1.a, which depicts the system architecture for the library system. This diagram, together with the design equation, is sufficient to describe the design of the software system.

Termination of Decomposition and a "Leaf"

When the decomposition process propagates down to lower levels, the design eventually reaches the leaf level, where one or more FRs can be fully satisfied (or controlled) by the selected set of DPs without further decomposition. The design process terminates when all of the lowest branches of the FR tree are leaves. In Figure E5.1.a, a leaf is shown as a bold-outlined box.

As defined in Section 4.6.2, a module is a row of the design matrix that generates an FR when it is multiplied by or supplied with a DP. Each FR leaf has one module of software that

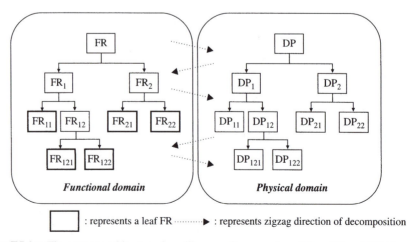

☐ : represents a leaf FR ········▶ : represents zigzag direction of decomposition

Figure E5.1.a The system architecture for a library software system. These FR and DP hierarchies, together with design equations, provide complete information on the software design.

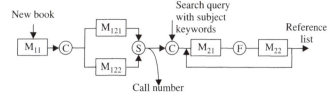

Figure E5.1.b The flow chart diagram for the library software. This representation of the system architecture is based on the decomposed hierarchies of FR and DP shown in Figure E5.1.a.

generates the FR given the input of DP. For example, FR_{121} in Figure E5.1.a is generated when DP_{121} is input to the module M_{121}. A module is an algorithm.

Based on this process of decomposition and development of uncoupled or decoupled designs, axiomatic design inherently ensures good modularity, because the design satisfies the Independence Axiom and the modules are generated from a diagonal or triangular matrix. Furthermore they contain less information than coupled designs.

It is often stated in the software field that "modularity" ensures a good software design. However, it should be noted that modularity itself does not guarantee a good software design if the Independence Axiom is violated.

A Flow Chart for the Library Software

Using the junction characteristics, the module-junction structure diagram can be easily converted to a flow chart diagram that shows the flow of data stream among modules and the sequence of execution. Figure E5.1.b shows the flow chart of the library software, which represents the system architecture of the software for the library system and has more information than the decomposed hierarchical trees of FRs and DPs shown in Figure E5.1.a. The flow chart diagram is based on the hierarchical tree structure of FRs and DPs and the design matrix. This diagram contains sufficient information to build the software system. Modules for FR leaves are program codes (i.e., algorithms) that can be constructed into a system to yield the specified FRs.

After the system architecture based on the axiomatic design of the software is completed, the programmers should then start creating program codes for each module. Then the modules can be combined to develop the main module that yields the final results upon execution.

Based on the flow chart shown in Figure E5.1.b, a software program can be developed using the object-oriented technique (OOT) discussed in Sections 5.4 and 5.5.

EXAMPLE 5.2 Design of Software for Design of a Rib in Injection Molding of Plastic Parts[7]

The axiomatic approach to software design for a library system has been described in the preceding section. To illustrate the concepts of axiomatic design for a manufacturing-

---

[7] From Kim and Suh (1987).

related system, the design of a software system for injection molding of plastic parts will be described in this section.

Design Process

Consider the design of a molded part. The primary shape of an injection-molded part is usually designed by an application engineer to satisfy a set of FRs, including the desired external appearance. The moldability of the part and its mechanical performance are also considered separately based on the product design, which must be provided by a mold designer. Supplementary features are then added to the primary shape by experienced mold engineers to reinforce the structure, to facilitate the melt flow, and to supplement the FRs with additional FRs. Ribs and bosses are the typical supplementary features that are added to the primary shape of the injection-molded parts.

The major role of a rib structure is to enhance the structural rigidity of a plastic part while maintaining uniform wall thickness and minimizing the cycle time of the molding process. A rib can also play the role of a secondary runner that quickly delivers the polymer melt to the remote region where the slow filling may cause defects such as short shots, flow marks, and voids. A bad design of a rib structure may deteriorate the quality of the product by causing sink marks, warpage, short shot, and difficulty in removing the part from the mold. Therefore, the design task for a rib structure consists of the specification of reinforcing requirements, the choice of shape parameters to meet the reinforcing requirements and moldability, and the check of the procedure for molding defect-free parts. A software system needs to be developed that can assist mold designers to make the primary shape of a plastic part moldable and mechanically sound.

STEP 1
Establishment of the Highest Level FRs and Mapping in the Physical Domain

The customer needs may be mapped into FRs in the functional domain. The highest level FRs are

$FR_1 = $ Specify the reinforcement requirement.
$FR_2 = $ Generate supplementary geometry.
$FR_3 = $ Check the performance of the reinforced structure by simulation.

The DPs that satisfy the FRs are selected as

$DP_1 = $ Primary shape of the plastic part
$DP_2 = $ Ribbed structure (core side)
$DP_3 = $ Applied loads

The design matrix at this level is a triangular matrix, which indicates that the solution is decoupled.

$$\begin{Bmatrix} FR_1 \\ FR_2 \\ FR_3 \end{Bmatrix} = \begin{bmatrix} X & 0 & 0 \\ X & X & 0 \\ X & X & X \end{bmatrix} \begin{Bmatrix} DP_1 \\ DP_2 \\ DP_3 \end{Bmatrix} \tag{a}$$

At this level, it is recognized that once DP_1 and DP_2 are determined, then FR_3 becomes an FR leaf in the functional domain. FR_3 can be readily generated using a module M_3 that is a commercially available structural analysis software package (ANSYS is selected as M_3 in this case study). The module M_3 constitutes the row of the design matrix.

FR_1 and FR_2 need to be decomposed further as the selected DPs cannot finalize the design.

STEP 2
Second-Level Decomposition

To decide if reinforcement is required, the structural performance of the given primary geometry must be quantified under certain loading conditions. However, the loading conditions of most plastic parts are not prespecified explicitly, but are given so as to avoid the maximum allowable deflection of the lid plate or the possible stress concentrations at corners and handles, and so forth. Therefore, the use of a finite-element program for the complete structural analysis is not warranted at this time.

FR_1 (Specify the reinforcement requirement) is decomposed as

FR_{11} = Characterize the structural performance of the primary geometry under implicit loading conditions.

FR_{12} = Calculate the reinforcement requirements.

Depending on the shape of the part, the regions in which the maximum deflection or stress concentration occurs are identified by the designer, based on his or her experience (i.e., database). Likewise, the complex shape of the primary geometry can be characterized by a set of geometric elements in which simple structural formulas for thin plates and beams can be applied as the initial guess for the reinforcing requirements. In this study, three elementary geometric elements for the rib reinforcement are formulated—for example, a rectangular plate, a circular plate, and a curved beam—for which appropriate structural formulas are available in handbooks (for example, Orr, 1977). The selected DPs are

DP_{11} = Elementary geometric elements

DP_{12} = Structural formulas from the handbook

The resulting design matrix is triangular.

$$\begin{Bmatrix} FR_{11} \\ FR_{12} \end{Bmatrix} = \begin{bmatrix} X & 0 \\ X & X \end{bmatrix} \begin{Bmatrix} DP_{11} \\ DP_{12} \end{Bmatrix} \tag{b}$$

FR_2 is decomposed as

FR_{21} = Generate ribs that do not adversely affect the manufacturability of the part.

FR_{22} = Generate ribs that can meet the reinforcement requirements.

FR_{21} can be satisfied only if both the DPs in the physical domain and the PVs in the process domain do not violate the Independence Axiom. Assuming that the PVs will be selected appropriately, the cross-sectional shape of the rib is the key input that controls FR_{21}. The number of ribs and their locations should be determined to meet the reinforcing requirements after the cross-sectional shape parameters are determined. This can be accomplished by determining the moment of inertia of the ribbed structure that is composed of a number of ribs of a certain cross-sectional shape to provide an equivalent stiffness to that of the reinforced structure of flat plate. This generates a decoupled design. The DPs that can control FR_{21} and FR_{22} are selected as

$DP_{21} =$ Cross-sectional shape of a rib structure
$DP_{22} =$ Number of ribs

The design equation is

$$\begin{Bmatrix} FR_{21} \\ FR_{22} \end{Bmatrix} = \begin{bmatrix} X & 0 \\ X & X \end{bmatrix} \begin{Bmatrix} DP_{21} \\ DP_{22} \end{Bmatrix} \qquad (c)$$

At this level, it seems that FR_{11}, FR_{12}, and FR_{22} can be satisfied by the selected set of DPs. These FRs form the terminal nodes of the FR hierarchy, from which the software modules for the FR leaves can be developed. However, FR_{21} needs to be decomposed further as DP_{21} is not sufficient to fulfill the design goal.

STEP 3
Third-Level Decomposition of FR_{21}

The manufacturability of ribs can be specified for ejectability, prevention of excessive warpage, and minimization of a sink mark that is a local dent formed on the surface opposite to the rib. FR_{21} (Generate ribs that do not adversely affect the manufacturability of the part) is decomposed as

$FR_{211} =$ Avoid warpage.
$FR_{212} =$ Ensure ejectability.
$FR_{213} =$ Minimize sink marks.

The cross-sectional shape of a rib should be based on the process characteristics of injection molding. Cross-sectional parameters include the height, the root thickness, the fillet radius, and the draft angle. The wall thickness and the material to be injected are predetermined together with the primary geometry, and will act as constraints (Cs). For a given material and wall thickness, the cross-sectional parameters have design ranges in order to prevent jamming, warping, or sink mark formation. By considering the major causality between the defects and cross-sectional parameters, the following DPs are selected:

$DP_{211} =$ Rib height
$DP_{212} =$ Draft angle
$DP_{213} =$ Root thickness

However, the design matrix becomes coupled if the allowable warpage (given in terms of tolerances) cannot be satisfied by the process.

$$\begin{Bmatrix} FR_{211} \\ FR_{212} \\ FR_{213} \end{Bmatrix} = \begin{bmatrix} X & 0 & X \\ X & X & 0 \\ X & X & X \end{bmatrix} \begin{Bmatrix} DP_{211} \\ DP_{212} \\ DP_{213} \end{Bmatrix} \qquad (d)$$

Injection molding is inherently a coupled process between the flow and the solidification of polymer melts (Kim and Suh, 1987). Therefore, it is very difficult to select DPs that can make the design matrix uncoupled or decoupled unless the tolerance is very large. In this case, the decomposition process must include zigzagging not only between the FR and DP domains but also between the DP and PV domains in order to yield the FRs through proper processing. Because of the coupled nature of the injection molding process, mold design for precision parts is done through trial-and-error processes. Often a compromised shape must be accepted as the Independence Axiom cannot be satisfied.

One way of handling this kind of problem (design for manufacture) is the use of the knowledge-based system. The Cs and expert knowledge are used to determine the cross-sectional parameters using an expert system module RIBBER (Jackson, 1975) and other equivalent programs. The expert system provides information on the cause of defects and acceptable cross-sectional geometry for the ribs. Such an approach will provide a compromised solution that may be acceptable in the absence of any uncoupled or decoupled solutions.

Module-Junction Structure Diagram

The hierarchical structures of FRs and DPs can be constructed based on the design Equations (a) through (d). The terminal nodes are formed when the selected DPs can satisfy the FRs. Figure E5.2 shows the module-junction structure diagram of the rib design software that represents both the hierarchical structures and design matrices.

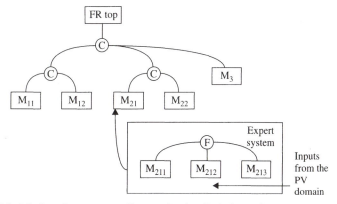

Figure E5.2 Module-junction structure diagram for the rib design software.

It should be noted that because the injection molding process cannot be completely uncoupled, an expert system was used to obtain the most acceptable solution for FR_{21}. This coupled nature of injection molding limits its applications to geometry with large tolerances, which was solved recenty by the use of microcellular plastics (see Chapter 7). Based on the module-junction structure diagram, we can construct a software system.

EXAMPLE 5.3 Software Development for Design Automation for Glass-Bulb Design[8]

The task of this project was to automate the design process for the design and manufacture of television tubes for a large consumer electronics company (Do, 1997; Do and Park, 1996).

---

[8] From Do and Park (1996).

The new software system by Do and Park has reduced the design cycle from 3 weeks to 6 hours by one designer. The secret behind their success was that they recognized a coupled process and developed a decoupled software system based on axiomatic design.

Design of the Software Architecture for Glass-Bulb Design

The glass bulb is a unit element of a TV tube, which is sometimes called a "Brown tube." The tube consists of a shadow mask, an electron gun, a band, and a glass-bulb that consists of a panel and a funnel. The panel is the front part and the funnel is the rear part of the glass bulb shown in Figure E5.3.a.

Figure E5.3.b shows the top and section view of the picture tube and design variables of the panel. In the original manual design process, the radii are supplied by the customer, but the position dimensions (in italics) are established by the designer through a trial-and-error process to satisfy geometric compatibility. Similarly, Figure E5.3.c shows the top and section view of the funnel section with design variables. As in Figure E5.3.b, the radii are supplied by customers, and the position dimensions (in italics) were determined by the designer through a trial-and-error process from a rough drawing. Therefore, in the original trial-and-error process, the designer needed a rough drawing, which was supplied by a CAD system.

The original manual design process consisted of several steps as shown in Figure E5.3.d. Step 1 is "Product Request," which consists of receiving an order from a customer who supplies the basic specifications or drawings. Based on this information, Step 2 consists

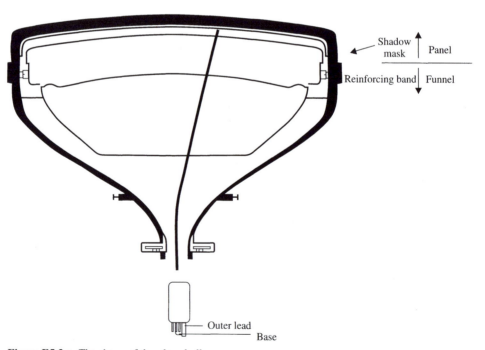

Figure E5.3.a The shape of the glass-bulb.

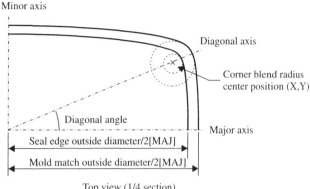

Top view (1/4 section)

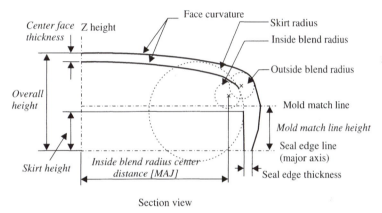

Section view

Figure E5.3.b Top and section view of the picture tube. Included are the design variables of PANEL for the initial design step.

of generating a rough drawing using CAD software, which requires some information on the final shape. To achieve this goal, an approximate final shape has to be generated by connecting tangential points between curvatures and lines, thickness of some parts, etc. In Step 3, the three-dimensional shape of the tube is generated using a software program, which uses the output of Steps 1 and 2 as well as part of information generated by Step 5—an iterative process. After the tube shape is generated, stress analysis is performed in Step 4. After the entire design specifications are determined, the final step (Step 5) generates the complete drawings. Some of this information is used in Steps 3 and 4, requiring iterations between Steps 2 through 5.

According to the original process of tube design, the designer gathers the basic design information, constructs the full product shape, conducts stress analysis, and generates the drawing. The functional requirements of this process are

$FR_1 =$ Construct the basic information of the product (Steps 1 and 2).
$FR_2 =$ Establish the product shape.
$FR_3 =$ Verify the mechanical characteristics of the product.
$FR_4 =$ Generate the product drawing.

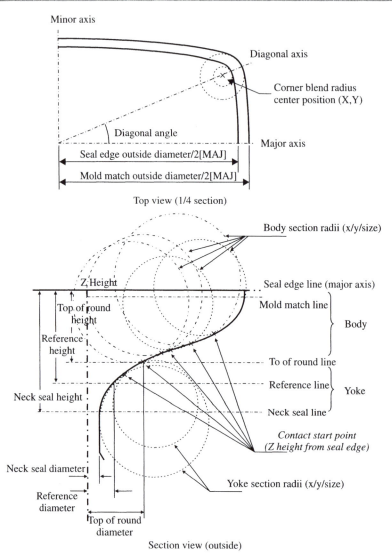

Figure E5.3.c Top and section view of the picture tube. Included are the design variables of FUNNEL for the initial design step.

If we analyze the original design process, the design parameters are

$DP_1 =$ A set of basic data—those supplied by the customer (Step 1 of Figure E5.3.d) and the rough drawing generated by CAD software (Step 2 of Figure E5.3.d)

$DP_2 =$ The three-dimensional shape structure (panel/funnel)

$DP_3 =$ Loading condition on panel and funnel

$DP_4 =$ A set of drawing data

The design equation for the highest level FRs is

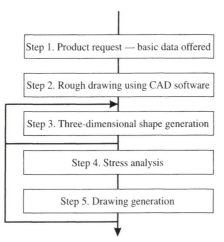

Step 1. Product request — basic data offered

Step 2. Rough drawing using CAD software

Step 3. Three-dimensional shape generation

Step 4. Stress analysis

Step 5. Drawing generation

Figure E5.3.d The design process for glass bulb before analysis.

$$\begin{Bmatrix} FR_1 \\ FR_2 \\ FR_3 \\ FR_4 \end{Bmatrix} = \begin{bmatrix} X & 0 & 0 & X \\ X & X & 0 & X \\ X & X & X & 0 \\ X & X & 0 & X \end{bmatrix} \begin{Bmatrix} DP_1 \\ DP_2 \\ DP_3 \\ DP_4 \end{Bmatrix} \qquad (a)$$

Equation (a) states that the design is a coupled design, requiring many iterations and approximations. FR_1 (construct the basic information of the product) is influenced by DP_1 (basic specification data supplied by the customer) and DP_4 (rough drawing). FR_2 (establish product shape) is affected by DP_1 (basic specification data), DP_2 (three-dimensional shape structure), and DP_4 (rough drawing data). FR_3 (verify the mechanical characteristics of the product) is similarly affected by DP_1, DP_2, and DP_3 (loading conditions on the panel and funnel). FR_4 is controlled by DP_1, DP_2, and DP_4. This design equation shows that the design process is very complex (or has a high degree of uncertainty).

To overcome this shortcoming of the complex manual process discussed above, Do and Park adopted the axiomatic design approach to develop an improved and "automatic" design process. The decomposition process for glass-bulb design is described below.

STEP 1
Mapping at the Highest Level FR → DP

First, FR_1 (construct the basic information about the product) was satisfied by the inputs from the customer (Step 1 shown in Figure E5.3.d). Thus, DP_1 is "a set of the specifications for the tube received from the customer." Then Step 2 was eliminated by fulfilling FR_2 (establish the product shape) using a numerical analysis program that automatically generated the information that used to be incorporated in the rough drawing and estimating the three-dimensional shape structure—the new DP_2. Then to satisfy FR_4 (generate the product drawing), DP_4 became only the accessory data needed to complete the set of drawings. By making these changes, the design process became decoupled. Do and Park also created a database that contained all the design data and the bulb shape information. This database is a PV in the process domain.

To satisfy the FRs at the highest level, the new DPs are

$DP_1 =$ A set of specifications supplied by the customer for the new product
$DP_2 =$ The three-dimensional shape structure of the panel and funnel
$DP_3 =$ Loading conditions on the panel and funnel
$DP_4 =$ A set of accessory drawing data

The design equation is

$$\begin{Bmatrix} FR_1 \\ FR_2 \\ FR_3 \\ FR_4 \end{Bmatrix} = \begin{bmatrix} X & 0 & 0 & 0 \\ X & X & 0 & 0 \\ X & X & X & 0 \\ X & X & 0 & X \end{bmatrix} \begin{Bmatrix} DP_1 \\ DP_2 \\ DP_3 \\ DP_4 \end{Bmatrix} \tag{b}$$

The first-level design is a decoupled design, because the set of accessory drawing data does not affect the construction of the database (i.e., FR_1), the three-dimensional shape of the glass-bulb (i.e., FR_2), and the mechanical strength of the Brown tube (i.e., FR_3).

STEP 2
Decomposition by Zigzagging DP → FR

Because the relationship between FRs and DPs at the highest level is not detailed enough to provide the desired output, decomposition of the FRs is required. For the design defined by the selected set of specifications (DP_1), FR_1 may be decomposed as

$FR_{11} =$ Assign an ID number to a new product.
$FR_{12} =$ Construct a set of data for a new product.

Similarly, FR_2 and FR_4 are decomposed as:

$FR_{21} =$ Check the curvature, panel flatness, and funnel axis profile.
$FR_{22} =$ Calculate the three-dimensional shape.
$FR_{23} =$ Consider the manufacturability.
$FR_{41} =$ Represent the shape of the product.
$FR_{42} =$ Display accessory of the drawing.

STEP 3
Second-Level Mapping FR → DP

Having established the second-level FRs, we need to map them in the physical domain to find DPs. For FR_{11} and FR_{12}, the second-level DPs are

$DP_{11} =$ Representative code of new product
$DP_{12} =$ A set of specific data for a new product

The design matrix is triangular, indicating that it is a decoupled design.

$$\begin{Bmatrix} FR_{11} \\ FR_{12} \end{Bmatrix} = \begin{bmatrix} X & 0 \\ X & X \end{bmatrix} \begin{Bmatrix} DP_{11} \\ DP_{12} \end{Bmatrix} \tag{c}$$

The lower level DPs for FR_{21}, FR_{22}, and FR_{23} are

$DP_{21} =$ Inside/outside curvature for the product
$DP_{22} =$ Characteristic geometric equation for the product
$DP_{23} =$ A set of data for mold manufacture

The design equation indicates that the design is decoupled.

$$\left\{ \begin{array}{c} FR_{21} \\ FR_{22} \\ FR_{23} \end{array} \right\} = \left[\begin{array}{ccc} X & 0 & 0 \\ X & X & 0 \\ X & X & X \end{array} \right] \left\{ \begin{array}{c} DP_{21} \\ DP_{22} \\ DP_{23} \end{array} \right\} \qquad (d)$$

For FR_{41} and FR_{42}, the DPs are

$DP_{41} =$ A set of data for product design
$DP_{42} =$ A set of data for accessory

This is an uncoupled design and therefore the design matrix is diagonal.

$$\left\{ \begin{array}{c} FR_{41} \\ FR_{42} \end{array} \right\} = \left[\begin{array}{cc} X & 0 \\ 0 & X \end{array} \right] \left\{ \begin{array}{c} DP_{41} \\ DP_{42} \end{array} \right\} \qquad (e)$$

STEP 4
Zigzagging Back to the Functional Domain DP → FR

FR_{23} is decomposed based on the selected DP_{23} as

$FR_{231} =$ Check the useful screen dimension for the panel.
$FR_{232} =$ Evaluate the ejectability.
$FR_{233} =$ Examine the deflection angle of a scanning line for the funnel.

STEP 5
Third-Level Mapping FR → DP

The selected DPs for the decomposed FR_{23} are

$DP_{231} =$ Distance of blending circle center position
$DP_{232} =$ Angle of the side wall
$DP_{233} =$ Inside curvature of the yoke part

This design is an uncoupled design.

$$\left\{ \begin{array}{c} FR_{231} \\ FR_{232} \\ FR_{233} \end{array} \right\} = \left[\begin{array}{ccc} X & 0 & 0 \\ 0 & X & 0 \\ 0 & 0 & X \end{array} \right] \left\{ \begin{array}{c} DP_{231} \\ DP_{232} \\ DP_{233} \end{array} \right\} \qquad (f)$$

Now the decomposition process is complete because the highest level FRs can be controlled by the DPs selected. The lowest level FRs of each decomposition branch are leaves of the hierarchical tree. Figure E5.3.e shows the FR and DP hierarchies and the decomposition process.

Module-Junction Structure Diagram

Based on the information provided by Figure E5.3.e and the design matrices of the glass-bulb presented in this section, the module-junction structure diagram may be constructed as shown in Figure E5.3.f. Each FR leaf in Figure E5.3.e can be transformed into an independent module. Each leaf can be executed independently and has separate input and output. FRs are the output of the software, and DPs are the key input or input files to the software that can control the corresponding FRs. The software code is a set of design matrices that transforms DPs to FRs at each level of the hierarchy. Each module can be

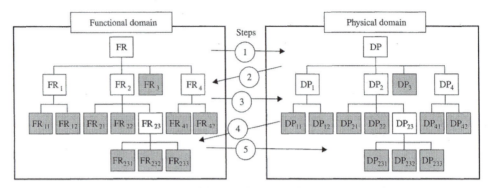

Figure E5.3.e The FR and DP hierarchies and the zigzagging process. The leaves are shown in shaded boxes.

utilized in different environments, which include operating systems, compilers, languages, and so forth. A module or a commercial software system can be programmed.

The Flow Chart Diagram for Software Architecture—Glass Bulb

Using the module-junction structure diagram of Figure E5.3.f, the flow diagram for the architecture of the glass-bulb design can be constructed as shown in Figure E5.3.g to illustrate the design sequence. As presented earlier, the design equation is represented at each node by either S for an uncoupled design or C for a decoupled design. Figure E5.3.g represents the complete software architecture developed using axiomatic design. Once Figure E5.3.g is finalized, the software engineer does not have to know anything about the development of the FR and DP hierarchies to develop the software. It should be noted that the larger module M_1 is constructed from M_{11} and M_{12}, etc. The final design objective is achieved through four modules, M_1, M_2, M_3, and M_4 that relate DP_1 to FR_1, DP_2 to FR_2, DP_3 to FR_3, and DP_4 to FR_4, respectively.

Software Development for a Glass Bulb

A system that can automatically design the glass-bulb system has been developed based on the flow chart diagram shown in Figure E5.3.g. The structure of the program is shown in Figure E5.3.h. The sequence of execution is controlled by the graphic user interface (GUI). Each major module (M_1, M_2, M_3, and M_4) is shown as a bubble in Figure E5.3.h. Each module can be a developed program, commercial software, or a combination of the two.

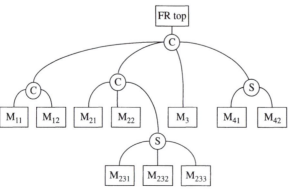

Figure E5.3.f Module-junction structure diagram for glass-bulb design.

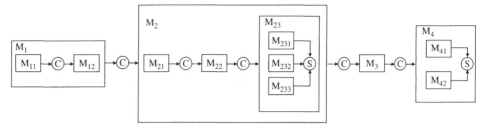

Figure E5.3.g The flow chart diagram for the glass-bulb design.

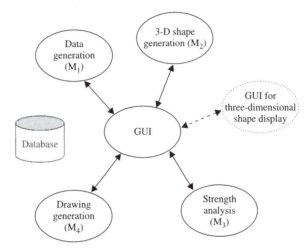

Figure E5.3.h Structure of the glass-bulb automatic design software.

The GUI controls the glass-bulb design process shown in Figure E5.3.h. Do and Park (1996) used an X-Window (MOTIF) system, which is a standard graphics tool in a UNIX operating system.

The M_1 module deals with data generation. To reduce the programming effort, Do and Park used commercial software called ORACLE. The SQL language is supported for the transaction commands in the ORACLE database.

Do and Park developed a module for three-dimensional shape generation. The shape-generation module uses geometric interrelations and some specific numerical analysis algorithms. From the design variables, the geometry of the glass bulb is constructed by module M_2.

The strength analysis module M_3 is used to determine the strength of the tube using fracture theory for brittle materials after the three-dimensional shape has been generated by M_2. Special mesh generation routine is used to compute the stress distribution in the bulb. Commercial software can be used for this purpose. If the results of the stress analysis cannot be accepted, the designer can go back to the 3-D shape-generation algorithm to repeat the process, which is indicated by the dotted line in Figure E5.3.h. Do and Park used ANSYS, a commercial algorithm, to determine the stress.

The last module is the drawing module M_4. A commercial package was used to produce the drawing, which could interface with other programs.

In addition to these modules, Do and Park also used a commercial graphic library to confirm the shape of the glass bulb during the design process.

▨ 5.4 AXIOMATIC DESIGN OF OBJECT-ORIENTED SOFTWARE SYSTEMS[9]

In this section, the development of a software system based on axiomatic design and the object-oriented programming technique is presented. This *Axiomatic Design of object-oriented Software Systems* (ADo-oSS) is a major new paradigm shift in the field of software engineering. It combines the power of axiomatic design with a popular software programming methodology called the object-oriented programming technique (OOT). The goal of ADo-oSS is to make software development a subject of science rather than an art and thus reduce or eliminate the need for debugging and extensive changes.

In software engineering, it is well known that software should be developed or written with independent modules.[10] However, a collection of modules can also lead to coupling of functions as will be evident when the details of software design methodology are presented later in this section. Furthermore, the decomposition of software is often done in the heads of the programmers, based on their experience, as they do not explicitly define FRs, or map, or zigzag, or write down design equations. Therefore, the use of separate modules does not in itself generate uncoupled software design. To have good software, the relationship between the modules must be designed to make them work independently and explicitly. These shortcomings can be overcome by designing software first based on axiomatic design.

The AD framework ensures that the modules are correctly defined and located in the right place in the right order. A "V model for software" will be used here to explain the concept of ADo-oSS.[11] The first step is to design the software, build the software hierarchy that follows the top-down approach of axiomatic design, and then generate the full design matrix table to define modules. The final step is to build the object-oriented model with a bottom-up approach, following the AD flow chart for the designed system. Figure 5.2 schematically represents the concept.

The right side of the V model shown in Figure 5.2 can be used with any software design methodology, e.g., SADT. However, the object-oriented method is used in most commercial software products, which exclusively use C++ or Java. Before we discuss the details of ADo-oSS, the essence of the object-oriented technique is reviewed in Section 5.4.1.

5.4.1 Object-Oriented Techniques[12]

In this section, OOT is introduced without any modification. In Section 5.4.2, OOT will be modified to make it simpler and consistent with axiomatic design. One of the difficulties associated with learning object-oriented techniques (OOT) is the definition of keywords that describe the method. Too many keywords are used to describe the same thing because OOT cannot deal with the hierarchical nature of design and the decomposition through zigzagging between the domains in a systematic manner. In this section, OOT will be

---

[9] This section is taken from Do and Suh (1999).

[10] Modules used in typical software may not meet the rigid definition of module used in axiomatic design.

[11] A "V model" concept was introduced to improve the quality in mechanical systems based on axiomatic design (El-Haik, 1999). This idea was adapted to show a "V model for software."

[12] The best-known methods for object-oriented designs are Rumbaugh et al.'s (1991) object modeling technique (OMT) and Booch's (1994) object-oriented design (OOD).

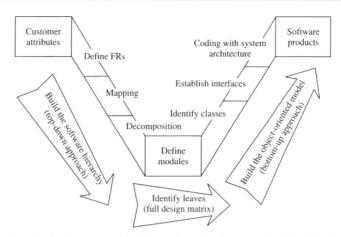

Figure 5.2 Axiomatic design process for object-oriented software system (the V model).

described by using the keywords used in OOT and also by providing equivalent terms used in axiomatic design, where appropriate, to facilitate the learning process.

Object-oriented methodology (OOM) is a conceptual process that is independent of any programming language until the final stage. This is a formal method of creating a software structure and abstraction, helping designers to express and communicate abstract concepts. The fundamental construct for the object-oriented method is the *object*,[13] which is equivalent to an FR. Object-oriented design decomposes a system into objects. Objects "encapsulate" both data (equivalent to DPs), and method (equivalent to relationship between FR_i and DP_j, i.e., module) in a single entity. An object retains certain information on how to perform certain operations, using the input provided by the data and the method imbedded in the object. In terms of axiomatic design, this is equivalent to saying that an object is $[FR_i = A_{ij} DP_j]$. A graphic representation of *object* is shown in Figure 5.3.

Object-oriented design generally uses four definitions to describe its operations: *identity, classification, polymorphism*, and *relationship*. Identity means that data—equivalent to DPs—are incorporated into specific objects. Objects are equivalent to an FR with a specified $[FR_i = A_{ij} DP_j]$ relationship, where DPs are data or input and A_{ij} is a method or

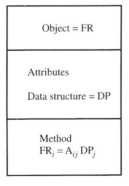

Figure 5.3 A graphic representation of an object.

---

[13] Words italicized in this section have specific definitions.

a relationship. In axiomatic design, the design equation explicitly identifies the relationship between FRs and DPs.

Classification means that objects with the same data structure (attributes) and behavior (operations or methods) are grouped into a class. The object is represented as an *instance* of a specific *class* in programming languages. Therefore, all objects are instances of some classes. A class represents a template for several objects and describes how these objects are structured internally. Objects of the same class have the same definition both for their operations and for their information structure. There are two types of object diagram: class diagram and instance diagram. A class diagram is a schema, pattern, or template for describing many possible instances of data. An instance diagram describes how objects relate to each other. A class diagram describes object classes. An instance diagram describes object instances. Figure 5.4 illustrates the concept graphically.

An object is linked to other objects through *messages*, which require an object to perform certain functions by executing a method (i.e., algorithm). The object performs the function using the internal operations that may not be apparent to other objects—equivalent to the elements of the design matrix that link specific FRs to specific DPs.

Sometimes an object is also called a tangible entity that exhibits some well-defined *behavior*. Behavior is a special case of FR. The relationship between objects and behavior may be compared to the decomposition of FRs in the FR hierarchy of axiomatic design. Object is the parent FR, relative to behavior, which is the child FR. That is, the highest level FR among the two layers of decomposed FRs is "object" and the children FRs of the object FR are "behavior."

Polymorphism means that the same operation may behave differently on different classes, and, therefore, two or more classes of objects may respond to the same message each in its own way. The move operation, for example, may behave differently on the Window and Chess-Piece classes.

Relationship describes the interactions between objects or classes. It can be of the following types, as shown in Figure 5.5 (Rumbaugh et al., 1991).

The relationships between objects and classes are established by means of *links* and *associations* (Rumbaugh et al., 1991; Booch, 1994). A link is a physical or conceptual connection between objects. An association represents a group of bidirectional links with common structure and common semantics.

Aggregation refers to the "part-whole" or "a-part-of" relationship in which objects represent the components of an object that represents the entire assembly (Rumbaugh et al., 1991; Booch, 1994).

Sharing similarities between classes while preserving their differences is encapsulated in *generalization* and *inheritance*. *Inheritance* allows the conception of a new class of objects as a refinement of another and abstraction of the similarities between classes [Wirfs-Brock, Wilkerson, and Wiener, 1990). *Generalization* defines the relationship between a

Person		(Person) Bob Powers 50	(Person) Derrick Tate 28
name: string age: integer			

Class diagram   Instance diagram

Figure 5.4 Class and instance of class. "Person" is a *class*; "Bob Powers" and "Derrick Tate" are *instances*, which are also called *objects*.

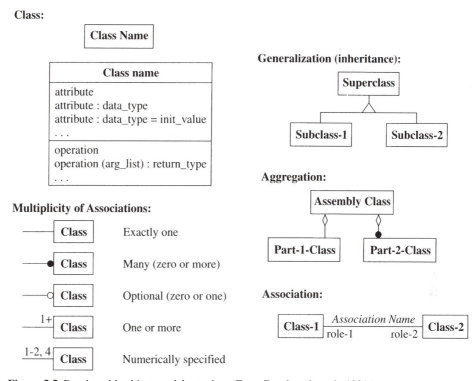

Figure 5.5 Rumbaugh's object model notation. (From Rumbaugh et al., 1991.)

class and one or more refined versions of it. The *superclass* is the class being refined, and a *subclass* is a class that inherits behavior from another class, which is a refined version of *superclass*. Attributes and operations common to a group of subclasses are attached to the superclass and shared by each subclass, which inherits the features of its superclass. Generalization and inheritance are transitive across an arbitrary number of levels.

The following software design sequence describes the object-oriented technique based on these concepts:

1. Define the problem or task.
2. Identify objects and their attributes and input data.
3. Identify operations that may be applied to objects.
4. Establish interfaces by showing the relationship between objects and operations.
5. Repeat Steps 1, 2, 3, and 4.

Steps 2 and 3 are summarized as class definition, sometimes called modeling or analysis. Step 4, called the design step, utilizes many diagrams, such as the data flow diagram and the entity relation (ER) diagram to realize the relationships. As shown in Figure 5.6, the object diagram is generated when all these steps are finished. It is really difficult to tell whether the design is good from an object diagram because no design-verification criterion exists, which is a shortcoming of OOT.

Figure 5.6 describes the following:

• Each box is a class or an object. The top section of the box (e.g., **Box** in the left top box of the figure) is the name of the class, the middle section of the box (e.g., **text** in the

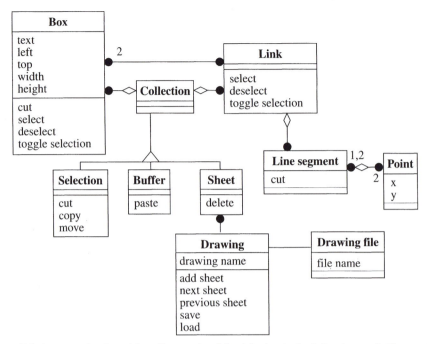

Figure 5.6 An example of an object diagram (explained further in the following text). The numbers indicate the number of associations. (From Rumbaugh et al., 1991.)

figure) describes the attributes of the class, and the bottom section of the box (e.g., **cut**) defines the operations of the class.

- An object is represented as an instance of a specific class (e.g., object **Box** is an instance of **Box** class). However, sometimes each box in Figure 5.6 may be called an object.
- Object **Link** has two relationships with the attributes of object **Box**.
- Object **Collection** is an aggregation of the attribute of **Box** and the operation of **Link** with many relationships. That is, object **Collection** can have many **Box** and **Link** objects.
- Object **Collection** is the superclass of subclasses **Selection, Buffer,** and **Sheet**.
- Object **Drawing**, which has a relationship with object **Drawing** file, can have many **Sheet** objects.
- Object **Link** is an aggregation of object **Line segment**, which in turn is an aggregation of object **Point**. That is, object **Line segment** is a part of object **Link** with many relationships.

5.4.2 Modified OOT for Compatibility with Axiomatic Design

The distinction between *superclass, class, object,* and *behavior* is necessary in OOT to deal with FRs at successive layers of a system design. In OOT, *class* represents an abstraction of *object*s and thus is at the same level as *object* in the FR hierarchy. However, *object* is one level higher than *behavior* in the FR hierarchy. The use of these keywords, although necessary in OOT, adds unnecessary complexity when the results of axiomatic design are to be combined with OOT. Therefore, we will modify the use of these keywords in OOT.

We will use one keyword "Objects with indices" to represent all levels of FRs, i.e., *class*, *object*, and *behavior*. For example, *class* or *object* may be called *Object i*, which is equivalent to FR_i. *Behavior* will be denoted as *Object ij* to represent a second-level FR_{ij}. Conversely, a third-level FR_{ijk} will be denoted as *Object ijk*. Thus, *Object i*, *Object ij*, and *Object ijk* are equivalent to FR_i, FR_{ij}, and FR_{ijk}, respectively, which are FRs at three successive levels of the FR hierarchy.

To summarize, the equivalence between the terminology of axiomatic design and that of OOT may be stated as follows:

- FR can represent an object.
- DP can be data or input for the object, FR.
- The product of a module of the design matrix and DP can be a method, i.e., $FR = A \cdot DP$.
- Different levels of FRs are represented as objects with indices, that is, FR_i, FR_{ij}, and FR_{ijk} are equivalent to *Object i*, *Object ij*, and *Object ijk*, respectively.

5.4.3 Basics of Axiomatic Design of Object-Oriented Software Systems

The axiomatic design of object-oriented software system (ADo-oSS) shown in Figure 5.2 involves the following steps:

STEP 1
Define FRs of the Software System

The first step in designing a software system is to determine the customer attributes in the customer domain that the software system must satisfy. Then the functional requirements (FRs) of the software in the functional domain and constraints (Cs) are established to satisfy the customer needs. FRs must satisfy the customer needs with fidelity.

STEP 2
Mapping between Domains and the Independence of Software Functions

The next step in axiomatic design is to map these FRs of the functional domain into the physical domain by identifying the DPs. DPs are the "how's" that satisfy specific FRs of the design. DPs must be chosen to be consistent with the constraints. During the mapping process, the design must satisfy the Independence Axiom, which requires that functional independence be maintained throughout the development of an uncoupled or decoupled design. A design matrix is used for this evaluation.

STEP 3
Decomposition of FRs, DPs, and PVs

The FRs, DPs, and PVs must be decomposed until the design can be implemented without further decomposition. These hierarchies of FRs, DPs, PVs, and the corresponding matrices represent the system architecture. The decomposition of these vectors cannot be done by remaining in a single domain, but can be done only through zigzagging between domains. This is illustrated in Figure 5.7. From FR_1 in the functional domain, the designers go to DP_1 in the physical domain. Then the designers return to the functional domain to generate

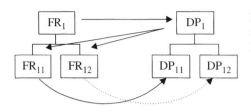

Figure 5.7 Zigzagging between the functional and physical domains to decompose FRs and DPs.

FR_{11} and FR_{12} that collectively satisfy the highest level FR_1 and the corresponding DP_1. Then the designers move to the physical domain to find DP_{11} and DP_{12}, which satisfy FR_{11} and FR_{12}, respectively. This process continues until the FRs can be satisfied without further decomposition. Each terminal FR and DP (those that are not further decomposed) is called a "leaf" in the tree structure.

As stated earlier, the software design methodology that was favored in the mid-1970s and 1980s uses similar decomposition with the top-down approach. As shown in Figure 5.8, structured analysis and design technique (SADT) is a well-known example (Ross, 1977, 1985; Marca and McGowan, 1988). However, SADT decomposition is conducted only in the physical domain from an axiomatic design viewpoint. Because of this nonzigzagging process, there is no guarantee of functionality or productivity. Therefore, those methods faded away as the requirements for software systems increased and the object-oriented method was introduced.

STEP 4
Definition of Modules—Complete Design Matrix

One of the most important features of the AD framework is the design matrix, which provides the relationships between the FRs and the DPs. For software creation, the design matrix provides two important bases. One important basis is that each element in the design matrix can be a method (or operation) in terms of the object-oriented method. The other basis is that each row in the design matrix represents a module to satisfy a specific FR when a given DP is provided. In most cases, the off-diagonal terms in the design matrix are important as the coupling comes from these off-diagonal terms. However, it is possible

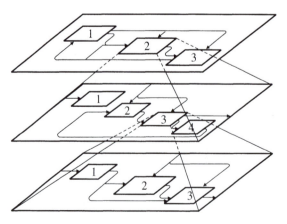

Figure 5.8 A concept of SADT diagram. The numbers indicate the flow of information and decisions at a given level of design. The layers indicate decomposition. This decomposition is done only in the physical domain without zigzagging. (From Ross, 1977, 1985; Marca and McGowan, 1988.)

to avoid coupling with a proper sequence of change if the design matrix is triangular—a decoupled design.

As illustrated in Figure 5.9, it is important to construct the complete design matrix based on the leaf-level FR–DP–A_{ij} to check for consistency of decisions made during decomposition. In Figure 5.9, the decomposition of $FR_{x.2}$ shows a relationship between $DP_{x.2.3}$ and $FR_{x.1}$, whereas before decomposition, it was decided that there is no relationship between $FR_{x.1}$ and $DP_{x.2}$. Designers must check for the consistency of every layer of the design hierarchy to make sure that the design decisions are made correctly at every step of decomposition (Baldwin, 1994; Tate, 1999; Kim, Suh, and Kim, 1991). If an off-diagonal term is shown in a lower level matrix, contrary to the higher level matrix, the designer must correct this inconsistency by checking the FRs, DPs, design matrices, and Cs.

STEP 5
Identify Objects, Attributes, and Operations

Because all DPs in the design hierarchy are selected to satisfy FRs, it is relatively easy to identify the objects. The leaf is the lowest level object in a given decomposition branch, but all leaf-level objects may not be at the same level if they belong to different decomposition branches. Once the objects are defined, the DPs (attributes or data) and products of modules with DPs (operations or methods) for the object should be defined to construct the object model. This activity should use the complete design matrix table.

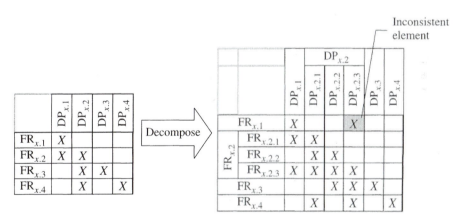

Figure 5.9 Complete design matrix deployment and consistency check. (From Baldwin, 1994.)

EXAMPLE 5.4 Software Design for a Simple Line Drawing Program[14]

Design a software program that can draw a straight line. Translate the resulting axiomatic design into the object-oriented method.

---

[14] Contributed by S.-H. Do (August 1999).

SOLUTION

The highest level FRs may be stated as

$FR_1 = $ Define line element.
$FR_2 = $ Detect the drawing location.
$FR_3 = $ Draw the line element through the window.

The corresponding DPs are selected as

$DP_1 = $ Line characteristic
$DP_2 = $ Mouse click information
$DP_3 = $ GUI for the drawing

The design matrix may be written as

$$\begin{Bmatrix} FR_1 \\ FR_2 \\ FR_3 \end{Bmatrix} = \begin{bmatrix} A_{11} & 0 & 0 \\ 0 & A_{22} & 0 \\ A_{31} & A_{32} & A_{33} \end{bmatrix} \begin{Bmatrix} DP_1 \\ DP_2 \\ DP_3 \end{Bmatrix}$$

The FRs and DPs may be decomposed as follows:

Second Level for FR_1

$FR_{11} = $ Define start.
$FR_{12} = $ Define end.

$DP_{11} = $ Start point
$DP_{12} = $ End point

The design matrix may be written as

$$\begin{Bmatrix} FR_{11} \\ FR_{12} \end{Bmatrix} = \begin{bmatrix} a & 0 \\ 0 & b \end{bmatrix} \begin{Bmatrix} DP_{11} \\ DP_{12} \end{Bmatrix}$$

Second Level for FR_2

$FR_{21} = $ Detect mouse push.
$FR_{22} = $ Detect mouse release.

$DP_{21} = $ Event for push
$DP_{22} = $ Event for release

The design matrix may be written as

$$\begin{Bmatrix} FR_{21} \\ FR_{22} \end{Bmatrix} = \begin{bmatrix} c & 0 \\ 0 & d \end{bmatrix} \begin{Bmatrix} DP_{21} \\ DP_{22} \end{Bmatrix}$$

Second Level for FR_3

$FR_{31} = $ Prepare the drawing environment.
$FR_{32} = $ Draw the line.

$DP_{31} = $ Window type
$DP_{32} = $ Graphics information

The design matrix may be written as

$$\begin{Bmatrix} FR_{31} \\ FR_{32} \end{Bmatrix} = \begin{bmatrix} e & 0 \\ f & g \end{bmatrix} \begin{Bmatrix} DP_{31} \\ DP_{32} \end{Bmatrix} \tag{d}$$

The above design can be put in OOT format as shown in Table E5.4. The top box of the four-tiered object box in Table E5.4 is for the name of the object for each of the FRs. The second boxes are for the second-level FRs, i.e., FR_{11}, FR_{12}, FR_{21}, etc. (in OOT, these were called behaviors). The data, which are DP_{11}, DP_{12}, and so on, are in the third-tier boxes. The fourth-tier boxes are for methods (or operations), which are a^*DP_{11}, b^*DP_{12}, and so on. We can keep the parent-level relationships (i.e., $FR_1-A_{11}-DP_1$, $FR_2-A_{22}-DP_2$, and $FR_3-A_{33}-DP_3$) in a library for future use; this ensures reusability as each one of these relations reveals the fundamental object structure.

TABLE E5.4 Two Different Levels of Objects

	Object 1 (for FR_1)		Object 2 (for FR_2)		Object 3 (for FR_3)	
Second-level objects (behavior in OOT)	Object 11 (for FR_{11}): Define start	Object 12 (for FR_{12}): Define end	Object 21 (for FR_{21}): Detect mouse push	Object 22 (for FR_{22}): Detect mouse release	Object 31 (for FR_{31}): Prepare the drawing environment	Object 32 (for FR_{32}): Draw the line
Attribute or data	DP_{11}: Start point	DP_{12}: End point	DP_{21}: Event for push	DP_{22}: Event for release	DP_{31}: Window type	DP_{32}: Graphics information
Operation or method	$a^* DP_{11}$	$b^* DP_{12}$	$c^* DP_{21}$	$d^* DP_{22}$	$e^* DP_{31}$	$f^* DP_{31} + g^* DP_{32}$

We will name the sets $(FR_1-A_{11}-DP_1)$, $(FR_2-A_{22}-DP_2)$, and $(FR_3-A_{33}-DP_3)$ as Object 1d for FR_1, Object 2d for FR_2, and Object 3d for FR_3, respectively. The "d" is added to the object number to denote the fact that they represent only the diagonal element.

We now have to take care of the off-diagonal terms. This can be done if we name $(FR_3-A_{31}-DP_1/FR_3-A_{32}-DP_2)$ as Object 3* for FR_3^*. The asterisk is used to denote the off-diagonal elements. Then FR_3 is "Object 3 for FR_3," which is an "aggregate" of Object 3d for FR_3 and Object 3* for FR_3^*. This can be represented in terms of the OOT notation, as shown in Figure E5.4.

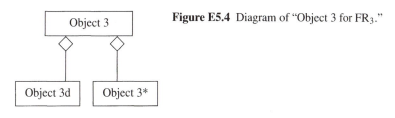

Figure E5.4 Diagram of "Object 3 for FR_3."

The complete design matrix with FRs and DPs can be translated into the OOT structure, as shown in Figure 5.10. This is illustrated further in Figure 5.11. The FRs are equivalent to objects and the DPs are equivalent to data.

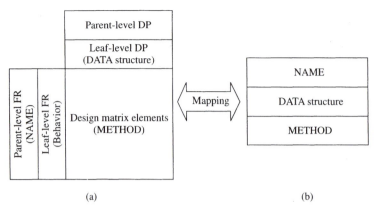

(a)   (b)

Figure 5.10 Correspondence between complete design matrix and OOT diagram. (a) Complete design matrix table. (b) Class diagram.

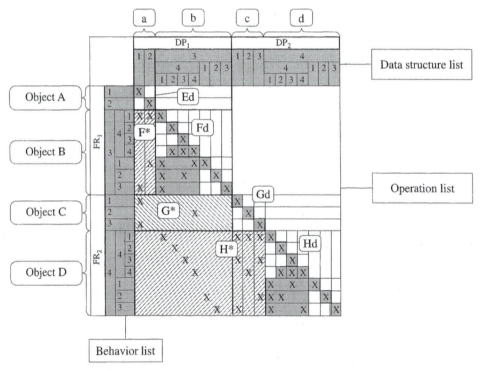

Figure 5.11 The OOT terms used to represent the complete design matrix.

STEP 6

Establish Interfaces by Showing the Relationships between Objects and Operations

Most efforts are focused on this step in the object-oriented method as the relationship is the key feature. However, it lacks information about functional relationships because the

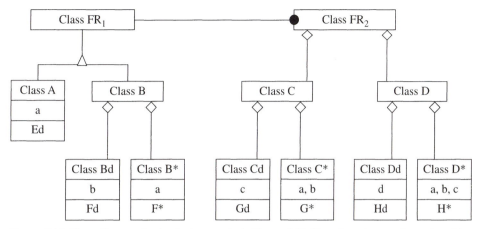

Figure 5.12 Class diagram for the design matrix in Figure 5.11. Note that *class* can be replaced by *object* without loss of generality.

relationship is represented only by lines between objects (or classes) as shown in Figure 5.12. To overcome this lack of information, the Axiomatic Design methodology presented in this chapter utilizes the off-diagonal elements in the design matrix as well as the diagonal elements at all levels. A design matrix element represents a link or association relationship between different FR branches that have totally different behavior. All of these concepts are illustrated in Figure 5.12 for the case of Figure 5.11.

STEP 7

Construct the System Architecture of the Software System

One of the advantages of the AD framework is that it represents the interrelationships between modules in the design hierarchy. To represent the properties of junctions at each level of decomposition, a "module-junction structure diagram" is defined (see Chapter 4). The module-junction structure diagram shown in Figure 5.13 is for the design given by Equation (5.4).

$$\begin{Bmatrix} FR_1 \\ FR_2 \end{Bmatrix} = \begin{bmatrix} X & 0 \\ 0 & X \end{bmatrix} \begin{Bmatrix} DP_1 \\ DP_2 \end{Bmatrix}$$

$$\begin{Bmatrix} FR_{11} \\ FR_{12} \end{Bmatrix} = \begin{bmatrix} X & 0 \\ X & X \end{bmatrix} \begin{Bmatrix} DP_{11} \\ DP_{12} \end{Bmatrix}$$

$$\begin{Bmatrix} FR_{21} \\ FR_{22} \\ FR_{23} \end{Bmatrix} = \begin{bmatrix} X & 0 & 0 \\ X & X & 0 \\ 0 & X & X \end{bmatrix} \begin{Bmatrix} DP_{21} \\ DP_{22} \\ DP_{23} \end{Bmatrix} \qquad (5.4)$$

$$\begin{Bmatrix} FR_{121} \\ FR_{122} \\ FR_{123} \end{Bmatrix} = \begin{bmatrix} X & 0 & 0 \\ X & X & 0 \\ X & 0 & X \end{bmatrix} \begin{Bmatrix} DP_{121} \\ DP_{122} \\ DP_{123} \end{Bmatrix}$$

$$\begin{Bmatrix} FR_{1231} \\ FR_{1232} \end{Bmatrix} = \begin{bmatrix} X & 0 \\ X & X \end{bmatrix} \begin{Bmatrix} DP_{1231} \\ DP_{1232} \end{Bmatrix}$$

It should be noted that the higher level modules are obtained by combining lower level modules. For example, by combining modules M_{121}, M_{122}, and M_{123} with a control junction (C), we obtain the next higher level module M_{12}. Then M_{11} and M_{12} are combined to obtain M_1 using a control junction (C) again. M_1 and M_2, although not shown because they are not leaves, are simply summed as indicated by the symbol "S."

When the AD framework is used, the entire process can be clearly shown by a flow chart that represents the system architecture, which was discussed in Chapter 4. Figure 5.14 shows the final flow diagram for the design represented by Equation (5.4).

The sequence of software development begins at the lowest level, which is defined as the leaves. To achieve the highest level FRs, which are the final outputs of the software, the development of the system must begin from the innermost modules shown in the flow chart that represent the leaves. Then move to the next higher level modules (next innermost box) following the sequence indicated by the system architecture; that is, go from the innermost boxes to the outermost boxes. In short, the software system can be developed in the following sequence.

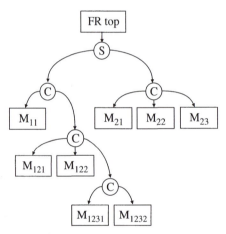

Figure 5.13 Module-junction diagram for design given by Equation (5.4).

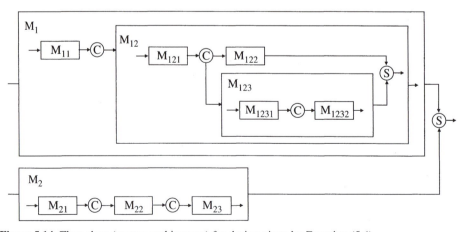

Figure 5.14 Flow chart (system architecture) for design given by Equation (5.4).

a. Construct the core functions using all diagonal elements of the design matrix.
b. Make a module for each leaf FR, following the sequence given in the flow chart that represents the system architecture.
c. Combine the modules to generate the software system, following the module-junction diagram.[15]

When this procedure is followed, the software developer can reduce the coding time as the logical process makes software construction a routine operation. It also guarantees minimal effort for maintenance or extension if either is required in the future. These points are further clarified through the case study described in Section 5.5.

▨ 5.5 AXIOMATIC DESIGN OF OBJECT-ORIENTED SOFTWARE SYSTEM FOR DESIGNERS: ACCLARO SOFTWARE[16]

5.5.1 Introduction

In Section 5.4, the basic concept of designing software based on axiomatic design of object-oriented software systems (ADo-oSS) was presented. In this section, a case study involving the design of a large, commercial software package based on ADo-oSS will be presented. The software—called Acclaro[17]—is an interactive, general-purpose software package for designers who practice axiomatic design. The Acclaro software, which means "make clear" in Latin, was developed very quickly because the (ADo-oSS) methodology was used in developing the software. The entire software package has many layers of decomposition with about several hundred FRs and DPs. For obvious reasons, only a limited portion of the software design will be discussed in this section.

The Acclaro software was designed to aid designers in their design tasks. The software can be used to design a variety of different things in many different fields, including hardware, software, organizations, strategic plans, processes, systems, and products. The ultimate output of the software is a design that satisfies the FRs and Cs. The designer provides inputs to the software, such as the FR, DP, and PV vectors, when prompted by Acclaro software. The designer then answers questions on the relationships between these characteristic vectors, again prompted by the software. Based on these inputs, Acclaro creates design matrices and helps the designer make correct design decisions. It also prompts the designer when a mistake is made, based on various theorems of axiomatic design. The final outcome of the software is the system architecture, which is represented in the form of the {FR}/{DP}/{PV} tree diagram, the module-junction diagram, and the flow diagram. The Acclaro software also generates the final design documentation.

---

[15] Coding of the software can be done using either the module-junction diagram or the flow chart. The module-junction diagram is complementary to the flow chart. The module-junction diagram is similar to the SADT structure, but has more information, such as combining sequences based on the design matrix. Software programmers familiar with the SADT structure may be more comfortable with the module-junction diagram than with the flow chart.

[16] This software was developed by Dr. Sung-Hee Do, which is included as part of Do and Suh (2000).

[17] Acclaro is the trademark of Axiomatic Design Software, Inc.

5.5.2 Axiomatic Design of Acclaro Software

Acclaro software was developed top-down as shown in Table 5.1. A constraint is that the software must be independent of specific operating systems and run on many computers without change or modification. Therefore, the Java language was chosen as the programming language. These top-level FRs and DPs are decomposed until the design task is completed.

The desired first-level FRs of the software are described in Table 5.2. A group of customer attributes, which are listed in Table 5.3, provides the basis for generating the first-level FRs.

TABLE 5.1 Top-Level Decomposition of Acclaro

FR_1	Make a decision-making tool which has impact on the design world
DP_1	Computerized system with the Axiomatic Design software
C	Independent of the operating system

TABLE 5.2 First-Level Decomposition of Acclaro

	Functional Requirements $FR_{1.x}$	Design Parameters $DP_{1.x}$
P	Make a decision-making tool that has impact on the design world	Computerized system with the axiomatic design software
1	Manage design workflow	Design roadmap
2	Provide decision-making environment	Decision-making criterion
3	Support ease of use while using software	Graphic user interface (GUI)
4	Provide efficient data I/O	Data manager
5	Provide utility function	Plug in software

TABLE 5.3 Some Customer Attributes for Acclaro

CA_1	Make design document
CA_2	User-friendly graphical user interface
CA_3	Indicate the user's mistake or guide
CA_4	Provide a collaborative design environment
CA_5	Provide efficient database
CA_6	Provide decision-making environment
CA_7	Manage the design activity
. . .	. . .

TABLE 5.4 Second-Level Decomposition for $FR_{1.2}$

	Functional Requirements $FR_{1.2.x}$	Design Parameters $DP_{1.2.x}$
P	Provide decision-making environment	Decision-making criterion
1	Provide design sequence in terms of axiomatic design	Decomposition roadmap
2	Maintain functional independence	Criterion for Independence Axiom
3	Make suggestions for better design	Criterion for Information Axiom and robust design

Equation (5.5) shows the first-level design equation. It is a decoupled design, although with the sequence shown, the matrix has Xs above and below the diagonal.

$$\begin{Bmatrix} FR_{11} \\ FR_{12} \\ FR_{13} \\ FR_{14} \\ FR_{15} \end{Bmatrix} = \begin{bmatrix} X & 0 & 0 & 0 & 0 \\ X & X & 0 & 0 & 0 \\ X & X & X & X & X \\ X & X & 0 & X & 0 \\ 0 & 0 & 0 & X & X \end{bmatrix} \begin{Bmatrix} DP_{11} \\ DP_{12} \\ DP_{13} \\ DP_{14} \\ DP_{15} \end{Bmatrix} \qquad (5.5)$$

Equation (5.6) shows the rearranged design matrix, indicating that the sequence of development or operation should follow a certain order, FR_{11}–FR_{12}–FR_{14}–FR_{15}–FR_{13}. The design matrix shows that the GUI is the most involved module in the first level because it is affected by all DPs.

$$\begin{Bmatrix} FR_{11} \\ FR_{12} \\ FR_{14} \\ FR_{15} \\ FR_{13} \end{Bmatrix} = \begin{bmatrix} X & 0 & 0 & 0 & 0 \\ X & X & 0 & 0 & 0 \\ X & X & X & 0 & 0 \\ 0 & 0 & X & X & 0 \\ X & X & X & X & X \end{bmatrix} \begin{Bmatrix} DP_{11} \\ DP_{12} \\ DP_{14} \\ DP_{15} \\ DP_{13} \end{Bmatrix} \qquad (5.6)$$

FR_2 (provide decision-making environment) may be decomposed into a sublevel with DP_2 (decision-making criterion). Table 5.4 shows the second-level decomposition for FR_2 and Equation (5.7) shows the relationships in the second level for FR_2.

$$\begin{Bmatrix} FR_{121} \\ FR_{122} \\ FR_{123} \end{Bmatrix} = \begin{bmatrix} X & 0 & 0 \\ 0 & X & 0 \\ 0 & X & X \end{bmatrix} \begin{Bmatrix} DP_{121} \\ DP_{122} \\ DP_{123} \end{Bmatrix} \qquad (5.7)$$

In this manner, the software system can be designed using the AD framework. The Acclaro software design has a nine-level hierarchy and well over 1000 leaves, the number of which may increase as other features are added.

5.5.3 Axiomatic Design on the FR_{1141} Branch

Figure 5.15 illustrates the complete design matrix table with module information for the FR_{1141} branch, which deals with the data structure for FRs, DPs, and design matrices in Acclaro software. Figure 5.16 also gives more detailed information for one of the modules to clarify the definition of modules. The FR_{11412} branch module for the data structure of FRs, which is located in the upper-left corner in Figure 5.15, is selected as an example. The off-diagonal term reveals the interrelationship between FRs and DPs located in different branches. It will guide the integration sequence between objects or classes that will be defined later.

5.5.4 Object-Oriented Model: Bottom-Up Approach

The representation of the design for the FR_{1141} branch based on the FR and DP hierarchies and the design matrix can be transformed into the OOT representation as shown in Figure 5.17. This is done by using indices for objects in a manner consistent with the indices for FRs and DPs, and also treating the object as consisting of a diagonal element whose index is suffixed by "d" and an off-diagonal element whose index is suffixed by a "*". Using this

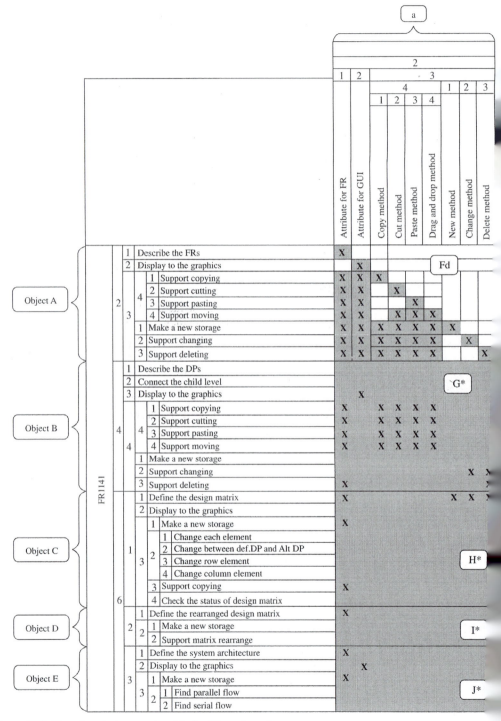

Figure 5.15 Complete design matrix table for FR$_{1141}$ branch.

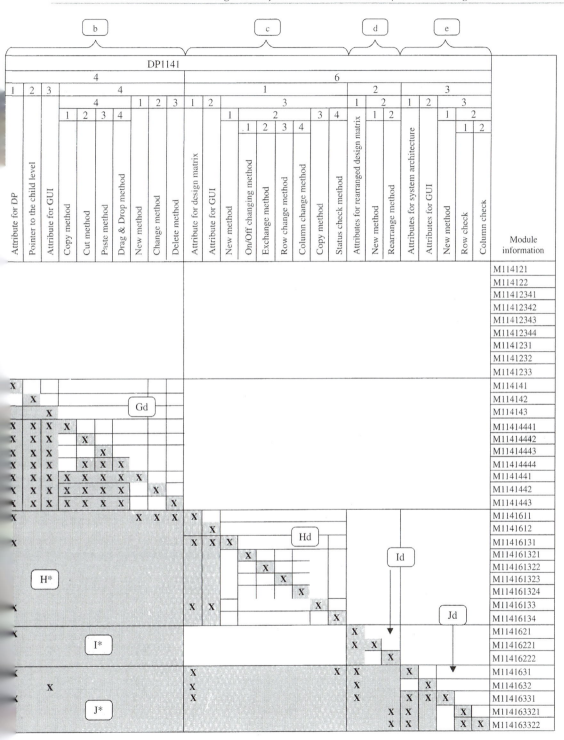

			2						
					3				
				4				1	2	3
			1	2	3	4			
	Attribute for FR	**Attribute for GUI**	**Copy method**	**Cut method**	**Paste method**	**Drag & drop method**	**New method**	**Change method**	**Delete method**
1 Describe the FRs	description, keyword..								
2 Display to the graphics		posx, posy,..							
4 1 Support copying	Relationship: If one of the attributes is changed, all the methods should be checked		Copy()						
2 Support cutting				Cut()					
3 Support pasting					Paste()				
4 Support moving		setPosition()		X	X	D&D()			
1 Make a new storage							Constructor		
2 Support changing				Clipboard				Delete()	
3 Support deleting									Change()

(Row labels at left: FR 1 1 4 1)

Figure 5.16 Detailed module definition for FR$_{11412}$ branch.

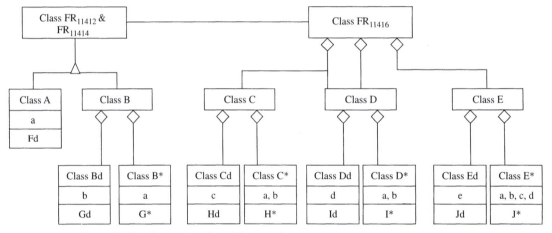

Figure 5.17 Object-oriented model generation.

system, the software designer can generate an object-oriented model such as that shown in Figure 5.17, which is derived from Figure 5.15.

5.5.5 Coding with the System Architecture

The system architecture for the FR$_{1141}$ branch may be illustrated as Figure 5.18. After the diagonal element in the full design matrix is coded, this flow diagram guides the coding sequence as well as maintenance sequences.

As discussed in Section 5.2.2, the flow chart shown in Figure 5.18 can be used in many different ways, such as

    a. Diagnosis of software failure
    b. Software change orders
    c. Job assignment and management of software-development team
    d. Distributed systems

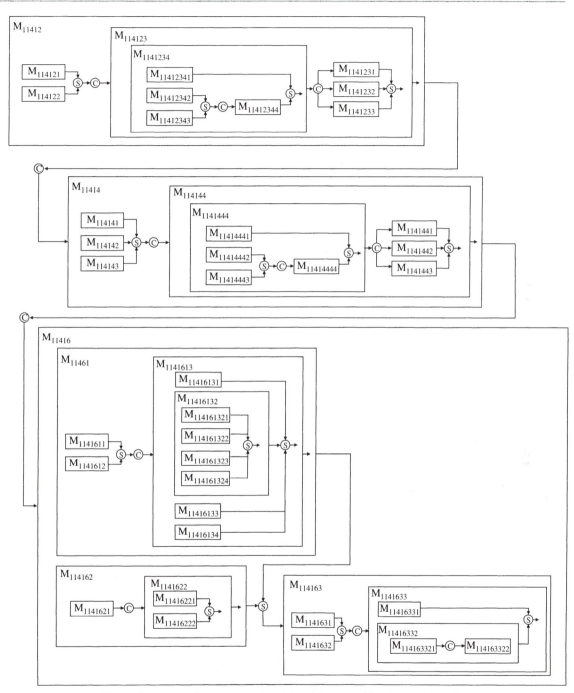

Figure 5.18 Flow diagram at the fourth-level decomposition for Acclaro software.

e. Software design through assembly of modules

f. System consisting of hardware and software

g. Interactive software program

Figure 5.19 shows an example screen of the Acclaro software.[18]

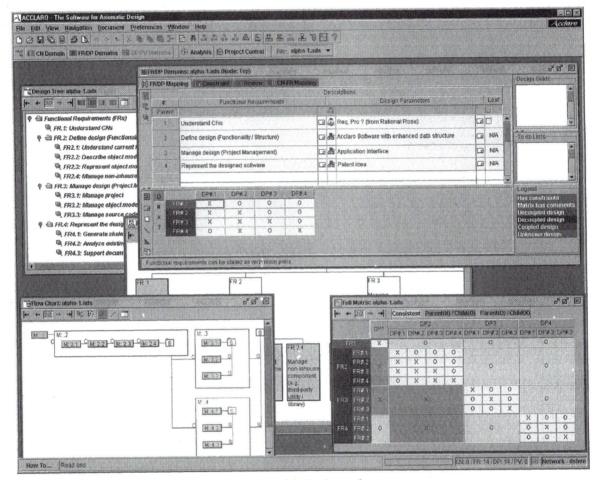

Figure 5.19 An example screen of the Acclaro software.

5.6 DESIGN OF RAPID-PROTOTYPING SOFTWARE FOR REAL-TIME CONTROL OF HARDWARE/SOFTWARE SYSTEM[19]

In an ideal software-design situation, we should be able to define the FR to be satisfied, select the DP we want to use to control the FR, and determine the design matrix and the modules

---

[18] An introduction to Acclaro can be found on the web page of Axiomatic Design Software, Inc., http://www.axiomaticdesign.com.

[19] From Igata (1996).

through modeling without resorting to any experimental verification. However, in some physical situations, we find that although the FR and the DP are chosen, the modules cannot be determined because the modeling technique is not sufficiently accurate to predict the FR/DP relationship. In this case, we have to depend on experimental results to determine the modules. The software must be so programmed that the modules can be determined based on test results.

An example of such a case is the design of the software for an antilock braking system (ABS). Igata (1996) investigated the use of axiomatic design theory for the specific case of developing software for the ABS, which cannot be fully specified from the beginning because of the complicated physical environment that the vehicle experiences during its lifetime. Although in theory, such a dynamic system can be modeled, the validity of the model must be independently verified through testing. Even then, it may be difficult to capture all possible events analytically. Therefore, in these situations, it may be quicker and easier to develop a small portion of the software to evaluate and modify through real-time testing until the desired software system is fully developed.

In the case of a system involving both hardware and software, a part of the hardware may constitute the module. The DP is input to this hardware. The product of the module and the DP generates the FR. The input DP can come from the lower level DPs, in accordance with the DP hierarchy and the associated design matrix. This is illustrated using the design of a software system for the ABS (Igata, 1996).

ABS was introduced in the 1970s to help automobile drivers to control and stop a vehicle in the shortest possible distance when driving on slippery roads. The idea behind ABS is that when the wheel completely locks up too quickly, the car will skid because there is not sufficient friction between the tire and the road. When the car skids, the driver loses control of the vehicle and the vehicle may move transversely as well as longitudinally. To prevent this skidding, the speeds of the four wheels are monitored and compared with the estimated vehicle speed to apply intermittent braking force to prevent a premature lock-up of the wheels and bring the vehicle to a stop in the shortest possible distance.

The functional requirements at the highest level of ABS are

$FR_1 =$ Minimize the stopping distance.
$FR_2 =$ Maintain stability and steerability of the vehicle.
$FR_3 =$ Inform driver of the vehicle condition.
$FR_4 =$ Maintain reliability.
$FR_5 =$ Generate self-diagnosis.

The C is the total system cost. In 1996, the price of the ABS offered to the original equipment manufacturer was in the range of $100 to $200. Therefore, the manufacturing cost must be substantially less than $100.

To minimize the stopping distance, the maximum resisting force to longitudinal motion is required. To have maximum stability, the resisting force to transverse motion must also be large. To have the maximum friction force along the longitudinal direction, there must be a very small slip between the tire and the road surface. That is, the tangential velocity of the tire must be slightly lower than the vehicle speed. However, the transverse friction force is at a maximum when there is no slip, and as the slip increases, the transverse force decreases. This is a well-known tribological phenomenon, i.e., the friction force is at a maximum along the direction of the resultant sliding velocity. This situation is schematically illustrated in Figure 5.20 in terms of the slip ratio S, which is defined as $(V_b - V_w)/V_b$, where V_b is the vehicle speed and V_w is the tangential velocity of the wheel.

In view of the tribological nature of the friction force between the tire and the smooth road, FR_1 and FR_2 may be combined to come up with a new FR_1. The new set of FRs may be stated as

$FR_1 =$ Minimize the stopping distance while maintaining stability and steerability.
$FR_2 =$ Inform driver of the vehicle condition.
$FR_3 =$ Maintain reliability.
$FR_4 =$ Generate self-diagnosis.

The corresponding set of DPs may be chosen as

$DP_1 =$ Slip ratio S
$DP_2 =$ Brake pedal vibration
$DP_3 =$ Redundant mechanism
$DP_4 =$ Self-diagnosis module

The design matrix is a diagonal matrix, indicating that at the highest level, the design is a completely uncoupled design.

As discussed earlier, one of the critical inputs for decision making is the absolute velocity of the vehicle, as the slip ratio S must be known. However, many of the current commercially available ABS products do not have an independent sensor that can measure the absolute vehicle velocity. In this case, the vehicle speed must be estimated from the speeds of the four wheels.

Figure 5.21 shows a method used to estimate vehicle speed, which is plots of the vehicle speed versus time. Figure 5.21a shows the case when the vehicle is running at constant speed, i.e., $S = 0$. Figure 5.21b shows a normal braking case when some brake force is applied and the vehicle stops at $t = t_1$. In this case, S is nonzero while braking because the wheel speed is slightly below the vehicle speed and we have maximum deceleration. Figure 5.21c is the case of excessive braking force, causing lock-up of the wheels while the vehicle is still moving forward. In this case, the lateral frictional force is very small and the vehicle is unstable and out of control. Figure 5.21d shows the case of the ABS system acting properly. The brake force is controlled to prevent wheel lock-ups by taking the maximum of the

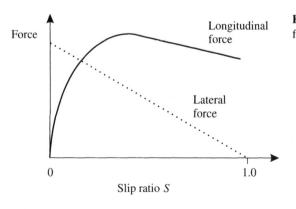

Figure 5.20 Schematic diagram of friction force versus slip ratio.

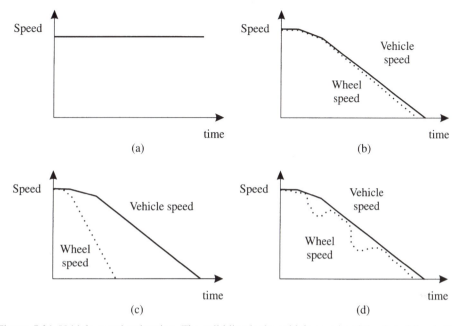

Figure 5.21 Vehicle speed estimation. The solid line is the vehicle speed and the dotted line is the wheel speed. (a) Steady speed. (b) Normal braking. (c) Excessive braking. (d) With ABS.

four wheel speeds to estimate the vehicle speed. When the speed of the four wheels drops synchronously, the deceleration can be calculated and compared with a reference value that is typically $1.0g$.

Based on the FR_1/DP_1 mapping, FR_1 can be decomposed in many different ways. Igata (1996) decomposed it as follows:

$FR_{11} =$ Detect the wheel speed.
$FR_{12} =$ Estimate vehicle speed.
$FR_{13} =$ Calculate the slip ratio S from the wheel speed and the vehicle speed.
$FR_{14} =$ Determine the output signal so that S is controlled at the target value.
$FR_{15} =$ Convert the output signal to brake force.

The corresponding DPs are:

$DP_{11} =$ Wheel speed sensor and calculation module
$DP_{12} =$ Vehicle speed estimation module
$DP_{13} =$ Slip ratio calculation module
$DP_{14} =$ Output signal calculation module
$DP_{15} =$ Brake force actuator

The design equation is

$$\begin{Bmatrix} FR_{11} \\ FR_{12} \\ FR_{13} \\ FR_{14} \\ FR_{15} \end{Bmatrix} = \begin{bmatrix} X & 0 & 0 & 0 & 0 \\ X & X & X & X & 0 \\ X & X & X & 0 & 0 \\ X & X & 0 & X & 0 \\ 0 & 0 & 0 & 0 & X \end{bmatrix} \begin{Bmatrix} DP_{11} \\ DP_{12} \\ DP_{13} \\ DP_{14} \\ DP_{15} \end{Bmatrix} \qquad (5.8)$$

This is a coupled design. The coupling is caused by the fact that we estimate the absolute velocity of the vehicle rather than measuring it directly. This problem can be overcome if we use an accelerometer to measure the deceleration directly, as is done in many four-wheel-drive vehicles, i.e., make DP_{12} = accelerometer. Then the design matrix becomes triangular, and the design becomes decoupled.

$$\begin{Bmatrix} FR_{11} \\ FR_{12} \\ FR_{13} \\ FR_{14} \\ FR_{15} \end{Bmatrix} = \begin{bmatrix} X & 0 & 0 & 0 & 0 \\ X & X & 0 & 0 & 0 \\ X & X & X & 0 & 0 \\ X & X & X & X & 0 \\ 0 & 0 & 0 & 0 & X \end{bmatrix} \begin{Bmatrix} DP_{11} \\ DP_{12} \\ DP_{13} \\ DP_{14} \\ DP_{15} \end{Bmatrix} \tag{5.9}$$

Some of the FRs at this level may require further decomposition. For example, FR_{14} must be further decomposed in order to achieve the objective of determining the output signal so that the slip ratio S can be controlled to a set value. Because DP_{14} is the output calculation module, FR_{14} may be decomposed as

FR_{141} = Estimate other vehicle condition parameters.
FR_{142} = Calculate the output signal.
FR_{143} = Correct the output signal based on other vehicle conditions.

The DPs at this level of decomposition are

DP_{141} = Input from sensors supplied to vehicle condition estimation module M_{141}
DP_{142} = Sensor output that feeds the output estimation module M_{142}
DP_{143} = Sensor output that feeds the signal correction module M_{143}

This is a decoupled design, but further decomposition may be necessary if some of the modules are not available in the marketplace or within the company.

The flow diagram for this ABS design is shown in Figure 5.22. If we proceed with further decomposition, there will be additional boxes inside of the innermost boxes shown in Figure 5.22.

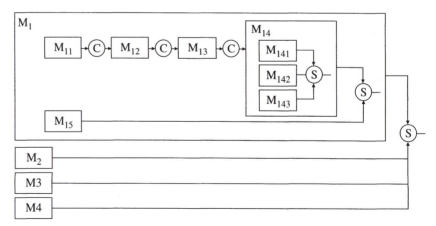

Figure 5.22 Flow diagram for the ABS system. The software modules are shown in boxes. The system can be controlled starting from the innermost boxes.

Some of the modules can be constructed based on test results such as those shown in Figure 5.22. Others can be determined through mathematical modeling. However, this is not done in this section.

5.7 AN IDEAL SOFTWARE SYSTEM

The Independence Axiom states that at a given level of the FR hierarchy, all of the FRs must be maintained independent when any DP is changed to alter its respective FR. This fact can be used to create a thinking design machine for products and processes (Suh, 1990; Suh and Sekimoto, 1990). This same concept can be used to develop a software design system that will considerably simplify the software design process.

For this purpose, consider a functional requirement hierarchy where the lowest level FRs, designated FR^Ls (i.e., "leaves"), do not require further decomposition. The higher level FRs can be constructed by means of the lowest level FR^Ls. Then we can construct a software database that consists of the product of modules and design parameters as follows:

$$FR_1^L \$ [MDP_1^a, MDP_1^b, MDP_1^c, \ldots, MDP_1^n]$$

$$FR_2^L \$ [MDP_2^a, MDP_2^b, MDP_2^c, \ldots, MDP_2^n] \tag{5.10}$$

$$\cdots$$

$$FR_x^L \$ [MDP_x^a, MDP_x^b, MDP_x^c, \ldots, MDP_x^m]$$

where MDP_i is the product of module M_i and DP_i. The $\$$ sign signifies the fact that any one of the MDP_1 modules can satisfy the functional requirement FR_1^L. For example, the first equation for FR_1^L states that any one of the modules (e.g., $MDP_1^a, MDP_1^b, MDP_1^c$, etc.) in the database can be used to control FR_1^L. Similarly, any one of the MDP_2^x can satisfy FR_2^L. An MDP may be subroutines, machine language, or any software that yields an $FR_{ijkl}(d)$, where $FR_{ijkl}(d)$ is obtained from the diagonal element of an FR_{ijkl}/DP_{ijkl} relationship discussed in Section 5.4.3. The FR_{ijkl}^* due to an off-diagonal element may be stored only for specific situations, because there are too many combinations of FR_i/DP_j relationships to store for unforeseen events. Equations (5.10) constitute a library of software modules.

Now suppose that we need to develop software for a problem consisting of the following three leaves, FR_1^L, FR_2^L, and FR_3^L. The best software package is the one that consists of DPs, each of which affects only one of the FR^Ls, i.e., there are no FR^{L*}. This can be expressed as

$$\begin{Bmatrix} FR_1^L \\ FR_2^L \\ FR_3^L \end{Bmatrix} = \begin{bmatrix} M_1 & 0 & 0 \\ 0 & M_2 & 0 \\ 0 & 0 & M_3 \end{bmatrix} \begin{Bmatrix} DP_1^s \\ DP_2^n \\ DP_3^b \end{Bmatrix} \tag{5.11}$$

In conceptualizing the design solution represented by Equation (5.11), only those DPs that affect just one of the FRs are chosen. The above software package is an uncoupled design, which satisfies the Independence Axiom. The flow chart diagram for the above design is simply a junction S connecting the three modules.

When we cannot find an uncoupled design, we may look for a decoupled solution, which also satisfies the Independence Axiom. For example, when we first look for a DP that can yield FR_1^L, it can be anything in the database, for example, DP_1^c or DP_1^h. However, when we then look for the software module that can yield FR_2^L, we must look for a DP_2^x that does not affect FR_1^L. Finally, when we look for a DP_3 that can satisfy FR_3^L after FR_1^L and

FR_2^L are taken care of, there are now two conditions to satisfy the Independence Axiom: the DP_3^x in the database must not affect either FR_1^L or FR_2^L. Then the design equation for the decoupled software may be written as

$$\left\{ \begin{array}{c} FR_1^L \\ FR_2^L \\ FR_3^L \end{array} \right\} = \left[\begin{array}{ccc} M_1 & 0 & 0 \\ ? & M_2 & 0 \\ ? & ? & M_3 \end{array} \right] \left\{ \begin{array}{c} DP_1^r \\ DP_1^i \\ DP_1^v \end{array} \right\} \tag{5.12}$$

An interesting question arises when the elements of the design matrix given by (?) are not known. Obviously, it is desirable to know them. However, even if they are not known, the software can execute the computation when the desired outputs, FR_1^L, FR_2^L, FR_3^L, are known and measurable. This can be done if the modules are computed in the sequence indicated in Equation (5.12), i.e., DP_1 first, then DP_2, and finally DP_3. Such a program will involve two control junctions in the flow chart diagram.

If the database (i.e., library of modules) is given for FRs at higher levels, the software can be constructed without going through the decomposition process. Whatever is in the library is for the lowest level FRs, as we do not have to decompose these modules. The higher level FRs can be constructed from the elementary-level FR^Ls through vertical integration.

The key to the creation of an idealized software package is the creation of the complete data (i.e., module) base and the creation of uncoupled or decoupled designs.

In many situations, the database is so extensive that we may be able to find another set of DPs that also satisfies FR_1^L, FR_2^L, FR_3^L independently. Then the question becomes: "Which set of DPs is a better solution for a given set of FRs?" To decide which is a better solution, we have to invoke the Information Axiom. Different software modules may satisfy FR^Ls with different degrees of accuracy. That is, we must evaluate how well the FRs are satisfied by the chosen DPs. The best solution, according to the Information Axiom, is the one with the minimum information content.

5.8 OTHER ISSUES RELATED TO SOFTWARE DESIGN

5.8.1 Reusability

One of the important issues in software engineering is reusability. The fact that a library of software knowledge exists means that when certain software is created, it can be stored in the library and retrieved at will. Thus, reusability is ensured. In describing software design in Section 5.5, it was noted that the best way of having a robust library of software for reusability is to create modules for FR_i and DP_i that involve only the diagonal elements. The additional component of the module for any off-diagonal elements can be treated ad hoc for each individual case and later added to the module for diagonal elements to create a complete module.

5.8.2 Extensionality

Whenever a new FR needs to be added at any level, the FR to be inserted must be consistent with the overall architecture of the software system. For example, an FR cannot be inserted at a mid-level of a system architecture if it cannot be obtained by decomposing a higher level FR to be consistent with the criterion for decomposition. The criterion is that the lower level FR must be consistent with its parent FR and must represent the FRs of the parent DP. If there is clear need for a certain FR but it cannot be generated from the higher level FRs, move up the design hierarchy to a level where it can be added. In the extreme case, a new FR can be added only at the highest level.

When a new FR is inserted, it should show in the flow diagram as a separate box that does not create coupling with other boxes. If the FR is decomposed further, there will be inner boxes inside this outer box.

5.8.3 Knowledge and Information Requirements in Software Design

The purpose of design is to achieve the goals articulated by the functional requirements. We design artifacts that satisfy the specified FRs and constraints. One question that comes up in all design situations is: "Do we have to have complete knowledge about a proposed design to come up with a viable design solution?" Or "What is the minimum knowledge that is necessary and sufficient to make design decisions given a set of DPs for a given set of FRs?" Obviously, it would be desirable and reassuring if complete knowledge about a given design were available. However, there are situations in which the available knowledge may not be complete.

The required knowledge about a design depends on whether the proposed design satisfies the Independence Axiom. In the case of a coupled design, which is not a good design, we need the knowledge that will enable us to formulate all elements of the design matrix precisely. The same is true of the decoupled design. However, in the case of a decoupled design, when there is a state change (i.e., when the FRs change from State A to State B), the effect of the state change must be isolated to minimize the information required for the decoupled design.

Consider an ideal design that consists of three FRs. For an uncoupled design, which is the simplest case, the design equation may be written as

$$\begin{Bmatrix} FR_1 \\ FR_2 \\ FR_3 \end{Bmatrix} = \begin{bmatrix} A_{11} & 0 & 0 \\ 0 & A_{22} & 0 \\ 0 & 0 & A_{33} \end{bmatrix} \begin{Bmatrix} DP_1 \\ DP_2 \\ DP_3 \end{Bmatrix} \tag{5.13}$$

A_{11}, A_{22}, and A_{33} relate FRs to DPs. They are constants in the case of linear design, whereas in the case of nonlinear design, A_{11} is a function of DP_1, etc. To set the values of FR_1, FR_2, and FR_3, we need to know only that the design matrix is a diagonal matrix for the chosen DP_1, DP_2, and DP_3. We do not need any other information, although to have a robust design, the magnitudes of the diagonal elements should be small.

Theorem Soft 1 (Knowledge Required to Operate an Uncoupled System)

Uncoupled software or hardware systems can be operated without precise knowledge of the design elements (i.e., modules) if the design is truly an uncoupled design and if the FR outputs can be monitored to allow a closed-loop control of FRs.

Again consider the three-FR case, but this time the design is a decoupled design given by the following design equation:

$$\begin{Bmatrix} FR_1 \\ FR_2 \\ FR_3 \end{Bmatrix} = \begin{bmatrix} A_{11} & 0 & 0 \\ A_{21} & A_{22} & 0 \\ A_{31} & A_{32} & A_{33} \end{bmatrix} \begin{Bmatrix} DP_1 \\ DP_2 \\ DP_3 \end{Bmatrix} \tag{5.14}$$

In the case of the decoupled design given by Equation (5.14), we need to know the diagonal elements A_{ii}. It is also desirable to know the off-diagonal elements A_{ij}. However, in the absence of complete information, we can proceed with the design and make design decisions even if we do not know the off-diagonal elements, especially if they are smaller in magnitude than the diagonal elements, i.e., $A_{ij} < A_{ii}$. For instance, we can vary DP_1 to set the value of FR_1 first, which will also affect FR_2. Then we may proceed to set the value of FR_2 by varying the value of DP_2. When we choose DP_2, we must be certain that it does not affect FR_1, but it is not necessary that we have any information on A_{21}, especially if DP_2 has the dominant effect on FR_2, i.e., $A_{22} > A_{21}$, and if FR_2 can be measured to allow a feedback control scheme. Similarly, as long as DP_3 does not affect FR_1 and FR_2, we can proceed with the design, even if we do not have any information on A_{31} and A_{32}.

The point is that we can proceed with design decisions even when we do not know the relationship between FR_i and DP_j precisely if the system is going to operate in a closed-loop mode. In this case, to make a correct decision, we must vary the DPs in the right sequence, i.e., we must know the design equation and the design matrix.

Theorem Soft 2 (Making Correct Decisions in the Absence of Complete Knowledge for a Decoupled Design with a Closed-Loop Control)

When the software system is a decoupled design, the FRs can be satisfied by changing the DPs if the design matrix is known to the extent that knowledge about the proper sequence of change is given, even though precise knowledge about the design elements may not be known, if the output FRs can be measured.

The design matrix must be truly triangular. Otherwise, Theorem Soft 2 does not hold true. Suppose that the upper triangular elements are not quite equal to zero but have very small values a_{12}, a_{13}, and a_{23} as shown in Equation (5.15).

$$\begin{Bmatrix} FR_1 \\ FR_2 \\ FR_3 \end{Bmatrix} = \begin{bmatrix} A_{11} & a_{12} & a_{13} \\ A_{21} & A_{22} & a_{23} \\ A_{31} & A_{32} & A_{33} \end{bmatrix} \begin{Bmatrix} DP_1 \\ DP_2 \\ DP_3 \end{Bmatrix} \tag{5.15}$$

The absolute magnitudes of the elements a_{ij} are much smaller than those of A_{ij}, i.e., $|a_{ij}| << |A_{ij}|$. In this case, FR_1 will be affected by large state changes of DP_2 and DP_3 and this effect may not be negligible as

$$\Omega FR_1 = A_{11}\, \Omega DP_1 + a_{12}\, \Omega DP_2 + a_{13}\, \Omega DP_3 \tag{5.16}$$

where Ω denotes a major change in the states of DPs and FRs. When DP_1 changes by ΩDP_1 for a linear design, FR_1 also changes by ΩFR_1. In this case, the effect of the state change (e.g., the change in FR_1 due to the change in DP_2 when the system goes from State A to State B) must be subtracted from the value of FR. For this to be possible, the magnitude of the a_{ij}'s and A_{ij}'s must be known. In this case, the decision making becomes much more difficult. The best thing to do is to develop a design such that all a_{ij}'s are zeros.

When the design is coupled, we must have complete knowledge of the design matrix to be able to satisfy FRs by changing DPs. Even a closed-loop control will not be useful, as a feedback control of one FR may negatively affect the outcome of other FRs. Thus, much more exact information is required to search for the unique operating point of a coupled system. When the system knowledge is incomplete, uncoupled designs cannot be operated with precision.

5.9 IMPLICATIONS OF THE INFORMATION AXIOM IN SOFTWARE DESIGN

One of the key issues in developing a software system is the evaluation of the quality of the system. In axiomatic design, it is done through the application of the Information Axiom and the measurement of the information content. The Information Axiom states that the information must be minimized, and the information content is a measure of the probability of success in achieving the stated FRs.

5.9.1 Qualitative Implementation of the Information Axiom

The Information Axiom can be applied both qualitatively as well as quantitatively. In a qualitative sense, it means any way the information content can be reduced by maximizing the probability of success in achieving the FRs. For example, the number of the lines of a computer code may be minimized; this assumes that by minimizing the length of the computer code, the probability that a computer program performs the desired functions correctly increases. Another way might be to minimize the number of computational steps by developing an efficient algorithm.

5.9.2 Quantitative Measure of the Information Content

In previous chapters, the information content was defined as the probability of success in achieving the FRs and was measured by determining the area of the common range, which is the area of the probability density function (pdf) of the system range that is inside the design range. There may be situations in which the design range and the system range for a software system can be defined quantitatively, so that the information content may be calculated by determining the area of the common range.

In some software systems, the design range may not be definable because the software gives either the right answer or a wrong answer. In this case, we may define the information content as

$$I = \log_2 \left(\frac{\text{Number of relevant hits}}{\text{Total number of executions}} \right) \tag{5.17}$$

If we have two software systems that satisfy the same set of FRs and the information contents are the same, but one goes through 20 times the computational steps, is one of these software systems more complex than the other? Since they achieve the same FRs with the same probability of success, they are equivalent but *not* identical. The difference between these two software systems is the *computational efficiency*, which is different from the information content. This subject will be revisited in Chapter 9 in relation to complexity.

5.10 SUMMARY

This chapter presents a framework for software design based on axiomatic design that provides a systematic way of designing software. Many case studies and examples are given to illustrate the axiomatic design of software.

The AD framework has been applied to the design and development of an object-oriented software system. In particular, axiomatic design for object-oriented software system (ADo-oSS) is presented as a specific methodology for developing software. The AD framework supplies a method to determine the viability of software systematically. It ensures that the modules are in the right place in the right order, when the modules are established as a row of the design matrix. Axiomatic design methodology for software development can help software engineers and programmers to develop effective and reliable software systems quickly.

A case study was presented to show how ADo-oSS can be designed and implemented using the Java language. The interactive software system called Acclaro has been developed for designers who design hardware, software, systems, and processes. It automatically generates the FR/DP/PV hierarchies and the flow chart to represent the system architecture. It makes suggestions for design improvements based on the design axioms. It also generates engineering documents automatically.

In summary, the axiomatic approach to software systems can be characterized in terms of the following features:

1. The software design begins with the specification of FRs from the perceived needs of the user.
2. The axiomatic approach recognizes the existence of independent domains and spaces in software design.
3. The design process requires mapping between these domains.
4. There are two design axioms that must be satisfied by the mapping process in order to develop an acceptable system.
5. Each domain has a characteristic vector that can be decomposed to establish a hierarchical tree. In software design, the outputs constitute the FR tree in the functional domain, while key inputs form the DP tree in the physical domain.
6. The decomposition process requires zigzagging between two adjacent domains.
7. The decomposition process terminates when all branches of the FR hierarchical tree (FR leaves) can be fully satisfied by the selected set of DPs.
8. Each FR leaf has one module of software that is a row of the design matrix that describes the relationship between the FR leaf and the selected DP.
9. Modularity of the software system does not guarantee a good software design. The modules must satisfy the Independence Axiom.
10. These modules are reusable in axiomatic design when the identical FR is to be satisfied by the same DP. These modules can be stored in a library of modules.
11. The module-junction structure diagram is devised to represent the vertical integration of the modules. Three types of junctions are defined based on the coupling property of the design matrix, i.e., summation, control, and feedback junctions.

REFERENCES

Baldwin, D. F. "Microcellular Polymer Processing and the Design of a Continuous Sheet Processing System," Ph.D. Thesis, Massachusetts Institute of Technology, 1994.

Booch, G. *Object-Oriented Analysis and Design with Applications*, 2nd ed., Benjamin/Cummings, San Francisco, CA, 1994.

Cox, B. J. *Object-Oriented Programming*, Addison-Wesley, Reading, MA, 1986.

DeMarco, T. *Structured Analysis and System Specification*, Prentice-Hall, Englewood Cliffs, NJ, 1979.

Do, S.-H., and Park, G.-J. "Application of Design Axioms for Glass-Bulb Design and Software Development for Design Automation," Presented at the CIRP Workshop on Design and Intelligent Manufacturing Systems, Tokyo, June 20–21, 1996.

Do, S.-H. "Application of Design Axioms to the Design for Manufacturability for the Television Glass Bulb," Ph.D. Thesis, Hanyang University, Seoul, Korea, 1997.

Do, S.-H. "Axiomatic Design Software," Unpublished paper of the MIT Axiomatic Design Group, August 1999.

Do, S.-H., and Suh, N. P. "Axiomatic Design of Software Systems," *CIRP Annals*, Vol. 49, No. 1, 2000.

El-Haik, B. "The Integration of Axiomatic Design in the Engineering Design Process," 11th Annual RMSL Workshop, May 12, Detroit, MI, 1999.

Igata, H. "Application of Axiomatic Design to Rapid-Prototyping Support for Real-Time Control Software," S.M. Thesis, Massachusetts Institute of Technology, May 1996.

Jackson, M. A. *Principles of Program Design,* Academic Press, San Diego, CA, 1975.

Kim, S. G., and Suh, N. P. "The Knowledge Synthesis System for Injection Molding," *International Journal of Robotics and CIM*, Vol. 3, No. 3, 1987.

Kim, S. J., Suh, N. P., and Kim, S.-K. "Design of Software Systems Based on Axiomatic Design," *Annals of the CIRP*, Vol. 40, No. 1, pp. 165–170, 1991 [also *Robotics & Computer-Integrated Manufacturing*, 3:149–162, 1992].

Marca, D. A., and McGowan, C. L. *SADT: Structured Analysis and Design Technique*, McGraw-Hill, New York, 1988.

Orr, K. *Structured Systems Development*, Yordon Press, New York, 1977.

Park, G. J. "Axiomatic Design vs. Software Engineering," NSF Sponsored Axiomatic Design Workshop for Professors, MIT, Boston, MA, June, 1998.

Pressman, R. S. *Software Engineering—A Practitioners Approach*, 4th ed., McGraw Hill, New York, 1997.

Ross, D. T. "Structured Analysis (SA): A Language for Communicating Ideas," *IEEE Transactions on Software Engineering*, Vol. SE-3, No. 1, pp. 16–34, 1977.

Ross, D. T. "Applications and Extensions of SADT," *IEEE Computer*, Vol. 18, No. 4, pp. 25–35, 1985.

Rumbaugh, J., Blaha, M., Premerlani, W., Eddy, F., and Lorensen, W. *Object-Oriented Modeling and Design*, Prentice-Hall, Englewood Cliffs, NJ, 1991.

Suh, N. P. *The Principles of Design*, Oxford University Press, New York, 1990.

Suh, N. P. "Design and Operation of Large Systems," *Journal of Manufacturing Systems,* Vol. 14, No. 3, pp. 203–213, 1995.

Suh, N. P. "Design of Systems," *Annals of CIRP*, Vol. 46, No. 1, pp. 75–80, 1997.

Suh, N. P. "Axiomatic Design Theory for Systems," *Research in Engineering Design*, Vol. 10, pp. 189–209, 1998.

Suh, N. P. and Sekimoto, S. "Design of Thinking Design Machine," *Annals of CIRP,* Vol. 39/1, pp. 145–148, 1990.

Tate, D. "A Roadmap for Decomposition: Activities, Theories, and Tools for System Design," Ph.D. Thesis, Massachusetts Institute of Technology, 1999.

Wirfs-Brock, R., Wilkerson, B., and Wiener, L. *Designing Object-Oriented Software*, Prentice-Hall, Englewood Cliffs, NJ, 1990.

HOMEWORK

5.1 A software system for automatic design of bolt and nuts is to be created. The design process is as follows:

a. The design specification is given through a menu.
b. Failure analysis is conducted.
c. The dimensions of the parts are determined.
d. Select a standard design by comparing the dimensions determined through the above process and the available database for standards bolts.
e. Generate a drawing using a commercially available CAD program.

All of the design processes are executed automatically except the generation of the menu for design specifications.

We want to design (not code this time) the software using the Independence Axiom. An example of the software, illustrated in Figures H.5.1.a and H.5.1.b (given on page 299), is developed based on AutoCad, a commercial CAD software.

In Figure H.5.1.a, ADME in the toolbar is an added item to the AutoCad menu. Figure H.5.1.b shows a blueprint of the bolt designed. The design specification is shown on the right-hand side of the drawing. Solve the following problems:

a. Design the software using the Independence Axiom.
b. Discuss the possibility of expanding the software for the design of other elements.
c. Discuss the application of the Information Axiom.

  Hints:

1. Define functional requirements of the software.
2. Define DPs with software modules.
3. Any commercial CAD software package may be used.

5.2 We want to develop an axiomatic design software package that supports a designer's decisions in terms of axiomatic design. The desired functional requirements of the software package are

a. Receive the designer's description of FR, DP, and PV with comments.
b. Analyze the proposed design by writing the design equations.
c. Draw the flow chart, tree structure, and the module-junction diagram.

Solve the following problems:

a. Design the axiomatic design software package, which is consistent with the Independence Axiom.
b. Draw the flow chart (system architecture) of the software.

5.3 One of the most important features in the AD framework for designing a system is the flow chart (system architecture), which can be used in many different applications. The flow chart is generated based on the FR/DP/PV tree diagram and the associated design matrices. Suppose we have the design matrix shown below.

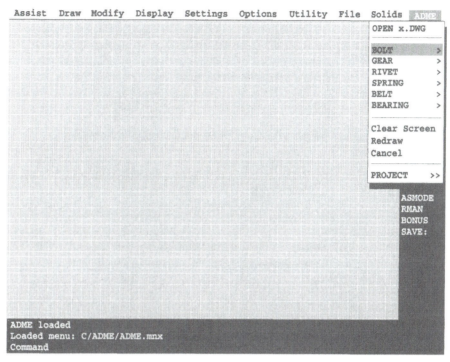

Figure H.5.1.a Initial menu.

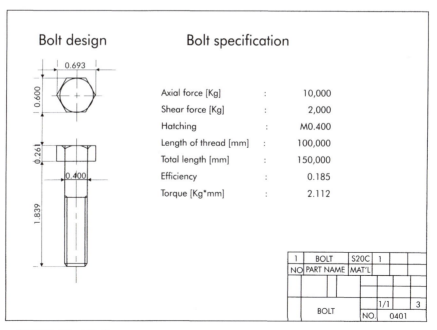

Figure H.5.1.b Final draft.

$$\begin{bmatrix} X & 0 & 0 & 0 & 0 & 0 & 0 \\ X & X & 0 & 0 & 0 & 0 & 0 \\ X & 0 & X & 0 & 0 & 0 & 0 \\ X & 0 & X & X & 0 & 0 & 0 \\ X & X & X & 0 & X & 0 & 0 \\ 0 & X & 0 & 0 & 0 & X & 0 \\ 0 & 0 & 0 & 0 & 0 & X & X \end{bmatrix}$$

Based on the above FR/DP design matrix, your task is to

a. Draw the flow chart of the proposed design matrix by hand.
b. Design the generalized software algorithm that can draw the flow chart using axiomatic design.
c. Compare your design with those of other students and select a good one among them.
d. Provide the rationale for your selection.

5.4 Design a software system for managing the sale of shoes on the Internet.

5.5 Design a software system that can translate English into another language.

5.6 Design a software system that can be used to compose poetry in English.

5.7 Use Acclaro to design a software system that can automatically schedule courses in your school so as to utilize existing classrooms most efficiently.

6
Axiomatic Design of Manufacturing Systems

6.1 INTRODUCTION

The United States is the world's largest market for manufactured products and thus is the testing ground for the competitiveness of industrial firms worldwide. Manufacturing firms in the United States have faced these challenges well, particularly in the 1990s. Many industrial firms are among the strongest in the world because they have been at the forefront in innovating new technologies, applying them to increase manufacturing productivity, and creating markets for new products.

The manufacturing sector of the U.S. economy contributes approximately 20% to the gross domestic product (GDP). More importantly, the manufacturing industry supports many service industries that depend on manufacturing firms for their business. Thus, manufacturing is important to the United States and will evolve continuously, as the relative cost of manufacturing goods in the United States vis-à-vis other countries will determine the domestic value-added content of the manufactured goods sold in the United States. One of the important factors that determine manufacturing competitiveness is the design of manufacturing systems.

The design and operation of production and manufacturing systems have important effects on manufacturing productivity, return on investment, and market share. The increase in American industrial productivity in the 1990s may largely be attributed to improvements in the design and operation of manufacturing systems in conjunction with better design of products. During this period, labor productivity, once thought to be the primary measure of productivity, has improved in U.S. industry, as a result of improvements in manufacturing system design and product design and secondarily, because of manufacturing process improvements. To be globally competitive, many American industrial firms have been attempting to adopt "lean" manufacturing techniques (Womack, Roos, and Jones, 1990).

To achieve the desired goals of a manufacturing enterprise, manufacturing systems must be designed to satisfy a specific set of functional requirements and constraints. Depending on the nature of the functional requirements and constraints, the final manufacturing system has different embodiments, which is the reason the manufacturing system evolved during the twentieth century from the mass-production techniques advanced by Henry Ford to the mass customization currently practiced by the personal computer industry.

Manufacturing systems can be designed based on axiomatic design theory. The functional requirements of a manufacturing system are to maximize the return on investment. This can be accomplished by maximizing the throughput rate given a capital investment and cost base. In general, there are two different kinds of manufacturing systems in terms of materials flow through the system: "pull" systems and "push" systems.

When an identical set of parts is subjected to the same set of manufacturing processes, we can design "push systems" that have the highest throughput rate. In this chapter, a "closed-form" solution for scheduling such a system is presented. This solution is deterministic rather than combinatorial; it is obtained by adding decouplers to satisfy the Independence Axiom. In this solution, an indefinite combinatorial approach to scheduling was replaced by a scheduling approach with a periodic function. This has enormous implications for the design of many engineering systems.

When a random set of parts of varying quantities is to be manufactured, a "pull" system is the best means of maximizing the throughput rate and minimizing inventory and work in progress. In this chapter, a hypothetical manufacturing system that has to produce a mix of products in large numbers in a highly competitive industry is designed to illustrate the methodology. This design, albeit hypothetical, may be an ideal and practical design for many manufacturing firms that make random sets of different parts, if they compete in consumer-oriented industries with worldwide overcapacity of manufacturing facilities.

What is a manufacturing system? What is an ideal manufacturing system? How should we design a manufacturing system? Can a manufacturing system be designed in a rational way based on scientific principles? Manufacturing systems range from a manufacturing cell to a large factory that produces a variety of products, services, and information. Manufacturing systems consist of machines, materials, people, and information. The goal of a manufacturing system is to improve customer satisfaction through improvements in the quality of products, short delivery time, and high labor productivity with a minimum of capital investment. Manufacturing systems must be designed to satisfy a specific set of functional requirements and constraints.

Manufacturing systems can be designed using the general design principles discussed for system design in Chapters 1, 3, and 4. This chapter applies the system-design theory presented in Chapter 4 to the design of manufacturing systems.

The design of an ideal manufacturing system depends on the selection of functional requirements (FRs) the system must satisfy within a given set of constraints (Cs). Therefore, an ideal manufacturing system design is not time invariant. It changes with the selection of specific sets of FRs and Cs. In many cases, these FRs and Cs have been selected implicitly and intuitively by experienced managers and engineers based on their understanding of available technologies and the cost structure of the manufacturing enterprise. Their decision is often affected by the performance measurements used by their company. The consequence is that in most cases, manufacturing systems have not been designed at all, they have evolved based on the way they are measured.

How have manufacturing systems evolved historically? History shows us how the FRs of a manufacturing system have changed to deal with the most costly—either real or perceived—item in manufacturing. At the beginning of the twentieth century, labor costs were thought to be the dominant cost item, and thus many efforts were made to minimize the labor content. F. W. Taylor influenced labor productivity by introducing the concept of division of labor. Improving the individual motions of human workers that were highly subdivided brought about productivity improvements. More efficient and automated machines were also introduced to eliminate the subdivided labor tasks. However, even today, in small batch manufacturing operations, labor cost is still one of the dominant cost items. This may be reduced through better product design, processes, and systems.

As the labor content in a manufactured product decreased, industrial firms have tried to reduce capital investment and operating cost, which has led to a variety of ideas, including capacity planning for machines, group technology, and flexible manufacturing systems. Some of these ideas still dominate the thinking of many manufacturing engineers.

In recent years, the materials costs—including externally purchased component cost, inventory cost, and work in process—has become the largest fraction (50% or more) of the manufacturing cost, especially in repetitive production. Therefore, large industrial firms engaged in mass production are driven to reduce materials costs, in addition to enhancing flexibility of manufacturing systems, quality of products, and ease of manufacturing. In the past, the changes and improvements in manufacturing systems have been made ad hoc based on empirical knowledge, because the design of products and manufacturing systems has been viewed as a subject that is outside of scientific treatment.

How do societal changes affect the design of manufacturing systems? The requirements of manufacturing systems have also changed with society and the standard of living. When Henry Ford was building his cars using a mass-production concept, the demand for these cars or the demand rate was much greater than the supply rate. Therefore, as long as Ford could produce cars inexpensively in large quantities, the market was there. Therefore, the emphasis was on mass production at low cost. Quality and customer satisfaction were not big issues as customers were happy just to have the product even if they had to take their cars to the garage for repair every month. However, when you have overcapacity of production facilities (as in the late 1990s, when it was estimated that there was 20% overcapacity of automobile production facilities), the requirements of a manufacturing system are different. Customer satisfaction in terms of quality, delivery time, customization, and short lead time is as important as the cost of manufacturing.

When customers with money look for merchandise, they have more options today than ever. The preferences of customers today also change quickly. Therefore, industrial firms must be able to respond to market forces quickly and have high-quality products that have functions the customers prefer. All these things must be done with a minimal investment in hardware.

Has manufacturing research helped industrial practice? Various theories for design and operation of manufacturing systems have been advanced to rationalize the design process

(Black, 1991; Cochran, 1994; Suh, Cochran, and Lima, 1998; Ham, Hitomi, and Yoshida, 1985; Sohlenius, 1998; Van Brussel, Peng, and Valckenaers, 1993; Yien, 1998). However, to date, many industrial firms design manufacturing systems based on empirical knowledge or simulations. Like many fields that have been led by technological development rather than by scientific discoveries or theories, the "science" for manufacturing systems has lagged behind the technological development by decades. This situation should change in the future with better understanding of the scientific approaches to the design of manufacturing systems.

Much of the research in the field of manufacturing systems has been concerned with increasing the efficiencies of existing operations and manufacturing processes and with optimizing inventory levels—through both hardware and software—rather than increasing total productivity through a rational design of manufacturing systems. In some instances, assumptions about the causality of manufacturing problems were ad hoc and, consequently, some of the research results have not made much impact on industry.

In the recent past, serious efforts were made to increase the efficiency of manufacturing systems by improving the management of information in a manufacturing enterprise. For example, IDEF0 has been advanced to integrate manufacturing operations by modeling the decisions and actions of systems for analysis and communication purposes (Marca and McGowan, 1993). IDEF0 is a descriptive model with few guidelines for design of systems (Yien, 1998). Lean manufacturing (essentially the renaming of the Toyota Production System) is a new paradigm that has made a major impact on the thinking and practice of manufacturing engineers in the design of manufacturing systems (Shingo, 1989).

What are typical manufacturing systems? Typically, manufacturing systems are classified in terms of the physical machine arrangement. A production job shop is a manufacturing system in which machines are grouped by function (e.g., turning department or milling department). Transfer lines, group technology cells, and lean, linked-cell manufacturing systems are established so that machines are arranged based on product flow (Ham, Hitomi, and Yoshida, 1985; Black, 1991; Cochran, 1994). The best known fixed manufacturing system is still the transfer line.

The cost of most manufacturing systems varies depending on the production volume, the degree of automation, labor cost, equipment costs, and location. To be competitive, these manufacturing systems must be larger than a certain minimum size to be able to process a minimum number of parts at a competitive cost. For example, in 1998, a typical automobile assembly plant that can produce 500,000 cars a year may cost about $1.5 billion, and a typical fabrication facility for semiconductors (often called a "Fab") may cost about $1.5 to $2 billion. On the other hand, a job shop may cost as little as $10 million. The smallest manufacturing system may be a stand-alone, multipurpose machine such as a computer-numerical-control machine (i.e., CNC machine or machining center) or a lithography machine for making IC chips. A CNC machine may cost about $1 million and a lithography tool may cost about $6 million per tool.

What is wrong with the way manufacturing systems have been designed in recent years? During the past two decades, the manufacturing industry and the research community have followed the promise of many hollow catch-phrases. They failed because there was no substance behind the buzzwords. A few stories illustrate the problem.

For a while, many companies tried to adopt the "just-in-time (JIT)" practice to reduce inventory. Because they ignored all other aspects of the manufacturing system issue, they simply forced their suppliers to build warehouses nearby, so that they could supply the part when it was needed. Neither the company nor the supplier benefited by this blind policy of emulating the JIT practice in Japan.

For a while the idea of using a flexible machining system (FMS) received much attention. The idea was to use the same set of computer-controlled machine tools to make a variety of different products. All kinds of predictions were made as to how FMS would take over factory operations by the early 1980s. After 30 years, only a limited number of FMS is in use. The reason is that FMS is not flexible. It takes a long time to program the machine and change the tools.

One of the largest automotive companies invested tens of billion of dollars to install robots and automate their manufacturing operations without a clear rationale as to how they would increase their profitability and quality using robots. Robots cannot replace human operators because human operators often supply the missing information to make the manufacturing facility function. Now many companies are making the same mistakes with information technology (IT) and various software tools. They must first define the functional requirements for their operations and then identify appropriate tools.

In the 1980s, Japan had a national project for unmanned factories. Although technically feasible, a completely unmanned factory does not make much sense because the cost of replacing human workers with machines and computers increases almost exponentially with the percentage of workers replaced. The cost of removing the last 5% of workers will be at least an order of magnitude greater than that of replacing the first 5% of the workers.

The lesson from these failures is that we must design the manufacturing system rationally from the system-design point of view. We should not start our thinking with an idea that sounds interesting—no matter how interesting—without considering all other requirements a system must satisfy.

How will the field of manufacturing evolve? Until recently, many manufacturing companies have emphasized the need to integrate their design and manufacturing operations—in contrast to throwing the design of products over the wall that divides the design section from the manufacturing section—and, thus, the emphasis on "concurrent engineering." This will continue to be the case in many businesses engaged in medium production volumes and highly specialized business. However, in the case of business in many consumer products and office equipment, the trend is to do the opposite—to separate the manufacturing part of the business from the soft side of the business such as design, marketing, sales, and distribution. As a result, custom manufacturers are beginning to emerge in many fields of manufacturing, including office equipment, DRAMs, and television sets. They specialize in manufacturing products for many different kinds of customers under contract. Even automobile companies will do less of component manufacturing, primarily concentrating on assembly operations. This trend is a result of many different economic and technological forces converging on industry at the same time. The efficiency of these specialized manufacturing companies can be higher than vertically integrated companies. To satisfy the stock market, companies try to minimize the capital investment so as to maximize return on investment. Because of the Internet and other technological advances, many companies operate globally—the global economy—and, therefore, they must distribute their manufacturing functions throughout the world near the market. Last but not least, the profit margin may be better when the companies in the industrialized nations concentrate on the soft side of business enterprise, i.e., design, marketing, sales, and distribution. It is clear that we have to reevaluate everything we—academia as well as industry—do in the field of manufacturing.

◼ 6.2 BASIC REQUIREMENTS OF A MANUFACTURING SYSTEM

A manufacturing system must be designed based on the design axioms to be most efficient and reliable. In Section 3.9,[1] it was shown that the scheduling and the flow of parts through a production system must be designed based on the Independence Axiom. It was demonstrated that when a large number of identical parts is to be processed at a given throughput rate, a "push" system can be designed that can maximize the production rate by using "decouplers" that provide a "queue" for the sequential steps and maintain the independence of functional requirements. *Decouplers* are any device (or means) that can separate the sequential steps of manufacturing processes so as to create a decoupled manufacturing system that satisfies the independence of functional requirements.

In a "push system" the parts are fed into the manufacturing system at a fixed interval for manufacturing through a series of processing steps. A classical example of a "push" system is the transfer line, such as those used in making automotive engines. Another example is a semiconductor-processing machine called "Track" that has to process silicon wafers through many different processes using robots as the transport system. In these machines, identical parts (i.e., silicon wafers) are processed sequentially by placing the wafer through a series of process modules. Robots move the wafers from module to module. Scheduling of the robot motion determines the throughput rate.

In the past, robots have been scheduled using the artificial intelligence technique, called an expert system, which creates "if–then" decision rules for the robot. However, because the number of combinations increases continuously with time, the system can come to a complete standstill when the appropriate "if–then" rule is missing. Often these machines crash for no apparent reason. These machines must be reinitialized and started, which can be very costly if the wafers are unusable, as wafers may already have several layers of circuits. Attempts were also made to solve these problems using Operations Research techniques with only limited success.

In a "push" system, we must be sure that the system is decoupled in order to maintain the independence of functional requirements. In a decoupled manufacturing system, "decouplers" are used to create independence of the functional requirements while the part is subjected to various processes in the manufacturing system. Decouplers are either the transport system or each workstation (module) where the parts may wait in a queue until the next machine in the sequence is ready to process the part. This subject, which was discussed in Section 3.9, is further discussed and analyzed in Section 6.4.

In a "pull" system, the production rate is determined by the rate at which the finished part leaves the manufacturing system. For manufacture of random sets of different types of parts that require a number of different processes, "push" systems cannot be made to

---

[1] Example 3.9, *Scheduling of Robots in Pharmaceutical Industry*, was used to solve the scheduling problem involving a robot serving many process modules. It was shown that an "exact" solution could be obtained for the scheduling of the robot motion, which has defied other approaches such as optimization techniques.

work unless there are large buffers between manufacturing steps, which may increase the manufacturing cost. For this type of manufacturing task, a "pull" system that satisfies the Independence Axiom can be designed. A "push system" can couple functional requirements when irregular parts are processed. This is the reason that a "pull" system, often called the lean manufacturing system, has been found effective when irregular parts are processed. An ideal "pull" system for such manufacturing operations is a cellular flexible manufacturing system. Such a system is designed in Section 6.5.

A "pull" system is inherently neither better nor worse than a "push" system—it is simply a matter of having to satisfy different sets of FRs. The foregoing reasoning may be stated as a theorem:

Theorem 25 ("Push" System vs. "Pull" System)

When identical parts are processed through a manufacturing system, a "push" system can be designed with the use of decouplers (that control queues) to maximize the productivity, whereas when irregular parts requiring different operations are processed, a "pull" system is the most effective system.

The Information Axiom provides a measure for choosing the best among the available options by comparing the effects of DPs and PVs on FRs. This is illustrated in this chapter, using a manufacturing system of a "pull" type.

6.3 ELEMENTS OF MANUFACTURING SYSTEMS

A manufacturing system is not unique from a design point of view. It is a subset of the general engineering systems discussed in Chapter 4 (Suh, 1995a, 1995c, 1997a). Therefore, the general methodologies developed for system design based on axiomatic design theory should apply equally well to manufacturing systems.

What are the key elements of a manufacturing system? A manufacturing system is a subset of an entire manufacturing enterprise. Manufacturing enterprises consist of people, "things," and information. People are deployed to perform various functions, such as marketing, design, purchasing, inventory control, inspection, machining, management, safety, service, and security. "Things" range from factories to machines, materials, transporters, computers, warehouses, vendors of components, and utilities. Information is related to marketing requirements, product design, manufacturing systems and operations, manufacturing processes, human resources, supply chain systems, and general management. All these elements constitute part of the manufacturing enterprise, and thus the design of manufacturing systems is regarded to be complex.

A manufacturing system—a subset of the production system—also consists of people, information, and "things." Therefore, when we design a manufacturing system, we must consider all these elements simultaneously to come up with the best system. We cannot design the hardware part of the system first and then think about information issues and the role of human operators at a later stage. A manufacturing system may be defined as the arrangement and operation of these basic elements (machines, tools, material, people, and information) to produce a value-added physical, informational, or service product whose success is specified by the highest functional requirements and constraints. The specific combination of a manufacturing system's elements is determined by the FRs placed on the manufacturing system.

For example, when the task is to machine various random metal parts on demand, the manufacturing system may consist of a set of flexible machine tools in a job shop arrangement. When the task is to produce millions of identical parts, the manufacturing system may consist of specialized, dedicated machines (such as transfer lines). When the task is to produce sets of different parts in large quantities, the best manufacturing system may be a cellular manufacturing system, a "flow line" type of manufacturing system. When the task is to build a limited number of large products such as lithographic machines for semiconductor manufacturing or locomotives, the manufacturing system may be a slowly moving or stationary site to which parts are brought in a specific sequence. When the task is to produce a variety of three-dimensional plastic parts by injection molding, the manufacturing system may consist of a large number of injection molding machines of various sizes with other accessory equipment for plastic storage and conveyance. When semiconductors are processed to make integrated circuit (IC) chips, the manufacturing system may consist of a series of dedicated (flexible or fixed) machines placed in a clean and controlled environment in which each wafer with a large number of dies (i.e., individual IC chips) can be processed.

To make all these machines and people work together for the common goals of the system, we must deal with information—generation, collection, processing, transmission, and interpretation of information. Information is typically embedded in hardware and software. It forms the interface between diverse elements of a manufacturing system and integrates the functions of the system. Without the proper management of information, a modern manufacturing system cannot function.

Two manufacturing systems can have identical sets of machines (i.e., "things") and yet have significantly different production rates and cost, depending on how the system is designed and operated, which, in turn, will affect how people and information are used. One factory may produce high-quality products at lower cost, whereas another similar factory may produce inferior products at higher costs, depending on the design and operation of the manufacturing system. Therefore, a productive manufacturing system must be *designed* so that the FRs—production rate, flexibility, and quality—of the manufacturing system can be satisfied at the lowest cost within regulatory and other constraints. The design task is to arrange various machines, people, materials, and information in a logical and rational manner so as to produce the products at the lowest cost to meet given demands.

To minimize the manufacturing cost, many diverse goals are often considered and pursued in developing a manufacturing system. Minimization of the initial capital investment and maximization of productivity are typical primary goals. There are other considerations: minimization of inventory and work in progress, reliability of the equipment with long mean times between failures (MTBF), and labor cost. Sometimes, the term cost of ownership is used to describe the life-cycle cost of using a machine or equipment. Another factor in manufacturing system design is the fact that customer orders will fluctuate in time and yet the ordered products must be delivered on time. Most of all, the manufacturing system must produce quality products so that the company can capture a large market share and charge a premium price.

◼ 6.4 AXIOMATIC DESIGN OF FIXED MANUFACTURING SYSTEMS FOR IDENTICAL PARTS

6.4.1 Highest Level Design of a Fixed Manufacturing System

We will first consider the simple case of making identical parts by processing them through a set of different processes. For example, it may be related to coating, curing, and developing a photoresist material[2] on silicon wafers for semiconductor manufacturing. In this case, the highest level FR may be selected as

FR_1 = Maximize the return on investment (ROI).

To maximize ROI, we have to produce the maximum number of coated wafers, sell them at the highest possible price, minimize the manufacturing cost, and minimize the capital investment. However, we will consider here only the task of maximizing the output of a dedicated, automated machine to illustrate the basic concept. Then, the design parameter may be written as:[3]

DP_1 = Dedicated automated machine that can produce the desired part at the specified production rate

The FRs of the dedicated and automated machine may be written as

FR_{11} = Process the wafers.
FR_{12} = Transport the wafers.

The corresponding DPs may be chosen to be

DP_{11} = Process modules
DP_{12} = Robots

The Cs are:

C_1 = Throughput rate
C_2 = Manufacturing cost
C_3 = Quality of the product
C_4 = Yield (production of acceptable products divided by total output)

What we have designed can be expressed by the design equation:

$$\begin{Bmatrix} FR_{11} \\ FR_{12} \end{Bmatrix} = \begin{bmatrix} X & 0 \\ X & X \end{bmatrix} \begin{Bmatrix} DP_{11} \\ DP_{12} \end{Bmatrix} \tag{6.1}$$

Equation (6.1) expresses the fact that in the proposed design, DP_{11} (the process modules) will affect FR_{12} (transport wafers), but DP_{12} (the robot) will not affect FR_{11} (process wafers). All subsequent decisions as we decompose these FRs and DPs must be consistent with this decision.

---

[2] "Photoresist" is a viscous, light-sensitive polymer that is used to create images on a silicon-wafer surface in integrated electric circuits.

[3] It should be noted again here that the FRs are written starting out with verbs, whereas the DPs start with nouns.

The design represented by Equation (6.1) states that given an arrangement of the modules, we must design a transportation system that will not affect the manufacturing process. This is a decoupled design.

Because this machine processes exactly identical parts, a "push" system may be designed, where the part will be supplied to the machine on a regular time interval, denoted as period T. The period T is equal to $(3600/m)$ seconds where m is the number of wafers supplied to the machine per hour. T is the cycle time during which the robot must pick up wafers from all modules at least once.

The number of modules needed for each process is related to the period T, because if the process time of a module is larger than T, it will take more than one module to be able to meet the required throughput rate. If the process time in Module i is denoted as t_i (seconds), the number of modules needed to meet the production requirement, n_i, is given by

$$n_i = \text{Int}\frac{t_i}{T} = \text{Int}\left[\frac{t_i}{\left(\frac{3600}{m}\right)}\right] \tag{6.2}$$

$\text{Int}[x]$ is a function that rounds x up to the next nearest integer. The total number of modules, M, required to process the wafers is

$$M = \sum_{i=1}^{N} n_i \tag{6.3}$$

In Equation (6.3), N is the number of tasks, i.e., processes. These process modules must be so arranged that the robots can serve all of these modules in the shortest possible time.

Within the cycle-time period T, the module for each process (or one of the modules when there is more than one module for a given process) completes its task. Therefore, within a given period T, the robot must pick up the wafers from the modules that have just completed their process cycles and transfer them to the next set of modules. The robot must also deliver a wafer from the supply cassette station to the first module as well as from the last module to the outgoing cassette station, all in a given period T. If it takes t_T for the robot to transport the wafer from one module to the next, then the number of moves the robot can make in time T is equal to T/t_T.

The sequence of the robot operation is as follows. The robot picks up a wafer from the supply cassette station and delivers it to Module 1 for Process 1. On completion of Process 1 in Module 1, the robot picks up the wafer from Module 1 and inserts it into Module 2 for Process 2, and from Module 2 to Module 3, and so on. When it is again time to pick up another new wafer from the supply cassette after an elapse of time T, the robot goes back to the supply cassette and loads another unprocessed wafer from the cassette to the first module for Process 1. If the first Module 1 is still processing a wafer, this new wafer is loaded into the second Module 1. This sequence of wafer transfer continues until the entire process is completed. In one period T, the robot must move all wafers that have just finished a prescribed process to the next module for another process [*Note*: there can be more than one module for each process as per Equation (6.2)]. In addition, it must load a new wafer from the supply cassette station and also deliver the finished wafer from the last module to the outgoing cassette station.

A conflict can arise in scheduling the robot motion if two processes are completed nearly at the same time (i.e., within the time required for single robot motion), as the robot has to be at two different places at the same time. This coupling of functional requirements can cause a system failure. In the past, this problem was tackled using an "if–then" algorithm for deciding which wafer the robot should pick up next. As discussed in Chapter 3 (Example 3.7), the "if–then" approach is unreliable because the number of combinations increases continuously, with each subsequent decision depending on the decisions made earlier. The number of possible combinations increases to Πn_i where n_i is the number of modules available for process i. Furthermore, when there is no appropriate "if–then" rule, the system breaks down.

This problem can be solved rigorously by introducing decouplers, i.e., by redesigning the system! The coupling occurs when two or more wafers complete the prescribed processes nearly simultaneously (within the transport time of the robot). We can decouple the pick-up functions by introducing "decouplers"—devices that store the wafers until the robot becomes available. The role of decouplers is to decouple functional requirements of the robot. The decouplers do not have to be separate physical devices. In this case, the modules can act as decouplers by letting the wafers stay in the modules longer. Decouplers provide queues between modules so that the wafers can be picked up in a predetermined sequence by the robot. The design task is to determine where the decouplers should be placed and how long their queue should be. Some modules cannot act as decouplers if the process time in the module is tightly controlled for chemical reasons.

When decouplers are introduced with queue q_i, the total cycle time T_C increases. T_C is defined as the sum of the process time, the queues, and the transport time, which may be expressed as

$$T_C = t_P + t_T + \sum_{i=1}^{N} q_i$$

where t_P is the total process time and t_T is the total transport time. As T_C increases, the number of modules may increase depending on the process time t_i of each module. Therefore, we have dual goals: decouple the process by means of the decouplers and minimize T_C by selecting the best set of q_i. Then FR_{12} (transport wafers) may be decomposed as

$FR_{121} = $ Decouple the process times.
$FR_{122} = $ Minimize the total process time T_C (or minimize the number of modules).

The corresponding DPs are

$DP_{121} = $ Decouplers with q_i
$DP_{122} = $ The minimum value of T_C

The design equation is

$$\left\{ \begin{array}{c} FR_{121} \\ FR_{122} \end{array} \right\} = \left[\begin{array}{cc} X & 0 \\ X & X \end{array} \right] \left\{ \begin{array}{c} DP_{121} \\ DP_{122} \end{array} \right\} \tag{6.4}$$

To minimize the number of modules M, we must satisfy the following two conditions:

$$\sum_{i=1}^{N} \frac{\partial M}{\partial q_i} = 0$$

$$\sum_{i=1}^{N} \frac{\partial^2 M}{\partial q_i^2} > 0 \tag{6.5}$$

where N is the number of processes.

6.4.2 Analytical Solution for Queues in Decouplers[4]

Having designed the manufacturing system, we must replace those Xs with mathematical expressions if they can be modeled. In this section, the queues q_i will be determined through modeling and analysis to determine the exact relationship between FR_{121} (decouple the process times) and DP_{121} (decouplers with q_i).

Let T_i denote the time the wafer has to be picked up on the completion of process j in Module i. Then T_i is the sum of the total process time t_P (up to and including the ith process) and the accumulated transport time t_T (up to the ith module), which may be expressed as

$$T_i = t_P + t_T \qquad (6.6)$$

T_i, t_P, and t_T are normalized with respect to the sending period T, i.e., actual time divided by T. Therefore, throughout this analysis, all of the times will be dimensionless, i.e., the actual time divided by the period T.

The total process time t_P is the sum of the individual process times t_{Pj}, and the accumulated transport time, t_T, is the sum of the all robot transport times, t_{Tj}. Therefore, Equation (6.6) may be expressed as

$$T_i = t_P + t_T = \sum_{j=1}^{i} t_{Pj} + \sum_{j=1}^{i-1} t_{Tj} \qquad (6.7)$$

The number of pick-up moves that the robot can make in a given period T is given by

$$n_R = \frac{T}{t_T} \qquad (6.8)$$

If there are N process steps, there are N modules with wafers that have gone through their respective processes and are ready to be picked up within a given period T (see Figure ex.3.7.b). Within this time period, the robot must pick up all wafers from the modules that have completed their processes. The robot must pick up a wafer at time τ_i, measured from the beginning of each period T, which may be expressed as

$$\tau_i = T_i - \text{Int}\left(\sum_{j=1}^{i} t_{Pj} + \sum_{j=1}^{i-1} t_{Tj}\right) = \sum_{j=1}^{i} t_{Pj} + \sum_{j=1}^{i-1} t_{Tj} - \text{Int}\left(\sum_{j=1}^{i} t_{Pj} + \sum_{j=1}^{i-1} t_{Tj}\right) \qquad (6.9)$$

where $\text{Int}(x)$ is a function that rounds x down to the next nearest integer. However, if the pick-up times are coupled because two or more processes are finished nearly at the same time, the robot cannot implement the schedule given by Equation (6.9).

We must modify the design to decouple the process by adding decouplers with queues q_i. For example, if Process 1 in Module 1 and Process 3 in Module 3 are finished within the transport time required $t_{T,1}$, then the robot cannot pick up both pieces at the same time. Therefore, in this case, we may add a "decoupler" to Module 1, which may be either a

---

[4] This robot-scheduling problem comes from SVG, Inc., which hired many consultants to solve the robot-scheduling problem without obtaining any satisfactory solution. Dr. Larry Oh, Vice President of SVG, Inc. and the author were waiting at an airport when we began to discuss this problem. The author suggested the use of decouplers and Dr. Oh (with the author's graduate student Tae-Sik Lee) came up with this elegant closed-form solution. A patent has been filed by SVG to protect this work.

physically separate device or just a queue in Module 1 to keep the wafer in longer. In this case, T_i given by Equation (6.7) is extended by q_i. Then the new time for pick-up T_i^* is given by

$$T_1^* - T_1 = q_1 \tag{6.10}$$

Similarly, extending it to the more general case:

$$T_2^* - T_2 = q_1 + q_2$$
$$T_3^* - T_3 = q_1 + q_2 + q_3$$
$$etc.$$

Substituting these relationships into Equation (6.9), we obtain the modified actual pick-up time. If we denote this modified time as τ_i^*, then $(\tau_i^* - \tau_i)$ may be expressed approximately as

$$\tau_i^* - \tau_i = \sum_{j=1}^{i} q_j = a_{ij} q_j \tag{6.11}$$

where the matrix a_{ij} is defined as:

$$a_{ij} = \begin{cases} 0 & \text{when} \quad i < j \\ 1 & \text{when} \quad i > j \end{cases}$$

We can determine τ_i^* approximately by solving Equation (6.9), by determining where the decouplers may be needed, and by approximating the values of queues.

Equation (6.11) may be expressed as

$$\left\{ \tau_i^* - \tau_i \right\} = [a] \left\{ q_j \right\} \tag{6.12}$$

Equation (6.12) may be solved for q_i as

$$\{q\} = [a]^{-1} \{\Delta\tau\} = \frac{1}{|a_{ij}|} [A] \{\Delta\tau\} = [A] \{\Delta\tau\} \tag{6.13}$$

where

$$\Delta\tau = \tau_i - \tau_i^*$$
$$[a] \equiv \text{matrix with elements } a_{ij}$$
$$|a_{ji}| \equiv \text{determinant of matrix } [a] = \prod_{i=1}^{N} a_{ii} = 1$$
$$[A] = Adj\, [a_{ij}] = [A_{ji}]$$
$$A_{ji} = (-1)^{i+j} M_{ji}$$
$$M_{ji} \equiv minor \ of \ a_{ji}$$

Equation (6.13) can be solved iteratively. To do so, we need to know $\{\Delta\tau\}$, which can be approximated by estimating reasonable values for τ_i^* and by solving Equation (6.9) for τ_i. The value for τ_j^* can be estimated by adding transport time to τ_i^* as $|\tau_i^* - \tau_j^*| > t_T$ where $j \neq i$. The solution can be improved by successive substitution of the improved values of τ_i^*. The determinant $|a_{ij}|$ of the triangular matrix $[a]$ is equal to the product of the diagonal elements.

Because the best solution is the one that makes the total cycle time T_C a minimum, we seek a set of values of q_i that yields a minimum value for the total queue, Σq_i. When the precise control of processing time is critical, the queue q_i associated with that process should be set equal to zero.

To solve Equation (6.12) for the best set of queues qs, Oh (1998) and Oh and Lee (1999) developed an optimization software program based on a genetic algorithm. Multiplying these q_i by T, we can obtain actual values of queues.

EXAMPLE 6.1 Determination of the Queues of a Fixed Manufacturing System That Processes Identical Parts

A manufacturing system is being developed for coating of wafers. To produce the final semiconductor product, wafers coated with photoresist must be subjected to various heating and cooling cycles at various temperatures for different duration before they can be shipped to the next operation. The manufacturing system is an integrated machine that consists of five (5) process steps involving five (5) different modules. A robot must place the wafer into these modules then take them out of the modules and transport them to the next process module according to a preset sequence. We want to maximize the throughput rate by using the robot and the modules most effectively. The desired throughput rate is 60 units an hour. A constraint is that we must use a minimum number of modules. The time it takes for the robot to travel between the modules is 6 seconds. The wafers are processed through the following sequence:

Steps	Modules	Temperature (°C)	Duration ± Tolerance (Seconds)
1	A	35	50 ± 25
2	B	80	45 ± 0
3	C	10	60 ± 20
4	D	50	70 ± 10
5	E	68	35 ± 0

The process times in Modules B and E must be precise because of the critical nature of the process. The cycle time is assumed to be the process time plus the transport time both for placement and pick-up of the wafer.

The robot must pick up the wafers from a supply bin (load-lock) and deliver it to Module A and when the process is finished, it must pick up the container from Module G and place it on a cassette. These operations take 6 seconds each.

SOLUTION

The number of modules is dependent on the process time T_C and the desired throughput rate. The required number of modules is as follows:

Modules	Number of Modules
A	2
B	1
C	2
D	2
E	1

Without any decouplers, there are simultaneous demands for the service of the robot, as shown in Figure E6.1.a, which shows the time the process is finished in each of the modules within a given period T. In the figure, the horizontal axis is dimensionless time—1 represents one period T. Because the transport time is equal to 6 seconds, i.e., $(T/10)$, the figure shows that the robot has a conflict. Similarly, Processes 1 and 3, and Processes 2, 3, and 5 are all finished nearly at the same time.

The solution is obtained by solving Equation (6.12) using the software program developed by Oh and Lee (2000). The best solution was obtained by finding a set of values that give the shortest cycle time T_C solving Equation (6.12) repeatedly using a genetic algorithm. The solution yields the following values for qs.

$$q_A = 19.7 \; seconds$$

$$q_B = 0 \; seconds$$

$$q_C = 9.3 \; seconds$$

$$q_D = 9.3 \; seconds$$

$$q_E = 0 \; seconds$$

The queues for B and E are zeros because the tolerance on these two modules is specified to be zero. Therefore, the queues of the other modules have been adjusted to make these two queues zero. The actual pick-up times at the completion of the processes of Modules A, B, C, D, and E are given in Figure E6.1.b.

There are other possible sets of solutions for q_i, but they may not give the minimum T_C and the minimum Σq_i.

One of the interesting results of this solution is that the number of combinations for

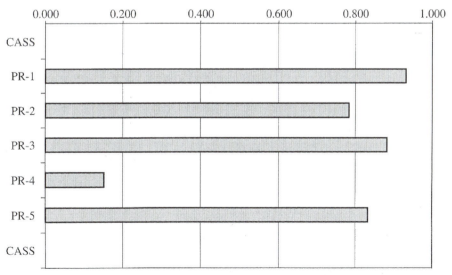

Figure E6.1.a The pick-up schedule in a period T without any decouplers. There are conflicts among the processes finished in Modules 1, 2, 3, and 5. This result is obtained using the software program developed by Oh (1998) and Oh and Lee (2000).

part flow reduces from several thousand to a few, because the parts flow through the manufacturing system along deterministic paths. What the concept of decouplers has done is to change a combinatorial problem into a periodic function that repeats itself with a given cycle that is deterministic.

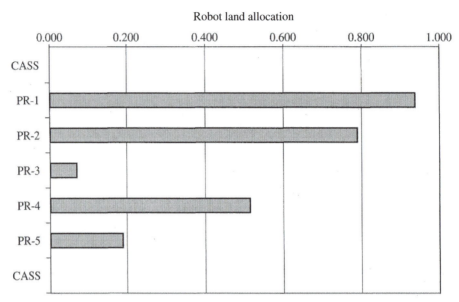

Figure E6.1.b The actual pick-up times of Modules A, B, C, D, and E. PR-1 is Process 1 that takes place in Module A, PR-2 is for Module B, PR-3 is for Module C, PR-4 is for Module D, and PR-5 is for Module E. This solution is obtained using the software program developed by Oh (1998) and Oh and Lee (2000).

What is the main message of this section? By decoupling a coupled design, we have reduced a problem that used to be treated as a combinatorial problem—thus, extremely intractable mathematically—to a deterministic problem that can be solved exactly! Also the number of possible combinations was reduced from several thousand to two! The throughput rate is also controlled precisely because the system has been designed with a given throughput in mind. The decoupled design of a manufacturing system that processes identical parts can be modified without much effort.

The design solution that led to a decoupled design and the accompanying analytical solution have profound implications on system design. The shortcomings of the traditional approach to these problems (e.g., Operations Research) can be overcome by developing designs that satisfy the Independence Axiom.

We considered the manufacture of identical parts. What we have shown here is that in this case, the best manufacturing system is a "push" system. A "push" system can be designed that maximizes the productivity if we can decouple the transport system from the manufacturing process.

The ideas presented in this section can be applied to a variety of system problems that involve design, analysis, and optimization. The important message is that if the system design violates the Independence Axiom, it is extremely difficult to achieve the functional requirements of a manufacturing system.

In the next section, we will consider the case of processing a random set of different parts with a varying lot size. In this case, the manufacturing system must be flexible to accommodate different shapes and different lot sizes.

6.5 AXIOMATIC DESIGN OF A FLEXIBLE MANUFACTURING SYSTEM FOR DIFFERENT TYPES OF PARTS[5]

In the preceding section, a manufacturing system (consisting of a robot and manufacturing modules) for manufacturing identical parts using the same process steps was considered. It was shown that a "push" type system could be designed. A robot, rather than people, was used in this system because the manufacturing environment for semiconductors must be clean and because the process times are so short that people cannot perform the transport function. In this section, we will design a manufacturing system that processes nonidentical parts and uses people as part of the system.

How should we design a manufacturing system for nonidentical parts that require different operations? Can a manufacturing system be designed in a rational way based on axiomatic design? An efficient manufacturing system must utilize "things," people, and information in a rational manner, consistent with basic principles. In this section, the methodology of axiomatic design is illustrated in the step-by-step design of a manufacturing system for machined parts of nonidentical shape. The principles and procedures illustrated here are the same regardless of the specific nature of the functional requirements a manufacturing system must satisfy. However, the specific choice of FRs for the manufacturing system will generate different manufacturing systems.

Manufacturing firms must satisfy their customers in order to survive. Customers of manufacturing plants desire specific products, the nature of which typically varies over the life cycle of the manufacturing system. The products must be of high quality and reliability—they must function well and be free of defects. The products must be delivered in a required volume on time at an agreed-on price. These customer attributes must be mapped into the functional domain and the FRs must be established.

In this example, it will be assumed that the task is to design a manufacturing system that must produce a mix of machined products during its lifetime.

STEP 1
Choose FRs in the Functional Domain

The first step in designing a manufacturing system is to determine the highest level FRs of the manufacturing system to be designed. There can be many different sets of FRs that we may wish to satisfy in designing a manufacturing system. The resulting manufacturing system will be different, depending on the specific set of FRs chosen for the design. Regardless of which set of FRs is chosen, a minimum set of FRs under a given circumstance must be chosen as per the definition of FRs (Suh, 1990).

---

[5] The issue of complexity is discussed further in Chapter 9.

Most manufacturing firms would state the highest FR of its manufacturing system as

FR_1 = Maximize the return on investment (ROI).

FR_1 is based on the interest of the owners or shareholders of a manufacturing company, which is profit. It is interesting to note that there are three very different stakeholders of manufacturing systems: the shareholders (owners), the employees (internal customers), and the external customers who purchase the products produced by the manufacturing system.

Although the following design starts with the point of view of the shareholders, as the decomposition is developed, the needs of the internal and external customers are included as lower level FRs.

It should be noted here that all FRs are stated starting out with verbs. This is a good way of distinguishing an FR from a DP, which should start with a noun, if possible.

STEP 2
Mapping of FRs in the Physical Domain

The second step is mapping from the functional domain to the physical domain to determine the DPs that can satisfy the FRs at this level. For a given set of FRs, there can be many different sets of DPs. The manufacturing system is designed when the DPs are chosen to satisfy the FRs. For example, to satisfy FR_1 of Step 1, we may choose one of the following two DPs as the corresponding DP_1:

DP_1^a = Manufacturing system to provide products at minimum cost

or

DP_1^b = Manufacturing system designed to provide a high-quality product that meets customer needs

The consequences of choosing DP_1^a rather than DP_1^b (or vice versa) can be quite significant as we develop the detailed design through decomposition.

STEP 3
Decompose FR_1 in the Functional Domain—Zigzagging between the Domains

We just defined FR_1 and DP_1 at the highest level. The next step in axiomatic design is to go back (i.e., "zigzag") to the functional domain from the physical domain, if the chosen DP_1 cannot be implemented without further detailed design (i.e., until we reach the leaf level). The lower level FRs must be determined by decomposing FR_1, which is equivalent to determining the functional requirements of the DP_1 chosen (i.e., either DP_1^a or DP_1^b). If the designer chooses DP_1^a, the corresponding FRs of the next level may be different from or the same as those if DP_1^b were chosen. For the particular DP_1^a or DP_1^b chosen, FR_1 may be decomposed as

FR_{11} = Increase the sales revenue.
FR_{12} = Minimize the manufacturing cost.
FR_{13} = Minimize manufacturing investment.

These FRs are derived from the formula that calculates ROI and, therefore, are the same for both DP_1^a and DP_1^b.

$$\text{ROI} = \frac{\text{Sales} - \text{Cost}}{\text{Investment}} \tag{6.14}$$

STEP 4
Find the Corresponding $\{DP_{1x}\}$ by Mapping $\{FR_{1x}\}$ in the Physical Domain

Now we have to find $\{DP_{1x}\}$ that correspond to FR_{11}, FR_{12}, and FR_{13}. These $\{DP_{1x}\}$ are also the decomposition products of DP_1^a or DP_1^b. Therefore, the $\{DP_{1x}\}$ may be different, depending on whether DP_1^a or DP_1^b was chosen in Step 2.

For DP_1^a, the decomposition of the next level of the DP hierarchy may be based on the manufacturing situation in 1913 when mass-production systems were evolving:

$DP_{11}^a = $ Maximization of production output
$DP_{12}^a = $ Unit cost minimization
$DP_{13}^a = $ Maximum machine utilization

It is clear that this design of the manufacturing system works well when there is unlimited demand for a new and innovative product (as was the case when low-cost automobiles first became available because of Henry Ford's car design and mass-production implementation).

For DP_1^b, the corresponding $\{DP_{1x}\}$ may be chosen as:

$DP_{11}^b = $ Design and manufacture products to maximize customer satisfaction
$DP_{12}^b = $ Target production cost
$DP_{13}^b = $ Investment in production facilities and machines based on a systems-thinking approach

This second type of manufacturing system is designed to increase the sales revenue while decreasing cost and investment by making products that customers want to have when they want to have them. This may be a good strategy when there is overcapacity of production facilities and when the design, delivery, quality, and price of products determine competitiveness.

STEP 5
Determine the Design Matrix

Having determined $\{FR_{1x}\}$ and $\{DP_{1x}\}$ for two different choices of DP_1, the design matrix must be created to determine whether the proposed design violates the Independence Axiom. The design equation and design matrix are

$$\begin{Bmatrix} FR_{11} \\ FR_{13} \\ FR_{12} \end{Bmatrix} = \begin{bmatrix} X & 0 & 0 \\ X & X & 0 \\ X & X & X \end{bmatrix} \begin{Bmatrix} DP_{11}^a \\ DP_{13}^a \\ DP_{12}^a \end{Bmatrix} \tag{6.15}$$

X signifies a strong relationship between the FRs and DPs. This design represents the mass-production system design for the automotive industry in the early 1900s. It was a decoupled design. The order of implementation was DP_{11}^a, DP_{13}^a, then DP_{12}^a. To increase sales revenue, the factory DP was to produce more products. The classic statement that "you can purchase any color vehicle as long as it is black" was actually the result of the tremendous pressure to decrease operation cycle time in the Ford factory (Arnold and Faurote, 1915) and black paint dried faster than any other paint color. Second, high machine utilization (DP_{13}^a) reduced manufacturing investment due to economies of scale. In addition, DP_{13}^a had a strong effect on manufacturing cost. The machines were designed to eliminate direct labor.

However, in terms of current dynamics of market forces, the design equation looks quite different for DP_{1x}^a. The new design matrix that reflects the current relationship between the FRs and DPs may be written as

$$
\begin{Bmatrix} FR_{11} \\ FR_{13} \\ FR_{12} \end{Bmatrix} = \begin{bmatrix} 0 & 0 & 0 \\ X & X & 0 \\ X & X & X \end{bmatrix} \begin{Bmatrix} DP_{11}^a \\ DP_{13}^a \\ DP_{12}^a \end{Bmatrix}
\tag{6.16}
$$

This design matrix illustrates that manufacturing system designs change over time. DP_{11}^a no longer satisfies FR_{11} as producing more products today does not guarantee that a product will be bought.

The second design, represented by $\{DP_{1x}^b\}$, satisfies $\{FR_{1x}\}$ better in the current market environment, where the customers are more demanding and the competitive pressure is much greater because of the overcapacity of manufacturing facilities worldwide.

$$
\begin{Bmatrix} FR_{11} \\ FR_{12} \\ FR_{13} \end{Bmatrix} = \begin{bmatrix} X & 0 & 0 \\ X & X & 0 \\ X & X & X \end{bmatrix} \begin{Bmatrix} DP_{11}^b \\ DP_{12}^b \\ DP_{13}^b \end{Bmatrix}
\tag{6.17}
$$

The analysis of the design matrices (6.16) and (6.17) illustrates that DP_{1x}^bs constitute a better design solution for manufacturing systems. We will pursue only the second design further in this section. Therefore, from now on, we will drop the superscript b from DP_1^b and use DP_1 for simplicity.

After the design is completed, the Xs can be replaced with precise expressions or constants through modeling of the physics or geometry of the design. As discussed in Chapters 4 and 5, the modeling is done for the lowest level (called "leaves") design equations, as the higher level design equations are made up of the lower level design parameters and matrices.

STEP 6
Decompose FR_{11}, FR_{12}, and FR_{13} by going from the Physical to the Functional Domain Again and Determine the Corresponding DPs (Level 3)

Step 6a. FR_{11}—Sales Revenue Branch

FR_{11} (Increase sales revenue) must be decomposed with DP_{11} in mind, which is "Design and production of products to maximize customer satisfaction." The sales revenue (SR) is equal to the product of the number of products sold and the price per unit, which may be expressed as

$$
SR = \sum_{i=1}^{n} (\text{Price}_i \times \text{Volume}_i)
\tag{6.18}
$$

where n is the number of types of products.

Then FR_{11} may be decomposed, with DP_{11} in mind, as

$FR_{111} =$ Sell products at the highest acceptable price.
$FR_{112} =$ Increase market share (volume).

The corresponding DPs may be chosen as

$DP_{111} =$ Customer perceived value of product
$DP_{112} =$ Broad product applications

The design matrix for the above set of FRs and DPs is

$$\left\{ \begin{array}{c} FR_{111} \\ FR_{112} \end{array} \right\} = \left[\begin{array}{cc} X & 0 \\ X & X \end{array} \right] \left\{ \begin{array}{c} DP_{111} \\ DP_{112} \end{array} \right\} \tag{6.19}$$

Naturally, to increase sales revenue, the objective is to sell products at the highest acceptable price (FR_{111}) now and in the future. That is why it becomes so important to maintain and improve the customer's perceived value of the product. DP_{111} also affects FR_{112} because satisfied customers contribute to market share in two ways: repeat purchases and product referrals (free advertising) to potential new customers. A specific action to increase market share (FR_{112}) is satisfied by a design policy to broaden product applications (DP_{112}). This design is a decoupled design and thus satisfies the Independence Axiom.

Step 6b.* FR_{12}—*Production Cost Branch

FR_{12} (Minimize the production cost) may be decomposed with DP_{12} (target production cost) in mind as

$FR_{121} =$ Reduce material costs.
$FR_{122} =$ Reduce operational activity costs.
$FR_{123} =$ Reduce overhead.

The elements of the production cost are the cost of raw materials and components, the direct cost, indirect cost, and administrative costs or overhead. The corresponding DPs may be stated as

$DP_{121} =$ Target price given to suppliers
$DP_{122} =$ Targeted performance of operational activities
$DP_{123} =$ Right size business processes

The FRs and DPs in italics do not need to be decomposed because the DPs can be implemented. These are leaves. However, other FRs and DPs must be decomposed.

The cost of raw materials and components purchased from suppliers is based on their price, which strongly affects the cost of the products and the profits. Without further decomposition to satisfy FR_{121}, the DP_{121}—target price given to suppliers—can be implemented.

The direct and indirect cost classification is useful for accounting purposes, but to reduce production costs, all activities must be analyzed. The direct cost is related to departments that perform value-added activities to the part, such as machining or assembly departments. The indirect cost is related to departments that perform non-value-added activities to the part, for example, maintenance or storage rooms. This classification is useless when we need to reduce costs. There are activities performed in the machining departments, such as transportation of parts and excessive material handling, that are non-value-added activities. Also there are activities performed in indirect-cost departments, such as preventive maintenance, which are extremely important to the value-added activities of the direct-cost departments. We must analyze what activities are performed throughout production, with a major focus on non-value-added activities, and identify what actions (DPs) are valid to reduce or eliminate them. Moreover, we have to focus on the improvement of value-added activities and on those non-value-added activities, such as maintenance and setup, that have a fundamental effect on production performance. The activities performed on the production environment will be called operational activities. The goal then is to "reduce total operational activity cost" stated as FR_{122}, and the DP_{122} that satisfies this FR is "targeted performance of operational activities."

The administrative cost, which is related to all administrative activities, will not be considered in the decomposition of this production system design. We will just state that to satisfy FR_{123} without further decomposition, the DP_{123} must state "right sized business processes." The design equation and matrices are

$$\begin{Bmatrix} FR_{121} \\ FR_{122} \\ FR_{123} \end{Bmatrix} = \begin{bmatrix} X & 0 & 0 \\ 0 & X & 0 \\ 0 & 0 & X \end{bmatrix} \begin{Bmatrix} DP_{121} \\ DP_{122} \\ DP_{123} \end{Bmatrix} \tag{6.20}$$

This design is an uncoupled design and thus satisfies the Independence Axiom.

Step 6c. FR_{13}—Capital Investment Branch

FR_{13} (minimize production investment) may be decomposed with DP_{13} (investment in production facilities and machines based on a systems-thinking approach) in mind as

> $FR_{131} =$ Acquire machines with cycle time less than or equal to the minimum TAKT time (i.e., the time interval between two consecutive parts coming out of the system).
> $FR_{132} =$ Ensure flexibility to accommodate capacity increments at lowest cost.
> $FR_{133} =$ Develop flexible tooling.
> $FR_{134} =$ Ensure flexibility to accommodate future products.

The corresponding DPs may be stated as

> $DP_{131} =$ Machine design focused on customer demand pace and value-added work
> $DP_{132} =$ Linked-cell manufacturing systems
> $DP_{133} =$ Flexible tooling design
> $DP_{134} =$ Movable machines and reconfigurable stations to enable new cell design

The acquisition of machines to satisfy FR_{131} adds a new dimension to minimizing investment in the production environment, while producing the parts at the pace that the customers want. Because in a number of cases there is the possibility of increasing the cycle time of the machine, it brings new opportunities for the manufacturing engineers to acquire simpler machines and reduce the required level of automation. DP_{132} addresses the answer for large and rapid increases in demand. The replication of manufacturing cells has a shorter decision and acquisition time and adds a smaller capacity increment than transfer lines, increasing the system flexibility to accommodate capacity increments at lowest cost

(FR$_{132}$). The investment in flexible tooling and reconfigurable equipment further increases the long-term flexibility.

The design equation and matrix is

$$\begin{Bmatrix} FR_{131} \\ FR_{132} \\ FR_{133} \\ FR_{134} \end{Bmatrix} = \begin{bmatrix} X & 0 & 0 & 0 \\ X & X & 0 & 0 \\ 0 & 0 & X & 0 \\ 0 & 0 & 0 & X \end{bmatrix} \begin{Bmatrix} DP_{131} \\ DP_{132} \\ DP_{133} \\ DP_{134} \end{Bmatrix} \tag{6.21}$$

This design is a decoupled design and, thus, satisfies the Independence Axiom.

STEP 7
Fourth-Level Decomposition

Step 7a. FR$_{11}$—Sales Revenue Branch

Functional requirement FR$_{111}$ (sell products at the highest acceptable price) must be decomposed with DP$_{111}$ (customer perceived value of product improved) in mind. FR$_{111}$ may be decomposed as

FR$_{1111}$ = Increase the appeal of products by providing desired functions and features.
FR$_{1112}$ = Increase the reliability of products.
FR$_{1113}$ = Provide on-time delivery (for a variety of products).
FR$_{1114}$ = Decrease variation of the delivery time.
FR$_{1115}$ = Provide effective after-sales service.

The corresponding DPs are

DP$_{1111}$ = Design of high-quality products that meet customer needs as specified by FRs and Cs
DP$_{1112}$ = Robust design of products
DP$_{1113}$ = Production based on actual demand
DP$_{1114}$ = Predictable production output
DP$_{1115}$ = Service network

The design equation is

$$\begin{Bmatrix} FR_{1111} \\ FR_{1112} \\ FR_{1113} \\ FR_{1114} \\ FR_{1115} \end{Bmatrix} = \begin{bmatrix} X & 0 & 0 & 0 & 0 \\ X & X & 0 & 0 & 0 \\ 0 & X & X & 0 & 0 \\ 0 & X & X & X & 0 \\ 0 & X & 0 & 0 & X \end{bmatrix} \begin{Bmatrix} DP_{1111} \\ DP_{1112} \\ DP_{1113} \\ DP_{1114} \\ DP_{1115} \end{Bmatrix} \tag{6.22}$$

FR$_{1113}$ is achieved by production based on actual demand. DP$_{1113}$ is fundamental to avoid the long throughput time of mass-production systems. In the traditional mass-production systems, production would start based on the sales forecast. But as the forecast is always wrong, the information related to sales volumes and mix of products will change. Production must be based on actual demand to avoid delays on due dates, overproduction, or high inventory levels. To be able to produce to actual demand, the throughput time of the production system must be radically reduced. FR_{1114} states the necessity to decrease variation of the delivery time. The corresponding DP addresses the importance of a predictable production output.

One important effect about enabling production to be based on actual demand is that when decisions are made to reduce or eliminate activities that cause delays, this not only reduces the throughput time but also eliminates sources of variation within the manufacturing system. For example, when a manufacturing cell is created with single piece-flow of parts, it not only eliminates transport and queues between machines, but also eliminates sources of time variation. This design is a decoupled design and thus satisfies the Independence Axiom.

Functional requirement FR_{112} [increase market share (volume)] must be decomposed with DP_{112} (broad product applications) in mind. FR_{111} may be decomposed as

$FR_{1121} =$ Develop niche (new or custom) products.
$FR_{1122} =$ Develop multiple solutions within the product line.

The corresponding DPs are

$DP_{1121} =$ Short product-development process
$DP_{1122} =$ Product variety

The design equation is

$$\begin{Bmatrix} FR_{1121} \\ FR_{1122} \end{Bmatrix} = \begin{bmatrix} X & 0 \\ X & X \end{bmatrix} \begin{Bmatrix} DP_{1121} \\ DP_{1122} \end{Bmatrix} \tag{6.23}$$

The development of niche products (FR_{1121}) and the development of product variety within the same product line (FR_{1122}) are the FRs to fulfill DP_{112} to broaden product applications. A short product-development process (DP_{1121}) is essential to the development of niche markets. The product development process also affects the ability to develop a variety of products (or options) within an existing product line. This design is decoupled and thus satisfies the Independence Axiom.

Step 7b. FR_{12}—Production Cost Branch

FR_{122} (Reduce operational activity costs) may be decomposed with DP_{122} (target production cost) in mind as

$FR_{1221} =$ Reduce transport costs.
$FR_{1222} =$ Reduce setup costs.
$FR_{1223} =$ Reduce costs of manual operations (machine load/unload, assembly, inspection).
$FR_{1224} =$ Reduce fabrication costs.
$FR_{1225} =$ Reduce maintenance costs.

The corresponding DPs are

$DP_{1221} =$ Product-flow-oriented layout
$DP_{1222} =$ Setup performed with reduced resources
$DP_{1223} =$ Effective use of the workforce
$DP_{1224} =$ Fabrication parameters based on TAKT time to increase tool life
$DP_{1225} =$ Total productive maintenance program

The objective is to reduce cost while increasing the performance of the main operational activities. One major difference in the modern production system is the separation of man and machine on the analysis of the activities. The development of autonomous machines provides opportunity for the worker to perform other activities while the machine is working. Therefore, FR_{1223} and FR_{1224} can be stated separately. FR_{1221}, FR_{1222}, and FR_{1225} are related to activities that we need to dramatically improve while reducing cost. Transport has to be done more frequently and in smaller quantities, the setup has to be done more frequently and much faster, and the maintenance should improve machine reliability.

The design equation is

$$\begin{Bmatrix} FR_{1221} \\ FR_{1222} \\ FR_{1223} \\ FR_{1224} \\ FR_{1225} \end{Bmatrix} = \begin{bmatrix} X & 0 & 0 & 0 & 0 \\ 0 & X & 0 & 0 & 0 \\ 0 & 0 & X & 0 & 0 \\ 0 & 0 & 0 & X & 0 \\ 0 & 0 & 0 & 0 & X \end{bmatrix} \begin{Bmatrix} DP_{1221} \\ DP_{1222} \\ DP_{1223} \\ DP_{1224} \\ DP_{1225} \end{Bmatrix} \tag{6.24}$$

Again the design is an uncoupled design.

STEP 8
Fifth-Level Decomposition

Step 8a. FR_{11}—Sales Revenue Branch

FR_{1112} (Increase the reliability of products) may be decomposed with DP_{1112} (Robust design of products) in mind as

> $FR_{11121} =$ Determine the lowest tolerable stiffness of the product.
> $FR_{11122} =$ Determine the design range for manufacturing tolerance.
> $FR_{11123} =$ Select manufacturing operations with a system range that is within the design range.

The corresponding DPs are

> $DP_{11121} =$ Mathematical model for stiffness determination
> $DP_{11122} =$ Mathematical model for derivation of design range for PVs
> $DP_{11123} =$ Selected machines with appropriate system range for PVs

The idea behind lowering the stiffness (i.e., reducing the magnitude of the diagonal design element) is central to robust design as discussed in Chapters 1 and 2 (Suh, 1995b). The stiffness should be reduced until the signal-to-noise ratio $\eta \geq$ its minimum value. The design equation is

$$\begin{Bmatrix} FR_{11121} \\ FR_{11122} \\ FR_{11123} \end{Bmatrix} = \begin{bmatrix} X & 0 & 0 \\ X & X & 0 \\ X & X & X \end{bmatrix} \begin{Bmatrix} DP_{11121} \\ DP_{11122} \\ DP_{11123} \end{Bmatrix} \tag{6.25}$$

The design matrix is triangular and thus the design is decoupled.

Functional requirement FR_{1113} (Decrease mean delivery time) must be decomposed with DP_{1113} (Production based on actual demand) in mind. FR_{1113} may be decomposed as

FR_{11131} = Produce at the customer-demand cycle time (or TAKT time).
FR_{11132} = Produce the mix of each part type demanded per time interval.
FR_{11133} = Be responsive to the downstream customer's demand time interval.

The corresponding DPs are

DP_{11131} = Linked-cell manufacturing system balanced to customer demand
DP_{11132} = Level production
DP_{11133} = Reduced response time across the production system

To satisfy FR_{11131}, the DP is "linked-cell manufacturing system balanced to customer demand." In a linked-cell manufacturing system, manufacturing processes and operations are designed with cycle times lower than the TAKT time to enable the whole system to produce parts at the customer-demand cycle time. Similarly, DP_{11132} satisfies FR_{11132} with the key elements of short setup time and an information system to level production. Figure 6.1 shows an illustrative linked-cell manufacturing system balanced to customer demand.

The response time is defined as the sum of the "administrative time" (information transfer), the "manufacturing throughput time," and the "shipping or transport time" (which is internal time in cases in which there is no inventory of final products or parts). This sum does not include the manufacturing throughput time when the parts are delivered from stock. For the internal customer, administrative time in a pull-system is the time required to send the information on the status of the manufacturing system (e.g., Kanban card) to

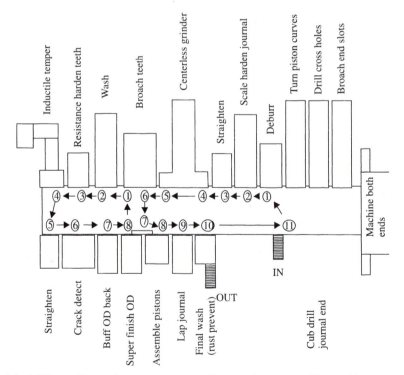

Figure 6.1 A linked-cell manufacturing system to illustrate the concept. The machines are arranged in a cellular structure. In this figure, two workers are moving in two loops opposite to the flow of the work piece. (From Cochran and Lima, 1998).

the upstream cell. The ideal situation is when the production subsystem can satisfy FR_{11133} by DP_{11133} with minimum or even without inventory. In the case in which it needs to have some inventory, if the replenishment time of the cell is short, then the inventory levels are low. The design equation and matrices are

$$\begin{Bmatrix} FR_{11131} \\ FR_{11132} \\ FR_{11133} \end{Bmatrix} = \begin{bmatrix} X & 0 & 0 \\ 0 & X & 0 \\ X & X & X \end{bmatrix} \begin{Bmatrix} DP_{11131} \\ DP_{11132} \\ DP_{11133} \end{Bmatrix} \tag{6.26}$$

When all machines or stations are running at the same TAKT time, buffers between machines can be reduced to one unit. This enables single-piece flow of parts, which eliminates lot delay and transport delay within the manufacturing subsystem (or cell). When the manufacturing system is producing the right mix of parts, it reduces or eliminates process delays and therefore reduces the throughput time and increases the ability of the system to "be responsive to the downstream customer's timing." A lot has been said about the importance of reducing throughput time. But sometimes the mechanisms to reach this goal are not understood. In the above design matrix, it is fundamental to understand the impact that DP_{11131} and DP_{11132} have as enablers to satisfy FR_{11133}. The above design matrix states this point very clearly. This design is a decoupled design and satisfies the Independence Axiom.

Functional requirement FR_{1114} (decrease variation of the delivery time) must be decomposed with DP_{1114} (predictable production output) in mind. FR_{1114} may be decomposed as

$FR_{11141} =$ Respond quickly to production problems.
$FR_{11142} =$ Produce with a predictable quality output.
$FR_{11143} =$ Produce with a predictable time output.

The corresponding DPs are

$DP_{11141} =$ Visual control system to provide rapid response
$DP_{11142} =$ Production with no defects and able to identify root cause
$DP_{11143} =$ Predictable production resources

The design equation and matrices are

$$\begin{Bmatrix} FR_{11141} \\ FR_{11142} \\ FR_{11143} \end{Bmatrix} = \begin{bmatrix} X & 0 & 0 \\ X & X & 0 \\ X & X & X \end{bmatrix} \begin{Bmatrix} DP_{11141} \\ DP_{11142} \\ DP_{11143} \end{Bmatrix} \tag{6.27}$$

The following factors characterize this decoupled design. The visual system requires the display of standard worksheets and inspection plans at every station, and also the correct display of all tools and devices to reduce the occurrence of mistakes among the workers, which improves quality. As the visual control system enables a faster response to undesired events, it has a beneficial impact on the ability of the production system to have a predictable time output. As the generation of scrap and rework is reduced, the system becomes more predictable in terms of time output. This design is decoupled and thus satisfies the Independence Axiom.

Step 8b. FR_{12}—Manufacturing Cost Branch

FR_{1223} (reduce costs of manual operations [machine load/unload, assembly, inspection]) may be decomposed with DP_{1223} (effective use of the workforce) in mind as

FR_{12231} = Reduce tasks that tie the operator to the machine.
FR_{12232} = Enable worker to operate more than one machine or station.
FR_{12233} = Plan resources to produce with different production volumes.

The corresponding DPs are

DP_{12231} = Machines and stations designed to run autonomously
DP_{12232} = Work-loops implemented in a cell layout
DP_{12233} = Standardized work-loops designed for different volumes

FR_{12231} and its corresponding DP_{12231} (machines and stations designed to run autonomously) have enabled the separation of the worker from the machine. One worker operates multiple kinds of machines to keep work moving in a continuous flow at TAKT time. At each machine, a part is already unloaded when the operator arrives to load a new part. The worker loads the new part and then restarts the machine and walks to perform a similar task in the next machine in the manufacturing cell. FR_{12232} and its corresponding DP_{12232} cover the cell layout, ergonomics, and organization to enable the implementation of work-loops. Once the elements for the establishments of work-loops are in place, the next step is to plan the configuration of the work-loops for different production volumes as a result of the need to accommodate fluctuations in demand, as stated by FR_{12233} and DP_{12233}. A manufacturing cell may be able to work with four, three, or two workers, depending on the actual customer demand cycle time or TAKT time, and, therefore, operational cost scales down when the customer demand decreases. The design equation is

$$\begin{Bmatrix} FR_{12231} \\ FR_{12232} \\ FR_{12233} \end{Bmatrix} = \begin{bmatrix} X & 0 & 0 \\ X & X & 0 \\ X & X & X \end{bmatrix} \begin{Bmatrix} DP_{12231} \\ DP_{12232} \\ DP_{12233} \end{Bmatrix} \tag{6.28}$$

The design is a decoupled design.

STEP 9
Sixth-Level Decomposition: FR_{11}—Sales Revenue Branch

FR_{11131} (produce at the customer-demand cycle time or TAKT time) must be decomposed with DP_{11131} (linked-cell manufacturing system balanced to customer demand) in mind. FR_{11131} may be decomposed as

FR_{111311} = Define customers, parts, and volumes for each subsystem or cell within production.
FR_{111312} = Design subsystem for a range of volume fluctuation.

The corresponding DPs are

DP_{111311} = Configuration of subsystems to enable flow at the ideal range of cycle times
DP_{111312} = Cell or subsystem designed to meet the minimum TAKT time

The design equation is

$$\begin{Bmatrix} FR_{111311} \\ FR_{111312} \end{Bmatrix} = \begin{bmatrix} X & 0 \\ X & X \end{bmatrix} \begin{Bmatrix} DP_{111311} \\ DP_{111312} \end{Bmatrix} \tag{6.29}$$

This design is decoupled and thus satisfies the Independence Axiom.

Functional requirement FR_{11132} (produce the mix of each part type demanded per time interval) must be decomposed with DP_{11132} (level production) in mind. FR_{11132} may be decomposed as

FR_{111321} = Produce in small run-sizes.
FR_{111322} = Convey in small and consistent quantities.
FR_{111323} = Produce and supply only the parts needed.

The corresponding DPs are

DP_{111321} = Setup performed in less than 10 minutes
DP_{111322} = Standard containers that hold small amounts of parts
DP_{111323} = Information system to produce only the parts needed (pull system)

Level production controls the mix of products produced and the production run size within a given time interval. A pull system sends an information signal for an upstream cell to supply the downstream cell with the required parts in standard quantities. Therefore, short setup times and standard containers are important enablers of level production. Level production enables the manufacturing system to supply in small quantities and, therefore, the system can supply only the parts needed. The design equation and matrices are

$$\begin{Bmatrix} FR_{111321} \\ FR_{111322} \\ FR_{111323} \end{Bmatrix} = \begin{bmatrix} X & 0 & 0 \\ X & X & 0 \\ X & X & X \end{bmatrix} \begin{Bmatrix} DP_{111321} \\ DP_{111322} \\ DP_{111323} \end{Bmatrix} \tag{6.30}$$

This design is a decoupled design and thus satisfies the Independence Axiom.

FR_{11133} (be responsive to the downstream customer's demand time interval) must be decomposed with DP_{11133} (reduced response time across the production system) in mind. FR_{11133} may be decomposed as

FR_{111331} = Reduce sub-system replenishment time to less than the customer demand interval.
FR_{111332} = Ensure that sufficient parts are available to satisfy the customer demand interval.

The corresponding DPs are

DP_{111331} = Elimination of wastes that cause excess lead time
DP_{111332} = Standard work-in-process (swip) quantity of parts

To satisfy FR_{111331}, the DP is to eliminate wastes that cause excess lead time. For repetitive production systems, the customer-demand interval is the time interval between transport to the downstream cell. The cell must be able to replace the parts to the standard inventory during the customer-demand interval.

To satisfy FR_{111332}, it is necessary to implement a standard inventory of parts (DP_{111332}). As the cell lead time decreases, the ability of the system to ensure that parts are available for the customer increases. Decreasing cell lead time also makes it possible to proportionally reduce the size of the standard inventory, as can be seen in the design equation shown below.

$$\begin{Bmatrix} FR_{111331} \\ FR_{111332} \end{Bmatrix} = \begin{bmatrix} X & 0 \\ X & X \end{bmatrix} \begin{Bmatrix} DP_{111331} \\ DP_{111332} \end{Bmatrix} \tag{6.31}$$

This design is decoupled and thus satisfies the Independence Axiom.

FR_{11142} (produce with a predictable quality of output) must be decomposed with DP_{11142} (production with no defects and with the ability to identify root cause) in mind. FR_{11142} may be decomposed as

$FR_{111421} =$ Ensure capable processes.
$FR_{111422} =$ Decrease sources of variation due to multiple flow paths.
$FR_{111423} =$ Prevent making defects throughout.
$FR_{111424} =$ Do not advance defects to the next operation.

The corresponding DPs are

$DP_{111421} =$ Capable machines, equipment, tools, and fixtures
$DP_{111422} =$ Single path through manufacturing system and external supplier (no parallel processing)
$DP_{111423} =$ Use of standards and devices to prevent defects
$DP_{111424} =$ Use of successive checks to detect defects if they do occur

In a continuous-flow factory, there must be no conflict between quantity and quality. Quality must always prevail. There should be no flow unless high-quality parts are being produced. This philosophy represents a major difference from the common practice of producing poor products and fixing them later. To achieve high-quality yields, machines and equipment must be capable (DP_{111421}), but, most importantly, defects must be prevented from happening (DP_{111423}), or detected early to prevent defective parts from flowing to the downstream operations (DP_{111424}). The design equation is

$$\begin{Bmatrix} FR_{111421} \\ FR_{111422} \\ FR_{111423} \\ FR_{111424} \end{Bmatrix} = \begin{bmatrix} X & 0 & 0 & 0 \\ X & X & 0 & 0 \\ X & 0 & X & 0 \\ 0 & 0 & 0 & X \end{bmatrix} \begin{Bmatrix} DP_{111421} \\ DP_{111422} \\ DP_{111423} \\ DP_{111424} \end{Bmatrix} \tag{6.32}$$

This design is decoupled and thus satisfies the Independence Axiom.

STEP 10
Seventh-Level Decomposition: FR_{11}—Sales Revenue Branch

FR_{111312} (design subsystem for a range of volume fluctuations) must be decomposed with DP_{111312} (subsystem designed to meet the minimum TAKT time) in mind. FR_{111312} may be decomposed as

$FR_{1113121} =$ Select appropriate manufacturing process.
$FR_{1113122} =$ Design manufacturing process cycle time at each station to meet minimum TAKT time.
$FR_{1113123} =$ Design station fixtures to enable minimum TAKT time.

The corresponding DPs are

$DP_{1113121} =$ Physics of the manufacturing process
$DP_{1113122} =$ Manufacturing process work content defined to be less than the minimum TAKT time

$DP_{1113123} =$ Fixture design to provide quick load/unload (within required tolerance)

The design equation is

$$\begin{Bmatrix} FR_{1113121} \\ FR_{1113122} \\ FR_{1113123} \end{Bmatrix} = \begin{bmatrix} X & 0 & 0 \\ X & X & 0 \\ X & X & X \end{bmatrix} \begin{Bmatrix} DP_{1113121} \\ DP_{1113122} \\ DP_{1113123} \end{Bmatrix} \qquad (6.33)$$

This is a decoupled design and thus satisfies the Independence Axiom.

6.6 MATHEMATICAL MODELING AND OPTIMIZATION OF DESIGN

In the preceding example, the design matrix was formulated in terms of X and 0. In some cases, it may be sufficient to complete the design at this level of definition. However, whenever it is possible, each X must be replaced with physical variables by modeling the relationship between the FR and the DP, based on physics and the laws of nature.

The modeling should be done at the leaf-level only as the higher level modules are a combination of the lower level modules dictated by the design matrices. When all these leaves are modeled and integrated in a single package, the complete package constitutes the model for the entire system.

If the design does not have coupling and satisfies the Independence Axiom, the exact design values for DPs can be determined through modeling. Even when the design is nonlinear, it is possible to find the design windows in which the design satisfies the Independence Axiom best. Then the design values for DPs can be determined. Designs that satisfy the design axioms are stable and controllable (Suh, 1997b). If the design were coupled at the highest level, it would be impossible to decompose the design to several levels, because any slight variations of any DP may create an unstable and uncontrollable design.

If the design is coupled, thereby violating the Independence Axiom, we may search for an optimum point in the design space. However, there is no guarantee that we can find an optimum point, especially for a nonlinear design. Even if we find a solution, it may be unstable and uncontrollable.

6.7 REPRESENTATION OF MANUFACTURING SYSTEM ARCHITECTURE

The manufacturing system designed in the preceding section has a system architecture—the hierarchical structure in the FR, DP, and PV domains with a precisely defined relationship given by the design equations and design matrices (Suh, Cochran, and Lima, 1998). The PVs, which are process variables that define how the selected DPs can be implemented, were not specified in this design illustration.

As discussed in Chapter 4, the system architecture can be represented in several different forms: the FR, DP, and PV hierarchies with the corresponding design equations and matrices, the module-junction diagram, and the flow diagram (Kim, Suh, and Kim, 1991; Suh, 1997a). Although all these different representations are equivalent, each emphasizes different aspects of the system architecture.

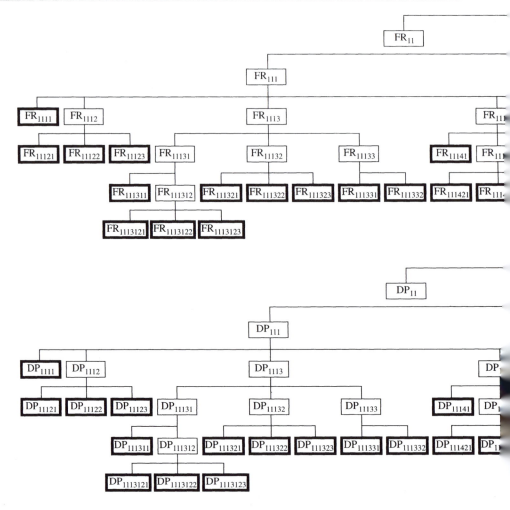

Figure 6.2. Functional requirement—design parameter decomposition hierarchy. (a) The FR hierarchy tree; (b) the DP hierarchy tree.

The hierarchical diagram gives the entire decomposition steps and the leaves. The module-junction diagram is created to show the hierarchical structure of modules. The flow diagram is a concise and powerful tool that provides a road map for implementation of the system design, which will be illustrated for the manufacturing system designed in this section. The flow diagram illustrates the design relationships and the precedence of implementation based on design matrices at each level of the design decomposition.

The tree diagrams of the FR and DP hierarchies are shown in Figure 6.2. In this representation, we must augment the diagram with the design matrices to indicate how the design must be implemented to satisfy the Independence Axiom because implementation of the design depends on whether the design is uncoupled, decoupled, or coupled. If the design matrix is triangular, the lower-level FRs must be satisfied in the sequence given by

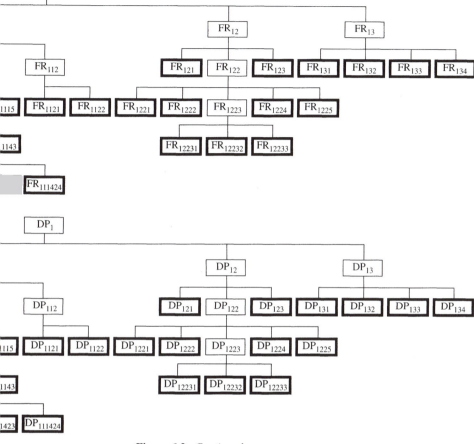

Figure 6.2. *Continued*

the design matrix in order to satisfy the higher-level FRs. Each design matrix represents a junction and shows how the lower-level modules must be assembled to yield the desired FR.

Each FR of the lowest layer of each branch (i.e., each FR that does not need further decomposition) is defined as the leaf. The leaves of the system shown in Figure 6.2 are FR_{121}, FR_{123}, $FR_{1113121}$, $FR_{1113122}$, $FR_{1113123}$, etc. The higher-level FRs are satisfied by combining the leaves according to the information contained in the design matrix. The thick lined boxes in Figure 6.2 show the leaves.

Normally, at each level of the design, the physical embodiment of the design should be shown. At higher levels, the physical embodiments do not have all the details. As the decomposition proceeds, the details are added to the physical embodiment.

The concept of "modules" was introduced in Chapter 4, Section 4.6.2. Modules relate FRs to DPs. A module is defined as the row of the design matrix that yields an FR when it is provided with the input of its corresponding DP.

Combining FR_{12231}, FR_{12232}, and FR_{12233} as per the design matrix can represent M_{1223}. However, M_{1223} does not appear in the design explicitly, as it is not the leaf or the terminal design. Only those DPs corresponding to leaves need to be brought in to execute the system architecture. Lower level modules are combined to obtain higher level modules that satisfy the higher level FRs. The design matrix at each level and at each junction provides guidance as to how the lower level modules can be combined to obtain the higher level modules.

In Section 4.6.2, the following symbols in circles were used to denote the control of the system flow chart. A circled **S** is for simple summation of FRs for an uncoupled design. It connects parallel modules of an uncoupled design. A circled **C** represents sequential control of DPs as suggested by the design matrix for a decoupled design. It links the modules sequentially. We must operate on the module on the left first and then move to the next module on the right. A circled **F** is for a coupled design, indicating that it requires feedback and violates the Independence Axiom. Figure 4.4, which is reproduced here as Figure 6.3, shows these control symbols and associated modules.

Having defined the modules for the system, the flow chart representation of the manufacturing system architecture for the system designed in Section 6.5 is shown in Figure 6.4. The flow diagram is generated using the axiomatic design software presented in Chapter 5, Section 5.4 (Do and Suh, 2000). This system architecture represents the entire manufacturing system, including all the modules and their interrelationships. Some of the modules represent hardware and some represent software. This diagram shows how any change in a given module will affect other modules. To achieve systems goals, all affected modules must be changed when one of them is changed. The system architecture shown in Figure 6.4 satisfies the Independence Axiom because there is no coupling. Although the DPs were chosen to minimize the information content, the information content of the system in its entirety has not been checked to this point.

Once the manufacturing system is designed, it can be operated using the flow chart of the system architecture shown in Figure 6.4. The sequence of operation begins either at the highest level (i.e., following the flow chart) or at the lowest level (i.e., the leaves). In some cases, the DPs at the leaf level must be determined to satisfy the given set of the highest level FRs. In this case, the top-down operation allocates the specific overall values to modules in the case of a decoupled design. However, when the lowest level DPs are given, a bottom-up approach may be appropriate. In this case, to achieve the highest level FRs, which are the system goals, the operation of the system must begin from the innermost modules that represent the lowest level leaves in each major branch. Then move to the next higher level modules (i.e., next innermost box) following the sequence indicated by the system architecture—from the innermost boxes to the outermost boxes. Therefore, when

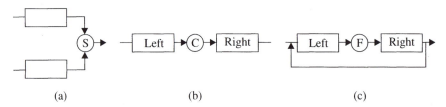

Figure 6.3 Representation of the design at each junction of a flow diagram. (a) Summing junction (uncoupled design), (b) control junction (decoupled design), (c) feedback junction (coupled design). (From Kim, Suh, and Kim, 1991.)

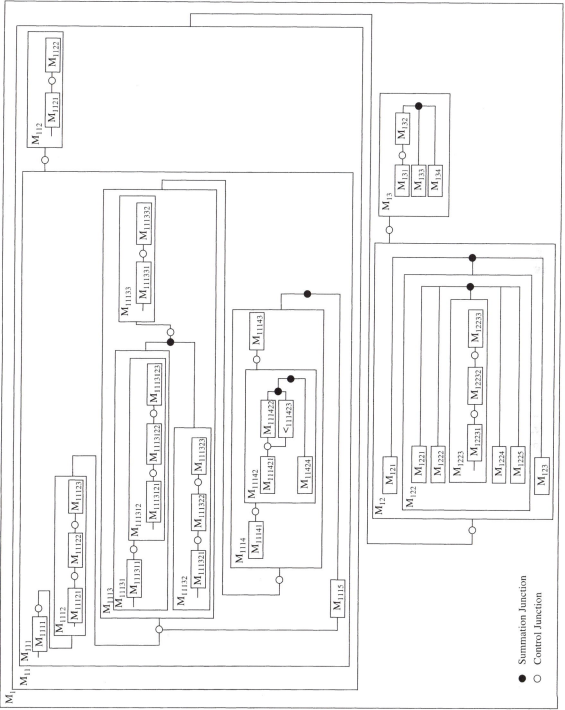

Figure 6.4 Flow chart of the manufacturing system design developed in Section 6.5.

the lowest level DPs are given, the rule for execution of the system architecture (the system command control—SCC) may be stated as follows:

1. From the input data for the lowest level DPs (i.e., DP leaves) and the corresponding elements of the design matrix, construct the lowest level modules of all branches. Multiply these modules with appropriate DPs (or make these modules operational by providing DPs as inputs) to obtain the low-level FRs corresponding to leaves.

2. Combining these low-level FRs based on the system architecture, obtain the next higher level FR and then proceed with the operation at the next higher level modules until the highest level module is obtained.

3. Execute this procedure simultaneously at all branches in which such an action can be achieved independently to minimize the operational time.

The system architecture represented by the flow chart can be used in dealing with the following aspects of manufacturing systems:

1. *Diagnosis*: The impending failure of a manufacturing system can be diagnosed based on the system architecture. When a certain FR is not being satisfied fully, the DPs that are causing the problem can be identified and fixed quickly based on the flow diagram. When the system fails, the flow diagram can be used to trace and find the source of the problem.

2. *Engineering changes*: During the development process of a manufacturing system, it is inevitable that changes will need to be made to the decisions made in the past. The system architecture and the flow diagram can be used to identify all the related modules that must also be changed when a single change in the system architecture is proposed.

3. *Job assignment and management*: The flow diagram can be used to organize management of the manufacturing system and to assign tasks to various participants.

4. *Distributed systems*: When a large manufacturing system is being developed, the lower level tasks are performed by many people located at distant places, sometimes involving subcontractors and subsubcontractors. The flow diagram of the system architecture will provide the road map by which the tasks can be coordinated and controlled.

5. *Software development*: The system architecture can be used in developing software to control the system operation.

6.8 SUMMARY

In this chapter, the axiomatic design theory for systems discussed in Chapters 1 through 4 is applied to manufacturing system design.

In the past, manufacturing systems have been designed ad hoc based on experience without a clear analysis of system needs. As a result, many manufacturing systems tried in the past have not been effective in increasing manufacturing productivity. Incremental improvements of an existing manufacturing system, although necessary, cannot provide a long-term strategy for manufacturing competitiveness.

Manufacturing systems must be designed rationally to achieve a set of functional requirements. The resulting manufacturing system will be different depending on the functional requirements chosen to satisfy the customer requirements. The functional requirements of a manufacturing system change over time. In this sense, there is no ideal, time-invariant manufacturing system, as the system design must evolve to deal with the most important manufacturing issues of a given era. In the future, it is highly conceivable that one of the important functional requirements of a manufacturing system may be related to environmental issues, which may change the design of manufacturing systems.

Two different kinds of manufacturing systems were considered: "push" type and "pull" type. In both cases, the design of these systems must satisfy the Independence Axiom. When identical parts are to be manufactured, a push system can be designed for a given throughput rate. To satisfy the Independence Axiom, the push system requires the use of decouplers with queues. In this chapter, the exact means of solving for the queues of a push system are given.

This chapter also presents a pull type of manufacturing system, which is an ideal manufacturing system design for manufacturing a mix of large numbers of similar-quality products over a period of a few weeks to years. This is often the case with many discrete consumer products. The system architecture for the designed manufacturing system is presented. The procedure for creating a quantitative mathematical model for the manufacturing system is outlined, which is based on the modeling of leaves and integrating these leaves into a system model for the entire system.

REFERENCES

Arnold, H. L., and Faurote, F. L. *Ford Methods and the Ford Shops*, J. J. Little and Ives Co., New York, 1915.

Black, J. T. *The Design of the Factory with a Future*, McGraw-Hill, New York, 1991.

Carrus, B. J., and Cochran, D. S. "Application of a Design Methodology for Production Systems," *Annals of the 2nd International Conference on Engineering Design and Automation,* Maui, HI, 1998.

Cochran, D. S. "The Design and Control of Manufacturing Systems," Ph.D. Thesis, Auburn University, 1994.

Cochran, D. S., and Lima, P.C. *The Production System Design Decomposition*, Version 4.2, unpublished report, MIT Production System Design Group, 1998.

Cochran, D. S., and Lima, P.C. *Production System Design: Theory, Evaluation and Implementation,* 2000 (in preparation).

Do, S.-H., and Suh, N. P. Axiomatic Design Software (Acclaro), Axiomatic Design Software, Inc., 2000.

Ham, I., Hitomi, K., and Yoshida, T. *Group Technology: Application to Production Management,* Kluwer Academic Publishers, Hingham, MA, 1985.

Kim, S. J., Suh, N. P., and Kim, S.-K. "Design of Software Systems Based on Axiomatic Design," *Annals of CIRP*, Vol. 40, pp. 165–170, 1991.

Marca, D. A., and McGowan, C. L. *IDEF0/SADT,* Eclectic Solutions, San Diego, CA, 1993.

Oh, H. L. *Synchronizing Wafer Flow to Achieve Quality in a Single-Wafer Cluster Tool,* Unpublished Report, Silicon Valley Group, Inc., 1998 (Patent Pending, 1998).

Oh, H. L., and Lee, T. S. "Synchronous Algorithm" presented at the First International Conference on Axiomatic Design, MIT, June 2000.

Shingo, S. *The Toyota Production System from an Industrial Engineering Viewpoint,* Productivity Press, Portland, OR, 1989.

Sohlenius, G. "The Productivity of Manufacturing through Manufacturing System Design," KTH working paper, 1998.

Suh, N. P. *The Principles of Design,* Oxford University Press, New York, 1990.

Suh, N. P. Axiomatic Design of Mechanical Systems, Special 50th Anniversary Design Issue, *Journal of Mechanical Design and Journal of Vibration and Acoustics, Transactions of the ASME*, Vol. 117, pp. 2–10, 1995a.

Suh, N. P. "Designing-in of Quality through Axiomatic Design," *IEEE Transactions on Reliability,* Vol. 44, No. 2, pp. 256–264, 1995b.

Suh, N. P. "Design and Operation of Large Systems," *Journal of Manufacturing Systems,* Vol. 14, No. 3, pp. 203–213, 1995c.

Suh, N. P. "Design of Systems," *Annals of CIRP,* Vol. 46, No. 1, pp. 75–80, 1997a.

Suh, N. P. *Is There Any Relationship Between Design Science and Natural Science?* The 1997 Ho-Am Prize Memorial Lecture given at Seoul National University, Published by the Ho-Am Prize Committee, Korea, 1997b.

Suh, N. P., Cochran, D. S., and Lima, P.C. "Manufacturing System Design" (Keynote paper presented at the CIRP General Assembly, Greece, August 1998), CIRP *Annals,* Vol. 47, No. 2, 1998.

Van Brussel, H., Peng, Y., and Valckenaers, P. "Modeling Flexible Manufacturing Systems Based on Petri Nets," *CIRP Annals*, Vol. 42, No. 1, pp. 479–484, 1993.

Womack, J. P., Roos, D., and Jones, D. *The Machine That Changed the World*, Rawson Associates, New York, 1990.

Yien, J. T. S. "Manufacturing System Design Methodology," Ph.D. Thesis, Hong Kong University of Science and Technology, 1998.

HOMEWORK

6.1 In many manufacturing firms, the assembly operation is very labor intensive. Assembly labor cost accounts for more than 25% of the total labor cost of manufacturing. To reduce the manufacturing cost, many efforts have been made during the past two decades to introduce flexible, automatic assembly systems without much success. In recent years, the problem has become much more difficult because of frequent change of products and small volumes of production to respond quickly to changing customer demands. Therefore, the goal of lowering the assembly cost through the use of flexible automatic assembly systems is only a dream in manufacturing firms. Your task is to design a lean assembly system for electrical hand drills sold to professionals. The specific details of the desired operation are as follows:

The assembly system must be able to assemble 12 kinds of drills with different horsepower. The demand rate is about 100,000 drills a year for each drill size—evenly distributed over the 12-month period. The price of drills is about 50% greater than comparable drills that are specifically made for home use—weekend amateur carpenters and electricians. The drill must last about 10 years.

6.2 A plastic part used in a consumer product must be processed in seven different process modules before the part is released from the production plant. A specially designed robot must be used for pick-up, transport, and placement of the part for safety reasons. The desired throughput rate is 100 parts/hour. The modules are cubic boxes with a dimension of 0.5 m. The part is picked up from an input cart and placed on Module A and finally leaves the machine when all the processes are completed. The time it takes for the robot to do these initial and final operations of picking up the part from the cart and from the last module to the outgoing cart is 5 seconds each. The process must be performed in the sequence given below:

The robot arm, which is 0.7 m long, can pick up the part from the modules in 3 seconds. It can place the part into the module in 5 seconds. The entire robot can also move on a rail at 5 m/second.

Your task is to

  a. Design the machine that can perform the desired functions, but you do not have to develop the individual process modules.
  b. Develop a strategy for robot motion.

Modules	Process Time ± Tolerance (seconds)
A	35 ± 5
B	20 ± 0
C	15 ± 2
D	55 ± 5
E	15 ± 0
A	35 ± 5
F	25 ± 10

6.3 A pharmaceutical company is producing bags that hold dialysis fluid and is considering modifying the existing manufacturing process. The most critical manufacturing issue involves the area of the bag that is to be punctured by the customers to add customer-specific doses of additives (medicines, etc.) to the fluid already inside. This is done by injecting a hollow needle through this area. To seal the contained fluid from the environment and prevent any bacterial infection, a rubber disk is positioned in this area to create a hermetically sealed interface. The needle is injected through the rubber disk and bag membrane, as show in Figure H.6.3.

The tube shown in figure H.6.3 encompasses the rubber disk. To perform its task adequately, the rubber disk must create a hermetic seal between the membrane and tube and thus prevent any bacterial passage.

The company designers have selected four production methods for the placement of the disk. In addition, they have asked you to use the Information Axiom to select the best solution.

The possible solutions are as follows:

  1. Mold the tube and membrane around the disk.
  2. Use a disk smaller in diameter than the tube, insert the disk, and then inject sealant between the disk and tube.
  3. Insert a hollow needle through the disk (to prevent air from being trapped in the disk/membrane interface), insert the disk, and then remove the hollow needle.
  4. Squeeze the disk and insert into place.

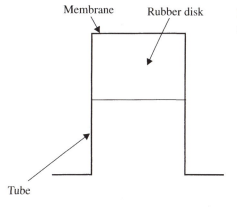

Membrane   Rubber disk

Tube

Figure H.6.3 Geometry of rubber disk and membrane (not to scale).

How would you manufacture this seal?

6.4 Typical automobile engines are produced using a transfer line, which consists of a series of specialized machine tools. An engine block is transferred from machine to machine, sometimes on a pallet. Each machine performs a specific task, such as drilling a set of tapped holes for bolts, boring cylinders, etc. The shortcomings of a transfer line are its high initial investment cost and the lack of flexibility. When the engine design changes, a new transfer line must be installed. Therefore, a transfer line is not economical when the production volume for a given engine is small. In recent years, attempts have been made to machine the engines using a flexible manufacturing system, which typically consists of machining centers that can perform multiple functions. The machining centers can be competitive with the transfer line when the production volume is small. Your job is to analyze the design of a transfer line using the design axioms and offer a more flexible manufacturing system.

6.5 Investigate how large products such as commercial airplanes are currently manufactured. Does it make sense? How would you improve it?

6.6 In the past, an attempt was made to increase the productivity of discrete-parts manufacturing by using the concept of Group Technology. The central idea behind Group Technology is to group similar parts and machine them together on the most appropriate machine tools. Analyze the effectiveness of Group Technology in terms of the design axioms.

6.7 Go to a small manufacturing company in your neighborhood and ask if you can study their manufacturing system for the purpose of improving it or devising an alternate manufacturing system to improve their manufacturing productivity. Define the FRs, DPs, and PVs. Apply the Independence Axiom to eliminate the coupling between various functions. After you have developed two different manufacturing systems, apply the Information Axiom to choose the better manufacturing system. MIT students have done this around Boston, with a significant improvement in manufacturing productivity of various companies.

Axiomatic Design of Materials and Materials-Processing Techniques

7.1 INTRODUCTION

Axiomatic design theory has been used to design new materials and materials-processing techniques. This chapter presents three examples. In the first two examples, two new materials—Mixalloy and Microcellular Plastics—that have been developed using the theory are presented to show how axiomatic design can be applied to the creation and development of new materials and processes. In these examples, FRs were developed to satisfy customer needs, new materials (in terms of DPs) were designed, and new industrial processes (in terms of PVs) were designed and commercialized. The third example, which deals with rapid prototyping and a layered manufacturing process, is presented to illustrate the process of creating new materials and processes.

New advanced materials have enabled us to meet the requirements of modern machines (e.g., gas turbines), consumer goods (e.g., carpets), airplanes (e.g., skins and structural parts), industrial equipment (e.g., cutting tools), computers (e.g., semiconductors), and electronic/communications equipment (e.g., fiber optics). These materials must be designed and processed. Axiomatic design has been applied to develop new materials and processing techniques. In this chapter, two of these materials and their processing techniques—Mixalloy and Microcellular Plastics—will be presented as case studies, which the author invented and developed with his co-workers.

In Chapter 1, it was shown that there are four domains in the design field: the customer domain, the functional domain, the physical domain, and the process domain. In the case of materials, the customer domain is where the desired *performance* of the materials is specified. Then, in the functional domain, the FRs are the *properties* of the materials that can provide the desired performance specified in the customer domain. These FRs are satisfied by choosing the *microstructure* (or morphology) as the DPs in the physical domain. Finally, the *process variables* (PVs) in the process domain specify how the desired microstructure can be created. Therefore, the sequence of mapping is from the desired performance to properties to structure to processing variables as shown in Figure 7.1.

The history of materials development has not followed this sequence until recently. Often an inverse mapping was done—from process to microstructure to properties to performance (or applications). Prehistoric discoveries indicate that humans discovered materials naturally produced from volcanic actions and found suitable uses for them. Gold was used for several thousand years because of its easy processing capabilities. Even in the case of plastics, processing techniques, such as the polymerization process and the incorporation of fibers in polymers, were done first, followed by characterization of material properties and microstructures. Even today, many materials scientists and engineers tend to view the materials field as a highly coupled "tetrahedron" whose four vertices (performance, properties, structure, and processes) are interconnected to each other, as shown in Figure 7.2. This traditional view of materials does not recognize the concepts of domains, mapping, zigzagging, and decomposition, i.e., the axiomatic design view shown in Figure 7.1.

During the past several decades, many new materials have been developed, most notably semiconductors, powder metallurgy products, and composites. Although they did

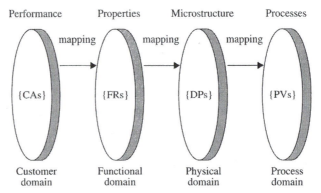

Figure 7.1 Four domains of the materials world: *Performance* of materials characterizes the customer domain. *Properties* characterize the functional domain. *Structure* characterizes the physical domain. *Process variables* characterize the process domain.

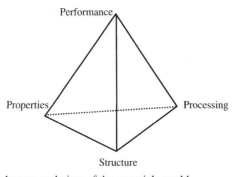

Figure 7.2 The traditional tetragonal view of the materials world.

not follow the axiomatic design process, the development of these materials—through extensive trial and error—was done by discovering or creating structures and processes that are decoupled. Indeed the development of complicated microelectro-mechanical systems (MEMs) is accelerated when it is done based on axiomatic design processes because the costly iterative process development cycle is reduced (Kim and Koo, 2000).

Two processes discussed in this chapter have been developed based on axiomatic design. One of these deals with metals and its alloys. The material is called Mixalloy and the process is called the Mixalloying process (Suh, 1981a, 1981b; Suh et al., 1982; Sanchez-Caldera et al., 1987, 1990, 1991a, 1991b). The other deals with polymers. The material is called Microcellular Plastics (Martini-Vvdensky et al., 1984; Baldwin et al., 1994a, 1994b; Suh, 1996). Microcellular plastics can be made from any starting polymers. Both of these materials were invented at MIT and commercialized by small new-venture firms. The history of the invention of microcellular plastics can be found in Suh (1996).

7.2 MIXALLOYS

What is a mixalloy? To create alloys that have high strength at high temperatures, high toughness, and pure matrix phase, mixalloy was designed and produced using a unique processing technique. The specific alloy made by the mixalloy process is a dispersion-strengthened copper alloy where nanoscale titanium diboride particles are dispersed throughout the metal matrix. The ceramic particles are generated through chemical reaction of metallic and nonmetallic elements. This is done by dissolving a small amount of titanium in molten copper in one stream and a small amount of boron in stoichiometric ratio in another stream of molten copper, and subjecting them to impingement mixing and rapid solidification. The Cu/TiB_2 mixalloys thus produced have the properties originally designed during the material design phase.

7.2.1 History of Mixalloys

The idea for mixalloys came about because the traditional view of the materials field—the inverse process of going from processes to properties to applications—did not make much sense from the axiomatic design point of view. Axiomatic design proceeds from the desired performance of materials specified in the customer domain to the desired material properties in the functional domain. Then, from the functional domain, we would proceed to the physical domain that must specify how the FRs (properties) will be satisfied by conceiving a microstructure and the associated design parameters that characterize the microstructure. Finally, having specified the design parameters (DPs) of the microstructure, we can proceed to the process domain and develop processing techniques (PVs) that will yield the desired DPs.

The idea for mixalloys was first conceived to make metals with fine microstructure, because the inventor of the original concept (Suh, 1981a, 1981b; Suh et al., 1982) was interested in grain-size effects on mechanical properties and fine-grained metals. Preliminary experiments were performed to show the feasibility of the idea, and based on the results, patents were obtained.

It was later realized that the process was better suited to making alloys, especially dispersion-strengthened materials. However, to be thoroughly tested, process required much larger facilities than were available at MIT. Therefore, an industrial firm—Sutek Corporation—was established to commercialize the process. Three new MIT Ph.D.s became the key leaders of this firm, who built the factory, developed the commercial process, created the market for this new product, and developed an industrial infrastructure. They did all this in less than 4 years—a remarkable achievement in the materials field, which normally spends 12 years to introduce new materials!

The mixalloy process was successfully developed and the final mixalloy products in the form of high-strength conductors and spot-welding electrodes were sold commercially to automotive companies in the United States, Korea, and Japan. Also a major defense contractor in the United States tested mixalloys for use in hypersonic aircraft, which showed that the microstructure does not change even after repeated heating to 1020°C. In spite of its technical success, after a few years of operation, Sutek Corporation had to abandon its commercial production of the mixalloys, because a larger competitor continued to lower the price of their products, eventually driving the small, start-up company out of the market. Although the investors in Sutek had been patient for many years, ultimately they decided to abandon their support of the mixalloy business for financial reasons. On average, it takes about 12 years or more to develop successful new materials on a commercial scale and succeed in the marketplace, which makes it difficult for small venture firms to compete.

7.2.2 Design of Dispersion-Strengthened Metals: Mapping from the Functional Domain to the Physical Domain

At the beginning of the mixalloy development project, the customer was identified to be automotive companies as they spot weld sheet-metal parts to manufacture car bodies. They use spot-welding electrodes in large numbers—about one per robot per 8-hour shift.

The desired performance characteristics are high strength at high temperatures (i.e., resistance to deformation), long life (i.e., resistance to wear), ease of formability, and toughness. Typically, automotive firms have used copper/chrome alloys or dispersion-strengthened copper with aluminum oxide particles. The copper/chrome alloy has low conductivity and low strength, but it is cheap. Because the copper/chrome alloy deforms readily, the welding tips made of this alloy must be dressed regularly and replaced frequently. The dispersion-strengthened Cu/Al_2O_3 alloy, which consists of submicroscale aluminum-oxide particles dispersed throughout the pure copper matrix, has high electrical and thermal conductivity as well as high strength and toughness. However, the internal oxidation process is expensive. Therefore, the competitive advantages of mixalloy were thought to be superior performance over the cheaper copper/chrome alloy and lower cost than the internally oxidized, Al_2O_3 dispersion strengthened copper.

The FRs of mixalloy may be stated as

FR_1 = Provide high strength at high operating temperatures = τ_s.
FR_2 = Provide high elongation and toughness = K_c.
FR_3 = Provide high electrical and thermal conductivity = σ.

We can readily design the microstructure of the desired material that can satisfy the above FRs. Figure 7.3 shows the microstructure of an alloy that consists of pure phase (matrix phase) and small ceramic particles that are thermally stable. To increase the flow strength of the metal at all temperatures relative to the flow strength of pure copper (or copper

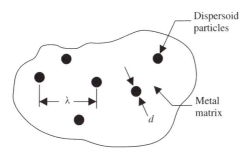

Figure 7.3 Microstructure of the alloy with hard ceramic particle in a metal matrix.

alloy), we can disperse a large number of thermally stable ceramic particles in copper so as to block the motion of dislocations that are responsible for plastic deformation.

For plastic deformation of the alloy to occur, the dislocations must be extruded between the hard particles. The extruded dislocation loops rejoin at the other side of particles and also form dislocation loops around the hard particles as shown in Figure 7.4. The force required for dislocation extrusion through the particles is greater than the force required to slide dislocations in pure copper. These extruded dislocation segments rejoin to form continuous dislocations on the other side of the hard inclusions, leaving dislocation loops around the particles. (Suh and Turner, 1975; Weertman and Weertman, 1964.)

If the distance between the dispersoids is λ, the shear strength τ is given by

$$\tau = 2Gb/\lambda \tag{7.1}$$

where G is shear modulus and b is the Burgers vector. At a given volume fraction of the solid particles, λ is inversely proportional to the diameter of the particle as $\lambda = [\pi/6V]^{1/3}d$, where V is the volume fraction of the hard particles and d is the diameter of the particles. Therefore, we can choose λ as DP_1.

From the fracture mechanics, it is known that if the particle size is less than a critical size d_c, which is of the order of 100 Å, cracks cannot nucleate around the hard particles, thus preserving the toughness of the pure matrix phase. Therefore, the diameter d of the particle may be used as DP_2.

To have the required high conductivity, the matrix phase of the material must be pure, free of alloying elements. Therefore, the purity of the matrix ϕ may be chosen as DP_3. Then the design equation for FR and DP may be written as

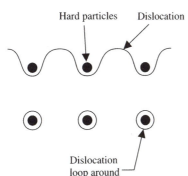

Figure 7.4 (Top) A dislocation being "extruded" between the hard second-phase particles, which require an increase in applied stress for deformation. (Bottom) The dislocation loop surrounding the hard particles after the dislocation has passed by the dispersoids.

$$\begin{Bmatrix} \tau_s \\ K_c \\ \sigma \end{Bmatrix} = \begin{bmatrix} X & 0 & x \\ 0 & X & 0 \\ 0 & 0 & X \end{bmatrix} \begin{Bmatrix} \lambda \\ d \\ \phi \end{Bmatrix} \tag{7.2}$$

The small x indicates that the alloying element has a relatively small effect on the strength of the matrix phase as the amount of metal and nonmetal elements of the ceramic phase should be in a stoichiometric ratio. After considering the solubility of various elements, we chose titanium and boron to make TiB_2 dispersion-strengthened copper.

Equation (7.2) suggests that we can set the value of particle size d first, before deciding on the distance between the particles. Because the manufacturing process often determines the size of the hard particles, the volume fraction of the hard phase can be used to control DP_2. Now we have designed the microstructure of the material in the physical domain.

7.2.3 Design of the Process: Mapping from the Physical Domain to the Process Domain

Having designed the material, we need to design a process that can generate the DPs by varying a suitable set of PVs. Many obvious techniques have been tried. Some of the following techniques are typical:

a. In one process, silicon carbide particles are dispersed into molten aluminum before casting it into various shapes. One of the problems is that the particles are typically of the order of 10 μm. Such large particles reduce the elongation and toughness of the material significantly. Furthermore, the machinability of the material is very poor because the hard silicon carbide phase wears out cutting tools and forming dies.

b. In a technique called mechanical mixing, alloys with second-phase particles are continuously deformed using two roll mills to break up the particles. It is very difficult and energy consuming to produce a microstructure with nanoscale particles. The final product is very expensive because of the time, energy, and human resources needed to make mechanical alloys.

c. In a process called the internal oxidation process, an alloy of copper with a small amount of aluminum is exposed to oxygen atmosphere to promote the diffusion of oxygen through the copper matrix. The oxygen then reacts with aluminum in copper, forming nanoscale particles. This process creates dispersion-strengthened copper alloys with small alumina (Al_2O_3) particles. The shortcoming of this process is the high process cost.

We wanted to come up with a different and better means of creating the microstructure, similar to that shown in Figure 7.3. The idea we have developed is the mixalloy process (Suh, 1981a, 1981b; Suh et al., 1982; Sanchez-Caldera et al., 1987, 1990, 1991a, 1991b). The idea is shown in Figure 7.5. It consists of two reservoirs of molten metals, for example, one with molten copper with a small amount of titanium and the other with molten copper with a stoichiometric ratio of boron. These molten metal streams are brought to an impingement mixing chamber at high speeds and mixed through the turbulent motion of the molten metals. Initially the mixing is done by macroscale mixing of turbulent eddies and then by the diffusion of titanium and boron. When these two elements meet, a chemical reaction occurs because the exothermic reaction of titanium and boron reduces the free energy by forming TiB_2.

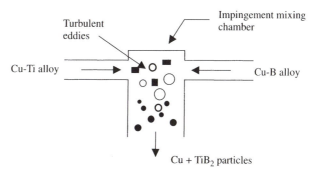

Figure 7.5 Mixalloy process for making dispersion-strengthened alloys. This figure shows impingement mixing of a Cu/Ti liquid solution with a Cu/B liquid solution to form a pure copper phase with nanoscale titanium diboride particles. (Courtesy of Sutek Corporation, Hudson, MA.)

Through a dimensional analysis, it can be shown that the size r of the turbulent eddies, to a first approximation, is related to the Reynolds number Re of the stream (see Tucker and Suh, 1980) as

$$\frac{r}{D} = [\text{Re}]^{-3/4} = [\rho V D/\mu]^{-3/4} \tag{7.3}$$

where D is the diameter of the nozzle, ρ is the density of the liquid, V is the velocity of the liquid, and μ is the viscosity of metals. For metals, ρ is larger and μ is less than that of water. Therefore, it is easy to attain a high Reynolds number and small eddy size. When Re = 20,000, r is equal to about 0.1 μm. Therefore, good mixing is not an issue.

To control the particle size, we must quench the material quickly so as to prevent the agglomeration and growth of TiB$_2$ particles in the molten copper.

The DP/PV design equation for the mixalloy process may now be written as

$$\begin{Bmatrix} \lambda \\ d \\ \phi \end{Bmatrix} = \begin{bmatrix} X & 0 & x \\ 0 & X & 0 \\ 0 & 0 & X \end{bmatrix} \begin{Bmatrix} \text{Solute Fraction of Ti and B} \\ \text{Rapid solidification} \\ \text{Metering of two streams} \end{Bmatrix} \tag{7.4}$$

The above design is a decoupled design. The metering ratio affects the volume fraction of TiB$_2$. Although the above process is designed well to this level, we must develop the process further to be sure that we can achieve the design as stipulated by Equation (7.4). We need to design the process of rapid cooling and the metering of two molten streams to obtain a pure matrix phase.

7.2.4 Further Development of the Process

7.2.4.1 Rapid Solidification

Through experimentation, we found that casting of the mixalloy in a conventional die or mold coarsens the hard particles through agglomeration of the small particles and diffusion of solutes to the nucleated particles. The physics is as follows. When the titanium atom and the boron atom in solution collide, some of them will react to nucleate TiB$_2$. Once the ceramic particles are nucleated in the melt, the solute atoms in the solution will diffuse to the nucleated site and the particles will grow, as free energy change can be minimized through a reaction at the solid surface. When the metal is in a liquid state, the diffusion rate will be fast. Through this process, the TiB_2 particles will grow. Some of these particles will also collide, agglomerating to form large particles, because of the fluid and Brownian motion. When solidification occurs, the diffusion rate of these atoms will be negligible. Therefore, the size of the particles is determined by the time available for diffusion before solidification.

Because the diffusion rate of elements in molten metals is almost as rapid as the thermal diffusion rate, slow cooling will result in coarsening and formation of large ceramic particles. Therefore, to prevent the coarsening of the particle beyond a certain size, the solidification time must be shorter than the agglomeration time. When the metal solidifies, the diffusion rate is slow and particle coarsening through particle collision cannot occur. Therefore, to have small ceramic particles, thermal diffusion must occur fast to solidify the alloy quickly.

Based on the conservation of mass, the distance Λ_m over which titanium atoms and boron atoms must diffuse to reach the nucleated site may be related to the maximum-allowable particle size, d_{TiB2}, as

$$\Lambda_m = \left(\frac{\rho_{TiB_2}}{2(\phi_{Ti}\rho_{Ti} + \phi_B\rho_B)} \right)^{1/3} d_{TiB_2} \qquad (7.5)$$

where ϕ_{Ti} and ϕ_B are the volume fractions of titanium and boron, respectively, and ρ_{TiB2}, ρ_{Ti}, and ρ_B are the mass density of titanium diboride, titanium, and boron, respectively. Then the distance Λ_T over which thermal effects can diffuse in the time period the particle reaches d_{TiB2} is given by

$$\Lambda_T = (\alpha_T/\alpha_m)\,\Lambda_m \qquad (7.6)$$

where α_T and α_m are thermal and mass diffusivities, respectively. When a series of experiments are performed, Λ_T was found to be of the order of a few millimeters.

Then, FR_2 (high elongation/toughness $= K_c$) of Equation (7.2) may be decomposed as

$FR_{21} =$ Control the amount of material to be cooled $= dm/dt$.
$FR_{22} =$ Rapid removal of heat.

The corresponding design parameter DP_2 (diameter of the ceramic particle $= d$) of Equation (7.4) may be written as

$DP_{21} =$ Thickness of ribbon or strip $= \delta$
$DP_{22} =$ Rapid quenching on rotating copper disk $= q$

The manufacturing process used to achieve DP_{21} and DP_{22} was the spin-casting process shown in Figure 7.6. The process variables (PVs) are

$PV_{21} =$ The ratio (metal flow rate/wheel speed) $= Q_m$
$PV_{22} =$ Water flow rate through the rotating copper disk $= Q_w$

The design equations may be written as

$$\begin{Bmatrix} FR_{21} \\ FR_{22} \end{Bmatrix} = \begin{bmatrix} X & 0 \\ X & X \end{bmatrix} \begin{Bmatrix} \delta \\ q \end{Bmatrix} = \begin{bmatrix} X & 0 \\ X & X \end{bmatrix} \begin{bmatrix} X & 0 \\ 0 & X \end{bmatrix} \begin{Bmatrix} Q_m \\ Q_w \end{Bmatrix} = \begin{bmatrix} X & 0 \\ X & X \end{bmatrix} \begin{Bmatrix} Q_m \\ Q_w \end{Bmatrix} \qquad (7.7)$$

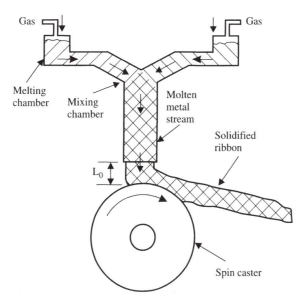

Figure 7.6 Spin casting that rapidly solidifies the molten copper/titanium diboride liquid stream from the mixalloy equipment. (Courtesy of Sutek Corporation, Hudson, MA.)

Equation (7.7) shows that rapid solidification is possible using the spin-casting process.

To obtain the exact quantitative information on the size of the caster, the flow rates of metal and the cooling water, and the speed of rotation, the Xs in Equation (7.7) should be replaced by modeling of the physical process based on the laws of nature.

7.2.4.2 Metering the Flow Rate of Two Streams

In Equation (7.4), a decision was made to control the composition of the matrix phase ϕ by means of the metering of two molten metal streams. If we denote the mass flow rate of the titanium-containing stream and the boron-containing steam as m_{Ti} and m_B, respectively, we can have a pure copper phase if $m_{Ti} = m_B$. If $m_{Ti} > m_B$, the matrix phase will contain a small amount of titanium, which will greatly reduce the electrical conductivity of the dispersion-strengthened alloy. If $m_{Ti} < m_B$, the electrical conductivity will also suffer, although to a lesser extent than when the matrix phase contains titanium. Therefore, it is extremely important that $m_{Ti} = m_B$.

To control the matrix phase chemical composition, the flow rates must be controlled. We chose a pipe of a given diameter and varied the pressure of the reservoir of molten metals to control the mass flow rate. To ensure that the cross-sectional area does not change during the ejection of molten metal because of freezing of copper at the wall, the pipes were kept at the melting point of copper, ensuring isothermal flow of molten metals. Otherwise, the flow rate is coupled to the temperature of the molten metal.

7.2.4.3 Final Processing

To make a solid billet of the mixalloy, the spin-cast mixalloy in a form of ribbon was chopped and put into a copper can, which was then compacted under hotisostatic pressure (HIP) and extruded.

7.2.5 Mixalloy Equipment

The final equipment thus designed and used in production is shown in Figure 7.7. It consists of two pressure vessels in which the crucibles for molten metals are stored. A pipe is connected from the crucible to the impingement-mixing chamber, which is located between the two pressure vessels. The mixing chamber is connected to a nozzle, which controls the shape of the molten stream that casts on a spin caster. The spin caster is made of pure copper disk with cooling channels inside to circulate cooling water to remove thermal energy from the caster. The caster rotates at high speeds. The cast mixalloy freezes on the wheel quickly and flies off the surface of the caster, continuously exposing new copper disk surface for continuous casting.

The first step in the process is to charge the crucible with copper and alloying element (either titanium or boron). The charges are weighed exactly to minimize the likelihood of having products that do not have the desired microstructure consisting of the pure copper phase and the ceramic phase. Then the metal is molten by induction heating. After the metal is completely molten, a small amount of sample is taken out of the crucible to check its chemistry using mass spectroscopy. Then the vessel is closed, and the air in the pressure vessel is replaced with an inert gas. The pressure in the vessel is set to the desired pressure for a given volume of charge. When the valve is opened, the molten copper alloy is pushed out of the vessel by the gas pressure in the vessel. A continuous strip of mixalloy is collected and sent out for consolidation by hotisostatic pressure extrusion.

7.2.6 Properties of Mixalloys: Dispersion-Strengthened Copper

A typical microstructure of mixalloy is shown in Figure 7.8. It consists of TiB_2 particles of about 0.05 μm in diameter, well dispersed throughout the copper matrix. (The diameter of the ceramic particles in the final product sold commercially was much smaller than these initial test runs.) The microstructure of mixalloys does not change even after they have been cycled to 1020°C. Its electrical conductivity and strength are given in Figure 7.9. This material has good conductivity and much higher strength than copper/chrome alloys, especially at high temperatures.

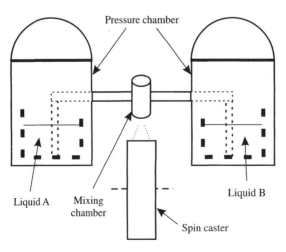

Liquid A   Mixing   Liquid B
            chamber

Spin caster

Figure 7.7 A schematic drawing of the mixalloying equipment. The metal was melted using induction heating. The equipment was about 15 feet high and occupied an area about 20 × 20 feet, excluding auxiliary equipment such as power supply, water tank, metal chip collectors, etc. (Courtesy of Sutek Corporation, Hudson, MA.)

Figure 7.8 Micrograph of the mixalloy with copper matrix and titanium diboride particles. (Courtesy of Sutek Corporation, Hudson, MA.)

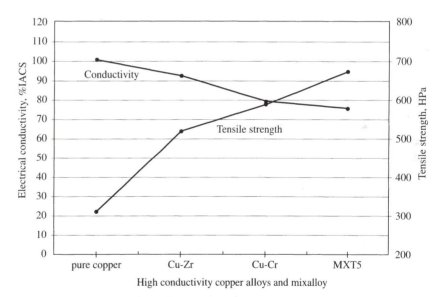

Figure 7.9 Electrical and mechanical properties of pure copper, copper zirconium alloy, copper chromium alloy, and mixalloy with 5% titanium diboride at room temperature. Pure copper, Cu–Zr, and Cu–Cr are all fabricated products manufactured by traditional methods. MXT5 is a 5 vol % Cu–TiB$_2$ mixalloy. (Courtesy of Sutek Corporation, Hudson, MA.)

Mixalloy has high elongation and high toughness. Even after significant cold work, it has high elongation as shown in Figure 7.10 by the stress–strain relationship of the alloy. This material has been forged to make welding tips without cracking. The ultimate tensile strength increases significantly through cold working.

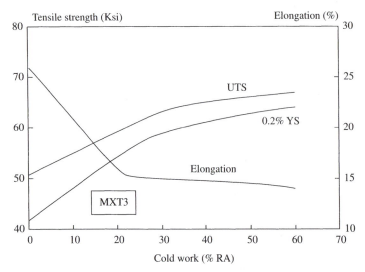

Figure 7.10 Effect of cold work on room temperature mechanical properties: ultimate tensile strength (UTS), elongation, and yield stress (YS) of mixalloy with 3% titanium diboride. (Courtesy of Sutek Corporation, Hudson, MA.)

7.3 MICROCELLULAR PLASTICS

7.3.1 Introduction to Microcellular Plastics

What is microcellular plastic? Microcellular plastics (MCP) are plastics that have a large number of uniform-sized, microscale bubbles of less than 30 μm in a polymeric matrix (Martini, Waldman, and Suh, 1982). A typical microcellular plastic has a myriad of small bubbles of a uniform size (0.1 to 30 μm in diameter) with a cell density of 10^9 to 10^{15} bubbles/cm^3. Figure 7.11 shows the microstructure of a typical MCP.

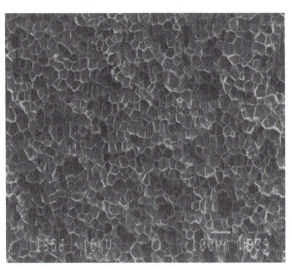

Figure 7.11 Microstructure of microcellular plastics. This particular micrograph shows an average cell size of about 30 μm, but they can be made to have a uniform cell size of 0.1 to 50 μm. Many polymers can be made into microcellular plastics.

MCP is a new class of polymers invented in 1979 at MIT [see Suh (1996) for a detailed historical account as well as a detailed review of microcellular plastics] in response to a customer's statement: "Most of our products are made of plastics. Is there any way we can reduce materials cost without sacrificing material properties, especially toughness?" In response to this request, a material microstructure was quickly designed, which was named microcellular plastics (Suh, 1996). The central idea behind microcellular plastics was to replace plastics with a large number of very small bubbles that are smaller than the preexisting inherent flaws. Small bubbles can blunt the crack-tips and act as crazing initiation sites, making the material tougher. They can be created in thermoplastics, thermosetting plastics, and elastomers.

Microcellular plastics are designed to reduce the consumption of polymers in polymeric products and to enhance certain physical properties. The designs of both the material and the manufacturing process are based on the principles of axiomatic design. Commercial production, which was also developed from axiomatic design theory, uses modified conventional extrusion and injection-molding machines. Almost any polymers can be used as the matrix material. Some of the microcellular plastics are tougher than the original materials by as much as 600% and have lower dielectric constant.

In the customer domain the need is the reduction of the manufacturing cost. This need has been mapped into the functional domain as the FR of reducing materials consumption without sacrificing the mechanical properties. The design parameter in the physical domain is the creation of microcellular plastics. Having decided on the DP, the highest level FR is then decomposed into the next-level functional requirements as

FR_1 = Reduce the amount of plastic used.
FR_2 = Increase the toughness of the plastic product.
FR_3 = Make three-dimensional geometrical shape.

The design parameters are

DP_1 = Number of cells
DP_2 = Cell size
DP_3 = Die design

The design equation for the product may be written as

$$\begin{Bmatrix} FR_1 \\ FR_2 \\ FR_3 \end{Bmatrix} = \begin{bmatrix} X & X & 0 \\ 0 & X & 0 \\ 0 & 0 & X \end{bmatrix} \begin{Bmatrix} DP_1 \\ DP_2 \\ DP_3 \end{Bmatrix} \tag{7.8}$$

Equation (7.8) indicates that the product design is a decoupled design.

Is there any relationship between cell size and cell density? In an ideal microcellular plastic, where spherical bubbles are packed in body-centered cubic structure, the bubble size can be directly related to the bubble density. In a 1-cm cube of foamed material, the number of cells is inversely proportional to the cube of the bubble diameter. Therefore, a microcellular plastic with 10-μm size bubbles has approximately 10^9 bubbles/cm^3 of unfoamed material, whereas microcellular plastics with 1-μm and 0.1-μm size bubbles have approximately 10^{12} and 10^{15} bubbles/cm^3 of unfoamed material, respectively.

Because the volume taken by spherical bubbles in an ideal, closely packed hexagonal or cubic structure is approximately 74%, the plastic occupying the interstitial space is 26%.

Therefore, the cell density of an ideal closely packed spherical microcellular plastic is equal to $(1/\text{cell size})^3$ times $(1/0.26)$. For a microcellular plastic with 1-μm cell diameter, the bubble density is 3.85×10^{12} cells/cm^3 of the solid plastic.

It should be noted again that in developing microcellular plastics, the design of the microstructure was done first to satisfy the customer's requirements before the processing techniques were developed. This is in contrast to the more typical approach of developing the process first and then examining the microstructure of the resulting foamed plastics.

7.3.2 Design of a Batch Process

How do you make microcellular plastics? Having now designed microcellular plastics, our next task is to figure out how such microcellular plastics can be manufactured economically. It took many years to develop processing techniques. First, batch processing and later, continuous processing were developed by the authors' students at MIT. Now Trexel, Inc. of Woburn, Massachusetts is commercializing microcellular plastics (trade name MuCell).

A batch process for microcellular plastics was first developed using a thermodynamic instability phenomenon (Martini-Vvedensky, Suh, and Waldman, 1984, Suh, 1996). The basic physics involved is as follows:

1. The plastic must be supersaturated with sufficient gas such as CO_2 to nucleate a large number of cells simultaneously. High pressure is needed because gas solubility increases with pressure.
2. The temperature must be controlled in order to control the deformability of the plastic matrix phase during cell growth.
3. A gas with a suitable solubility and diffusivity for the given plastic must be selected.
4. Homogeneous nucleation must dominate the nucleation process in order to create a large number of microcells even when heterogeneous nucleation sites are available.

The processing technique consists of forming a polymer/gas solution and then suddenly inducing a thermodynamic instability by either lowering the pressure or raising the temperature to change the solubility S. The solubility is a function of two thermodynamic properties, temperature and pressure:

$$S = S(p, T) \tag{7.9}$$

The change in the solubility can be expressed as

$$\Delta S = \frac{\partial S}{\partial p}\Delta p + \frac{\partial S}{\partial T}\Delta T \tag{7.10}$$

The $\partial S/\partial p$ term of Equation (7.10) is positive, whereas the $\partial S/\partial T$ term is negative. Therefore, to decrease the solubility and induce the thermodynamic instability, either the pressure must be decreased (i.e., $\Delta p < 0$) or the temperature must be increased (i.e., $\Delta T > 0$). Furthermore, regardless of whether the process is continuous or batch type, the thermodynamic instability must be induced quickly so that cells will nucleate simultaneously before significant diffusion of gas has taken place. Therefore, the higher the temperature of the polymer, the quicker nucleation has to occur because the diffusion of the gas occurs faster at higher temperatures. Such simultaneous cell nucleation will ensure a uniform cell size distribution.

What is the physics behind the continuous extrusion of microcells? It was shown through extensive research at MIT that the number of cells nucleated is a function of the supersaturation level—the higher the supersaturation level, the greater the number of cells nucleated. Furthermore, because the amount of dissolved gas that fills the nucleated cells is finite, and because all the cells are nucleated nearly at the same time, the gas distributes more or less evenly among all these cells. The final bubble size is then determined by the total gas per bubble and the flow characteristics of the polymer at the nucleation temperature.

Micrographs of several different kinds of microcellular plastics processed in batch processes, i.e., extruded or injection-molded parts foamed after saturation with CO_2 gas, are similar to that shown in Figure 7.11. Figure 7.12 is a plot of cell density of various plastics, and Figure 7.13 is a plot of the cell size of various plastics plotted in Figure 7.12. It should be noted that the plastics with a large cell density have smaller cell diameters. In fact, the products of average cell volume and cell density are nearly equal when the amount of gas dissolved in the plastic is the same.

7.3.3 Design of Continuous Process

To create a continuous process, we must be able to design a process and associated equipment to perform the following functions:

1. Rapid dissolution of gas into molten flowing polymer to form a solution
2. Nucleation of a large number of cells
3. Control of the cell size
4. Control of the geometry of the final product

The designer must fully understand each one of these processes in order to achieve the task.

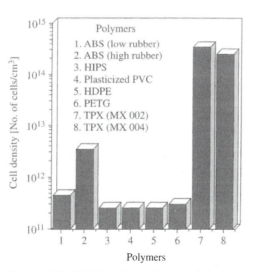

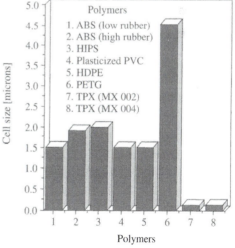

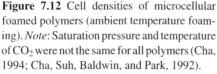

Figure 7.12 Cell densities of microcellular foamed polymers (ambient temperature foaming). *Note*: Saturation pressure and temperature of CO_2 were not the same for all polymers (Cha, 1994; Cha, Suh, Baldwin, and Park, 1992).

Figure 7.13 Cell sizes of the microcellular foamed polymers (ambient temperature foaming). *Note*: Saturation pressure and temperature of CO_2 were not the same for all polymers.

7.3.3.1 Brief Introduction to the Physics of the Process

Gas Diffusion and Formation of Polymer/Gas Solution To produce the microcellular plastics at an acceptable production rate through a continuous process, we must dissolve the gas in polymers quickly despite the slow diffusion rate.

When a block of polymer is suddenly exposed to high-pressure gas at temperature T, the polymer will approach a thermodynamic equilibrium by gas diffusion into the polymer if the polymer and gas are miscible and by thermal conduction toward an equilibrium temperature. The driving force is the free energy of the polymer/gas system. The diffusion phenomenon can be expressed with a partial differential equation as

$$\nabla(\alpha \, \nabla c) = \frac{\partial c}{\partial t} \tag{7.11}$$

where c is the concentration of the gas, t is the time, and α is the diffusivity, which may be a function of temperature and pressure. In general, the diffusivity increases with temperature by an Arrhenius relationship:

$$\alpha = \alpha_0 \exp(-\Delta G / kT) \tag{7.12}$$

where ΔG is the activation energy, k is Boltzmann's constant, and T is the absolute temperature. The solution to Equation (7.11) for the case of constant gas diffusivity has been obtained. An important implication of the solution is that the time for gas diffusion is proportional to the thickness of the plastic ℓ as

$$t \propto \frac{\ell^2}{\alpha} \tag{7.13}$$

Table 7.1 shows the diffusivity D of CO_2 and N_2 in various plastics, and Table 7.2 lists the diffusion times estimated by Park (1993) for various striation thicknesses of plastics as a function of diffusivity. It should be noted that it takes a long time to diffuse gas into a polymer at room temperature and that the diffusivity of CO_2 and N_2 are nearly the same. For example, the diffusivity of CO_2 in most thermoplastics at room temperature is in the range of 5×10^{-8} cm^2/second and the diffusion time estimated using Equation (7.12) is approximately 14 hours when ℓ is 0.5 mm. The diffusivity at 200°C is three to four orders of

TABLE 7.1 Estimated Diffusion Coefficients of Gases in Polymers at Elevated Temperatures

Polymer	D of CO_2 (cm^2/sec)		D of N_2 (cm^2/sec)	
	At 188°C	At 200°C	At 188°C	At 200°C
PS	—	1.3×10^{-5}	—	1.5×10^{-5}
PP	4.2×10^{-5}	—	3.5×10^{-5}	—
PE	—	2.6×10^{-6}	—	8.8×10^{-7}
HDPE	5.7×10^{-5}	2.4×10^{-5}	6.0×10^{-5}	2.5×10^{-5}
LDPE	—	1.1×10^{-4}	—	1.5×10^{-4}
PTFE	—	7.0×10^{-6}	—	8.3×10^{-6}
PVC	—	3.8×10^{-5}	—	4.3×10^{-5}

From Durril and Griskey (1960, 1969).

TABLE 7.2 Estimated Diffusion Time at Various Striation Thicknesses and Diffusion Coefficients

Striation Thickness (s) (μm)	Diffusion Coefficient (D)			
	10^{-5} cm^2/sec	10^{-6} cm^2/sec	10^{-7} cm^2/sec	10^{-8} cm^2/sec
1	1×10^{-3} sec	0.01 sec	0.1 sec	1 sec
10	0.1 sec	1 sec	10 sec	100 sec
50	2.5 sec	25 sec	4 min	42 min
100	10 sec	100 sec	17 min	3 hr
250	63 sec	10 min	2 hr	17 hr
500	4 min	42 min	7 hr	3 days
750	9 min	94 min	16 hr	7 days
1 mm	17 min	3 hr	28 hr	12 days

From Park (1993).

magnitude greater than that at room temperature. Even at high temperatures, the diffusion rate is still the rate-limiting step in continuous processes.

To accelerate the diffusion rate and shorten the time for the formation of gas/polymer solutions, we must raise the temperature and shorten the diffusion distance. Deforming the two-phase mixture of polymer and gas through shear distortion, as shown in Figure 7.14, can decrease the diffusion path. This type of deformation occurs in an extruder under laminar flow conditions. The bubbles are stretched by the shear field of the two-phase mixture and eventually break up to minimize the surface energy when a critical Weber number is reached (Taylor, 1934). The Weber number We is defined as

$$\text{We} = \frac{\text{shear forces}}{\text{surface forces}} = \frac{(d\gamma/dt)d_b \, \eta_p \, f(\lambda)}{2\sigma} \tag{7.14}$$

where λ = viscosity ratio(η_g/η_p)

$\quad f(\lambda) = (19\lambda + 16)/(16\lambda + 16)$

$\quad \eta$ = dynamic viscosity

$\quad \gamma$ = shear strain

$\quad d_b$ = bubble diameter

$\quad \sigma$ = surface tension

Using Equation (7.14), the disintegrated bubble size is calculated to be about 1 mm and the initial striation thickness after bubble disintegration to be about twice the bubble

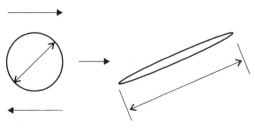

Figure 7.14 Deformation of a spherical bubble in a shear field to form an ellipsoid. The distance between the ellipsoids (measured perpendicular to the major axis of the ellipsoids) is the striation thickness. The dissolution rate of gas increases as the striation thickness becomes smaller and as the interfacial area of gas/polymer increases.

diameter (Park, 1993). This striation thickness decreases with further shear, and the gas diffusion occurs faster as a result of the increase in the surface area and the decrease in striation thickness. The striation thickness in an extruder is estimated to decrease to about 100 mm. At this thickness, the diffusion time is in the range of 1 minute in PET, from 10 to 20 seconds in polystyrene (PS), polyvinylchloride (PVC), and high-density polyethylene (HDPE), and in the range of a few seconds in low-density polyethylene (LDPE).

At thermodynamic equilibrium, the gas concentration in the polymer is related to pressure and temperature as

$$c = H(p, T) \tag{7.15}$$

where H is known as Henry's law constant. At low pressures and low concentrations, H is constant. At high pressures, H depends on both pressure and temperature. The temperature dependence follows the Arrhenius type rate equation. At low pressures, the gas solubility is low but Henry's constant is high. At higher pressures, the gas solubility is high, but the rate of the weight increase with pressure decreases with pressure.

The solubility of gas in polymers decreases with an increase in temperature. Table 7.3 lists the estimated gas solubility of CO_2 and N_2 in various thermoplastics. It should be noted that the solubility of N_2 is considerably less than that of CO_2. Because the amount of gas that can be dissolved is a function of the saturation pressure and because the gas diffusion rate is the rate-limiting process, we can use CO_2 at a critical state to enhance the solubility and diffusion rate. Supercritical fluid is neither gas nor liquid in a certain temperature and pressure regime higher than the critical pressure and critical temperature of the fluid. In this state, it has both gas-like and liquid-like properties. CO_2 is supercritical at pressures and temperatures greater than 7.4 MPa and 31.1°C.

With dissolution of a large number of gas molecules in polymers, the glass transition temperature and viscosity decrease with the increase in gas concentration. The change in the glass transition temperature is quite substantial at high gas concentrations. These changes affect the processibility of polymers.

Nucleation The key idea in the formation of microcellular plastics is the nucleation of a large number of bubbles (cells). Cells can nucleate either homogeneously or heterogeneously. Homogeneous nucleation must overcome a larger activation energy barrier than heterogeneous nucleation. Homogeneous nucleation occurs throughout the matrix. On the other hand, heterogeneous nucleation occurs at an interface between two or more different materials or phases where the interfacial energy is high. When the driving force is very high, such as when the degree of supersaturation of the gas in the polymer is large, the difference in the activation level is so much smaller than the driving force that both homogeneous and heterogeneous nucleation can occur simultaneously.

TABLE 7.3 Estimated Gas Solubility in Polymers at 200°C and 27.6 MPa (4000 psi)

Polymer	CO_2 Weight Gain (%)	N_2 Weight Gain (%)
PE	14	3
PP	11	4
PS	11	2
PMMA	13	1

From Park (1993).

When the gas is homogeneously dissolved throughout the molten plastic by diffusion, the polymer is ready for nucleation. For nucleation to occur, a finite energy barrier has to be overcome. The energy barrier depends on two competing factors: (1) the energy available in the gas diffused into the embryo of the cell and (2) the surface energy that must be supplied to form the surface of the cell. There is a critical cell size beyond which the cell becomes stable and grows, and below which the cell embryo collapses. Typically the cell nucleation rate is expressed as

$$\frac{dN}{dt} = N_0 \, f e^{-\Delta G / kT} \tag{7.16}$$

where N = the number of cells

N_0 = the number of available sites for nucleation

f = the frequency of atomic or molecular lattice vibration

ΔG = the activation energy barrier

k = the Boltzmann constant

T = the absolute temperature

The activation energy term can be related to the pressure in the bubble and the surface energy (Martini, 1981; Colton, 1985) as

$$\Delta G = \frac{16\pi \gamma}{3\Delta p_0} \, g(\theta) \tag{7.17}$$

where γ is the surface energy of the polymer and Δp_0 is the pressure difference between the cluster of gas molecules in the bubble and the surrounding pressure, which is assumed to be equal to the saturation gas pressure. The function $g(\theta)$ depends on the wetting angle between the polymer, gas, and the second phase particle. It is equal to 1 in the case of homogeneous nucleation and is less than 1 for heterogeneous nucleation. The critical radius is given by

$$r_c = \frac{2\gamma}{\Delta p} \tag{7.18}$$

In homogeneous nucleation there are many possible nucleation sites, the most prominent of which are the free volume sites. Also, in the case of semicrystalline polymers, the interfaces between the amorphous region and the crystalline region could be the nucleation sites (Colton, 1985; Colton and Suh, 1987a, 1987b, 1990, 1992). Depending on the gas supersaturation level, all or part of these nucleation sites will be activated. At this time, we cannot quantitatively delineate the importance of these various nucleation sites at a given saturation pressure. From the foregoing discussion, it can be seen that even homogeneous nucleation is not truly homogeneous in terms of the activation energy levels involved for nucleation. At high pressures, both homogeneous and heterogeneous nucleation occur because of the driving force.

Cell Growth Immediately after the cells are nucleated, the pressure in the bubble is equal to the saturation pressure. Therefore, the cells will try to expand if the polymer matrix is soft enough to undergo viscoelastic plastic deformation. The cell expands until the final pressure inside the cell is equal to the pressure required to be in equilibrium with the surface forces and the stress in the viscoelastic cell wall.

Unlike in conventional foaming, in the case of microcellular plastics, there are so many cells nucleated and the diffusion length is so short that the diffusion of the gas to the cell growth stops relatively quickly when there is no more dissolved gas to diffuse into the nucleated cells.

In practice, the temperature of the surface of the extrudate changes as a result of heat transfer, and thus the expansion of the cell is constrained by the outer stiff layer. Also some of the gases from the cells near the surface escape, reducing the tendency to expand.

When the temperature is very high and the cell expands beyond the plastic instability point (i.e., equivalent to the ultimate tensile strength point in a uniaxial tensile test where necking begins), the cells will rupture and an open-cell microcellular structure will result.

Cell Density and Cell Size The cell density is a function of both the concentration of dissolved gas and the pressure drop rate. During the cell nucleation stage, there is competition for gas between cell nucleation and cell growth if the cells do not nucleate instantaneously. When some cells nucleate before others, the gas in the solution will diffuse to the nucleated cells to lower the free energy of the system. As the gas diffuses to these cells, low gas concentration regions where nucleation cannot occur are generated adjacent to the stable nuclei. As the solution pressure drops further, the system will either both nucleate additional microcells and expand the existing cells by gas diffusion or only expand the existing cells. Therefore, when the pressure drop occurs rapidly, the gas-depleted region where nucleation cannot occur will be smaller and a more uniform cell distribution will result. It has been determined experimentally that a drop rate of 2 GPa/second is the minimum pressure drop rate required for microcellular plastics processing.

The competition between cell nucleation and cell growth can be understood using two dimensionless groups (Baldwin, 1994). For nucleation to be completed in a time that is short compared to the diffusion time, the dimensionless group giving the ratio of these times must be small.

$$\frac{\text{Characteristic nucleation time}}{\text{Characteristic diffusion time}} \ll 1$$

or

$$\frac{\alpha}{(dN/dt)d_C^5} \ll 1 \tag{7.19}$$

where α is the gas diffusivity, dN/dt is the nucleation rate, and d_C is the cell diameter.

The second dimensionless group can be obtained by considering the length associated with the nucleation and growth of cells over finite processing times. The competition between microcell nucleation and cell growth is negligible when

$$\frac{\text{Characteristic gas diffusion distance}}{\text{Characteristic spacing between stable nuclei}} \ll 1$$

or

$$2\rho_c^{1/3}(\alpha t_D)^{1/2} \ll 1 \tag{7.20}$$

where ρ_c is the cell density, α is gas diffusivity, and t_D is the diffusion time. For a continuous process to work, Equations (7.19) and (7.20) must be satisfied.

7.3.3.2 Design of a Continuous Process

It was decided that a continuous process would be designed by modifying a conventional single-screw extruder. High-pressure CO_2 gas is introduced into the extruder barrel by metering the exact amount of CO_2 at pressures greater than 2000 psi. The flow rate of CO_2 into the extruder can be controlled using a metering pump. The gas forms a large bubble in the extruder as the flow of the gas is briefly interrupted whenever the screw flight wipes over the barrel. Then the gas in the bubble must be diffused quickly in the molten plastic by increasing the polymer/gas interfacial area and decreasing the striation thickness (i.e., the average distance between the elongated bubbles) of polymers between the gas bubbles, as the diffusion time is proportional to the thickness squared. By elongating the bubble in the barrel through the shear deformation of the two-phase mixture of the polymer and gas, the area-to-volume ratio of the gas bubble is increased and the striation thickness is reduced, promoting rapid diffusion of gas into the molten polymer.

When the striation thickness is in the range of 100 μm, the diffusion time is 100 seconds when the temperature of PET is 200°C, 1 second for low-density polyethylene at 200°C, and 10 seconds for polystyrene at 200°C. These numbers provide the approximate residency time required for diffusion and solution formation in the extruder.

The FRs and DPs were given in the preceding section as

FR_1 = Reduce the amount of plastic used.
FR_2 = Increase the toughness of the plastic product.
FR_3 = Make three-dimensional geometric shape.

DP_1 = Microcellular plastics (uniform cell distribution in large numbers)
DP_2 = Diameter of microcells
DP_3 = Die shape

The PVs for the process described that can satisfy the DPs given are

PV_1 = Supersaturation of the plastic with a large amount of gas and sudden pressure change (dp/dt)
PV_2 = Temperature of the molten polymer to control the expansion of cells at the die
PV_3 = Cross-sectional dimensions

The design equation for the extrusion process may be written as

$$\begin{Bmatrix} DP_1 \\ DP_2 \\ DP_3 \end{Bmatrix} = \begin{bmatrix} X & 0 & 0 \\ X & X & 0 \\ 0 & 0 & X \end{bmatrix} \begin{Bmatrix} PV_1 \\ PV_2 \\ PV_3 \end{Bmatrix} \tag{7.21}$$

Equation (7.21) shows that the process design is also a decoupled design. Therefore, each design satisfies the Independence Axiom. However, for concurrent engineering to be possible, the product of the product and process design matrices must also be diagonal or triangular. Because $FR_i = A_{ij}DP_j$ and $DP_j = B_{jk}PV_k, FR_i = C_{jk}PV_k$, where $C_{jk} = A_{ij}B_{jk}$. However, the process design matrix with elements given by C_{jk} is neither diagonal nor triangular. This means that the foam density and toughness of the plastic part cannot be independently controlled by means of the PVs chosen unless the tolerances on the density variation or the toughness variation are large enough to change one of the two design matrices. Experiments support this conclusion.

DP_1 and PV_1 can be further decomposed as

DP_{11} = Large number of nucleated cells
DP_{12} = Uniform-size cells

PV_{11} = The level of supersaturation of CO_2
PV_{12} = Rapid pressure drop dp/dt

The design matrix for this design may be represented as

$$\begin{Bmatrix} DP_{11} \\ DP_{12} \end{Bmatrix} = \begin{bmatrix} X & x \\ X & X \end{bmatrix} \begin{Bmatrix} PV_{11} \\ PV_{12} \end{Bmatrix} \tag{7.22}$$

Equation (7.22) states that DP_{11} and DP_{12} are coupled slightly in that if the pressure drop rate is really slow, we cannot get a large number of cells and uniform-sized cells. In most cases, the effect of dp/dt on the number of cells is negligible. A typical pressure profile in a single screw extruder is shown in Figure 7.15.

DPs have been decomposed further until the designs of the process and the equipment were completed by a number of the author's students at MIT (Park, 1993; Baldwin, 1994; Sanyal, 1998). Park investigated the continuous extrusion of microcellular plastics, in particular, the dissolution of gas at an acceptable production rate and the use of a rapid pressure drop nozzle as a nucleation device. He designed equipment to demonstrate the continuous extrusion of microcellular fiber as part of his doctoral thesis work (Park, 1993). Baldwin investigated the microcellular plastic structure in amorphous plastics and crystalline plastics using polyethylene terephthalate (PET) and polyolefin-modified polyethylene terephthalate (CPET) and also the continuous extrusion process, especially the shaping of the flat and tubular extrudates as part of his doctoral thesis (Baldwin, 1994). The work of Park and Baldwin was important as the original batch process had limited applicability because of the slow diffusion of gas into a solid plastic [Park, Baldwin, and Suh, 1995, 1996). Another doctoral student investigated the use

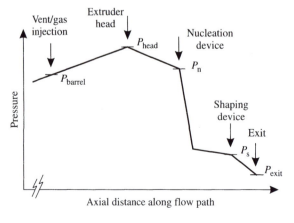

Figure 7.15 Representative pressure profile along the polymer flow field in the extruder and die. (From Baldwin, Park, and Suh, 1996, 1997.)

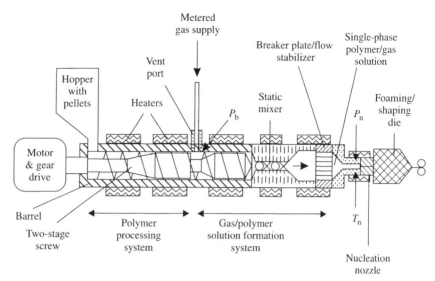

Figure 7.16 Schematic of the microcellular extrusion system used for the shaping and cell growth control experiments. (From Baldwin, Park, and Suh, 1997.)

of supercritical fluids, especially CO_2, to dissolve the gas faster and to increase the amount of dissolved gas to create "super" microcellular plastics. With supercritical fluids, the bubble density was increased from 10^9 bubbles/cm^3 to as high as 10^{15} bubbles/cm^3 (Cha, 1994).

The final design of the equipment that can provide the PVs is shown in Figure 7.16. This equipment consists of the CO_2 injection system, a shearing section for the two-phase mixture, a diffusion section to ensure the homogeneity of the polymer/gas solution, and a nucleation section. The morphology of the microcellular plastics extruded by the laboratory equipment is shown in Figure 7.17.

The commercially made MuCells have much more uniform and equiaxial shape as shown in Figure 7.18. The micrograph shows the cross section of polystyrene sheet extruded

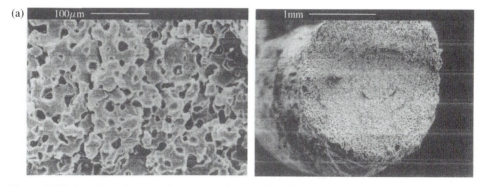

Figure 7.17 Scanning electron microscope micrographs of (a) extruded microcellular polystyrene thick filament and (b) extruded microcellular polystyrene sheet using the experimental apparatus. (From Baldwin, Park, and Suh, 1997.)

(b)

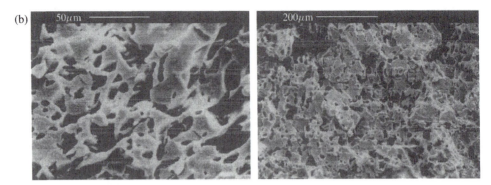

Figure 7.17 *Continued*

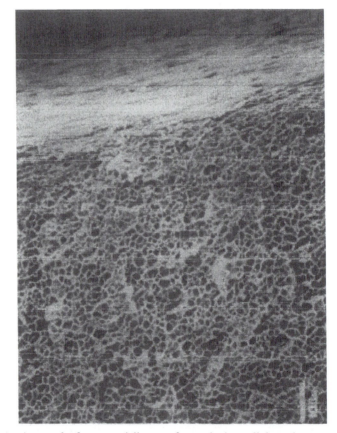

Figure 7.18 A micrograph of commercially manufactured microcellular polystyrene.

commercially (roughly 52 inches wide and 0.25 inches thick). These cells are closed cells with an average diameter of approximately 20 μm. Figure 7.19 shows some of the extruded commercial products and Figure 7.20 shows some of the injection molded products by MuCell technology.

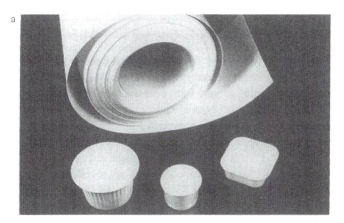

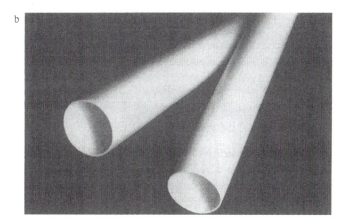

Figure 7.19 (a, b) Extruded commercial products of MuCell microcellular plastics.

Figure 7.20 Injection molded MuCell microcellular plastics.

EXAMPLE 7.1 Extrusion of Wire Coating with Microcellular Plastics

One hundred and fifty thin coaxial cables are bundled together to transmit many electrical signals in making ultrasound imaging machines. The bundle must have a small diameter for flexibility in bending around tight corners, etc. Therefore, the diameter of each cable must be 1 mm. The customer is looking for a coaxial cable bundle that is highly flexible and that has certain electrical characteristics. The highest level FR that can satisfy this customer need is "produce insulation of desired size and electrical performance." The highest level DP is "insulated wire with low dielectric constant and proper electrical performance." The highest level PV is "the extrusion coating of microcellular plastics on a thin wire." The design task is to develop DPs and PVs for this coated coaxial wire.

SOLUTION

The highest level FR may be decomposed as

FR_1 = Produce insulation of desired dielectric constant.
FR_2 = Produce foam insulation of a desired electric breakdown strength.

The DPs that can satisfy these FRs are:

DP_1 = Volume fraction of the void = V_F
DP_2 = Thickness of the insulation = t

The constraint is that the cell size must be very small and uniformly distributed as these microcoaxial cables require a very thin layer of electric insulation. The material must be either a Teflon family of polymer or a polyethylene. Neglecting the second order effects of voids on the electric breakdown strength, the design equation is a triangular matrix given by

$$\begin{Bmatrix} FR_1 \\ FR_2 \end{Bmatrix} = \begin{bmatrix} X & X \\ 0 & X \end{bmatrix} \begin{Bmatrix} DP_1 \\ DP_2 \end{Bmatrix} \tag{a}$$

Now we have to determine appropriate PVs that can satisfy the DPs. We may choose PVs as

PV_1 = Foaming variables
PV_2 = Flow rate of the foam, Q, relative to the speed of the wire, V

The design equation may be written as

$$\begin{Bmatrix} DP_1 \\ DP_2 \end{Bmatrix} = \begin{bmatrix} X & 0 \\ 0 & X \end{bmatrix} \begin{Bmatrix} PV_1 \\ PV_2 \end{Bmatrix} \tag{b}$$

The product design matrix given in Equation (a) is triangular (i.e., a decoupled design), and the process design matrix in Equation (b) is diagonal (i.e., an uncoupled design). Up to this stage of the design, both the product design and the process design satisfy the Independence Axiom, and thus concurrent engineering is shown to be possible.

It should be noted that PV_1 must be decomposed further as it cannot be implemented as stated. We need to decompose FR_1, DP_1, and PV_1 by means of zigzagging. All subsequent decomposition must be such that the design intent expressed by Equations (a) and (b) is satisfied.

Decomposition of FR_1, DP_1, and PV_1

FR_{11} = Control the cell size.
FR_{12} = Control the number of cells.

DP_{11} = Uniform cell size
DP_{12} = Cell density

The design matrix for the FR/DP relationship is a diagonal matrix.

PV_{11} = Pressure drop rate across die = dp/dt
PV_{12} = Amount of the gas dissolved

The maximum amount of the gas that can be dissolved, PV_{11}, is a function of the lowest pressure for maintaining the single-phase polymer/gas system, which is the die inlet pressure. The pressure drop is determined by the die geometry. In the case of a slit die with a constant cross section along the die length, the pressure drop rate, dp/dt, is a function of the length of the die lip. The design matrix for the DP/PV relationship may be represented as follows:

$$\begin{Bmatrix} DP_{11} \\ DP_{12} \end{Bmatrix} = \begin{bmatrix} X & X \\ x & X \end{bmatrix} \begin{Bmatrix} PV_{11} \\ PV_{12} \end{Bmatrix} \tag{c}$$

The microstructure of the foam—void fraction, cell size, and cell density—determines the performance of the insulation. Using the rule of mixture, the dielectric constant of the insulation may be expressed as

$$K \approx K_p - (K_p - 1)V_F \tag{d}$$

K_p is the dielectric constant of plastic, and the dielectric constant of gas is assumed to be equal to 1. V_F is the volume fraction of the cells. The volume fraction of the voids can be related to the average cell radius r and the cell density N as

$$V_F = \frac{\pi r^3 N}{\frac{3}{4} + \pi r^3 N} \tag{e}$$

Decomposition of FR_2, DP_2, and PV_2

The decomposition of FR_2 (produce foam insulation of a desired electric breakdown strength), DP_2 (thickness of the insulation), and PV_2 (flow rate of foam, Q, relative to the speed of the wire, V) will now be considered.

Assuming that microcellular plastic with desired cell size and cell density can be created, the electric breakdown of the insulation will depend on the elctric field strength around the wire. Therefore, the control of the insulation thickness may be the easiest means of satisfying FR_2. The thickness can be controlled by changing the flow rate of plastics at a given wire velocity at a given diameter of the die opening.

The volume flow rate of the foam, Q, will determine the final insulation thickness at a given velocity of the wire, V. The conservation of mass yields the following:

$$Q = \frac{\pi}{4}(D^2 - d^2)V \qquad \text{(f)}$$

where D is the outside diameter of the insulated wire and d is the bare conductor diameter. Therefore, we can treat DP_2 and PV_2 as leaves. However, the specific choice of process variables for PV_2 must be determined.

The flow rate, Q, can be changed in many different ways at a fixed production rate of the insulated wire. For example, the die diameter, the rotational speed, ω, of the extruder, and the temperature, T_p, of plastic in the die will affect the flow rate. The die diameter should be based on the ratio of the desired density of the insulation material after and before foaming. The temperature of the plastic is normally kept as low as possible to generate stable foam structure when the gas diffuses out of the polymer/gas solution. In light of these considerations, we may choose the rotational speed of the screw as the final PV_2 that can control the thickness of the foam.

Since the second order effects are ignored in this preliminary design of the production system and since our ability to model the process exactly depends on many assumptions, the final design of the system must be checked out through actual testing of the production system.

7.3.4 Performance of Microcellular Plastics

Microcellular plastics have many advantages over conventional (macrocellular) foamed plastics. First, the cells in conventional foams are so large that thin films and sheets cannot be foamed, whereas for microcellular foams, when the cell size can be 0.1 μm, 20-μm-thick sheets can be successfully foamed with good physical integrity. Similarly, thin microcellular fibers can be also made.

Some microcellular plastics also have exhibited better physical properties (Seeler and Kumar, 1992). Many microcellular plastics are tougher and have longer fatigue life. Also the specific mechanical strength is much better than conventional large-cell foamed plastics. Microcellular plastics can have densities as low as 0.03 g/cm^3, although the limit of the lowest density is not known at this time. At such low densities, thermal insulation properties of microcellular plastics are excellent because of the small cell size.

Because the cell size is extremely small, the cells cannot be seen by the naked eye. Therefore, the foamed plastic resembles a solid plastic, having a good physical appearance. At conventional cell sizes, they are opaque without the need to introduce pigments such as titanium dioxide. A very exciting prospect is the potential manufacture of "transparent supermicrocellular foams" with cell sizes less than 0.05 μm.

Because about 70% of the cost of foamed plastic goods is the material cost and up to 50% weight reduction is possible for many applications, the cost of plastic parts can be reduced by as much as 35%—a substantial cost saving. This reduction in material cost is achieved while enhancing physical properties of the plastic part. Because of the reduction in part weight, further savings are possible in associated costs, such as transportation.

Microcellular plastics can be made with closed cells or open cells, depending on the specific application. Open-cell microcellular plastics have been made from homopolymers and from blends of two incompatible polymers.

Microcellular plastics are environmentally acceptable because they are processed using carbon dioxide (CO_2) or nitrogen (N_2), instead of hydrocarbons or fluorinated materials. Because smaller amounts of plastics are used in a given product, there is less material to recycle or dispose. Furthermore, less raw material and energy are used to make the same plastic article.

Microcellular plastics should find many applications in housing and construction, sporting goods, vehicles, electrical and electronic products, chemical and biochemical applications, and the textile and apparel industry. They can be used in siding, pipes, electrical wire, automotive seats and other parts, airplane parts, filters, shoe soles, office equipment housing, artificial paper, food containers, polishing cloth, thermal insulation around pipes, and others. Lightweight apparel can be made using hollow fibers. Also the microporous structure enables production of materials that are water repellent and yet transmit moisture.

7.3.5 Other Advantages of the MuCell Process

Although they were not a subset of the original FRs, Microcellular plastics provide several additional advantages. This is a result of work done by the engineers and managers at Trexel, Inc.—the exclusive licensee of the Microcellular plastics technology. They found the following:

• *The productivity increases when the MuCell process is used to make microcellular plastics.* The viscosity of polymer/gas solutions decreases when gases dissolve in molten polymers. Therefore, the extrusion rate of Microcellular plastics increases—as much as by a factor of two. This means that when the MuCell process is used, the production capacity can be increased by a factor of two.

• *Large parts can be injection molded using small machines.* It takes much less pressure to injection mold microcellular plastics, again because of the reduced viscosity. Therefore, using the same clamping tonnage, an injection molding machine can make parts about twice as large as parts normally made by injection molding machines. Table 7.4 shows typical results obtained by Mar-Lee Companies, a licensee of Trexel, Inc.

• *Injection molded parts cool more rapidly.* MuCell plastics are processed at lower processing temperatures, because the glass-transition and melting temperatures and the viscosity of plastics decrease with increases in dissolved gas. As the gas is formed in the bubble during the nucleation and cell growth phase, the viscosity and the melting temperature of plastics increase, reverting back to the original state. Therefore, Microcellular plastics "solidify" much more quickly and injection-molded parts can be taken out of the mold quickly. See Table 7.4.

• *Injection mircrocellular plastic parts are precise.* Precision plastic parts can be injection molded using the microcellular plastics, as plastic parts made by the MuCell process do not shrink—the shrinkage is taken up by the expansion of the bubble. These

TABLE 7.4 Comparison of Injection-Molded Microcellular Plastics and Conventional Plastics

	Conventional	MuCell	%
Air bag canister made of 33% glass-filled nylon			
Part weight	365 g	252 g	30.9
Cycle time	45 sec	35 sec	22.2
Tonnage	150 tons	15 tons	90
Connector made of polycarbonate			
Part weight	48.8 g	42.9 g	10.6
Cycle time	17.5 sec	15.9 sec	9.1
Tonnage	140 tons	20 tons	85.7
Battery cover made of poypropylene			
Part weight	201 g	159 g	20.8
Cycle time	60 sec	37 sec	38.3
Tonnage	200 tons	15 tons	92.5

Courtesy of Mar Lee Mold Company, Inc. Leomister, MA.

parts also do not have the so-called "sink marks"—slight dips on the surface of molded parts—that occur as a result of the secondary flow when adjacent sections of the plastic part do not cool at the same rate.

• *Good for environment!* Microcellular plastics and the MuCell processes are good for the environment for several reasons: less materials, non-hydrocarbon-based gas, and less energy. The total impact that can be made by microcellular plastics on the environment is staggering. Since the annual consumption of plastics exceeds 100 billion pounds, we will save more than 10 billion pounds of plastics if 30% of all plastic parts are made by this process.

7.4 LAYERED MANUFACTURING PROCESSES FOR RAPID PROTOTYPING

During the past two decades, rapid prototyping of physical objects has become an area of intense research and development. The goal is to produce solid bodies that can be used as a prototype in product development, or as a master that can be replicated by other techniques such as casting, to make a large number of parts. Since then, the technology has evolved to making composite parts with different ingredients at different places to satisfy specific needs. It is now also used to make parts that can deliver medication by occluding it inside a career.

The specific technologies developed for rapid prototyping can be classified as "layered manufacturing processes." There are a variety of specific techniques: three-dimensional printing (Sachs et al., 1993), resin-based systems, laser welding of powders, and others. Some of these techniques satisfy the Independence Axiom. In this section, the functional requirements of a layered manufacturing process will be stated and a rapid prototyping process will be developed. Instead of describing one of the well-established processes, we will develop a new process as we write this chapter, although the designed process may be unrealistic. This is done to illustrate the process without any preconceived notion of what kind of process we will end up developing.

7.4.1 Design of Layered Manufacturing Processes

The highest functional requirement of rapid prototyping may be stated as

FR = Produce three-dimensional parts rapidly.

The design parameter is

DP = Rapid prototyping machine

The process variable may be stated as

PV = Layered manufacturing process

From now on, we will zigzag among the functional domain, the physical domain, and the process domain to obtain the FR, DP, and PV hierarchy. The highest level FR may be decomposed to define the FRs of the DP (rapid prototyping machine) that perform the PV (layered manufacturing process) as

FR_1 = Provide two-dimensional information for each layer of a three-dimensional body.
FR_2 = Deposit a thin layer of materials.
FR_3 = Bond the materials into a continuum.
FR_4 = Remove unwanted materials.
FR_5 = Build up the body layer by layer.

The design parameters are

DP_1 = Digitized two-dimensional information
DP_2 = Thin layer of photoresist (resin)
DP_3 = Light beam
DP_4 = Vacuum suction
DP_5 = Table motion

The process variables are

PV_1 = Computer memory for matrix table of $n \times m$ pixels
PV_2 = Resin dispenser and horizontal table
PV_3 = On–off intense light beam that scans in two dimensions
PV_4 = Suction tube that scans in two dimensions
PV_5 = Step motor with ball screw for motion in vertical direction

The product and the process conceived by the above FRs, DPs, PVs constitute a layered manufacturing process in which thin photoreactive resin (i.e., photoresist) is deposited on a table that is scanned by a light source to cross-link the resin based on the information supplied by the computer. Unreacted resin is removed by a vacuum suction tube that also rasters the resin surface after the reaction is completed.

The product design matrix for the FR/DP relationships is

$$\begin{Bmatrix} FR_1 \\ FR_2 \\ FR_3 \\ FR_4 \\ FR_5 \end{Bmatrix} = \begin{bmatrix} X & 0 & 0 & 0 & 0 \\ 0 & X & 0 & 0 & 0 \\ 0 & 0 & X & 0 & 0 \\ 0 & 0 & 0 & X & 0 \\ 0 & 0 & 0 & 0 & X \end{bmatrix} \begin{Bmatrix} DP_1 \\ DP_2 \\ DP_3 \\ DP_4 \\ DP_5 \end{Bmatrix} \qquad (7.23)$$

The product (i.e., the machine) is an uncoupled design. The design matrix for the DP/PV relationships is

$$\begin{Bmatrix} DP_1 \\ DP_2 \\ DP_3 \\ DP_4 \\ DP_5 \end{Bmatrix} = \begin{bmatrix} X & 0 & 0 & 0 & 0 \\ 0 & X & 0 & 0 & 0 \\ X & 0 & X & 0 & 0 \\ 0 & 0 & 0 & X & 0 \\ 0 & 0 & 0 & 0 & X \end{bmatrix} \begin{Bmatrix} PV_1 \\ PV_2 \\ PV_3 \\ PV_4 \\ PV_5 \end{Bmatrix} \tag{7.24}$$

The process is a decoupled design. Each one of the first-level FRs, DPs, and PVs must now be decomposed. This will be left as a homework problem (Homework 7.6).

7.4.2 Information Content of Layered Manufacturing Processes

The information content of layered manufacturing processes may be less than that of typical manufacturing processes such as casting and machining, because the information required for layered manufacturing processes is two dimensional. In the case of three-dimensional printing, regardless of the complexity of the three-dimensional part, the manufacturing process is reduced to a cyclic, two-dimensional process of depositing the powder, followed by selective bonding of particulates. The two-dimensional process repeats itself over and over again until the final part is manufactured. Thus the information content associated with the manufacturing process is greatly reduced if the dimensional tolerance is comparable to the particle size and if the part does not have to be subjected to any additional manufacturing processes to meet other functional requirements. If the part has to be heat treated or finished to final accuracy, a layered manufacturing process may require more information.

7.5 SUMMARY

In this chapter, axiomatic design theory is applied to the design of materials and materials-processing techniques. Three case studies are presented: one dealing with metal processing, another dealing with polymer processing, and the third on a layered manufacturing process. In the first two examples, FRs are created based on industrial needs, and new processes have been created and commercialized. In the third example, the process of innovation was illustrated by describing how a new rapid prototyping process might be developed.

In the first two examples, it is shown that the materials can be designed first, and then the process for creating the designed materials can be designed logically and systematically. In creating these materials and processes, the Independence Axiom was employed explicitly throughout, and the Information Axiom was applied implicitly. The development of these materials and processes based on axiomatic design shortens the time of development and increases the probability of success. Both of these materials have been commercialized with varying commercial success. However, technical success does not necessarily guarantee commercial success, which is a sobering thought for engineers and designers.

REFERENCES

Baldwin, D. F. "Microcellular Polymer Processing and the Design of a Continuous Sheet Processing System," Ph.D. Thesis, Department of Mechanical Engineering, Massachusetts Institute of Technology, Cambridge, MA, January 1994.

Baldwin, D. F., Suh, N. P., Park, C. B., and Cha, S. W. "Super-Microcellular Foamed Materials," U. S. Patent 5,334,356, 1994a.

Baldwin, D. F., Tate, D. E., Park, C. B., Cha, S. W., and Suh, N. P. "Microcellular Plastics Processing Technology," *Journal of the Japan Society of Polymer Processing*, Vol. 6, pp. 187–194 and 245–256, 1994b.

Baldwin, D. F., Park, C. B., and Suh, N. P. "An Extrusion System for the Processing of Microcellular Polymer Sheets: Shaping and Cell Growth Control," *Polymer Engineering and Science*, Vol. 36, No. 10, pp. 1425–1435, 1996.

Cha, S. W. "Foaming of Super-Microcellular Plastics," Ph.D. Thesis, Department of Mechanical Engineering, Massachusetts Institute of Technology, Cambridge, MA, 1994.

Cha, S. W., Suh, N. P., Baldwin, D. F., and Park, C. B. "Microcellular Thermoplastic Foamed with Supercritical Fluid," U.S. Patent 5,158,986, October 27, 1992.

Colton, J. S. "The Nucleation of Thermoplastic Microcellular Foam," Ph.D. Thesis, Department of Mechanical Engineering, Massachusetts Institute of Technology, Cambridge, MA, 1985.

Colton, J. S., and Suh, N. P. "The Nucleation of Microcellular Thermoplastic Foam with Additives: Part II: Experimental Results and Discussion," *Polymer Engineering and Science*, Vol. 27, pp. 493–499, 1987a.

Colton, J. S., and Suh, N. P. "Nucleation of Microcellular Foam: Theory and Practice," *Polymer Engineering and Science*, Vol. 27, pp. 500–503, 1987b.

Colton, J. S., and Suh, N. P. "Microcellular Semi-Crystalline Thermoplastic Foam," U. S. Patent 4,922,082, April 1990.

Colton, J. S., and Suh, N. P. "Microcellular Foams of Semi-Crystalline Polymeric Materials," U.S. Patent No. 5,160,674, November 3, 1992.

Durril, P. L., and Griskey, R. G. *AIChE Journal*, Vol. 12, p. 1147, 1960 and Vol. 15, p. 106, 1969.

Kim, S.-G., and Koo, M. K. "Design of Microactuator Array against Coupled Nature of Microelectromechanical (MEMS) Processes," *Annals of CIRP*, Vol. 49, No. 1, 2000.

Martini, J. "The Production and Analysis of Microcellular Foam," S.M. Thesis, Department of Mechanical Engineering, Massachusetts Institute of Technology, Cambridge, MA, January 1981.

Martini, J., Suh, N. P., and Waldman, F. A. "The Production and Analysis of Microcellular Thermoplastic Foam," *Society of Plastics Engineers Technical Papers*, Vol. 28, pp. 674–676, 1982.

Martini-Vvedensky, J., Waldman, F. A., and Suh, N. P. "Microcellular Closed Cell Foams and Their Method of Manufacture," U.S. Patent 4,473,665, September 25, 1984.

Park, C. B. "The Role of Polymer/Gas Solutions in Continuous Processing of Microcellular Polymers," Ph.D. Thesis, Department of Mechanical Engineering, Massachusetts Institute of Technology, Cambridge, MA, May 1993.

Park, C. B., Baldwin, D. F., and Suh, N. P. "Effect of the Pressure Drop Rate on Cell Nucleation in Continuous Processing of Microcellular Polymers," *Polymer Engineering and Science*, Vol. 35, pp. 432–440, 1995.

Park, C. B., Baldwin, D. F., and Suh, N. P. "Axiomatic Design of a Microcellular Filament Extrusion System," *Research in Engineering Design*, Vol. 8, No. 3, pp. 166–177, 1996.

Sachs, E. M., Cima, M., Cornie, J., Brancazio, D., Bredt, J., Lee, J., and Michaels, S. "Three Dimensional Printing: the Physics and Implications of Additive Manufacturing," *Annals of CIRP*, Vol. 42, No. 1, pp. 257–260, 1993.

Sanchez-Caldera, L. E., Suh, N. P., Chun, J.-H., Lee, A. K., and Blackall IV, F. S. "Mixing and Casting Apparatus," U. S. Patent 4,706,730, November 17, 1987.

Sanchez-Caldera, L. E., Lee, A. K., Chun, J.-H., and Suh, N. P. "Mixing and Cooling Techniques," U.S. Patent 4,890,662, January 2, 1990.

Sanchez-Caldera, L. E., Lee, A. K., Suh, N. P., and Chun, J.-H. "Dispersion Strengthened Materials," U.S. Patent 4,999,050, March 12, 1991a.

Sanchez-Caldera, L. E., Lee, A. K., Suh, N. P., and Chun, J.-H. "Dispersion Strengthened Materials," U.S. Patent 5,071,618, December 10, 1991b.

Sanyal, Y. "Synthesis and Analysis of a Microcellular Plastics Extrusion System for Insulation of Fine Wires," Ph.D. Thesis, Department of Mechanical Engineering, Massachusetts Institute of Technology, Cambridge, MA, June 1998.

Seeler, K. A., and Kumar, V. "Tension-Tension Fatigue of Microcellular Polycarbonate: Initial Results," *Journal of Reinforced Plastics and Composites*, Vol. 12, pp. 359–376, 1992.

Suh, N. P. "Method for Forming Metal, Ceramic, or Polymer Compositions Application," U.S. Patent 4,278,622, July 14, 1981a.

Suh, N. P. "Orthonormal Processing of Metals. Part I: Concept and Theory," *Journal of Engineering for Industry, Transactions of ASME*, Vol. 104, No. 4, pp. 327–331, 1981b.

Suh, N. P. *The Principles of Design*, Oxford University Press, New York, 1990.

Suh, N. P. "Microcellular Plastics," in *Innovation in Polymer Processing: Molding*, J. Stevenson, ed., SPE Books of Hanser Publishers, New York, 1996.

Suh, N. P., and Turner, A. P. L. *Elements of the Mechanical Behavior of Solids*, McGraw-Hill, New York, 1975.

Suh, N. P., Tsuda, N., Moon, M.G., and Saka, N. "Orthonormal Processing of Metals. Part II: Mixalloying Process," *Journal of Engineering for Industry, Transactions of ASME,* Vol. 104, No. 4, pp. 332–338, 1982.

Suh, N. P., Baldwin, D. F., Cha, S. W., Park, C. B., Ota, T., Yang, J., and Shimbo, M. "Synthesis and Analysis of Gas/Polymer Solutions for Ultra-Microcellular Plastics Production," *Proceedings of the 1993 NSF Design and Manufacturing Systems Grantees Conference*, Charlotte, NC, pp. 315–326, January, 1993.

Taylor, G. I. "The Formation of Emulsion in Definable Fields of Flow," *Proceedings of the Royal Society*, *London*, Vol. 146A, p. 501, 1934.

Tucker, C. L. III, and Suh, N. P. "Mixing for Reaction Injection Molding I. Impingement Mixing of Liquids," *Polymer Engineering and Science*, Vol. 30, pp. 875–886, 1980.

Waldman, F. A. "The Processing of Microcellular Foam," S.M. Thesis, Department of Mechanical Engineering, Massachusetts Institute of Technology, Cambridge, MA, 1982.

Weertman, J., and Weertman, J. R. *Elementary Dislocation Theory*, MacMillan, New York, 1964.

Youn, J. R., and Suh, N. P. "Processing of Microcellular Polyester Composites," *Polymer Composites*, Vol. 6, pp. 175–180, 1985.

HOMEWORK

7.1 A company is interested in developing a micromirror of 100 μm by 100 μm (dimension of each side) that can be controlled to reflect light to different directions for illumination. The idea is to use the lithography technique of putting patterns on a silicon wafer, etching away materials, and building up desired materials by deposition. The central idea is to use the lithography technique to build piezoelectric actuators behind the mirror as an integral part. The deflection of the piezoelectric actuators will move the mirror about its pivot point to bounce off the light. Your job is to design a process that can do the job.

7.2 The design project is to create a machine that can polish the surface of silicon wafers after each deposition of metallic or oxide layers on the silicon wafer to make integrated circuits.

The integrated circuit (IC) manufacturing process consists of putting a thin layer of photoresist on the silicon wafer and imprinting a circuit path or insulation path on the photoresist using a lithography technique (i.e., camera). Then the unexposed (or exposed, depending on whether the photoresist is positive or negative) photoresist is removed. Subsequently, a layer of metal or oxide is deposited on the wafer surface by a physical vapor deposition technique (or chemical vapor deposition technique). A thin layer of photoresist is again put on this wafer with a metallic or oxide layer, which is exposed to lithography to create the next layer of integrated circuit or insulation elements. After the unexposed photoresist is removed, the metallic or oxide layers are removed by an etching technique. Between each deposition of metal or oxide layers, the surface has to be planarized, as the surface must be planar to be able to print the circuit patterns using the lithography technique. This process of putting many layers continues until the manufacture of microprocessors or DRAMs is completed.

The process of planarization is becoming increasingly important as the critical dimension (CD), i.e., line width of the circuit printed on the silicon wafer by lithography, becomes smaller than 0.25 μm. As the CD becomes smaller, we have to use a light source or laser beam with increasingly shorter wavelengths. Unfortunately, the wavelength of the beam has a correspondingly smaller useful depth of focus (UDOF). Therefore, the "bumps and hills" created by the deposition of a new layer of oxides or metals must be removed to make the surface planar, maintaining a uniform thickness of the circuit elements. Otherwise, the next layer of photoresist cannot be exposed to the light beams at the right depth of focus.

The current industrial process used for planarization is called chemical mechanical polishing (CMP). This process is very similar to the conventional lapping process except that it uses a composite pad made of polyurethane and fibers rather than a hard metallic surface. The silicon wafers are attached to a chuck, which rotates. The polyurethane pad is mounted on a rotating table to create a uniform polishing action on the entire wafer surface. Fine abrasive particles suspended in aqueous solution are supplied between the wafer and the polishing pad. The pH of the solution is controlled to create chemical interaction. The pH is about 2 for metallic layers and about 9 for oxide layers.

Your job is to create either a CMP machine that is better than those currently available commercially or a new process.

7.3 Develop a process of making three-dimensional (3-D) parts by means of a layered manufacturing technique. In a typical process, thin layers of materials are deposited layer by layer until the 3-D shape is created. State the functional requirements for such a process and design a process that can make transistors.

7.4 Design a process for making "solder bumps," which are spheres of lead/tin alloy for use in bonding electrical components on a printed circuit board.

7.5 Design a process that is different from the mixalloy process for making dispersion-strengthened copper with ceramic particles.

7.6 Complete the design of the layered manufacturing process in Section 7.4.

7.7 Design a process that can make perfect spherical particles of 2 μm in diameter starting from a liquid tin alloy.

7.8 One of the exciting ways of manufacturing a light emitting diode (LED) is to deposit a galium arsenide layer on top of a silicon wafer. One of the problems is the lattice mismatch between the substrate material and the material being deposited. One approach is to introduce an intermediate gradient layer (such as a mixture of galium, arsenide, and silicon) to deal with the lattice mismatch. However, dislocations are generated in these intermediate gradient layers due to lattice mismatch that impedes the motion of electrons. When the strain energy becomes very large, dislocations burst out and create a large number of undesirable additional dislocations. Design a process that will generate a dislocaton-free galium arsenide layer on top of the silicon wafer.

8

Product Design

8.1 INTRODUCTION

This chapter deals with the application of axiomatic design to product design and product development. Product design is a system design issue, consisting of the design of hardware and software as a system. Products range from relatively simple items such as toasters to highly complex machines such as photolithography machines. Some products such as cars are mass produced and others such as airplanes are custom made based on orders. These products have a finite life cycle, and new products must be introduced in a timely manner to be competitive. In some industries, products become obsolete so rapidly that the product development cycle—the lead time for product development—must become shorter and shorter to keep up with competition and customer demand. This requires a much more systematic approach to product design.

Some products are capital intensive and thus only large companies can produce them. The worldwide consolidation of automobile companies through mergers and acquisitions is an example. Some products are technology intensive. Often, new-venture companies are formed to capitalize on proprietary technologies with the backing of venture capitalists. In recent years, these new companies have enjoyed extraordinary market capitalization, which has enabled them to acquire capital for further expansion. However, in many cases, large companies acquire successful new-venture companies when there is a strategic match between the technology and business opportunities that require significant capital investment.

There are always exceptions. The market capitalization of some of the traditional capital-intensive companies is so low that they can easily be acquired by successful new-venture companies that have enjoyed rapid growth, as exemplified by AOL's acquisition of Time Warner. The market capitalization of Microsoft is larger than that of all of the automotive companies in the world combined! General Motors has revenue of about $200 billion, but their market capitalization is about $40 billion, while corresponding numbers for Microsoft are $20 billion for revenue and $400 billion for market capitalization. What differentiates these companies is product and growth rate. It is clear that companies that do not grow—both in revenue and income—will not be valued highly by the stock market. This clearly creates an unstable equilibrium, which will invite changes in investment patterns and industrial infrastructure in the future.

Product development cost is high relative to the overall revenue the product can generate over its life cycle. Companies that develop noncompetitive products cannot survive in a highly competitive marketplace. Notwithstanding the importance of product development, in many companies, it is often done haphazardly by a group of experienced managers and engineers through trial and error. Management is sometimes at the mercy of the promises made by technical people and often finds that the actual development cost is twice the original planned cost, that the development time has to be extended, and that when the product is finally introduced, it is no longer competitive. This situation arises because of unexpected problems—many of them due to the coupling of functions—that require redesign and remanufacture. This situation is exacerbated when management does not understand product development.

Product development must follow a logical path. This cannot be done based on experience alone, although experience is essential. It must be based on logic and reasoning if the typical iterative process is to be avoided or eliminated. It is difficult to understand why it cost $2 billion to develop a new car when the new car is not much different from the old. When the development cost is so much, a typical management solution is to merge with or acquire other companies to spread the R&D, marketing, and administrative cost over a larger production base. Unless the merger is driven by capacity expansion or by the desire to create new markets based on new technologies, it cannot solve problems that have been generated by inefficiency and mismanagement.

In this chapter, the factors that affect the success of a new product are discussed. The importance of nontechnical factors—market size, management, marketing, personnel, and service—is emphasized. The process of designing a product based on axiomatic design is illustrated by means of a real-time design of an automobile power plant. An industrial case study is presented to show how axiomatic design has been used in designing new products. Many products have been designed or modified in industry based on axiomatic design.

Finally, this chapter presents the approach used by four graduate students to develop an industrially competitive machine at a fraction of the normal development cost. These students developed the specifications for the product based on a market study, designed the machine, which consists of hardware, control logic, and software, had the parts manufactured by vendors, and assembled the machine. The machine is a testimony to how a few bright people (in this case with little experience), working hard and using scientific reasoning, can create a rather complex machine at a reasonable cost in a relatively short period of time.

8.1.1 Important Questions to Ask before Developing a New Product

What is the most important question a firm, especially a new venture firm, must ask before commencing on new product development? Is it the level of technological sophistication? The functionality of the product? The price? Or the market size? Many engineers start their own business when they have a new invention or an idea for a new product. It is exciting to begin a new business and see one's idea turning into a product. It is even more exciting to think about all the money that will begin to pour into one's coffer. Most of all, engineers are so enamored of technology that they forget to ask perhaps the most important question: "How big is the market size?" Many people and companies have made wrong business decisions, because they ignored this simple question before embarking on development of a new product!

If the total worldwide market size for the proposed product is of the order of $10 million per annum, it is best not to waste money trying to develop the product—unless, of course, government is going to finance it. If the market size is only a few hundred million dollars per year worldwide, careful market research should be conducted before investing money to develop the product, especially if there are many competitive products in the marketplace already. If the potential market size is over $1 billion, it is worth venturing in, even if there are strong competitors, because the probability of success improves with the market size. When the potential market is very large, it is easier to get venture capital, penetrate the market, distribute, sell, and service the product, and survive!

Unfortunately, it is difficult to predict the market size. Even the best among us make mistakes in gauging the market size. For example, when the digital computer was first developed at a major computer company, the market size was estimated to be very small. Obviously, someone made a wrong estimate! The same company made a costly mistake in anticipating the growth of the personal computer market and practically gave away its jewel.

On the other hand, another computer company overestimated the market for its mini-computers and, as a result, it is no longer an independent company. In the 1970s, how many people could have predicted the market size for PCs to be as large as it is today? Many people have limited vision and can foresee the future only through their narrow prism, which is often a linear extrapolation of their own past experience. It is always easier to oppose a new idea than to promote it, as the probability that a new idea will become successful is low. However, this kind of attitude will ultimately bankrupt a company.

Notwithstanding these uncertainties in estimating the market size correctly, it is important to estimate the size of the potential market before undertaking product development. Otherwise, the initial euphoria and excitement that come with a new start-up effort can be replaced with frustration, agony, and disappointment.

What is the second most important thing in product development? Although we—the technologists—do not wish to admit it, the second most important thing is the *management* of the company, especially the top management. Management provides the vision for the product and oversees the process of product development, manufacture, financing, marketing, sales, and service.

Top executives of a company should provide the vision, develop business strategy, take care of financing, and get the job done through their organization and management of employees. One of the most critical jobs of top executives is the hiring and management of people. It is the people who make an enterprise succeed or fail. If management is weak and lacks leadership, even the best product may fail in the marketplace. Management must create an environment so that their employees can work smartly, improve their performance continuously, respond to their customer needs, and be innovative. Top management must be able to both generate long-term strategies and implement short-term goals.

For business to succeed, both the macro- and microissues must be considered and implemented effectively. Establishment and attention to macrostrategic goals provide a vision for the future, and careful implementation and attention to details of microissues enable achievement of the vision. These are prerequisites of a successful human endeavor.

Is technology important? Technology is very important. Technology provides the basis for developing a new product. Many modern products are technology driven. Furthermore, engineers thrive and have fun developing new technologies. However, a business should not be developed based on technology alone.

Technology is only one element, albeit a very important element, of a business enterprise. To be a successful developer of new products, technology must be considered in a systems context. Product development should not be undertaken solely centered around an interesting idea without evaluating it in a systems context. Technologists must work with everyone in the company—collaboratively, cooperatively, and harmoniously—to develop a consensus and a systems approach. This is a simple idea, but it is hard to make it work in practice.

What else should one consider before undertaking new product development? Once the market size is estimated to be sufficiently large, there are many other questions a budding entrepreneur must ask:

- What will the return on investment (ROI) be?
- How strong are our intellectual property rights (IPR)?
- How should we market our products—directly or indirectly through distributors?
- Should we manufacture the parts or should we have vendors make them, with us simply assembling the final product?
- How do we provide service after the products are sold?
- What kinds of sales force do we need?

Most business people are always taught to think in terms of ROI, as was discussed in Chapter 6. ROI must always be estimated before developing a new product. The venture capital community will not invest their money unless the ROI is very high—they want to get five times their investment back in 3 to 5 years! It is much more difficult to realize the estimated ROI when the need for the product is not clearly established.

Patent protection is absolutely necessary if the product is innovative and is easy to reproduce. Patents must be applied for globally, although it is expensive to do so. One basic patent is not sufficient protection; the technology must be protected by layers of related patents. The cost of creating strong IPR protection is not trivial.

If the product must be customized for unspecified needs of each potential customer, reconsider the venture. It is difficult to be in the business of selling highly customer-specific products—those that require special calibration and development. The after-sales service cost can be prohibitive.

It is difficult to manufacture products that require continuous customer service— consulting and hand-holding—unless service itself is the business. New venture firms should not use their meager capital in building their own buildings and buying equipment to make all their parts. They should utilize outside vendors for parts and, sometimes, even for assembly.

8.1.2 Basic Requirements of Product Manufacture

There are six fundamental factors to consider in product development: functions, lead time, quality, reliability, value added, and cost. Products must have functions that customers want and are willing to pay for (even though customers sometimes do not know they want these functions until after the purchase!). The lead time must be short because the company that introduces the "hit" product first can make the most money by charging more. The price of any product tends to come down as the competition among products becomes intense. The quality and reliability of products are basic prerequisites for a successful new product. Products must also give the customer the feeling that the product has a high value. Of course, these goals must be achieved at the lowest possible cost for development and manufacturing.

The United States has always led the world in developing innovative new products. It is a result of the unique American culture—a strong tradition and emphasis on innovation, willingness to invest in R&D, and educational emphasis on independent thinking. However, in recent decades, the issue of product quality has become a national concern in the United States as the domestic automotive companies began to lose market share to Japanese automobile manufacturers. Since then, much progress has been made in U.S. corporations, making the United States again the most productive and competitive nation in the world. However, in a number of other relatively mature industries (e.g., machine tools, consumer electronics, cameras, and watches), many foreign firms have become highly competitive, taking market share away from U.S. firms.

As stated earlier, to be competitive, always remember the six factors that determine the competitiveness of a product:

- FRs of the product (What do customers want?)
- Lead time (Remember that World War II lasted only 4 years. How long should your product development process last?)
- Cost (Can it be made cheaper? Why is the materials cost more than 50% of the manufacturing cost? Why is the gross margin[1] less than 50%? Why is the direct labor cost more than 7%?)
- Quality of products (Have you made rational design decisions?)
- Reliability of the product (Is it going to work all the time?)
- Intrinsic value of the product (Is it worth the money?)

8.1.3 How Should Companies Avoid Making Mistakes during Product Development?

When a company attempts to execute poor design concepts and wrong design decisions made during the design stage, it can cost the company a fortune! Unfortunately, this situation exists in many companies and should be corrected.

---

[1] Gross margin is the difference between the revenue from product sales and the direct manufacturing cost.

i. *Importance of Defining the FRs First.* One effective way of promoting innovation is to require that designers define FRs first without any regard to how such products can be made. When FRs are unambiguously stated, designers will know if the proposed design is good or bad. Once FRs are defined, the designers should develop basic ideas for products based on basic principles (i.e., laws of nature and design axioms), making sure that the chosen DPs satisfy the FRs and the Independence Axiom. They should write out the design equations to check whether the Independence Axiom is satisfied and then model the design decisions based on laws of nature to reduce the information content.

Without carefully stated FRs at all levels of decomposition, the quality of products—a minimum requirement—cannot be measured, as a high-quality product is one that satisfies the FRs at all times. Even benchmarking cannot be done without clearly stated FRs. Benchmarking the existing product against its competitors can deal only with DPs, not with FRs, unless the FRs are stated.

Without a clear definition of the FRs of a design, the participants of a product development team cannot communicate among themselves and work together as a team to achieve the ultimate goals of product development.

ii. *Coupled Design.* Once the decision is made to develop a certain kind of product, the most serious mistake many companies make in product development is caused by introduction of coupled designs—inadvertently or intentionally. The situation is made even worse by randomly decomposing these coupled designs to create FR/DP/PV hierarchies. Many companies do not define FRs and do not even know that coupling of FRs is bad—that it causes delays in product development and that the resulting bad products cannot be improved by optimization. Products that violate the Independence Axiom are not good products in terms of quality, reliability, and functional robustness. These products will fail more frequently.

iii. *System Integration.* The second most common mistake made in product development is related to system integration. Many companies often develop the hardware first and then try to introduce software to integrate the system functions and operate the product. In this sequential process of developing hardware first and then software, software programmers must struggle to understand the FRs of the hardware and develop the rationale behind the hardware design before they can design the software and integrate the system. Furthermore, much debugging time is necessary before the product can be introduced to the marketplace. This sequential process is costly and time consuming. Introduction of new products is often delayed because of this ad hoc serial process of designing hardware first, followed by software design and system integration.

As we discussed in preceding chapters, products should be developed based on axiomatic design, which enables the design of the entire system. It allows simultaneous consideration of the hardware and software issues from the beginning as the FRs, DPs, and PVs are defined and decomposed systematically. Such a process should improve the probability of introducing a creative new product on schedule and within the original budget. This kind of logical and rational thinking should replace the need to copy existing products.

iv. *Innovative Products versus "Me-Too" Products.* What is surprising is that most companies are organized to make the same product they used to make, rather than to introduce innovative products and to become the market leader. This is caused partly by the urge to produce prototypes quickly and test them instead of spending time and effort to do a rational and thorough job during the design stage. The entire system should be designed and tested on paper first before the actual construction of prototypes. A design that is well planned and executed on paper has a much higher probability of becoming a product that meets the original functional requirements. It is important that all interested parties in a company (i.e., marketing, design, engineering, manufacturing, and service) agree on the FRs of the product in a solution-neutral environment, know how to create uncoupled or decoupled designs, and analyze the proposed design through modeling and analysis.

In many companies, there is much internal resistance to a new way of conducting product development. They fear the introduction of any changes because their production technologies and development processes have been developed empirically and by trial-and-error processes after many years of hard work. Many people are worried that changes will disrupt their operations and that the product may not work as well. This is sometimes called "inertia" (i.e., "we have always done it this way") or "NIH" (i.e., not invented here) syndrome. They know only too well how difficult it was to reach the current steady state of operation—many old timers have scars to show for it! They also do not trust new theories because so many of them were not effective in improving their operations. To make the cultural changes that allow new ideas to be accepted more easily, the entire product development process must be more firmly based on science—a rational basis for decision making.

Benchmarking is an important thing for companies making competitive products to do, but it is not sufficient when new features and innovations are to be introduced in their products. Many large firms do not have an innovative culture. In these firms, employees are rewarded for not making mistakes, which indirectly creates a culture that discourages innovation and risk taking. This is one of the reasons why large firms end up buying innovative small companies. Furthermore, the lack of scientific decision-making tools forces many engineers to do engineering based on their "gut feeling," which frequently produces wrong results! Therefore, a large number of engineers is needed to produce products because trial and error is often the method used in new product development.

8.1.4 What Have Universities Done in This Area?

Notwithstanding the fact that product development involves both synthesis and analysis, often in an iterative feedback loop, most academic researchers concentrate on the analysis of design results or design processes. In some ways, this emphasis on improvement of design through analysis is not surprising, because engineering education has emphasized the "engineering science of analysis"—a derivative of reductionism of science—in contrast to "engineering science of synthesis." Consequently, academic researchers have conducted research on optimization techniques and ad hoc theories and methodologies that may lack a scientific base. The issues addressed in some research papers are often related to the lowest level DPs in the physical domain, often ignoring the relationship between functions and physical parameters as well as the importance of hierarchical decomposition through a systematic zigzagging between the domains in creating innovative products. This kind of academic research lies in the middle of the research spectrum shown in Figure 8.1, which tends to have a limited impact on technology innovation and fundamental knowledge generation.

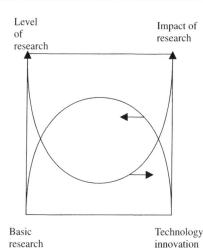

Level of research

Impact of research

Basic research

Technology innovation

Research spectrum

Figure 8.1 Level of research effort and potential impact as a function of the nature of the research. The impact is greatest at both ends of the research spectrum.

Universities do a reasonably good job of teaching students how to model well-defined problems, but do not teach them how to define the task and what to model. Many engineering schools do not teach their students to write design equations and to check their validity in terms of the design axioms. The fact that few industrial engineers bother to model and write governing equations after they graduate from engineering schools is not surprising, as they have not been rigorously educated in defining the problem. In axiomatic design, we emphasize the need to clearly state the FRs and Cs and then to write down the design equations that unambiguously define the problem that must be modeled—the relationship between FRs and DPs—and solved.

What should change? Product design should be based on science, e.g., axiomatic design and the natural sciences. Science is needed to develop designs that are rational and thus fulfill the FRs and Cs precisely. We should eliminate the idea that we will debug a product after prototypes are made and tested. We should also eliminate the idea that we will optimize a poor design by improving one FR at the expense of other FRs. Furthermore, the hardware and the software should not be treated as separate entities—they together constitute the system that is the product. These things can be done within a framework established by science.

A scientific base for design also enables us to document the design process and design results systematically. The decomposition process based on zigzagging between domains produces a design document that clarifies the complicated relationship between all FRs and DPs. The system architecture is a form of documentation that can be followed logically and easily during the design, manufacture, and maintenance of a product.

After the FR/DP relationships are clearly established, we can create and structure the design knowledge as modules, which can be stored in a knowledge library. Such systematized knowledge can be used in many other applications, enabling faster product development.

8.1.5 Customization of Products to Satisfy Individual Customers

Competition has intensified in a global economy where products made anywhere in the world are becoming available in all countries. Mass customization is a means of remaining competitive in the global market. In the era of electronic purchasing of products, the competition for market share will become even more intense than ever before.

To meet this challenge, products are increasingly becoming customized—to a limited degree—because customers are no longer satisfied with a standard product. They want to have special features. The challenge to manufacturers is how to use standard components and yet offer customized features. This customization is taking place now in many fields, such as the PC and automotive fields. Recently, a personal computer manufacturer has revolutionized the PC business by taking orders from individual customers directly for specially configured computers through the Internet. This trend is spreading to many other businesses, including the automobile industry. This is sometimes called "mass customization."

The goal of customization of products is to increase the market share. Customization is becoming possible as a result of the availability of enabling technologies (e.g., communications technologies). Customization is often driven by overcapacity in production facilities, the intense competitiveness of the marketplace, and the overall sophistication of consumers for quality products that accompany a high standard of living.

8.1.6 Total Quality Management

Even after the product is introduced, there must be a continuing effort to improve quality, productivity, and profit through continuous improvement of all aspects of the manufacturing operation. Total quality management (TQM) has been a movement to discover the source of inefficiencies and defects and to improve the performance of the company as a manufacturer and marketer of the product. Companies form quality circles involving people from different functional groups within the company and sometimes including vendors. This TQM movement has yielded significant dividends to many companies, making some of them the most productive companies in their field of business.

This chapter presents the design of products based on axiomatic design theory. Products may be hardware, hardware/software, or software. For ease of discussion, hardware will be used as examples. We will illustrate the product design process by actually designing a product that does not exist at the present time. This is done to illustrate the design process in real time, as this chapter is being written, so that readers can also follow the axiomatic design process. Because of the ad hoc nature of this design exercise, some of the details may be incorrect, but the pedagogical goal of illustrating the axiomatic design process should still be served by this case study.

8.2 MAPPING FROM THE CUSTOMER DOMAIN TO THE FUNCTIONAL DOMAIN

Mapping of customers' needs into FRs in the functional domain is one of the most important elements of the design process. It is a difficult process for several reasons. Often the customer needs are poorly defined, and, in many cases, it is tempting to define the FRs by recalling an existing product rather than in a solution-neutral environment. Also customer needs are diverse—there may be as many opinions as there are customers. Furthermore, Arrow's Impossibility Theorem (Example 1.2) showed that the individual preferences of a group of people do not directly translate into the preference of the group.

Even when the customer needs are established, mapping them into FRs is not an easy task. It often takes many months of discussion and analysis. It also requires many talented and knowledgeable people who have the following characteristics:

a. Strong engineering backgrounds with clear understanding of fundamental principles
b. Experience in knowing what can be done and what cannot be done
c. Creative ideas—the ability to think in terms of FRs and out of the box
d. Sixth sense of knowing what customers *really* want and are willing to pay for
e. An understanding of the market
f. The ability to think logically

There are substantial differences in the development and design processes used when the goal is to improve an existing product rather than to create a complete new product. Many methodologies have been proposed to deal with existing products (e.g., Clausing, 1994; Ulrich and Eppinger, 1995; Wood and Otto, 2000; and Shiba, Graham, and Walden, 1993).

8.2.1 For Existing Products

Many companies begin their product development process after a thorough market study and produce a document called Marketing Requirement Specification (MRS) as the basis for new product development. The MRS is typically a compilation—a very thick stack of papers—of detailed requirements the marketing department, with the help of some of their key customers, has generated for the next-generation product. In many cases, it is a mixture of FRs, DPs, and PVs—a random collection of wish-lists. These documents are not ideal for use by designers because they overspecify the requirements.

The marketing people, in collaboration with designers and engineering staff, should specify FRs only, not DPs and PVs. Customer specification of DPs and PVs tends to limit design options and also to force the designer to come up with a product that is similar to an existing one.

Benchmarking is an important step in comparing various existing products with one's own product. Price, functionality, reliability, and the cost of ownership are compared to determine the relative merit of one's own product against the competitors' products. The reliability is measured using the mean time between failures (MTBF), mean time for repair (MTFR), and the total life of the product. The cost of ownership includes all the things the customer must pay (e.g., consumables) to manufacture their products, which is a measure used in the semiconductor industry. Sometimes companies use a "spider" chart to show the comparison, which is illustrated in Figure 8.2.

All companies use the method of "reverse engineering" to a varying degree to copy the best design features of their competitors' products without infringing on patent rights. Companies obtain their competitors' product (sometimes very difficult to do in capital equipment business) and study its features. It is not uncommon to see parts from Lexus, Mercedes Benz, and other well-known automobiles at most engineering shops of automobile manufacturers. However, what we can learn from reverse engineering is limited. It provides dimensions of the part, the material used, and particular design features and their performance, but it is difficult to capture the design intent and all of the functional requirements. Also if the design is a decoupled design, it is difficult to reconstruct the design matrix. Example 8.1 illustrates this point.

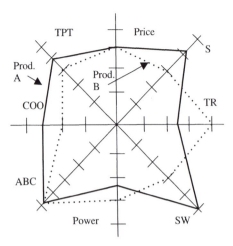

Figure 8.2 A spider diagram comparing various features and performance of one's product to those of competitors. Each axis represents one of the features or functions being compared.

EXAMPLE 8.1 Did They Steal Company Secrets?

In the late 1970s, the manager of a major division of a computer company (will be referred to as the BBM Corporation) and his two key engineers quit BBM to set up their own company (called CN). They raised venture capital—about $20 million—from HB Ventures to produce a product very similar to the product that they were responsible for at BBM. They set up a plant near the BBM plant in San Jose, California. Within a year, CN was able to manufacture a product very similar to the BBM product and had started to take BBM's market share away. BBM became very suspicious of the way CN got the technology, as the BBM product was a result of a $120 million R&D investment BBM had made over a 10-year period. Because of the major technical advances incorporated in this product, BBM has dominated the market, making handsome profits. Every competitor, including many in Japan and Europe, tried to come up with competitive products without any success—that is, until CN came along.

BBM Corporation retained a famous New York Law firm, CMS, and sued CN, Inc., claiming that their three former employees stole the technologies for the product from BBM and used them to produce the CN product. CN denied the allegations saying that it used only the knowledge publicly available, i.e., BBM's publications, videotapes shown by BBM to tour groups visiting BBM, and the information available in patents, to produce the CN product.

The law firm hired a professor of electrical engineering from a university on the West Coast and a mechanical engineering professor from an East Coast technical university. These professors' job was to demonstrate that CN could not have produced the product in such a short period of time without the information that they had access to at BBM, which they had misappropriated when they were with BBM.

How would you prove that BBM's claim is correct and CN's explanations are not credible? You must remember that although you are paid by BBM, you are legally, ethically, and morally bound to be objective.

SOLUTION

At first, the professors had to read many boxes of documents. The CMS lawyers—about 50 lawyers worked on this case—produced boxes and boxes of depositions taken from BBM engineers and managers as well as those from CN's employees and executives. (CMS also hired some local lawyers from California law firms.) The professors were given tours of BBM's facilities and all relevant technical details. They were impressed by the thoroughness of the lawyers, who knew what they were doing. The professors were told not to take any notes because the CN lawyers might demand to see them. The professors were treated royally—of course, at BBM's expense. (It seemed that the CMS lawyers are used to that kind of life-style.) After many hours of reading and thinking, the expert witnesses met BBM engineers to ask questions for clarification and further technical details.

The mechanical engineering professor looked over the design of the product and the manufacturing processes. The product was a small sensor made by plating layers of circuits on the surface of ceramic material, requiring about 50 different process steps. Many of these sensors were made at the same time and then each sensor (sometimes called die) was separated from other sensors by slicing the ceramic disk. The sensors made by this process were very small, about $3 \times 3 \times 2$ mm.

When the FRs of the device and the manufacturing processes were analyzed and the DPs identified, it was determined that they satisfied the Independence Axiom! (The lawyers had to learn axiomatic design quickly, especially the meaning of the independence of FRs.) The lawyers also learned that when there are so many steps and so many functions, it is difficult to design such a device and develop manufacturing processes in a short period of time—especially because BBM engineers did not know anything about axiomatic design in the late 1970s—and come up with almost identical processes.

From the axiomatic design point of view, the more interesting thing was that CN processes were not only similar to the BBM process, but they too satisfied the Independence Axiom. At one place where the CN process deviated from the BBM process, CN had a coupled design. If an attempt is made to develop such a multistep manufacturing system through a trial-and-error process, it cannot be done in 1 year because most people make many mistakes and must try alternative ideas before they get it right.

When the BBM and CN drawings were checked, it was amazing that the tolerances CN used were identical to those of BBM. Reverse engineering cannot generate identical tolerances! Measurements of nominal dimensions can be made, but the tolerances cannot be obtained from the measurements of a limited number of samples. Clearly, they had access to BBM's drawings, which were somehow made available to CN. They also bought machines and supplies from the same vendors BBM had been using.

For a year, the consultants and BBM lawyers put in hard work, traveling back and forth between the West Coast and the East Coast. Ultimately, CN lost the suit and closed its doors.

The law firm made a lot of money and the consultants were reasonably compensated. BBM could protect its market position. It may be safe to assume that HB Ventures lost its investment.

Some companies, especially those engaged in mass production, use a method called quality function deployment (QFD), which was imported from Toyota—a company known worldwide as the company that produces high-quality (but not the most advanced) products. The idea is to identify engineering characteristics and determine how different design features impact them. Furthermore, they benchmark their own products against competitors' products. From such an evaluation, engineering characteristics (some of which are equivalent to FRs) are developed that are important for the product both qualitatively and quantitatively. This information is used as the basis for their product development. Those interested in the topic of QFD should read the book by Clausing (1994).

Many companies end up copying the best product available on the market. When their competitors come up with a new feature, they try to introduce a similar—but slightly improved—product quickly at a lower price. They try to learn from their competitors' mistakes. Their marketing people go around the world, taking pictures of their competitors' products. Sometimes, it is much cheaper and less risky to copy ideas from successful products than to develop something from scratch, especially if the company has a strong marketing group with ample financial resources.

8.2.2 For New Innovative Products

When completely new innovative products are designed, we may or may not know customer needs. In this case, it is difficult to create the MRS. Often the customer needs are present right in front of us, but we do not recognize them. For example, people who worked with wet mimeographs knew that the chemical odor and the wet process were undesirable, but most people had assumed that there is no alternative. It took many years for the inventor of xerography to convince others in the industry that his dry electrostatic process had merits. Many people working in the mimeograph machine companies were so busy making better mimeograph machines that they did not have time to think of a better machine that is based on a radically different concept. Sometimes, we become slaves of our own creations and bad ideas!

Once we know what "customers" may wish to have, it is important to define FRs in a solution-neutral environment, which is not an easy thing to do. Defining FRs is not a trivial task. But perhaps the biggest problem has been that most designers and engineers do not even attempt to define FRs. In some situations, FRs can best be defined if we have a strong technical background. However, in some cases, technical understanding is less important than the understanding of human beings and nontechnical issues for defining FRs.

Chapter 7 discussed the development of FRs in a solution-neutral environment for microcellular plastics, which has resulted in a completely new class of polymers with microscale bubbles in large numbers, and for mixalloy, which has created dispersion-strengthened alloys in a new cost-effective way.

We will illustrate the creation of FRs in two different ways: one based on purely technical considerations and the other based on nontechnical considerations.

i. *Definition of FRs Based on Technical Understanding.* Consider the following question most engineers ought to be able to answer:

If we make the basic assumption that we will be driving a car powered by an engine that burns petroleum, what kinds of engines will we (i.e., the customers in this case) need in the twenty-first century?

Most people would say the following:

   a. An engine that provides all the acceleration we want when we need it
   b. An engine that burns a minimum amount of fuel
   c. An engine that does not harm nature—no pollution
   d. An engine that most people can afford to have
   e. An engine for a vehicle that can go long distances

Although it is possible to come up with a different set of customer needs, let us assume that we are satisfied with the above set. Then, what are the FRs that can satisfy these customer needs? One thing we must do is not think about the existing diesel or gasoline engines. To come up with a specific set of FRs in a solution-neutral environment that satisfies the above set of customer needs, we must resort to our technical understanding of the issues that are relevant to converting chemical energy to mechanical energy.

We may list the following FRs:

FR_1 = Supply the fuel.
FR_2 = Evaporate the liquid fuel into a vapor phase.
FR_3 = Deliver high power for acceleration.
FR_4 = Mix the fuel molecules with oxidizer.
FR_5 = Induce chemical reaction between the fuel molecules and oxidizer.
FR_6 = Convert the chemical energy into electrical energy.
FR_7 = Convert the combustion product into harmless molecules.
FR_8 = Exhaust the combustion product.

These FRs are created by thinking about the basic chemical and physical steps involved in converting chemical energy into either mechanical or electrical energy. Indeed it is interesting to note that as we write down these FRs, we begin to see how we can come up with an engine that may indeed revolutionize the engine business, although we have not yet considered actual hardware and the DPs that can do this job. We may find that we do not know how to satisfy one or more of the FRs because of the lack of scientific and engineering knowledge or because of the unavailability of suitable technologies.

The FRs defined above are based on technical factors. We will consider the creation of FRs that are not technically based, at least at the highest level of the FR hierarchy.

   ii. *Definition of FRs Based on Nontechnical Factors.* Consider the customer needs of a marketing organization that can sell automobiles through the Internet rather than through the usual dealerships. What are the FRs?

We may list the following FRs:

FR_1 = Create a web-based information system.
FR_2 = Devise a mean of identifying the customers who log in.
FR_3 = Create a set of typical questions customers ask about cars.
FR_4 = Develop a system that can give an equivalent feel for the car's performance even though the customer cannot drive the real car.
FR_5 = Develop a competitive pricing system.
FR_6 = Create a network of banks for low-cost financing for customers.
FR_7 = Take care of registration and insurance.

$FR_8 =$ Create a service network.

$FR_9 =$ Deliver cars.

In this case, the FRs were created without any regard to technical issues, although some of the DPs that we choose to satisfy these FRs may bring in technical factors. Here we have depended on our understanding of what customers go through and need in buying cars. However, we have not thought about how it is done currently. It was easy to be in a solution-neutral environment for this design because of the author's lack of prior experience or knowledge of Internet commerce. [A wise man once said: *"Sometimes ignorance is a blessing!"* Unfortunately, it is not always true.]

Is inspiration needed to develop FRs? What is inspiration? Inspiration is defined in many different ways in Webster's dictionary, one of which states that it is "a divine influence or action on a person held to qualify him to receive and communicate sacred revelation." In engineering, inspiration sometimes comes suddenly to a designer, revealing the question and providing answers. It may be that when curiosity or a questioning mind resonates with an observation or external stimulus, we get the inspiration that leads to insight and/or answers. Therefore, to be able to come up with a good set of FRs, a broad knowledge base should help because the probability of having resonance between our own knowledge and the external stimuli should increase with our knowledge base and the quality of external input. This process of generating ideas through inspiration depends on the richness of the knowledge base that can be acquired either through experience or education.

8.3 MAPPING FROM FRs TO DPs

In the preceding section, the process of defining FRs was illustrated by mapping from the customer domain to the functional domain. In this section, we will attempt to map the FRs into the physical domain to develop DPs. The task of designing a new power plant for an engine will be used as an exercise for product design.[2] The FRs were stated in Section 8.2.2 as

$FR_1 =$ Supply the fuel.

$FR_2 =$ Evaporate the liquid fuel into a vapor phase.

$FR_3 =$ Deliver high power for acceleration.

$FR_4 =$ Mix the fuel molecules with oxidizer.

$FR_5 =$ Induce chemical reaction between the fuel molecules and oxidizer.

$FR_6 =$ Convert the chemical energy into electrical energy.

$FR_7 =$ Convert the combustion product into harmless molecules.

$FR_8 =$ Exhaust the combustion product.

There are many constraints, some of which may be stated at this stage as

$C_1 =$ The engine must be portable in vehicles.

$C_2 =$ It should not be bigger than the V-6 engine currently used in mid-size cars.

$C_3 =$ It should cost less than \$2000 (in 1999 U.S. dollars) to manufacture.

---

[2] A power plant will be designed for automobiles here as this chapter is being written to illustrate the axiomatic design process. This may not be the best design, but the purpose of this exercise is to show how a product can be designed.

C_4 = The fuel will be gasoline.

C_5 = The engine must last 250,000 miles or for 10 years of service.

The first thing we must do now is to conceptualize the design solution by considering all of these FRs in aggregate and each individual FR in isolation. Then, we have to think of DPs for each FR and integrate these DPs to produce an integral product. Sometimes, it may be better to conceptualize the integral solution first and then identify the individual DPs within the integrated design embodiment.

The DPs may be stated as

DP_1 = Fuel pump

DP_2 = Fuel injection into a combustion chamber

DP_3 = Energy storage for use when peak power is needed

DP_4 = Injection of the vaporized fuel with compressed air (turbocharger)

DP_5 = Spark ignition (spark plug) in cylinder/piston

DP_6 = Electric generator (a permanent magnet piston moving in an electric coil)

DP_7 = Catalyst

DP_8 = Exhaust port

The conceptual design of the engine is schematically illustrated in Figure 8.3. The figure shows a free-floating piston engine (Galitello, 1989), but with many new features. It shows a two-piston/cylinder engine with its pistons linked together by a mechanical shaft.

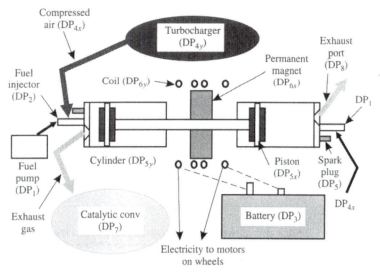

Figure 8.3 Free-floating piston engine. It has two cylinders and a direct fuel-injection system that uses compressed air to mix the fuel with air and deliver it into the combustion chamber. The electricity generator consists of a permanent magnet mounted on the shaft that connects two pistons and moves inside a coil. It treats the exhaust gas with a catalytic converter, stores energy in a battery for fast acceleration, and uses conventional spark plugs for ignition. Note that the DPs are shown on the drawing. (The concept of the free-floating piston engine with linear electric generator was introduced to the author by Galitello, 1989.)

A permanent magnet is mounted on this shaft. When this magnet moves back and forth inside the coil, electricity is generated. In place of a conventional fuel injector that typically uses a mechanism that involves a piston-type mechanical valve to inject the fuel at high pressure into the combustion chamber, this fuel injector uses compressed air coming in at high velocity from a high-pressure source (something similar to turbocharger). The exhaust gas is treated with a catalytic converter.

Figure 8.3 shows the DPs. DP_{4x}, DP_{4y}, etc. are shorthand notation for the next-level DPs of DP_4, although they have not been formally decomposed.

We have to write the design equations to be sure that the design satisfies the Independence Axiom. The design matrix for this design is shown in Table 8.1. From Table 8.1 it appears that FR_2 and FR_4 are coupled. But further analysis of FR_2 and FR_4 shows that they are essentially the same FRs. Therefore, one of these two FRs can be eliminated. All the FRs the designer could think of were put down without analyzing the details of the design, and yet the construction of the design matrix pointed out the mistake the designer had made! Indeed reexamination of Figure 8.2 shows that DP_2 and DP_4 are the subelements of the fuel-injection system.

Based on these results, FR_4 will be eliminated and FR_2 will be restated as

$FR_2 = $ Deliver mixture of the oxidizer and the fuel in gaseous phase.

Then DP_2 may be written as

$DP_2 = $ Compressed air activated fuel injector system

Then the design matrix may be modified as shown in Table 8.2. The drawing must also be revised, as shown in Figure 8.4.

TABLE 8.1 Design Matrix for Original Design of a Power Plant for Automobiles

	DP_1	DP_2	DP_3	DP_4	DP_5	DP_6	DP_7	DP_8
FR_1	X	0	0	0	0	0	0	0
FR_2	X	X	0	X	0	0	0	0
FR_3	0	0	X	0	0	0	0	0
FR_4	X	X	0	X	0	0	0	0
FR_5	0	X	0	0	X	0	0	0
FR_6	0	0	0	X	0	X	0	0
FR_7	0	0	0	0	X	0	X	0
FR_8	0	0	0	0	0	0	0	X

TABLE 8.2 Design Matrix for the Modified Power Plant for Automobiles

	DP_1	DP_2	DP_3	DP_5	DP_6	DP_7	DP_8
FR_1	X	0	0	0	0	0	0
FR_2	X	X	0	0	0	0	0
FR_3	0	0	X	0	0	0	0
FR_5	0	X	0	X	0	0	0
FR_6	0	0	0	0	X	0	0
FR_7	0	0	0	X	0	X	0
FR_8	0	0	0	0	0	0	X

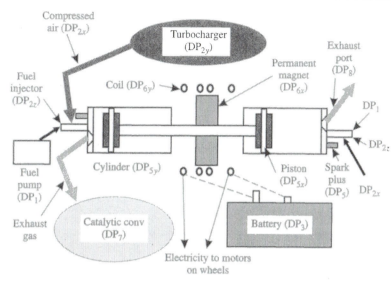

Figure 8.4 Modified drawing of the free-floating piston engine shown in Figure 8.3. This figure is different from Figure 8.3 in that DP_4s are changed to DP_2s.

8.3.1 Decomposition of FR_2 and DP_2

The next step in the design process is the decomposition of these highest level FRs and DPs to develop the detailed design for these major components. We will consider here one of the important FRs for decomposition to illustrate the process: FR_2 (deliver mixture of the oxidizer and the fuel in gaseous phase) and DP_2 (compressed air activated fuel-injector system). The decomposition of the rest of the FRs will be left to the reader as a homework problem (Homework 8.2).

FR_2 may be decomposed as

$FR_{21} =$ Meter the fuel.
$FR_{22} =$ Deliver the fuel into high-pressure chamber.
$FR_{23} =$ Atomize and vaporize the fuel.
$FR_{24} =$ Mix the vaporized fuel with oxidizer.
$FR_{25} =$ Supply enough oxidizer (air) to the combustion chamber.

The DP_{2x} that can satisfy the FR_{2x} may be stated as[3]

$DP_{21} =$ Axial position of Plunger A
$DP_{22} =$ Axial motion of Plunger A
$DP_{23} =$ Nozzle design
$DP_{24} =$ Channel and nozzle for compressed-air delivery through Plunger A
$DP_{25} =$ Valve-opening time through rotation of Plunger A to line up the channels with the compressed air supply hole on the cylinder wall

---

[3] It should be noted that this preliminary design is done without much thinking in about 1 hour including the typing and drawing time. As we decompose further, other ideas will come along, which will help in improving this design.

The above set constitutes the decomposition of DP_2 (compressed-air-activated fuel-injection system). The fuel-injection system meters the fuel when Plunger A retracts to a predetermined axial position with the rotational position of the plunger such that the compressed-air supply is sealed off. When the time for fuel injection comes, the plunger rotates to line up the air supply line and the plunger is pushed downward. As the compressed air flows out of the nozzle with the fuel, the fuel breaks up into microdroplets and vaporizes in the combustion cylinder. Each FR_{2x} and DP_{2x} must be decomposed to develop further detail.[4] However, before we proceed with decomposition we must check the design by writing the design matrix at this level of design hierarchy, which is shown in Table 8.3. It is a decoupled design and thus satisfies the Independence Axiom. The lower case x implies a weak relationship.

A fuel injector that can perhaps fulfill the above set of FR_{2x} with the proposed set of DP_{2x} is schematically sketched in Figure 8.5. More details will be worked out as we decompose further and the geometry will be later finalized through modeling.

How do we conceptualize and generate DPs? This is not an easy question to answer. The designer must have a database in his or her brain or in a machine (e.g., computer). Some people, especially in Germany, use "morphological tables" that show many different means of achieving certain design goals. Some people use "brainstorming sessions" to generate ideas. Thomas Edison used analogy to come up with new ideas—from phonographs to

TABLE 8.3 Design Matrix for a Fuel Injector

	DP_{21}	DP_{22}	DP_{23}	DP_{24}	DP_{25}
FR_{21}	X	0	0	0	0
FR_{22}	0	X	0	0	0
FR_{23}	0	X	X	0	0
FR_{24}	0	X	x	X	0
FR_{25}	X	0	0	0	X

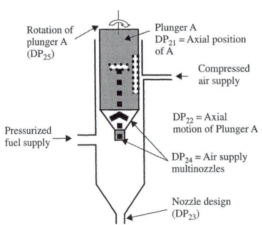

Figure 8.5 A schematic of a conceptual design of the fuel-injection system. The axial position of Plunger A determines the amount of fuel to be delivered into the combustion chamber, the axial motion of the plunger pushes the fuel into the combustion chamber, and the compressed air mixes with the fuel and aids in breaking up the fuel. The compressed air is turned on and off by turning Plunger A.

[4] As we decompose to lower and lower levels, we may find that we have to come back to the higher level design hierarchy and make some changes or corrections at the highest level. It is much cheaper to make corrections on paper than with hardware.

telephones, etc. The idea for the Mixalloy process of dispersion-strengthened metals came partly from the work done in reaction injection molding of polyurethane, which is a form of analogy. Sometimes simple experiments can shed a great deal of light that leads to ideas. Russian inventor Altshuller (1996) came up with a set of rules (algorithms) for developing DPs. One of the rules is to use contradiction when a certain DP violates a constraint. One of the most effective ways of identifying the need for a different DP is to identify coupled FRs and think of ways of eliminating the coupled design features.

Integration of DPs to create a synthesized solution is an important element of the design process. Designers have to depend on their experience and basic understanding of engineering science and the natural sciences to achieve this integration task. As of now, it is difficult to let a computer perform this function, especially when the design involves a completely new design.

What happens to constraints? As we continue to decompose, the number of constraints increases as all the higher level decisions made cannot be violated. Also the higher level constraints must be satisfied at all levels of the decomposed hierarchy. For example, to be consistent with C_5 that deals with the service life of the engine, we may have to design the surface of the plunger shown in Figure 8.5 as an undulated surface in order to prevent galling and roughening up of the surface. Further details on undulated surfaces can be found in Suh (1986).

8.4 APPLICATION OF THE INFORMATION AXIOM

8.4.1 General Criteria

In designing the power plant for automobiles in Section 8.3, we did not discuss the implications of the Information Axiom. However, we implicitly did several things that are consistent with the Information Axiom:

1. A minimum number of FRs was chosen.
2. Simple DPs were chosen rather than complicated and convoluted designs.
3. In selecting DPs, the rules for coming up with a robust design (e.g., lower stiffness) were considered.

As we continue to decompose and become more quantitative and analytical through detailed modeling at the leaf level, we can try to impose the Information Axiom in a rigorous manner.

In Chapters 2 and 3, it was stated that when the system range is determined for a given design range, we can find the minimum information value of the design—the best operating region for the given system—by moving the system range along the FR axis (horizontal axis) until it is fully contained within the design range. Because the FR is a function of the chosen DP, we can determine the best operating point for the design by differentiating the information content[5] with respect to DP as

---

[5] These information contents constitute real complexities. In some cases, imaginary complexities may be present because of lack of understanding of the proposed design or because one is not aware of axiomatic design. These issues are discussed in Chapter 9 in great detail.

$$\sum_{i=1}^{N} \frac{\partial I_i}{\partial DP_i} = 0$$

(8.1)

$$\sum_{i=1}^{N} \frac{\partial^2 I_i}{\partial DP_i^2} > 0$$

When the design is uncoupled, the above equation can be satisfied by setting each term of Equation (8.1) as

$$\frac{\partial I_1}{\partial DP_1} = \frac{\partial I_2}{\partial DP_2} = \cdots = \frac{\partial I_N}{\partial DP_N} = 0$$

(8.2)

$$\frac{\partial^2 I_1}{\partial DP_1^2} > 0 \cdots \frac{\partial^2 I_N}{\partial DP_N^2} > 0$$

In the case of a decoupled design, the order of differentiation should follow the order of DPs given by the triangular matrix. In the case of a coupled design, Equation (8.1) still must be satisfied, i.e., the summation of the slope of I_i with respect to DP_i must be equal to zero and the summation of the curvature of I_i with respect to DPi must be greater than zero. However, many coupled designs may not be able to satisfy these conditions.

8.4.2 Error Budgeting

Error budgeting is a concept of allocating tolerances to different components (i.e., DPs in the physical domain) of a system so that when the entire system is assembled together, the tolerance specified for the DPs at the highest level can be satisfied. Error budgeting replaces the need to keep track of the large number of errors that can occur when the six degrees of freedom are specified to define the relative motion of one rigid body with respect to another.

Error budgeting was an idea originally used in the optics industry. This concept was first introduced to mechanical systems by Donaldson (1980) when he was developing a precision machine called the single-point diamond turning machine at the Lawrence Livermore National Laboratory. He allocated errors such that each component was made within its error allocation (Slocum, 1992). This concept of allocating tolerances to different parts of a system can be applied to all designs. Unfortunately, traditional error budgeting is done in the physical domain, which is not consistent with axiomatic design.

In axiomatic design, error budgeting should be done in the functional domain first, followed by the tolerance specification for DPs in the physical domain. This will make it possible to apply robust-design concepts (such as lower stiffness) to come up with the largest possible tolerances for DPs for given specified ranges of FRs. Then the issue becomes how to deal with the propagation and allocation of tolerances on FR during the decomposition of FRs and DPs. Suffice it to say that the tolerances of children FRs must be consistent with the tolerance of the parent FR.

Suppose the design equations at the highest level of FRs, FR_1 and FR_2, are

$$FR_1 \pm \Delta FR_1 = A_{11} (DP_1 \pm \Delta DP_1)$$

(8.3)

$$FR_2 \pm \Delta FR_2 = A_{22} (DP_2 \pm \Delta DP_2)$$

The goal here is to develop a design such that ΔDP_1 and ΔDP_2 can be maximized for a given ΔFR_1 and ΔFR_2. It is clear that as long as FR_1 and FR_2 are independent, the allocation of tolerances to DP_1 and DP_2 can be done independently.

When the design requires further decomposition of FR_1 and DP_1 because DP_1 is not a leaf, the question of the allocation of tolerances becomes an issue. Consider a hypothetical case in which FR_1 and DP_1 are decomposed as

$$FR_{11} \pm \Delta FR_{11} = \alpha \, (DP_{11} \pm \Delta DP_{11})$$

$$FR_{12} \pm \Delta FR_{12} = \beta \, (DP_{12} \pm \Delta DP_{12})$$

(8.4)

where α and β are the elements of the design matrix. If the design at this second level is an uncoupled design, FR_{11} and FR_{12} are completely independent from each other.

How are the design ranges ΔFR_{11} and ΔFR_{12} related to ΔFR_1? The first requirement is that the design range of the children FRs must be *consistent* with that of the parent FR. In some cases, the tolerance on children FRs is exactly the same as that of the parent, i.e., $\Delta FR_1 = \Delta FR_{11} = \Delta FR_{12}$. To illustrate this case, let us reconsider Example 1.4 (Refrigerator Design).

EXAMPLE 8.2 Error Allocation for the Refrigerator (Example 1.4)

The highest functional requirements were

$FR_1 =$ Freeze food for long-term preservation.
$FR_2 =$ Maintain food at cold temperature for short-term preservation by keeping the food at between 2°C and 3°C (or keep the food at the temperature of 2.5°C $\pm$ 0.5°C).

To satisfy these two FRs, a refrigerator with two compartments is designed. Two DPs for this refrigerator were stated as

$DP_1 =$ The freezer section
$DP_2 =$ The chiller (i.e., refrigerator) section

Let us consider the decomposition of FR_2 and DP_2, which was stated with their specified design ranges as

$FR_{21} =$ Control temperature of the chiller section within 2.5C $\pm$ 0.5C.
$FR_{22} =$ Maintain a uniform temperature throughout the chiller section within 2.5C $\pm$ 0.5C.

In this case, both FR_{21} and FR_{22} simply inherited the design range of the parent FR_2, (i.e., ± 0.5°C). This, in turn, will determine the tolerances on DP_{21} and DP_{22}. Because DP_{21} was the fan for the chiller section, the volume of the air recirculated should have a tolerance (probably a minimum fan speed) to satisfy FR_{21}. Similarly, because DP_{22} was the vent to circulate air, the vent must be designed to allow air circulation independent of the amount of food stored in the chiller section.

In some situations, the design ranges on children FRs, in addition to being related to the parent FR design range, may also have a relationship between themselves. For this purpose, consider the design of the parking mode of the automatic transmission discussed in Example 3.5 (Parking Mode of Automatic Transmission).

EXAMPLE 8.3 Parking Mode of Automatic Transmission (Example 3.5)

The automatic transmission of automobiles is designed to prevent accidental engagement of the transmission in the park mode while the vehicle is still in motion. Example 3.5, which was to redesign the mechanism of the park mode to make it easier to disengage the gear from the park mode, also had an FR for preventing accidental engagement. Their design is shown here again in Figure E8.3.

To illustrate how the tolerances must be allocated and propagated, FR_3 will be restated in this example and its tolerance propagation will be investigated here.

FR_3 and the corresponding DP_3 are

FR_3 = Prevent the accidental engagement of the park mode when the vehicle is moving at a speed greater than 3 mph.

DP_3 = The tooth profile of the sprocket wheel and the profile of the pawl "teeth"/Spring A/tension spring

FR_3 was decomposed into the following children FRs:

FR_{31} = Control the force that pushes the pawl into sprockets to be less than 6 N ± 1 N (so that the pawl cannot be engaged at a speed greater than 3 mph).

FR_{32} = Create a reaction force at the sprocket/pawl interface so that the force transmitted to DP_{31} is greater than 8 N ± 1 N if the sprocket is turning.

At this time, FR_{31} and its design range ΔFR_{31} are arbitrarily defined based on a common understanding of what would be an acceptable force level. Also, the magnitudes of FR_{32} and ΔFR_{32} are established a priori based on the observation that the force exerted by the linkage to the pawl must always be less than the reaction force at the interface between the pawl and the sprocket.

The corresponding DPs are

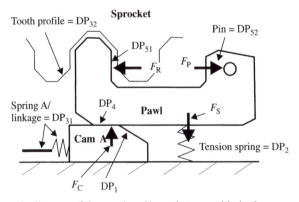

Figure E8.3 Schematic diagram of the cam/pawl/sprocket assembly in the engaged position, which shows the physical parts and DPs. The forces acting on the pawl are shown by thick arrows.

DP_{31} = Spring A that connects the linkage to the cam and the displacement of the
linkage
DP_{32} = Sprocket tooth profile/pawl profile

These DPs must be chosen so that they can satisfy FR_{31} and FR_{32} within their specified
design range. What we need to do as a designer is to lower "the stiffness" between the FRs
and DPs so that the tolerance on DP is large for a given value of the FR design range.

The design matrix is a triangular matrix.

$$\begin{Bmatrix} FR_{31} \\ FR_{32} \end{Bmatrix} = \begin{bmatrix} X & X \\ 0 & X \end{bmatrix} \begin{Bmatrix} DP_{31} \\ DP_{32} \end{Bmatrix}$$

This is a decoupled design.

8.5 CASE STUDY—DEPTH CHARGE[6]

8.5.1 Case Study Background

The Navy uses depth charges with explosives (i.e., warheads) to damage enemy submarines
during unfriendly encounters. One of the important elements of the system is the initiator,
which triggers a series of events that ultimately results in the detonation of the warhead
near the enemy submarine. The design task is to design an initiator that sends a signal to the
detonator only when the depth charge hits a target and is intended to explode the warhead.
The customer requires a unit that is cheaper and more reliable than the existing one.

A schematic drawing of an initiator is shown in Figure 8.6. The initiator requires the
following inputs before signaling to the detonator:

- Electrical energy
- Three independent arming conditions (ACs)
- Ignition signal

When all three are present, the detonator will detonate the warhead. The functional requirement of the design is to provide the initiator with these signals to detonate the depth
charge.

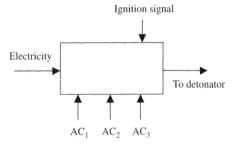

Figure 8.6 Schematic diagram of the operational
features of an initiator.

---

[6]This design project was conducted by Anders Swenson (Swenson, 1994), an engineer with Saab
Missiles AB, Sweden, with the support of Mats Nordlund. This section has been adapted from a
section in Dr. Nordlund's thesis, which was done in part at MIT (Nordlund, 1996).

Safety regulations mandate that at least one of the ACs should be satisfied by a state of the depth charge (e.g., underwater). Once it is set, the state should be constant until detonation. This means that the system will prevent detonation if the state changes or disappears before the cycle is completed. Furthermore, the system's sensors must not erroneously react to humidity, electromagnetic radiation, darkness, vibrations, accelerations, or temperature.

8.5.2 Effectively Searching for Potential Design Solutions

The Information Axiom states that the designer should select the design with the minimum information content (i.e., with the maximum probability of success). This axiom has been interpreted in many different ways. One common qualitative interpretation is that the simplest design is the best, whereas another interpretation is that the minimum number of parts is the best (provided that the Independence Axiom is still satisfied). However, the design with the minimum number of parts is not the best design if the total information content in the new system is higher than it was in the old system.

8.5.3 Design of the Depth Charge Initiator [7]

8.5.3.1 Problem Definition

Based on the general problem description, a more concrete description of the customer requirements or desired customer attributes (CAs) was developed to help focus the thinking process. Three CAs were established as

CA$_1$ = Lower cost
CA$_2$ = Simpler concept (lower part count if information content is reduced, as per Axiom 2)
CA$_3$ = More reliable concept

8.5.3.2 Highest Level FRs and DPs

The highest level FRs are to initiate detonation of the warhead and convey the driving gas pressure in the barrel to the entire depth charge causing it to accelerate and begin its ballistic trajectory. DPs were chosen as an electrical system and a launcher of depth charge. Thus, FRs and DPs may be stated as

FR$_1$ = Initiate detonator.
FR$_2$ = Launch the depth charge.

DP$_1$ = Electrical system
DP$_2$ = Launcher

The design matrix is diagonal.
FR$_2$ may be decomposed as

---

[7] Because of the nature of this product, specific numbers as well as some specific requirements and constraints that were part of the original design report cannot be included. However, this does not make any difference in the conclusions presented here.

FR_{21} = Provide force to launch device.
FR_{22} = Send the device in the desired direction.
FR_{23} = Convey force to the entire device.

The DP_{2x}s are chosen as

DP_{21} = Propellant
DP_{22} = Barrel
DP_{23} = Chassis

The design equation is diagonal, as shown.

$$\begin{Bmatrix} FR_{21} \\ FR_{22} \\ FR_{23} \end{Bmatrix} = \begin{bmatrix} X & 0 & 0 \\ 0 & X & 0 \\ 0 & 0 & X \end{bmatrix} \begin{Bmatrix} DP_{21} \\ DP_{22} \\ DP_{23} \end{Bmatrix} \tag{8.5}$$

The design equation shows that the Independence Axiom is satisfied, i.e., the independence of the FRs is maintained and so the design process continues. However, the design of the launcher (including the explosive, barrel, and chassis) is already complete and will not be further decomposed here.

At this level, the following constraints are introduced that will apply to all DPs that may be chosen later in the design process.

C_1 = Safety
C_2 = Weight
C_3 = Position of the center of gravity
C_4 = Outside measures (geometry) have to fit within chassis
C_5 = Environmental endurance

8.5.3.3 Decomposing the Initiator (FR_1)

FR_1 must now be decomposed into the lower level FRs. The decomposition must take into consideration that an electrical system was chosen as DP_1.

FR_{11} = Provide electricity.
FR_{12} = Activate arming condition 1 (AC_1).
FR_{13} = Activate arming condition 2 (AC_2).
FR_{14} = Activate arming condition 3 (AC_3).
FR_{15} = Send signal when the depth charge hits the target.

To determine the DPs, we must understand the environment within which the depth charge will be used in practice. There are seven states of the launching cycle, during which the depth charge must satisfy FR_{11} through FR_{15}.

State 1: Storage and transport
State 2: Loaded in launcher
State 3: Launching
State 4: Air trajectory
State 5: Penetrating water
State 6: Sinking
State 7: Hitting target

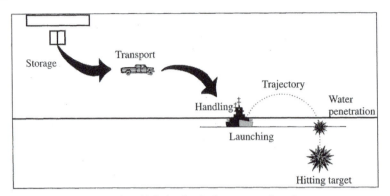

Figure 8.7 The environment within which the depth charge must function.

Figure 8.7 shows the seven states.

Swenson (1994) decided that the armament conditions should be satisfied during States 2 through 6 and that the initiation of the warhead activation must be satisfied during State 7. We must choose certain DPs that will satisfy the FRs as the depth charge goes through these states. These DPs must be independent of each other and unique, as the initiation of the depth charge must be based on recognition of a unique set of events that identifies the states the launcher has gone through starting from the launch of the depth charge. Each of the states and events will be examined in detail below.

> *Loaded in Launcher and Launching:* States 2 and 3 characterize a unique condition of the depth charge. These states cannot be used as a DP because they are not present just before detonation. On the other hand, an important event occurs when the depth charge leaves the launcher through the muzzle.

> *Air Presence:* At States 1, 2, 3, and 4, the common environmental condition is the presence of air. The detection of air is not a useful DP as it cannot be used to identify a specific state.

> *Dynamic Air Pressure:* When the depth charge is launched, dynamic air pressure will exist. However, this dynamic air pressure will be of the same order of magnitude as dynamic water pressure, which means that dynamic pressure cannot be used to distinguish between dynamic water pressure and dynamic air pressure. Therefore dynamic air pressure cannot be used as a DP in this design.

> *Gas Pressure (in the Launcher's Barrel):* The gas pressure due to the deflagration of solid propellants exists only when launching, which means it is a unique and independent event.

> *Water Presence:* The presence of water clearly distinguishes between the air and the water phases, making it suitable as a potential DP. However, we must be able to distinguish water due to rain and the body of water in the sea, which surrounds the depth charge everywhere.

> *Water Pressure:* Water pressure clearly distinguishes between the air and the water phases.

Dynamic Water Pressure: Dynamic water pressure is of the same order of magnitude as dynamic air pressure, which means that dynamic pressure cannot distinguish between dynamic water pressure and dynamic air pressure. Therefore dynamic water pressure cannot be used as a DP in this design.

Rotation: The launcher is not rifled; hence launching generates no rotation. However, rotation is generated when penetrating water as well as during storage and transport, and therefore rotation cannot be used as an environmental factor.

Hitting the Target: When the depth charge hits the target, there will be some negative acceleration that can be detected as an event.

Time: Time is not a good factor, as the time required for each phase will be different for each situation.

Based on the above analysis, the following environmental factors were determined to be unique and independent and thus could be used as the basis for formulating DPs to trigger the ACs (the FRs) and set off the ignition signal (IS) to the detonator:

DP_{11} = Gas pressure
DP_{12} = Leaving the launcher muzzle (Event 1)
DP_{13} = Entering a body of water (Event 2)
DP_{14} = Water pressure (state)
DP_{15} = Hitting target (Event 3)

The design equation for this system at this level of decomposition is given by Equation (8.6).

$$\begin{Bmatrix} FR_{11} \\ FR_{12} \\ FR_{13} \\ FR_{14} \\ FR_{15} \end{Bmatrix} = \begin{bmatrix} X & 0 & 0 & 0 & 0 \\ X & X & 0 & 0 & 0 \\ X & 0 & X & 0 & 0 \\ X & 0 & 0 & X & 0 \\ X & X & X & X & X \end{bmatrix} \begin{Bmatrix} DP_{11} \\ DP_{12} \\ DP_{13} \\ DP_{14} \\ DP_{15} \end{Bmatrix} \tag{8.6}$$

In this type of application, safety must be of paramount importance. Therefore, the system range must always be inside the design range so that the information content is zero.

8.5.3.4 Design of Subsystems

FR_{11} through FR_{15} must be decomposed to develop a detailed design of each subsystem to ensure their satisfaction. Only the decomposition of FR_{11}, FR_{12}, and FR_{15} can be disclosed in this case study. However, the approach to decomposing FR_{13} and FR_{14} is no different from the approach followed in decomposing the other FRs.

i. *Decomposition of* FR_{11} *(Provide Electricity) and* DP_{11} *(Gas Pressure).* To provide electricity (FR_{11}), gas pressure will be used to activate a battery. Then the decomposition of FR_{11} yields

FR_{111} = Sense launching event.
FR_{112} = Supply electrolyte.

A design concept for these FRs is shown in Figure 8.8, which uses a battery that can be activated when an electrolyte in an ampoule is supplied to a chamber with electrodes. Gas pressure enters the rear end of the depth charge. The gas is led to a chamber where an impact piston is forced to hit one end of a battery. This impact should suffice to break a glass ampoule containing an electrolyte. When the electrolyte comes into contact with the electrodes, the battery becomes active.

The DPs may be stated as

DP_{111} = Gas pressure-activated mechanical motion
DP_{112} = Mechanical impact to break the ampoule

The design equation may be written as

$$\left\{ \begin{array}{c} FR_{111} \\ FR_{112} \end{array} \right\} = \left[\begin{array}{cc} X & 0 \\ X & X \end{array} \right] \left\{ \begin{array}{c} DP_{111} \\ DP_{112} \end{array} \right\} \tag{8.7}$$

The design is decoupled design.

ii. *Decomposition of* FR_{12} *(Generate Arming Condition 1) and* DP_{12} *(Leaving the Launcher Muzzle).* What we need to do is to decompose FR_{12} in a way that can describe the functional requirements of DP_{12}. These lower level FRs may be stated as

FR_{121} = Sense launch.
FR_{122} = Activate the circuit after it leaves the barrel.

The concept proposed is the use of a rod that senses the presence of the barrel, and when the depth charge leaves the barrel, it closes an electric switch. Then the DPs may be stated as

DP_{121} = Rod sensing the presence of the barrel
DP_{122} = Electric switch activated by the rod

The resulting design equation may be written as

$$\left\{ \begin{array}{c} FR_{121} \\ FR_{122} \end{array} \right\} = \left[\begin{array}{cc} X & 0 \\ X & X \end{array} \right] \left\{ \begin{array}{c} DP_{121} \\ DP_{122} \end{array} \right\} \tag{8.8}$$

It is a decoupled design.

iii. *Decomposition of* FR_{121} *(Sense Launch) and* DP_{121} *(Rod Sensing the Presence of the Barrel).* The lower level FRs that describe the functions of DP_{121} may be written as

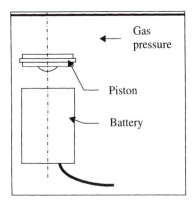

Gas pressure

Piston

Battery

Figure 8.8 Supplying Electricity

FR_{1211} = Push the rod toward the barrel.
FR_{1212} = Extend the rod when the depth charge leaves the barrel.
FR_{1213} = Prevent the rod from moving back after launch.

The corresponding DPs were chosen as

DP_{1211} = Piston
DP_{1212} = Expanding gas
DP_{1213} = Latch mechanism

The design equation may be written as

$$\begin{Bmatrix} FR_{1211} \\ FR_{1212} \\ FR_{1213} \end{Bmatrix} = \begin{bmatrix} X & 0 & 0 \\ X & X & 0 \\ 0 & 0 & X \end{bmatrix} \begin{Bmatrix} DP_{1211} \\ DP_{1212} \\ DP_{1213} \end{Bmatrix} \quad (8.9)$$

This is a decoupled design and thus satisfies the Independence Axiom.

A mechanism that integrates the three DPs is shown in Figure 8.9. It has a piston that separates the higher pressure side from the lower pressure side, which exerts pressure on the pin to contact the barrel. When the device leaves the barrel, the high-pressure gas behind the piston expands, pushing the rod farther out. This in turn closes the electric circuit. In order to ensure that the switch does not open after the depth charge leaves the barrel, a latch mechanism must be introduced to hold it in place. This is not shown in the figure. This latch mechanism could be either mechanical or electrical.

iv. *Decomposition of* FR_{15} *(Provide Initiation Signal) and* DP_{15} *(Hitting Target).* The decomposition of FR_{15} to generate FRs that can satisfy DP_{15} may be stated as

FR_{151} = Sense the impact with the target.
FR_{152} = Send the signal to the detonator.

The DPs may be chosen as

DP_{151} = Accelerometer
DP_{152} = Switch activated by the accelerometer

These DPs are the leaves as they can be made from off-the-shelf components. The design matrix is a lower triangular matrix. Thus this design satisfies the Independence Axiom.

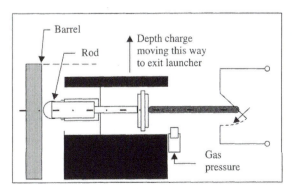

Figure 8.9 Generating arming condition 1.

8.5.3.5 Final Comments on the Case Study

The result of this case study—a commercially successful design of an initiator—was a more reliable and robust system, with the part count reduced from more than 350 parts to fewer than 100 parts. This system is now in serial production. Someday, when the information is declassified, the design of DP_{13} and DP_{14} may be made public. In the meantime, it is assigned as Homework 8.8.

8.6 CHEMICAL-MECHANICAL PLANARIZATION (CMP) MACHINE[8]

Integrated circuits (IC) for microprocessors and memory chips (DRAMS) are made on silicon wafers using the following process steps. A thin layer (less than 1 μm thick) of a photoresist (i.e., light-sensitive polymeric resin) is spin coated on a silicon wafer (typically 200 mm—soon to be 300 mm—in diameter). An electrical circuit is printed on the photoresist layer using a lithography machine (a huge and expensive precision camera). The unexposed photoresist is then etched (or dissolved) away, which is followed by the deposition of layers of conductors and insulators using a chemical vapor deposition (CVD) or sputtering process. Undesired deposited materials are then etched away. This process is repeated until several layers of the circuits are deposited on the wafer to make the IC chip.

To increase the memory density of DRAMS and the speed of microprocessors, the line-width of the circuits (sometimes called the critical dimension or CD) must be made as small as possible. The linewidth has been made smaller and smaller over the past 20 years. The latest lithography machine uses laser beams of ultraviolet wavelength (e.g., 193 nm) with a very high numerical aperture (VHNA of 0.6 and higher), projection optics, and phase shift masks.

To make the linewidth of integrated circuits smaller and smaller (0.1 μm or less), each time a new layer of electronic circuit elements is added, it must be planarized by polishing the surface flat and smooth within atomic dimensions. Circuits with a small CD cannot be printed on a "wavy and rough" surface using the ultraviolet beam, because the "useful depth of focus" of the incident beam decreases as the wavelength decreases. Therefore, the surface must be planarized to atomic flatness after deposition of a new layer. This planarization is done using a chemical and mechanical polishing (CMP) process, which uses slurries in a liquid carrier.

A market study of the semiconductor manufacturing business indicates that the demand for CMP machines will increase rapidly in the early twenty-first century as the demand for faster microprocessors and high-density memory chips increases, following the so-called "Moore's law." The MIT Manufacturing Institute has developed a CMP machine at the request of an industrial sponsor, which is interested in introducing a competitive machine to the semiconductor manufacturing equipment market.

---

[8]This section presents the work done by two graduate students at MIT. Jason Melvin, who designed the central part of the system as part of his Ph.D. thesis, prepared Section 8.6.1. Kwang-duk Douglass Lee prepared Section 8.6.2 on the design of the control system for this CMP machine. Other graduate students, Amir Torkaman and Jamie Nam, also designed and implemented other parts of the CMP machine, which are not presented in this section because of the lack of space. These students were supervised by Professor Jung-Hoon Chun, Dr. Nannaji Saka, and the author. An industrial firm sponsored this project. Several patent applications have been filed on this machine.

The CMP machine has many elements, including a polishing head, an *in situ* sensor for monitoring of the polishing process, a cleaning station, a gantry-type spindle-head carrier, etc. A completed machine is about 6 feet long, 4 feet wide, and 10 feet high, not including the electronic and computer control panels. A commercial CMP machine sells for $1.5–$2 million in 2000.

8.6.1 Design of the Mechanical System

8.6.1.1 Design through Decomposition

In this section, the design of the central section of the MIT CMP machine—the spindle, wafer holder, polishing platen, etc.—is presented. One of the goals of this section is to show how the highest FRs can be decomposed to leaf levels to implement the design. Some of the details of the machine are not presented to make the length of the chapter manageable as well as to protect the proprietary information of the industrial sponsor.

The highest level FR and DP. The top level FR and DP for the CMP machine are as follows:

Functional Requirements (FRs)	Design Parameters (DPs)
Maximize ROI	CMP Machine Design

Return on investment (ROI) is defined as

$$\text{ROI} = \frac{(\text{Value added} - \text{COO})(\text{Net wafers per hour})(\text{Machine life})}{\text{Capital investment}} \tag{8.10}$$

Decomposition of the highest FR and DP, design matrix, and constraints. The highest level FR and DP may be decomposed as shown in Table 8.4. The resulting FRs and DPs of the CMP machine design are as follows:

$$\begin{Bmatrix} \text{FR}_1 \\ \text{FR}_2 \\ \text{FR}_3 \\ \text{FR}_4 \\ \text{FR}_5 \end{Bmatrix} = \begin{bmatrix} X & 0 & 0 & 0 & 0 \\ X & X & 0 & 0 & 0 \\ X & X & X & 0 & 0 \\ X & X & X & X & 0 \\ X & X & 0 & 0 & X \end{bmatrix} \begin{Bmatrix} \text{DP}_1 \\ \text{DP}_2 \\ \text{DP}_3 \\ \text{DP}_4 \\ \text{DP}_5 \end{Bmatrix} \tag{8.11}$$

Description of DPs

- DP_1: Leading-edge technology extends the machine's life in the quickly progressing production environment.

TABLE 8.4 Highest Level FRs and DPs

	Functional Requirements (FRs)	Design Parameters (DPs)
1	Maximize machine life	Leading-edge technology
2	Maximize value added	Flexible, integrated system
3	Minimize cost of ownership (COO)	Target COO
4	Minimize investment	Production-optimized machine design
5	Maximize number of wafers produced	Maximized output

- DP_2: The value added to the wafer by the CMP process is difficult, if not impossible, to quantify. Through the design of the device, CMP becomes an essential step in fabrication. Therefore, the requirement to maximize value added is satisfied through customer perception. The current perception of value in CMP tools is the integration of polish, cleaning, and metrology in one station.
- DP_3: The target COO is the cost of ownership, expressed in dollars per wafer pass.
- DP_4: Production-optimized machine design is the design refinement process that begins with a functionally complete machine, and evolves the design using cost reduction methods to reduce the manufacturing cost of the tool, leading to reduced selling price.
- DP_5: Maximized output from the machine ensures that for the given process, useful product is made at the highest rate.

Decomposition of FR_2 and DP_2. FR_2 (Maximize Value Added) and DP_2 (Flexible, Integrated System) may be decomposed to the third level as shown in Table 8.5.

The design matrix is as follows:

$$\begin{Bmatrix} FR_{21} \\ FR_{22} \\ FR_{23} \\ FR_{24} \\ FR_{25} \end{Bmatrix} = \begin{bmatrix} X & 0 & 0 & 0 & 0 \\ X & X & 0 & 0 & 0 \\ X & X & X & 0 & 0 \\ X & X & X & X & 0 \\ X & X & X & X & X \end{bmatrix} \begin{Bmatrix} DP_{21} \\ DP_{22} \\ DP_{23} \\ DP_{24} \\ DP_{25} \end{Bmatrix} \qquad (8.12)$$

The constraints are many, which are given in Table 8.6.

Description of FR_{2x} and DP_{2x}

- DP_{21}: The removal module is the key competency of the machine. It enables the process around which the machine is designed, leading to the primary position in the decomposition. The process is defined by the fabrication requirements as a removal process; other parameters will be dealt with in the further decomposition. DP_{21} affects the following FRs.
  - FR_{22}: The removal module creates the materials that must be cleaned from the wafer.
  - FR_{23}: The manner in which the wafer handler interfaces with the removal module affects the wafer handler's design.
  - FR_{24}: The parameters available for user control and the possible range for control must be determined by the removal module.
  - FR_{25}: Any necessary support systems will be determined by the design of the removal module.
- DP_{22}: The cleaning module returns the wafers to their preprocess level of contam-

TABLE 8.5 Decomposition of FR_2 and DP_2

	Functional Requirements (FRs)	Design Parameters (DPs)
21	Process wafer	Removal module
22	Clean wafer	Cleaning module
23	Transport wafers	Wafer handler
24	Enable user control	User interface software
25	Support machine operation	Support subsystems

TABLE 8.6 Constraints C_{2x}

Constraint Table		Impacts: FR_{2x}				
Index	Description	1	2	3	4	5
Critical Performance Specifications						
C_{21a}	Polish output quality	—				
C_{21b}	Polish repeatability	—				
C_{22a}	Cleaner output quality		—			
Operational Constraints						
C_{24a}	Allow flexible user interface				—	
C_{24b}	Allow automated operation				—	
Global Constraints						
C_{21c}	Minimize costs (design, manufacturing, operational, maintenance, etc.)	—	—	—	—	—
C_{21d}	Maximize throughput	—	—	—	—	
C_{21e}	Do not damage wafers	—	—	—	—	
C_{21f}	Maximize availability/reliability (minimize MTBM and MTBF)	—	—	—	—	—
C_{21g}	Make tool serviceable (easy access for maintenance)	—	—	—	—	—
C_{21h}	Make tool "user-friendly" (ergonomics and software interfaces)	—	—	—	—	—
C_{21i}	Minimize footprint	—	—	—		—
C_{21j}	Conform to industry and safety standards	—	—	—	—	—
C_{21k}	Integrate maximum amount of existing technology (minimize redesign of proven components, use off-the-shelf equipment whenever possible)	—	—	—	—	—

ination. The requirement to clean the wafers is partially created by the choice of removal process. The cleaning module must meet the minimum throughput defined by the removal modules.

- DP_{23}: The wafer handler is a transport device used to move the wafers from one stage of their processing to the next. It allows the use of multiple removal modules and cleaning modules to meet a global throughput constraint.
- DP_{24}: The user interface software is the software that is common to all other software modules. This includes any interface with outside information or manual input. It is the normal operating display of the machine interface.
- DP_{25}: The support subsystems for the machine allow the implementation of the above design parameters. There is a sufficient role in providing those services that are common to multiple parts of the machine to necessitate a separate requirement.

Decomposition of FR_3 and DP_3. FR_3 (Minimize COO) and DP_3 (Target COO) may be decomposed as shown in Table 8.7.

The cost of ownership (COO) is defined as

$$COO = Materials + Operational\ Activities\ (OA) + Overhead$$

where

TABLE 8.7 Children of FR_3s and DP_3s

	Functional Requirements (FRs)	**Design Parameters (DPs)**
31	Minimize material costs	Optimized consumable use
32	Minimize operational activity costs	Target performance of operational activities
33	Reduce overhead	Reduced footprint design

Materials = Slurry, Pad
OA = Transport, Manual Operations, Setup
Overhead = Footprint

The design equation is given by

$$\begin{Bmatrix} FR_{31} \\ FR_{32} \\ FR_{33} \end{Bmatrix} = \begin{bmatrix} X & 0 & 0 \\ 0 & X & 0 \\ 0 & 0 & X \end{bmatrix} \begin{Bmatrix} DP_{31} \\ DP_{32} \\ DP_{33} \end{Bmatrix} \tag{8.13}$$

Description of DP_{3x}

- DP_{31}: Optimized consumable use is the method for reducing the use of consumables during machine operation to the minimum level required to meet performance specifications.
- DP_{32}: Target performance of operation activities is the result of actions taken to reduce OA costs. These will be detailed in the next level of decomposition.
- DP_{33}: Reduced footprint design provides the smallest possible footprint for the machine.

Decomposition of FR_5 and DP_5. FR_5 (Maximize Net Wafers per Hour) and DP_5 (Maximized Output) are decomposed as shown in Table 8.8.

TABLE 8.8 Decomposed FR_5 and DP_5

	Functional Requirements (FRs)	**Design Parameters (DPs)**
51	Maximize throughput	Process cycle time
52	Maximize availability	Reliable, robust design
53	Maximize yield	Scrap prevention mechanisms

where

Net wafers per hour = Availability × Throughput × Yield

Availability = Scheduled (MTBM, MTTM) + Unscheduled (MTBR, MTTR)

The design matrix is given by

$$\begin{Bmatrix} FR_{51} \\ FR_{52} \\ FR_{53} \end{Bmatrix} = \begin{bmatrix} X & 0 & 0 \\ 0 & X & 0 \\ 0 & 0 & X \end{bmatrix} \begin{Bmatrix} DP_{51} \\ DP_{52} \\ DP_{53} \end{Bmatrix} \tag{8.14}$$

Description of DP_{5x}.

- DP_{51}: Process cycle time is the total processing time, a combination of removal, cleaning, and transport.

- DP_{52}: A robust, reliable design is able to operate for extended periods without breaking down or requiring maintenance.
- DP_{53}: Accurate end-point detection prevents overpolished wafers, the most significant factor in yield loss.

Decomposition of FR_{21} *and* DP_{21}. FR_{21} (Process Wafer) and DP_{21} (Removal Module) may be decomposed as shown in Table 8.9.

The design equation is given as follows:

$$\begin{Bmatrix} FR_{211} \\ FR_{212} \\ FR_{213} \\ FR_{214} \end{Bmatrix} = \begin{bmatrix} X & 0 & 0 & 0 \\ 0 & X & 0 & 0 \\ X & X & X & 0 \\ X & 0 & 0 & X \end{bmatrix} \begin{Bmatrix} DP_{211} \\ DP_{212} \\ DP_{213} \\ DP_{214} \end{Bmatrix} \tag{8.15}$$

The constraints are shown in Table 8.10.

Description of DP_{21x}

- DP_{211}: The ARP is a process that removes material in a manner consistent with its constraints. Primary in the constraints is planarization, or the ability of the process to make the surface flat. The flat surface should be created as quickly as possible, to allow for a broader range of remaining thickness. DP_{211} affects the following FRs.
  - FR_{213}: The abrasive removal process (ARP) will determine which parameters are available for control of the remaining thickness, as well as the possible ranges for parameters.
  - FR_{214}: The ARP defines the interface that the wafer must be loaded into and out of.
- DP_{212}: Multiple ARP design is the inclusion of at least two independent platens for polishing. The second platen may enable the use of two-step polishing slurries or of a buffing operation after the main polish.
- DP_{213}: The ARP controller is the software implementation of a process control

TABLE 8.9 Third Level FRs and DPs

	Functional Requirements (FRs)	Design Parameters (DPs)
21.1	Remove material	Abrasive removal process (ARP)
21.2	Enable two-step processes	Multiple ARP design
21.3	Control remaining thickness	ARP controller
21.4	Enable wafer load/unload	Load/unload station

TABLE 8.10 Constraints C_{211x}

Constraint Table		Impacts: FR.		
Index	Description	1	2	3
	Critical Performance Specifications			
C_{211a}	Planarize surface—die scale flatness	—		
C_{211b}	Surface quality—scratches and roughness	—		
C_{213a}	Wafer-to-wafer variation $< 2\%$—SiO_2			—
C_{213b}	100% of land area cleared			—
C_{213c}	Minimal overpolish			—

scheme. It may contain the necessary end-point detection methods and/or *in situ*/in-process metrology.

- DP$_{214}$: The load/unload station is positioned in the center of the machine, and is responsible for positioning the wafer in a manner that will enable it to be picked up.

Several different planarization requirements are shown in Figure 8.10, which shows the surface of wafers before and after planarization. In the top figure, the oxide layer, which isolates the lower layer from the layer to be placed on top, is polished. In the middle figure, the metal layer is removed to the original surface of silicon. In the bottom figure, trenches must be filled with oxide and covered with silicon nitrides.

Decomposition of FR$_{24}$ *and* DP$_{24}$. FR$_{24}$ (Enable User Control) and DP$_{24}$ (User Interface Software) are decomposed as shown in Table 8.11.

The design equation may be written as

$$
\begin{Bmatrix} FR_{241} \\ FR_{242} \\ FR_{243} \\ FR_{244} \\ FR_{245} \\ FR_{246} \end{Bmatrix} = \begin{bmatrix} X & 0 & 0 & 0 & 0 & 0 \\ 0 & X & 0 & 0 & 0 & 0 \\ X & X & X & 0 & 0 & 0 \\ 0 & 0 & 0 & X & 0 & 0 \\ 0 & 0 & 0 & X & X & 0 \\ X & X & X & X & 0 & X \end{bmatrix} \begin{Bmatrix} DP_{241} \\ DP_{242} \\ DP_{243} \\ DP_{244} \\ DP_{245} \\ DP_{246} \end{Bmatrix}
\tag{8.16}
$$

The constraints are shown in Table 18.12.

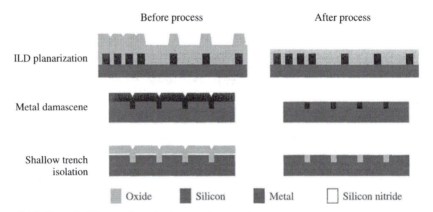

Figure 8.10 Several different planarization requirements.

TABLE 8.11 Children of FR$_{24}$ and DP$_{24}$

	Functional Requirements (FRs)	**Design Parameters (DPs)**
24.1	Allow process "recipes" to be processed	Flexible processing of machine state sequences
24.2	Interface with measurement data	Metrology module
24.3	Track wafer processing	Wafer database
24.4	Control subsystem operation	Subsystem control software
24.5	Allow machine setup and calibration	Setup and calibration module
24.6	Allow flexible machine operation	Machine operation interface

TABLE 8.12 Constraints C_{24x}

Constraint Table		Impacts: FR_{24x}					
Index	Description	1	2	3	4	5	6
Critical Performance Specifications							
C_{244a}	Speed—the subsystem control must be able to process the servo loops as quickly as necessitated by the respective subsystem. This constraint may lead to the use of two computers—one for the user interface and high-level machine operations, and one for low-level control loops and algorithms				—		
Operational Constraints							
C_{242a}	Nova compatibility—due to widespread use of the Nova 210/420 measurement system, compatibility is desired		—				
C_{243a}	Parameters to track include the wafer ID number along with any associated wafer metrology and information on any polishing processes that have been run on the wafer			—			
C_{246b}	Provide control of all machine functions						—
C_{246c}	Display/record all process parameters						—

Description of DP_{24x}

- DP_{241}: Flexible processing of machine states is the manner in which the software deals with process recipes. To run a particular process on a wafer, the machine must cycle through a series of states, each of which is a combination of machine parameters. Each state may also have duration associated with it. Flexible processing of machine states (DP_{241}) affects the following FRs:

   FR_{243}: The manner in which the machine sequences are created and used will affect how the wafer processing is tracked.
   FR_{246}: The manner in which the machine sequences are created and used will affect how they are accessed by the machine operation interface.

- DP_{242}: The metrology module is responsible for interfacing with any available method for determining wafer-coating thickness. This may include a Nova metrology module or off-line metrology data. The metrology module (DP_{242}) affects the following FRs:

   FR_{243}: The manner in which the metrology module deals with data will affect how those data are tracked by the wafer database.
   FR_{246}: The manner in which the metrology module deals with data will affect how those data are displayed by the operation interface.

- DP_{243}: The wafer database is a software module designed to track all parameters relating to an individual wafer. This may include wafer ID number, metrology data, and process data. The wafer database is coupled to the following FR:

FR_{246}: The manner in which the database tracks wafer information will affect how this information is displayed in the operation interface.

- DP_{244}: The subsystem control software is the layer of interface from individual axes to the machine operation interface. It consists of those functions that are common to multiple machine elements, and so may be shared. The subsystem control software affects the following FR:

FR_{246}: The parameters available for control and associated value ranges affect the interface to the operation interface.

- DP_{245}: The setup and calibration module is a part of the software designed to facilitate tool installation and maintenance. It will allow all software parameters to be adjusted, providing correct scaling and zeroing of sensor values.
- DP_{246}: The machine operation interface is the screen with which the operator has direct contact. Therefore, it must provide access to all other software modules. The interface must also display all critical machine status information for the operator to review. The interface must enable the use of all other software modules.

Decomposition of FR_{25} *and* DP_{25}. We may decompose FR_{25} (Support Machine Operation) and DP_{25} (Support Subsystems) as shown in Table 8.13.

The design equation is given by

$$\begin{Bmatrix} FR_{251} \\ FR_{252} \\ FR_{253} \\ FR_{254} \\ FR_{255} \\ FR_{256} \end{Bmatrix} = \begin{bmatrix} X & 0 & 0 & 0 & 0 & 0 \\ X & X & 0 & 0 & 0 & 0 \\ 0 & 0 & X & 0 & 0 & 0 \\ 0 & 0 & 0 & X & 0 & 0 \\ X & X & X & X & X & 0 \\ X & X & X & X & X & X \end{bmatrix} \begin{Bmatrix} DP_{251} \\ DP_{252} \\ DP_{253} \\ DP_{254} \\ DP_{255} \\ DP_{256} \end{Bmatrix} \tag{8.17}$$

The constraints are given by the following table:

Constraint Table		Impacts: FR_{25x}				
Index	Description	1	2	3	4	5
	Operational Constraints					
C252a	The materials used by the machine should be compatible with existing fabrication plant services	—				
C253a	Simple operation to allow use by gloved and otherwise physically encumbered operators	—				

Description of FR_{25x} *and* DP_{25x}

- DP_{251}: The motion system hardware is the hardware that is common to any of the machine motion modules. DP_{251} affects the following FRs:

FR_{252}: The raw materials that must be supplied are determined in part by the motion system hardware. These materials may include electrical and fluidic power.

FR_{255}: The particular access requirements must be determined by the design of the motion system hardware.

TABLE 8.13 Children of FR_{25} and DP_{25}

	Functional Requirements (FRs)	Design Parameters (DPs)
251	Enable motion system	Motion system hardware
252	Provide raw materials	Material supply systems
253	Enable user interface	User interface hardware
254	Dispose of waste	Waste disposal system
255	Allow physical access	Physical configuration
256	Provide mechanical support	Machine structure

FR_{256}: The hardware selected for the motion system may place demands on the machine structure for physical space as well as loads.

- DP_{252}: The material supply systems are those systems that supply the raw material and power needed by any other machine subsystem. This includes electrical, fluidic, and chemical systems. DP_{252} affects the following FRs:

    FR_{255}: The particular access requirements must be determined by the design of the material supply systems.

    FR_{256}: The space required by the supply system must be accounted for in the machine structure.

- DP_{253}: The operator interface hardware includes those devices that are necessary for operator contact with the machine. This may be a separate interface PC. This DP_{253} affects the following FRs:

    FR_{255}: The particular access requirements must be determined by the design of user interface hardware.

    FR_{256}: The user interface hardware requires support for its mass.

- DP_{254}: The waste disposal system must be capable of maintaining any necessary separation of materials as well as dealing with the abrasive and chemical nature of the materials. DP_{254} affects FR_{255} and FR_{256} in the following way:

    FR_{255}: The particular access requirements must be determined by the design of the waste disposal system.

    FR_{256}: The space requirements must be accounted for in the design of the machine structure.

- DP_{255}: The physical configuration of the machine is defined as the selection and layout of the machine components with which the operator and fab must interface to facilitate whichever tasks are required. DP_{255} affects FR_{256}.

    FR_{256}: The physical configuration of the machine highly influences the machine structure. The overall shape of the structure is determined by requirements from the physical configuration.

- DP_{256}: The machine structure is the base of the machine, which supports the various systems of the machine. It needs to support the static loads from gravity as well as process induced loading and dynamic loading.

Decomposition of FR_{31} *and* DP_{31}. FR_{31} (Minimize Material Costs) and DP_{31} (Optimized Consumables Use) may be decomposed as shown in Table 8.14

TABLE 8.14 Children of FR$_{31}$ and DP$_{31}$

	Functional Requirements (FRs)	Design Parameters (DPs)
311	Minimize slurry consumption	Optimized slurry delivery
312	Minimize pad wear	Optimized wafer motion

where

$$\text{Material cost} = \text{Slurry} + \text{Pad} + \text{DIW} + \text{N}_2 + \text{Electricity} + \text{etc.}$$

The design equation may be written as

$$\begin{Bmatrix} \text{FR}_{311} \\ \text{FR}_{312} \end{Bmatrix} = \begin{bmatrix} X & 0 \\ 0 & X \end{bmatrix} \begin{Bmatrix} \text{DP}_{311} \\ \text{DP}_{312} \end{Bmatrix} \tag{8.18}$$

Description of DP$_{31x}$

- DP$_{311}$: Optimized slurry delivery is a dispensing method that allows the highest efficiency of slurry usage.
- DP$_{312}$: Optimized wafer motion is the oscillation of the wafer from its nominal offset. This uses a greater fraction of the pad area, extending the pad life.

Decomposition of FR$_{32}$ *and* DP$_{32}$. FR$_{32}$ (Minimize Operational Activities costs) and DP$_{32}$ (Target Performance of Operational Activities) may be decomposed as shown in Table 8.15 where

$$\text{OAC} = \text{Transport} + \text{Setup} + \text{Manual operations}$$

$$\begin{Bmatrix} \text{FR}_{321} \\ \text{FR}_{322} \\ \text{FR}_{323} \end{Bmatrix} = \begin{bmatrix} X & 0 & 0 \\ 0 & X & 0 \\ 0 & 0 & X \end{bmatrix} \begin{Bmatrix} \text{DP}_{321} \\ \text{DP}_{322} \\ \text{DP}_{323} \end{Bmatrix} \tag{8.19}$$

Description of DP$_{32x}$

- DP$_{321}$: Product flow oriented layout concerns the layout of machine elements to reduce transport time into and out of the tool as well as part flow within the tool.
- DP$_{322}$: Automated wafer handling is the use of robotics to move wafers out of the cassettes into the polishing process, and then through a cleaning cycle back to the output cassette.
- DP$_{323}$: A cached load/unload station acts as a decoupler between process stages, eliminating any delay between the head unloading a processed wafer and loading a new one.

Decomposition of FR$_{53}$ *and* DP$_{53}$. FR$_{53}$ (Maximize Yield) and DP$_{53}$ (Scrap Prevention Mechanisms) may be decomposed as shown in Table 8.16.

Yield is defined as

TABLE 8.15 Children of FR$_{32}$ and DP$_{32}$

	Functional Requirements (FRs)	Design Parameters (DPs)
321	Reduce transport costs	Product flow oriented layout
322	Reduce manual operations	Automated wafer handling
323	Reduce setup costs	Cached load/unload station

TABLE 8.16 FR_{535} and DP_{535}

	Functional Requirements (FRs)	Design Parameters (DPs)
531	Minimize overpolish—ξ (die %)	Accurate EPD
532	Minimize breakage—κ (wafer %)	Wafer retention sensor

$$\text{Yield} = (1 - \xi)(1 - \kappa)$$

The design equation is given by

$$\begin{Bmatrix} FR_{531} \\ FR_{532} \end{Bmatrix} = \begin{bmatrix} X & 0 \\ 0 & X \end{bmatrix} \begin{Bmatrix} DP_{531} \\ DP_{532} \end{Bmatrix} \tag{8.20}$$

Description of DP_{53x}

- DP_{531}: Accurate end-point detection is the method used to stop polishing. By quickly and accurately determining when the wafer is finished polishing, the number of overpolished dies may be reduced.
- DP_{532}: The wafer retention sensor is an optical sensor placed to indicate the release of a wafer from the wafer carrier during polishing, so the machine motions may be stopped before the wafer is broken.

Decomposition of FR_{211} *and* DP_{211}. FR_{211} (Remove Material) and DP_{211} (Abrasive Removal Mechanism) may be decomposed as shown in Table 8.17.

The DP_{21x} are shown in Figure 8.11. The figure shows the design embodiment at this stage of decomposition with DP_{21x} indicated. Some of these DPs need to be decomposed further.

The design equation is given by

$$\begin{Bmatrix} FR_{2111} \\ FR_{2112} \\ FR_{2113} \\ FR_{2114} \\ FR_{2115} \\ FR_{2116} \\ FR_{2117} \\ FR_{2118} \end{Bmatrix} = \begin{bmatrix} X & 0 & 0 & 0 & 0 & 0 & 0 & 0 \\ X & X & 0 & 0 & 0 & 0 & 0 & 0 \\ 0 & X & X & 0 & 0 & 0 & 0 & 0 \\ X & X & 0 & X & 0 & 0 & 0 & 0 \\ X & 0 & X & X & X & 0 & 0 & 0 \\ X & X & X & X & X & X & 0 & 0 \\ X & X & X & X & X & X & X & 0 \\ X & X & X & X & X & 0 & 0 & X \end{bmatrix} \begin{Bmatrix} DP_{2111} \\ DP_{2112} \\ DP_{2113} \\ DP_{2114} \\ DP_{2115} \\ DP_{2116} \\ DP_{2117} \\ DP_{2118} \end{Bmatrix} \tag{8.21}$$

The constraints are given in Table 8.18.

TABLE 8.17 Children of FR_{211} and DP_{211}

	Functional Requirements (FRs)	Design Parameters (DPs)
211.1	Create relative velocity	Rotary velocity system
211.2	Supply abrasive	Abrasive slurry
211.3	Hold wafer	Wafer retention system
211.4	Provide polishing surface	Polishing pad—porous PU
211.5	Apply normal pressure	Pressure application system
211.6	Control removal parameters	ARP controller
211.7	Support ARP	ARP subsystems
211.8	Maintain process temperature	Temperature control system

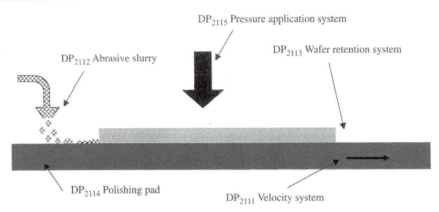

Figure 8.11 Design of material removal mechanism.

TABLE 8.18 Constraints C_{211}s

Constraint Table		Impacts: FR_{211x}				
Index	Description	1	2	3	4	5
	Critical Performance Specifications					
C_{2111a}	Uniform velocity profile—this requirement provides three options for motion: rotary, linear, and orbital	—				
C_{2111d}	Velocity range—there is some evidence that higher speed polishing leads to some better output characteristics. Therefore, the velocity system must be able to support speeds up to approximately 1000 feet/minute	—				
C_{2112a}	Minimize consumption of slurry		—			
C_{2113a}	Do not damage wafer			—		
C_{2113b}	Hold reliably—releasing the wafer from the carrier unintentionally is a severe hindrance to machine operation, and as such must be minimized			—		
C_{2113c}	Conform to wafer shape—incoming wafer shape is a variable that should be accommodated			—		
C_{2114a}	Force range—the force application system must be capable of delivering the necessary loads for polishing. Current predictions estimate a polishing pressure of up to 10 psi. For a 300-mm wafer, the required polishing force would therefore be 1100 lbf				—	
C_{2114b}	Force uniformity—the force application to the wafer should nominally be uniform across the wafer surface. This will lead to uniform removal over the wafer surface				—	
	Operational Constraints					
C_{2111b}	Vary velocity orientation with respect to wafer—this is necessary to prevent patterns forming in the polished surface from pad imperfections and a "smearing" tendency in the polishing process	—				
C_{2111c}	Polishing area—the wafer should use all available pad area to maximize pad life and prevent polishing patterns due to pad nonuniformity	—				

continued

TABLE 8.18 Continued

Constraint Table		Impacts: FR_{211x}				
Index	Description	1	2	3	4	5
	Critical Performance Specifications					
C_{2112b}	Compatibility—the abrasive system should support the use of commercially available compositions	—				
C_{2112c}	Flexibility—the abrasive system should be capable of supplying several different abrasive mixtures independently	—				
C_{2113d}	Contamination rejection—the wafer carrier should have means for preventing the contamination from polishing slurry. If the slurry is allowed to dry on a surface, it has a tendency to agglomerate and cause problems when released			—		
C_{2114c}	Force uniformity control—the ability to cause the force profile to deviate from the nominal pressure allows the machine to compensate for other parameters that may cause nonuniform polishing. Most of these influences are realized as radial variations across the wafer surface, so this degree of control is primary, allowing concentric rings of uniform pressure			—		
C_{2115a}	Compatibility with preceding DPs—the specific polishing pad must fit on the velocity system selected and be able to use the slurry selected					—
C_{2115b}	Lifetime—the number of wafers possible to polish on a pad affects the machine availability					—

Description of DP_{211x}

- DP_{2111}: Relative velocity must be applied between the wafer surface and the polishing surface or pad. The primary constraint in applying velocity to the interface is the uniformity of velocity profile. The rotary velocity system (DP_{2111}) affects the following FRs:

  FR_{2112}: The manner in which the abrasive is introduced to the polishing interface is affected by the velocity system chosen.

  FR_{2114}: The material used for a polishing surface must be compatible with the velocity system.

  FR_{2115}: The manner in which the force is supported by the polishing surface is affected by the velocity system. A rotary or orbital system may be supporting a large platen while a linear system supports a thin belt.

  FR_{2116}: Parameters available for control are determined by the design of the velocity system.

  FR_{2117}: Required subsystems are determined by the design of the velocity system.

  FR_{2118}: The means for removing heat generated during polishing depend on the configuration of the velocity system

- DP_{2112}: The abrasive slurry is the medium used to supply abrasive to the polishing process. The abrasive system (DP_{2112}) affects the following FRs:

FR$_{2114}$: The polishing pad must be compatible with the chemistry and particle content of the abrasive slurry.

FR$_{2118}$: The amount of heat generated during the process and the amount of heat removed by the slurry depend on the slurry used and its initial temperature.

- DP$_{2113}$: The wafer retention system is the device that holds the wafer for processing. In this application, it must be capable of automatically receiving the wafer, reliably holding it for processing, and then passing the wafer to the wafer handler. The wafer retention system (DP$_{2113}$) affects the following FR:

FR$_{2118}$: The ability to add/remove heat through the wafer carrier is determined by the design of the wafer carrier.

- DP$_{2114}$: The polishing pad is defined as the upper surface of any such pad—that part that makes contact with the wafer surface. The polishing pad affects the following FR:

FR$_{2118}$: The ability to add/remove heat through the platen surface depends on the specifications of the polishing pad.

- DP$_{2115}$: The pressure application system is the manner in which normal force is created on the wafer surface to enable removal. This means some manner of loading the wafer and a manner of supporting the polishing pad. The pressure application system (DP$_{2115}$) affects the following FR:

FR$_{2118}$: The ability to add/remove heat through the wafer backside depends on the design of the force application system.

- DP$_{2116}$: The ARP controller coordinates the various axes of the process to ensure that the desired process recipe is followed.
- DP$_{2117}$: ARP subsystems are those systems required to run the removal process.
- DP$_{2118}$: The cooling system is the system designed to remove the heat of polishing from the polishing area. It is likely to use a chilled water heat exchanger and a fluid circulating through an internal cavity of the platen.

Decomposition of FR$_{212}$ *and* DP$_{212}$. FR$_{212}$ (Enable Two-Step Processes) and DP$_{212}$ (Multiple ARP Design) may be decomposed as shown in Table 8.19.
The design equation is given by

$$\begin{Bmatrix} FR_{2121} \\ FR_{2122} \end{Bmatrix} = \begin{bmatrix} X & 0 \\ 0 & X \end{bmatrix} \begin{Bmatrix} DP_{2121} \\ DP_{2122} \end{Bmatrix} \tag{8.22}$$

Description of DP$_{212x}$

- DP$_{2121}$: The second polishing platen allows the use of multistep processes, where each of the polishing pads must maintain a separate and different chemistry.

TABLE 8.19 Children of FR$_{212}$ and DP$_{212}$

	Functional Requirements (FRs)	**Design Parameters (DPs)**
2121	Provide multiple removal processes	Second polishing platen
2122	Clean wafer between steps	Wafer rinse station

- DP_{2122}: The wafer rinse station prevents the contamination of polishing stages from earlier stages, rinsing the bulk of slurry particles and chemistry off the wafer.

Decomposition of FR_{251} *and* DP_{251}. FR_{251} (Enable Motion System) and DP_{251} (Motion System Hardware) may be decomposed as shown in Table 8.20.

The design equation may be written as

$$
\begin{Bmatrix} FR_{2511} \\ FR_{2512} \\ FR_{2513} \\ FR_{2514} \\ FR_{2515} \\ FR_{2516} \end{Bmatrix} = \begin{bmatrix} X & 0 & 0 & 0 & 0 & 0 \\ X & X & 0 & 0 & 0 & 0 \\ X & 0 & X & 0 & 0 & 0 \\ X & 0 & 0 & X & 0 & 0 \\ X & 0 & 0 & 0 & X & 0 \\ X & 0 & 0 & 0 & 0 & X \end{bmatrix} \begin{Bmatrix} DP_{2511} \\ DP_{2512} \\ DP_{2513} \\ DP_{2514} \\ DP_{2515} \\ DP_{2516} \end{Bmatrix}
\tag{8.23}
$$

Description of DP_{251x}

- DP_{2511}: DSP-based architecture describes the product selected for the control system computer. The product is a Kiethley Metrabyte ADWIN Pro system.
- DP_{2512}: The data acquisition board is a plug in module designed to sample up to 8 channels at 16 bits with a sample rate up to 100 kHz.
- DP_{2513}: The digital input channels are those channels of the digital interface board that are configured for input. There are a total of 32 digital lines, each of which may be input or output.
- DP_{2514}: The counter input board is a collection of timers and counters intended to take repetitive digital inputs such as optical encoders and provide position or count information to the central computer.
- DP_{2515}: The analog output board is a card with 8 channels of 16 bit analog output.
- DP_{2516}: The digital output channels are those channels of the digital interface board that are configured for output. There are a total of 32 digital lines, each of which may be input or output.

Decomposition of FR_{252} *and* DP_{252}. FR_{252} (Provide Raw Materials) and DP_{252} (Material Supply System) may be decomposed as shown in Table 8.21.

The design equation is as follows:

$$
\begin{Bmatrix} FR_{2521} \\ FR_{2522} \\ FR_{2523} \\ FR_{2524} \\ FR_{2525} \end{Bmatrix} = \begin{bmatrix} X & 0 & 0 & 0 & 0 \\ 0 & X & 0 & 0 & 0 \\ 0 & 0 & X & 0 & 0 \\ 0 & 0 & 0 & X & 0 \\ 0 & 0 & 0 & 0 & X \end{bmatrix} \begin{Bmatrix} DP_{2521} \\ DP_{2522} \\ DP_{2523} \\ DP_{2524} \\ DP_{2525} \end{Bmatrix}
\tag{8.24}
$$

TABLE 8.20 Children of FR_{251} and DP_{251}

	Functional Requirements (FRs)	**Design Parameters (DPs)**
251.1	Run control software	DSP-based architecture
251.2	Acquire analog signals	Data acquisition board
251.3	Acquire digital signals	Digital input channels
251.4	Acquire and interpret encoder signals	Counter input board
251.5	Output analog command signals	Analog output board
251.6	Output digital command signals	Digital output channels

TABLE 8.21 Children of FR_{252} and DP_{252}

	Functional Requirements (FRs)	Design Parameters (DPs)
252.1	Supply electrical power	Electrical distribution system
252.2	Supply clean water	Water filtration and distribution system
252.3a	Supply clean pressurized gas	Compressed N_2 distribution system
252.3b		Compressed air filtration and distribution system
252.4	Supply subatmospheric pressure	Vacuum system
252.5	Supply abrasive slurry	Slurry distribution system

Description of $DP_{252}s$

- DP_{2521}: The electrical distribution system is the main power line of the machine, likely a 220 V ac supply. It includes the distribution of this power to the machine systems as 220 V ac, 110 V ac, and various voltage levels of dc.
- DP_{2522}: The water filtration and distribution systems must include the connections to a water supply. Depending on the conditions of that supply, the water may have to be filtered and pressurized. The distribution to the machine systems should happen through some network of tubing. The axis drive unit handles individual switching, which might be a solenoid valve for water distribution.
- DP_{2523a}: The machine may use compressed nitrogen in its pneumatic systems. The use of nitrogen would require the inclusion of a main pressure regulator. Individual drive units handle control to systems, which may be a proportional pressure valve or a solenoid valve.
- DP_{2524}: Vacuum is used in machine systems, and must be distributed for use in a manner similar to the pressure system. The vacuum has the additional requirement of removing any acquired moisture or particle content from the vacuum to prevent them from entering the main supply system since the flow of material is away from the machine.
- DP_{2525}: The slurry distribution system must handle the machine's interface with external slurry. In a production machine, this would be via bulkhead connectors to the fab distribution service.

Decomposition of FR_{255} *and* DP_{255}. FR_{255} (Allow Physical Access) and DP_{255} (Physical Configuration) may be decomposed as shown in Table 8.22.

The design equation may be written as

$$\begin{Bmatrix} FR_{2551} \\ FR_{2552} \\ FR_{2553} \\ FR_{2554} \\ FR_{2555} \end{Bmatrix} = \begin{bmatrix} X & 0 & 0 & 0 & 0 \\ 0 & X & 0 & 0 & 0 \\ 0 & 0 & X & 0 & 0 \\ 0 & X & 0 & X & 0 \\ 0 & 0 & 0 & 0 & X \end{bmatrix} \begin{Bmatrix} DP_{2551} \\ DP_{2552} \\ DP_{2553} \\ DP_{2554} \\ DP_{2555} \end{Bmatrix} \tag{8.25}$$

Description of DP_{255x}

- DP_{2551}: The cassette holder is a device responsible for holding a wafer cassette in a rigid position so the wafer handler may load and unload wafers.
- DP_{2552}: The touchscreen allows a gloved operator to use the machine without the flat

TABLE 8.22 Children of FR_{255} and DP_{255}

	Functional Requirements (FRs)	**Design Parameters (DPs)**
255.1	Provide cassette interface	Cassette holder
255.2	Allow GUI input	Touchscreen
255.3	Allow data input	Keyboard
255.4	Supply machine information	Front panel display
255.5	Allow easy pad change	"Kinematic" platen interchange system

horizontal surface necessary for a mouse, and simplifies the operation of the software. The touchscreen is coupled to the following FR:

FR_{2554}: The use of a touchscreen requires a compatible display device. Therefore, the touchscreen must be selected first, so that improper display selection prevents the use of a touchscreen.

- DP_{2553}: The keyboard is a common, easy to use data input device.
- DP_{2554}: The front panel display is a computer screen to display machine information to the user. This would be a flat panel display in a production machine to help the machine meet footprint requirements, but will be a CRT in the alpha machine to keep costs down.
- DP_{2555}: In a rotary system, it may be desirable to use different pads before the pad life has expired. Also, to reduce the time involved with pad change, the procedure may be done off-line on a secondary platen, which is just swapped for the removed platen.

Decomposition of FR_{2111} *and* DP_{2111}. FR_{2111} (Create Relative Velocity) and DP_{2111} (Rotary Velocity System) may be decomposed as shown in Table 8.23.

The design equation may be written as

$$\begin{Bmatrix} FR_{21111} \\ FR_{21112} \\ FR_{21113} \\ FR_{21114} \end{Bmatrix} = \begin{bmatrix} X & 0 & 0 & 0 \\ 0 & X & 0 & 0 \\ 0 & 0 & X & 0 \\ X & X & X & X \end{bmatrix} \begin{Bmatrix} DP_{21111} \\ DP_{21112} \\ DP_{21113} \\ DP_{21114} \end{Bmatrix} \tag{8.26}$$

The constraints are given in Table 8.24.
Descriptions of $DP_{2111}s$

- DP_{21111}: The platen rotation system is the means for rotating the polishing pad relative to the machine frame. The platen rotation system (DP_{21111}) affects the following FR:

FR_{21114}: The parameters available for control and the range of values are determined by the design of the platen rotation system.

TABLE 8.23 Children of FR_{2111} and DP_{2111}

	Functional Requirements (FRs)	**Design Parameters (DPs)**
2111.1	Rotate platen	Platen rotation system
2111.2	Rotate wafer	Wafer rotation system
2111.3	Set offset	Radial X-axis
2111.4	Control parameters	Velocity control software

TABLE 8.24 Constraints C_{21113}

Constraint Table			Impacts: FR_{211x}			
Index	Parent	Description	1	2	3	4
		Critical Performance Specifications				
C_{21113a}		Acceleration		—		
C_{21113b}		Velocity		—		
C_{21113c}		Resolution		—		
		Operational Constraints				
C_{21113d}		Support vertical load $\sim$1000 lb		—		
C_{21113e}		Support lateral load $\sim$500 lb		—		

- DP_{21112}: The wafer rotation system is the means for rotating the wafer relative to the machine frame. The wafer rotation system (DP_{21112}) affects the following FR:

  FR_{21114}: The parameters available for control and the range of values are determined by the design of the wafer rotation system.

- DP_{21113}: The radial X-axis is the means for moving the wafer in the radial direction relative to the polishing pad. In the proposed design, the wafer carrier remains centered on the axis between the polishing pads. Therefore, motion of the wafer is always radial with respect to the pad. This DP affects the following FR:

  FR_{21114}: The parameters available for control and the range of values are determined by the design of the X-axis.

- DP_{21114}: The velocity control software is the means for coordinating the velocity of the wafer, the polishing pad, and the offset between the two, so as to create a relative velocity between wafer and pad.

Decomposition of FR_{2112} *and* DP_{2112}. FR_{2112} (Supply Abrasive) and DP_{2112} (Abrasive Slurry) may be decomposed as shown in Table 8.25.

The design equation may be written as

$$\begin{Bmatrix} FR_{21121} \\ FR_{21122} \\ FR_{21123} \end{Bmatrix} = \begin{bmatrix} X & 0 & 0 \\ X & X & 0 \\ X & X & X \end{bmatrix} \begin{Bmatrix} DP_{21121} \\ DP_{21122} \\ DP_{21123} \end{Bmatrix} \tag{8.27}$$

Description of DP_{2112x}

- DP_{21121}: The slurry chemistry is whatever chemical composition the slurry has, used to affect the wafer-coating surface in a desired manner. The slurry chemistry (DP_{21121}) affects the following FRs:

TABLE 8.25 Children of FR_{2112} and DP_{2112}

	Functional Requirements (FRs)	Design Parameters (DPs)
2112.1	Chemically treat wafer surface	Slurry chemistry
2112.2	Remove wafer material	Abrasive particles
2112.3	Transport particles	Liquid viscosity

FR$_{21122}$: The chemistry of the slurry affects how wafer material is removed.

FR$_{21123}$: The ability to transport particles may be affected by the chemical nature of the slurry.

- DP$_{21122}$: The abrasive particles used in the slurry make up the third body in the removal process. They are responsible for the mechanical material removal, and may be used to optimize this part of the removal process. The abrasive particles (DP$_{21122}$) are coupled to the following FR:

FR$_{21123}$: The ability to transport the particles depends on the particles being transported. The primary particle is the abrasive used for removal. Other particles in the system may be worn pad material and worn wafer-coating material.

- DP$_{21123}$: The liquid viscosity may be used to affect the slurry's ability to transport particles. The viscosity will directly affect the thickness of any fluid film in the polishing interface. It is this fluid film that will be used to transport the particles.

Decomposition of FR$_{2113}$ *and* DP$_{2113}$. FR$_{2113}$ (Hold Wafer) and DP$_{2113}$ (Wafer Retention System) may be decomposed as shown in Table 8.26.

The DP$_{2113x}$ are shown in Figure 8.12.

The design equation may be written as

$$\begin{Bmatrix} FR_{21131} \\ FR_{21132} \\ FR_{21133} \\ FR_{21134} \\ FR_{21135} \\ FR_{21136} \end{Bmatrix} = \begin{bmatrix} X & 0 & 0 & 0 & 0 & 0 \\ X & X & 0 & 0 & 0 & 0 \\ X & 0 & X & 0 & 0 & 0 \\ 0 & 0 & 0 & X & 0 & 0 \\ X & 0 & 0 & 0 & X & 0 \\ 0 & X & X & 0 & X & X \end{bmatrix} \begin{Bmatrix} DP_{21131} \\ DP_{21132} \\ DP_{21133} \\ DP_{21134} \\ DP_{21135} \\ DP_{21136} \end{Bmatrix} \tag{8.28}$$

Description of DP$_{21131}$

- DP$_{21131}$: The retaining ring is a device for surrounding the wafer and trapping it between the polishing pad and the carrier film, so that polishing pressure may be applied. The retaining ring (DP$_{21131}$) is coupled to the following FRs:

FR$_{21132}$: The load cycle depends on how the wafer handler must interface with the retaining ring.

FR$_{21133}$: The unload cycle depends on how the wafer handler must interface with the retaining ring

FR$_{21135}$: The amount of motion needed to gain access is influenced by the retaining ring design.

TABLE 8.26 Children of FR$_{2113}$ and DP$_{2113}$

	Functional Requirements (FRs)	**Design Parameters (DPs)**
2113.1	Prevent wafer translation	Retaining ring
2113.2	Load wafer	Bladder loading configuration
2113.3	Eject wafer	Bladder unloading configuration
2113.4	Prevent wafer rotation relative to carrier	Carrier film surface
2113.5	Allow access to wafer	Z-axis travel length
2113.6	Control parameters	Wafer retention control software

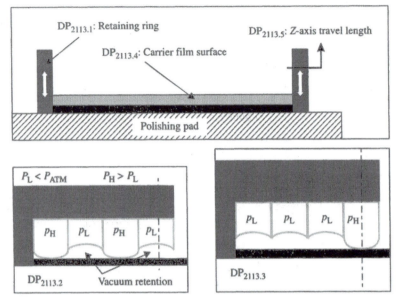

Figure 8.12 Design of the wafer retention system that consists of DP_{2113x}.

- DP_{21132}: The bladder-loading configuration is shown in a schematic. It is the means of creating multiple areas of suction on the wafer's top surface so that a net vertical force may be applied. The loading configuration (DP_{21132}) is coupled to the following FR:

  FR_{21136}: The control software must be designed to correctly control the pressure/vacuum configuration.

- DP_{21133}: The bladder-unloading configuration is shown in a schematic. It is the means of pushing the wafer out of the wafer carrier. The unloading configuration (DP_{21133}) affects FR_{21136} as follows:

  FR_{21136}: The control software must be designed to correctly control the pressure/vacuum configuration.

- DP_{21134}: The compliant carrier film is a porous material designed to provide a high friction with the wafer back surface. This friction will prevent the wafer rotation.
- DP_{21135}: The Z-axis travel length is a vertical axis of motion designed to allow the wafer to be lifted for loading and unloading. The unloading configuration (DP_{21135}) affects the following FR:

  FR_{21136}: The control software must be designed to correctly control the wafer carrier vertical position.

- DP_{11136}: The control software is responsible for coordinating the actions of all the wafer carrier systems and adjusting available parameters to meet the machine needs.

Decomposition of FR_{2114} *and* DP_{2114}. FR_{2114} (Provide Polishing Surface) and DP_{2114} (Polishing Pad—Porous Polyurethane) may be decomposed as shown in Table 8.27.

TABLE 8.27 Children of FR_{2114} and DP_{2114}

	Functional Requirements (FRs)	Design Parameters (DPs)
2114.1	Maintain slurry flow in polishing interface	Surface topology
2114.2	Create local force variation	Device scale pad stiffness
2114.3	Maintain uniform characteristics	Pad conditioning scheme

The design equation may be written as

$$\begin{Bmatrix} FR_{21141} \\ FR_{21142} \\ FR_{21143} \end{Bmatrix} = \begin{bmatrix} X & 0 & 0 \\ X & X & 0 \\ X & X & X \end{bmatrix} \begin{Bmatrix} DP_{21141} \\ DP_{21142} \\ DP_{21143} \end{Bmatrix} \tag{8.29}$$

The constraints are as shown in Table 8.28.

Description of DP_{2114x}

- DP_{21141}: The surface topology of the pad consists of those features used to draw slurry flow into and out of the polishing interface. This flow should reach all parts of the wafer evenly. The surface topology (DP_{21141}) affects the following FRs:

    FR_{21142}: The means for maintaining the surface of the pad depends on what that surface is. Therefore, any pad conditioning parameters must be selected after the pad surface.

    FR_{21143}: The geometry of the pad surface will play a role in its stiffness.

- DP_{21142}: The device scale pad stiffness is what creates a local force variation on the wafer-coating surface. The device-scale pad stiffness (DP_{21142}) affects the following FR:

    FR_{21143}: The desired pad stiffness may affect the strategy for maintaining a uniform surface. Therefore, the conditioning system should be designed after the pad surface is secured.

- DP_{21143}: The conditioning scheme is the means for controlling the characteristics of the pad. The characteristics that must be maintained are summarized in the constraint Table 8.28.

Decomposition of FR_{2115} *and* DP_{2115}. FR_{2115} (Apply Normal Pressure) and DP_{2115} (Pressure Application System) may be decomposed as shown in Table 8.29.

Figure 8.13 shows the design of the pressure application system with DP_{2115x}.

TABLE 8.28 Children of C_{2114x}

Constraint Table		Impacts: FR._		
Index	Description	1	2	3
	Critical Performance Specifications			
C_{21141a}	Chemical inertness	—		
C_{21142a}	Pad life		—	

TABLE 8.29 Children of FR$_{2115}$ and DP$_{2115}$

	Functional Requirements (FRs)	Design Parameters (DPs)
2115.1	Provide pressure	Nominal compartment pressure
2115.2	Comply to wafer thickness variation	Pad/wafer/membrane compliance (mm length scale)
2115.3	Comply to wafer out-of-flatness	Pad/wafer/membrane compliance (cm length scale)
2115.4	Control radial pressure variation	Active membrane
2115.5	Control edge effects	Retaining ring pad compression
2115.6	Control parameters	Profile adjustment software module
2115.7	Maintain wafer-pad orientation	Polishing load support structure

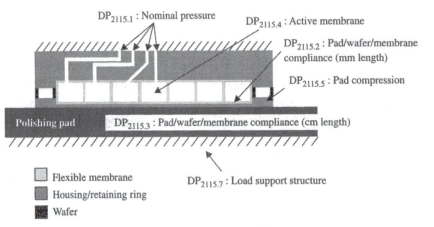

Figure 8.13 The design of the pressure application system with DP$_{2115x}$.

The design equation is given by

$$\begin{Bmatrix} FR_{21151} \\ FR_{21152} \\ FR_{21153} \\ FR_{21154} \\ FR_{21155} \\ FR_{21156} \\ FR_{21157} \end{Bmatrix} = \begin{bmatrix} X & 0 & 0 & 0 & 0 & 0 & 0 \\ 0 & X & 0 & 0 & 0 & 0 & 0 \\ 0 & X & X & 0 & 0 & 0 & 0 \\ 0 & 0 & X & X & 0 & 0 & 0 \\ 0 & 0 & 0 & 0 & X & 0 & 0 \\ X & 0 & 0 & X & X & X & 0 \\ X & 0 & 0 & 0 & X & 0 & X \end{bmatrix} \begin{Bmatrix} DP_{21151} \\ DP_{21152} \\ DP_{21153} \\ DP_{21154} \\ DP_{21155} \\ DP_{21156} \\ DP_{21157} \end{Bmatrix} \qquad (8.30)$$

Description of DP$_{2115x}$

- DP$_{21151}$: Compartment pressure is the pressurized gas supplied to the bladder compartments. This pressure is controlled with an E/P valve, using a control loop within the valve. Compartment pressure (DP$_{21151}$) is coupled to the following FRs:

  FR$_{21156}$: The range of control offered to the control software will depend on the control valve used.

  FR$_{21157}$: The range of loads applied by the compartment pressure determines what loads must be supported by the support structure.

- DP$_{21152}$: Membrane compliance allows for thickness variation in the wafer to have minimal effect on the front side pressure. Small length-scale stack compliance (DP$_{21152}$) affects the following FR:

FR_{21153}: The longer length-scale stack compliance is affected by its compliance on a shorter scale.

- DP_{21153}: Increased wafer-scale pad compliance has been shown to contribute to decreased within-wafer nonuniformity. Therefore, the pad and its support will be designed with this compliance in mind. The wafer-scale pad compliance affects the following FR:

FR_{21154}: The compliance of the pad affects how the pressure profile may be controlled. If the pad is overly compliant, the stiffness of the wafer may dominate the stack, preventing the control of the pressure distribution.

- DP_{21154}: The active membrane compartments are a means for applying a pattern of displacement in concentric rings to the wafer front surface. With this displacement, the wafer front side will see a variation in normal pressure due to the compression of the polishing pad. The membrane compartments (DP_{21154}) affect the following FR:

FR_{21156}: Parameters available for control and the available values are determined by the design of the active membrane.

- DP_{21155}: The retaining ring pad compression is defined as the deformation of the polishing pad as it enters the polishing interface. The retaining ring/pad contact (DP_{21155}) affects the following FRs:

FR_{21156}: Parameters available for control and the available values are determined by the design of the ring.

FR_{21157}: The retaining ring pad compression is affected by the orientation of the wafer and pad, and so the requirements for this orientation are partially set by the requirements of the retaining ring compression.

- DP_{21156}: The profile adjustment software module is responsible for varying the necessary parameters to meet the machine operation requirements.
- DP_{21157}: The polishing load support structure is the collection of necessary features that enables the machine to support the applied normal load as well as induced frictional loads.

Decomposition of FR_{2117} *and* DP_{2117}. FR_{2117} (Support ARP) and DP_{2117} (ARP Support Subsystems) may be decomposed as shown in Table 8.30.

The design equation is given by

$$\begin{Bmatrix} FR_{21171} \\ FR_{21172} \end{Bmatrix} = \begin{bmatrix} X & 0 \\ 0 & X \end{bmatrix} \begin{Bmatrix} DP_{21171} \\ DP_{21722} \end{Bmatrix} \quad (8.31)$$

Description of DP_{2117x}

- DP_{21171}: The slurry dispensing system is the means for moving slurry from storage containers to the point of use at the polishing process.

TABLE 8.30 Children of FR_{2117} and DP_{2117}

	Functional Requirements (FRs)	Design Parameters (DPs)
21171	Deliver slurry	Slurry dispensing system
21172	Prevent slurry accumulation	ARP rinsing system

- DP$_{21172}$: The ARP rinsing system is the means for supplying water to the polishing process, primarily when not polishing, to prevent slurry from drying.

Decomposition of FR$_{21112}$ *and* DP$_{21112}$. FR$_{21112}$ (Rotate Wafer) and DP$_{21112}$ (Wafer Rotation System) may be decomposed as shown in Table 8.31.

The design equation may be written as

$$
\begin{Bmatrix} FR_{211121} \\ FR_{211122} \\ FR_{211123} \\ FR_{211124} \\ FR_{211125} \end{Bmatrix} = \begin{bmatrix} X & 0 & 0 & 0 & 0 \\ 0 & X & 0 & 0 & 0 \\ 0 & 0 & X & 0 & 0 \\ X & X & X & X & 0 \\ X & X & X & 0 & X \end{bmatrix} \begin{Bmatrix} DP_{211121} \\ DP_{211122} \\ DP_{211123} \\ DP_{211124} \\ DP_{211125} \end{Bmatrix} \tag{8.32}
$$

Description of FP$_{21112x}$ *and* DP$_{21112x}$

- DP$_{211121}$: Hydrostatic bearings provide smooth motion that may benefit the polishing process. Also, the pocket pressures in a hydrostatic bearing could be used to monitor the loads on the wafer rotation axis, and therefore on the wafer itself. These loads are useful in controlling the polishing process. The hydrostatic bearing (DP$_{211121}$) affects the following FRs:

  FR$_{211124}$: Parameters available for control and the available values are determined by the bearing design.
  FR$_{211125}$: The shaft must be designed to interface with the chosen bearings.

- DP$_{211122}$: The wafer rotation motor is a servomotor capable of supplying the necessary torque and speed to the wafer rotation axis. The wafer rotation motor affects the following FRs:

  FR$_{211124}$: Parameters available for control and the available values are determined by the motor selection.
  FR$_{211125}$: The shaft must be designed to interface with the chosen motor.

- DP$_{211123}$: A rotary encoder is used to measure the rotational speed of the wafer. The rotary encoder (DP$_{211123}$) affects the following FRs:

  FR$_{211124}$: The data supplied to the control software depend on the selection of the encoder.
  FR$_{211125}$: The shaft must be designed to interface with the chosen encoder.

- DP$_{211124}$: The wafer rotation software module is responsible for varying the necessary parameters to meet the machine operation requirements.

TABLE 8.31 Children of FR$_{2112}$ and DP$_{2112}$

	Functional Requirements (FRs)	Design Parameters (DPs)
21112.1	Allow wafer rotation	Rotary motion bearing
21112.2	Provide rotation torque	Wafer rotation motor
21112.3	Measure rotation speed	Rotary encoder
21112.4	Control parameters	Wafer rotation software module
21112.5	Transmit torque to wafer carrier	Upper spindle shaft

- DP_{211125}: The upper spindle shaft is designed to accommodate the requirements of the rest of the rotation system. It must fit the selected bearings and motor and encoder.

Note: The platen rotation system (DP_{21111}) is nearly identical to the wafer rotation system, including the same components at this level, so is not detailed.

Decomposition of FR_{21131} *and* DP_{21131}. FR_{21131} (Prevent Wafer Translation) and DP_{21131} (Retaining Ring) may be decomposed as shown in Table 8.32.

The design equation is given by

$$\begin{Bmatrix} FR_{211311} \\ FR_{211312} \\ FR_{211313} \end{Bmatrix} = \begin{bmatrix} X & 0 & 0 \\ 0 & X & 0 \\ 0 & X & X \end{bmatrix} \begin{Bmatrix} DP_{211311} \\ DP_{211312} \\ DP_{211313} \end{Bmatrix} \qquad (8.33)$$

The constraints at this level are shown in Table 8.33.

Figure 8.14 shows the design of the retaining ring with relevant DP_{21131}s indicated.

Description of $DP_{211131x}$ *and* $FR_{211131x}$

- DP_{211311}: The ring ID is the inner surface of the retaining ring, which contacts the edge of the wafer. It is this surface that provides the support to prevent wafer translation.
- DP_{211312}: The ring flexure is a continuous ring of material that will support the frictional loads of polishing while minimally influencing the contact pressure. The ring flexure (DP_{211312}) affects the following FRs:

FR_{211313}: The influence of the loads and other factors on the contact pressure is affected by the design of the ring flexure.

- DP_{211313}: The contact pressure consists of the interface conditions around the bottom surface of the ring. To maintain contact with the pad, the contact pressure must be maintained above a certain value.

Decomposition of FR_{21143} *and* DP_{21143}. FR_{21143} (Maintain Uniform Pad Characteristics) and DP_{21143} (Pad-Conditioning Scheme) may be decomposed as shown in Table 8.34.

TABLE 8.32 Children of FR_{21131} and DP_{21131}

	Functional Requirements (FRs)	Design Parameters (DPs)
21131.1	Provide barrier	Ring ID—compliant
21131.2	Support loads	Ring flexure
21131.3	Maintain contact with pad	Ring contact pressure

TABLE 8.33 Constraints C_{211313}

Constraint Table		Impacts: FR._	
Index	Description	1	2
	Critical Performance Specifications		
$C_{211313a}$	The contact pressure should be of sufficient magnitude to prevent the retaining ring lifting off the pad surface enough to allow the wafer to leave the carrier		
$C_{211313b}$	The contact pressure should be uniform over the contact area of the ring, to prevent any low-pressure areas that may allow the wafer to escape		

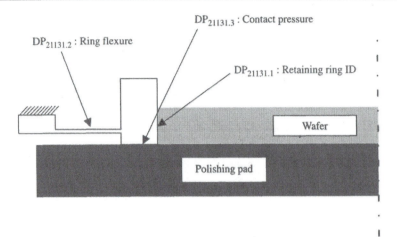

Figure 8.14 Design of the retaining ring with relevant DP_{11131x}.

TABLE 8.34 Children of FR_{21143} and DP_{21143}

	Functional Requirements (FRs)	Design Parameters (DPs)
21143.1	Maintain consistent surface	Rotary pad conditioner
21143.2	Control pad wear shape	Wafer offset oscillation
21143.3	Control parameters	Pad-conditioning system software

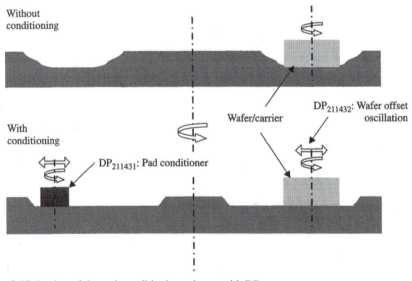

Figure 8.15 Design of the pad-conditioning scheme with DP_{21143x}.

Figure 8.15 shows the design of the pad-conditioning scheme with relevant DPs. The design equation is expressed as

$$\begin{Bmatrix} FR_{211431} \\ FR_{211432} \\ FR_{211433} \end{Bmatrix} = \begin{bmatrix} X & 0 & 0 \\ X & X & 0 \\ X & X & X \end{bmatrix} \begin{Bmatrix} DP_{211431} \\ DP_{211432} \\ DP_{211433} \end{Bmatrix} \qquad (8.34)$$

Description of DP_{21143x} *and* FR_{21143x}

- DP_{211431}: The pad conditioner is an abrasive mechanism used to roughen up the surface of the pad so it may carry slurry efficiently and consistently. The pad conditioner (DP_{211431}) affects the following FRs:

  FR_{211432}: The pad wear shape is controlled partially by the pad conditioner. As material is removed to maintain the surface, it will have some affect on the wear shape. The wafer-offset oscillation may be used to further influence this parameter.
  FR_{211433}: Parameters available for control and the available values are determined by the design of the pad conditioner.

- DP_{211432}: The wafer-offset oscillation is motion of the wafer in the radial pad direction. This is done to even the wear on the pad and prevent abrupt edge transitions. The wafer-offset oscillation (DP_{211432}) is coupled to the following FR:

  FR_{211433}: Parameters available for control and the available values are determined by the design of the oscillation mechanism.

- DP_{211433}: The pad-conditioning system software module is responsible for varying the necessary parameters to meet the machine operation requirements.

Decomposition of FR_{21154} *and* DP_{21154}. FR_{21154} (Control Radial Pressure Variation) and DP_{21154} (Active Membrane) may be decomposed as shown in Table 8.35.
  Figure 8.16 shows the design of active membrane with DP_{21154x}.
  The design equation for FR_{21154} and DP_{21154} is expressed as

$$\begin{Bmatrix} FR_{211541} \\ FR_{211542} \\ FR_{211543} \\ FR_{211544} \\ FR_{211545} \end{Bmatrix} = \begin{bmatrix} X & 0 & 0 & 0 & 0 \\ X & X & 0 & 0 & 0 \\ X & 0 & X & 0 & 0 \\ X & 0 & 0 & X & 0 \\ X & X & X & 0 & X \end{bmatrix} \begin{Bmatrix} DP_{211541} \\ DP_{211542} \\ DP_{211543} \\ DP_{211544} \\ DP_{211545} \end{Bmatrix} \qquad (8.35)$$

Description of DP_{21154x} *and* FR_{21154x}

- DP_{211541}: Membrane compartments are the annular sections of the bladder that define zones of the wafer where a particular pressure is to be applied. Membrane compartments (DP_{211541}) affect the following FRs:

TABLE 8.35 Children of FR_{21154} and DP_{21154}

	Functional Requirements (FRs)	Design Parameters (DPs)
21154.1	Apply normal force to annular areas of the wafer	Membrane compartments
21154.2	Create pressure variation	Compartment pressure distribution
21154.3	Control pressure at discontinuities	Bias pressure
21154.4	Smooth applied pressure profile	Front membrane thickness
21154.5	Control parameters	Active membrane control software

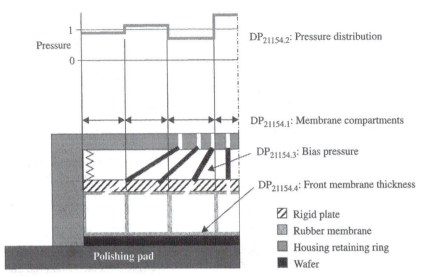

Figure 8.16 Design of active membrane with DP_{21154x}.

FR_{211542}: The ability to control pressure variation depends on how the total area is segmented. The dimension of a compartment may determine how it is used to control wafer scale effects.

FR_{211543}: The discontinuities are defined by the compartment design.

FR_{211544}: The degree of smoothing required for the force profile will depend on the spacing/pitch of the bladder compartments and the associated divisions.

FR_{211545}: The parameters available for control and the ranges of adjustment must be determined before the control code is developed.

- DP_{211542}: The compartment pressure distribution is the pressure supplied to a particular bladder compartment to load the respective area of the wafer. The compartment pressure distribution affects the following FR:

    FR_{211545}: The parameters available for control and the ranges of adjustment must be determined before the control code is developed.

- DP_{211543}: The bias pressure is the pressure applied to the membrane as a whole and determines the relative loading on the sidewalls of the compartments relative to the center of the compartments. The bias pressure affects the following FR:

    FR_{211545}: The range of control as well as the method must be selected before the software is designed.

- DP_{211544}: The front membrane thickness may be used to smooth the pressure distribution from the loading rings as it is transmitted to the wafer back surface.

- DP_{211545}: The active membrane control software is the control code responsible for adjusting the necessary parameters to achieve the desired force profile.

Decomposition of FR_{21171} *and* DP_{21171}. FR_{21171} (Deliver Slurry) and DP_{21171} (Slurry Dispensing System) may be decomposed as shown in Table 8.36.

TABLE 8.36 Children of FR_{21171} and DP_{21171}

	Functional Requirements (FRs)	Design Parameters (DPs)
21171.1	Control flow rate	Peristaltic pump
21171.2	Position dispensing point	Through-the-pad slurry delivery
		Overpad drip
21171.3	Transport to point-of-use	Slurry distribution plumbing
21171.4	Control parameters	Slurry dispensing software

The design equation may be written as

$$\begin{Bmatrix} FR_{211711} \\ FR_{211712} \\ FR_{211713} \\ FR_{211714} \end{Bmatrix} = \begin{bmatrix} X & 0 & 0 & 0 \\ 0 & X & 0 & 0 \\ X & X & X & 0 \\ X & X & X & X \end{bmatrix} \begin{Bmatrix} DP_{211711} \\ DP_{211712} \\ DP_{211713} \\ DP_{211714} \end{Bmatrix} \tag{8.36}$$

The constraints at this level of this branch are as shown in Table 8.37.
Description of DP_{21171x} *and* FR_{21171x}

- DP_{211711}: The peristaltic pump is used to pump the slurry because the only wetted part with this type of pump is the flexible tubing, which is easily changed. The peristaltic pump (DP_{211711}) affects the following FRs:

  FR_{211713}: The tubing and coupling used depend on the size of the pump chosen.
  FR_{211714}: The parameters available for control and the ranges of control must be determined before the software is written.

- DP_{211712}: The through-the-pad slurry delivery is the mechanism of supplying slurry through a hole in the pad. The slurry delivery (DP_{211712}) affects the following FRs:

  FR_{211713}: The plumbing to transport the slurry depends on the desired dispensing point.
  FR_{211714}: The specific control requirements may depend on the configuration of the dispensing point, as it may change the flow dynamics.

- DP_{211713}: The distribution plumbing is the set of components required to get the slurry where it is needed. The distribution plumbing (DP_{211713}) affects the following FR:

  FR_{211714}: The specific control requirements may depend on the configuration of the plumbing, as it may change the flow dynamics.

- DP_{211714}: The slurry dispensing software is the code responsible for controlling the delivery components, such as pump and valves.

TABLE 8.37 Constraints C_{211711}

Constraint Table			Impacts: FR._		
Index	Parent	Description	1	2	3
		Critical Performance Specifications			
$C_{211711a}$		Flow rate 50 to 250 ml/minute	—		—
		Prevent atmospheric exposure	—	—	—
		Maintain suspension	—	—	—

Decomposition of FR_{211313} *and* DP_{211313}. FR_{211313} (Maintain Contact with Pad) and DP_{211313} (Ring Contact Pressure) may be decomposed as shown in Table 8.38.

Figure 8.17 shows the mechanism that controls the ring contact pressure with $DP_{211312x}$. The design equation is given by

$$
\begin{Bmatrix} FR_{2113131} \\ FR_{2113132} \\ FR_{2113133} \\ FR_{2113134} \\ FR_{2113135} \end{Bmatrix} = \begin{bmatrix} X & 0 & 0 & 0 & 0 \\ 0 & X & 0 & 0 & 0 \\ 0 & X & X & 0 & 0 \\ 0 & 0 & X & X & 0 \\ X & X & X & X & X \end{bmatrix} \begin{Bmatrix} DP_{2113131} \\ DP_{2113132} \\ DP_{2113133} \\ DP_{2113134} \\ DP_{2113135} \end{Bmatrix} \qquad (8.37)
$$

Description of $DP_{211313x}$ *and* $FR_{211313x}$

- $DP_{2113131}$: The flexure tip-tilt compliance acts as the "gumball" action to keep the retaining ring aligned with the pad surface. The flexure tip-tilt compliance affects the following FR:

  $FR_{2113135}$: The relationship between vertical position and retaining ring pressure is affected by the compliance of the flexure, so the control must accommodate this.

- $DP_{2113132}$: Strain gauges on the ring flexure allow the influence of the flexure on the retaining ring to be measured and compensated. Strain gages affect the following FRs:

  $FR_{2113133}$: The ability to measure force on the ring flexure directly affects the ability to minimize such force.

TABLE 8.38 Children of FR_{211313} and DP_{211313}

	Functional Requirements (FRs)	**Design Parameters (DPs)**
211313.1	Maintain alignment of ring with pad	Ring flexure tip-tilt compliance
211313.2	Measure force from ring flexure	Strain gages on flexure
211313.3	Minimize force from ring flexure	Z-axis position
211313.4	Apply normal force to ring	Retaining ring bladder pressure
211313.5	Control parameters	Retaining ring software module

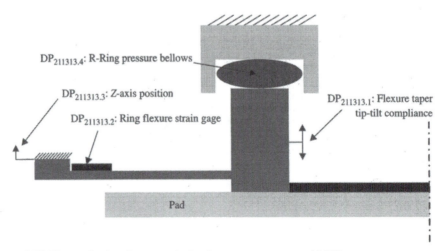

Figure 8.17 The mechanism that controls the ring contact pressure with $DP_{211313x}$.

FR$_{2113135}$: The design of the software is affected by the strain gauge specifics.

- DP$_{2113133}$: The Z-axis position is used to minimize the influence of the ring flexure on the retaining ring force. The relationship between the position and the force exerted on the retaining ring will depend on the design of the ring flexure. The Z-axis position affects the following FRs:

FR$_{2113134}$: The force on the retaining ring is the combination of the force from the flexure and the force from the pressure bellows.

FR$_{2113135}$: The control of the retaining ring pressure must account for the Z-axis position to correctly set pressure.

- DP$_{2113134}$: The retaining ring bladder pressure creates a normal force on the retaining ring by using pneumatic pressure to load the ring. The retaining ring bladder pressure affects the following FR:

FR$_{2113135}$: Parameters available for control and the available values are determined by the selection of the pressure control device.

- DP$_{2113135}$: The retaining ring software module is responsible for varying the necessary parameters to meet the machine operation requirements.

Decomposition of FR$_{211431}$ *and* DP$_{211431}$. FR$_{211431}$ (Maintain Consistent Surface) and DP$_{211431}$ (Rotary Pad Conditioner) may be decomposed as shown in Table 8.39.

The design equation is given by

$$
\begin{Bmatrix} FR_{2114311} \\ FR_{2114312} \\ FR_{2114313} \\ FR_{2114314} \\ FR_{2114315} \\ FR_{2114316} \end{Bmatrix} = \begin{bmatrix} X & 0 & 0 & 0 & 0 & 0 \\ X & X & 0 & 0 & 0 & 0 \\ 0 & 0 & X & 0 & 0 & 0 \\ 0 & 0 & 0 & X & 0 & 0 \\ 0 & 0 & 0 & 0 & X & 0 \\ 0 & X & X & X & X & X \end{bmatrix} \begin{Bmatrix} DP_{2114311} \\ DP_{2114312} \\ DP_{2114313} \\ DP_{2114314} \\ DP_{2114315} \\ DP_{2114316} \end{Bmatrix} \tag{8.38}
$$

Decomposition of FR$_{2114312}$ *and* DP$_{2114312}$. FR$_{2114312}$ (Apply Normal Force) and DP$_{2114312}$ (Overarm Force System) may be decomposed as shown in Table 8.40.

The design is shown in Figures 8.18 and 8.19.

The design equation is given by

$$
\begin{Bmatrix} FR_{21143121} \\ FR_{21143122} \\ FR_{21143123} \\ FR_{21143124} \\ FR_{21143125} \end{Bmatrix} = \begin{bmatrix} X & 0 & 0 & 0 & 0 \\ 0 & X & 0 & 0 & 0 \\ 0 & 0 & X & 0 & 0 \\ 0 & 0 & X & X & 0 \\ X & 0 & 0 & 0 & X \end{bmatrix} \begin{Bmatrix} DP_{21143121} \\ DP_{21143122} \\ DP_{21143123} \\ DP_{21143124} \\ DP_{21143125} \end{Bmatrix} \tag{8.39}
$$

TABLE 8.39 Children of FR$_{211431}$ and DP$_{211431}$

	Functional Requirements (FRs)	**Design Parameters (DPs)**
211431.1	Allow interpad travel	Vertical pad clearance when retracted
211431.2	Apply normal force	Overarm force system
211431.3	Control radial offset	Conditioner position system
211431.4	Abrade pad surface	Conditioner surface texture
211431.5	Rotate conditioner	Conditioner rotary drive system
211431.6	Control parameters	Conditioner control software

TABLE 8.40 Children of $FR_{2114312}$ and $DP_{2114312}$

	Functional Requirements (FRs)	Design Parameters (DPs)
2114312.1	Apply force	Bellows pressure
2114312.2	Prevent force variation from drag loads	Vertical offset of pivot point from conditioning point
2114312.3	Prevent pressure distribution from drag loads	Colocation of conditioner head pivot and conditioning point
2114312.4	Prevent pressure distribution from misalignment	Tip/tilt compliance of conditioner head
2114312.5	Control force	Overarm force system control software

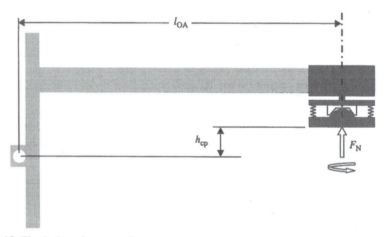

Figure 8.18 The design of overarm force system.

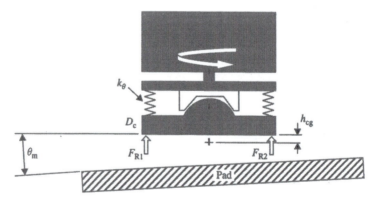

Figure 8.19 Detailed drawing of the overarm force system.

Decomposition of $FR_{2114313}$ *and* $DP_{2114313}$. $FR_{2114313}$ (Control Radial Offset) and $DP_{2114313}$ (Conditioner Position System) are decomposed as shown in Table 8.41.

Decomposition of $FR_{2114315}$ *and* $DP_{2114315}$. $FR_{2114315}$ (Rotate Conditioner) and $DP_{2114315}$ (Conditioner Rotary Drive System—CRDS) may be decomposed as shown in Table 8.42.

TABLE 8.41 Children of $FR_{2114313}$ and $DP_{2114313}$

	Functional Requirements (FRs)	Design Parameters (DPs)
2114313.1	Allow radial motion wrt platen	Linear guides
2114313.2	Provide force	Belt drive system
2114313.3	Measure position	CPS rotary encoder
2114313.4	Control force	CPS control software

TABLE 8.42 Children of $FR_{2114315}$ and $DP_{2114315}$

	Functional Requirements (FRs)	Design Parameters (DPs)
2114315.1	Allow rotation	Duplex pair angular contact bearings
2114315.2	Provide torque	CRDS motor and amp
2114315.3	Measure rotational speed	CRDS rotary encoder
2114315.4	Control torque	CRDS control software

Decomposition of $FR_{21143132}$ *and* $DP_{21143132}$. $FR_{21143132}$ (Provide Force) and $DP_{21143132}$ (Belt Drive System) may be decomposed as shown in Table 8.43.

The design equation is given by

$$\begin{Bmatrix} FR_{211431321} \\ FR_{211431322} \\ FR_{211431323} \\ FR_{211431324} \end{Bmatrix} = \begin{bmatrix} X & 0 & 0 & 0 \\ X & X & 0 & 0 \\ 0 & X & X & 0 \\ 0 & 0 & X & X \end{bmatrix} \begin{Bmatrix} DP_{211431321} \\ DP_{211431322} \\ DP_{211431323} \\ DP_{211431324} \end{Bmatrix} \tag{8.40}$$

The maximum rotational speed and the torque are given as follows:

$$w_{max} = V_{max}/R_{pulley} = 1.0/19.1e - 3 = 52 \text{ rad/second} = 500 \text{ rpm}$$

$$t_{max} = T_{max} \times R_{pulley} = 76.5 \text{ kgf} \times 19.1 \text{ mm} = 14.3 \text{ N-m}$$

Motor and gearbox specification: 1.43 N-m at 5000 rpm with 10:1 reduction.

8.6.1.2 Master Design Matrix (Complete Design Matrix)

When the design created through the foregoing decomposition process was checked by constructing the full (Complete) master matrix, it was found that unintended coupling was introduced during the decomposition process. This was recognized as being a problem and most of the mistakes were corrected by making the machine decoupled.

8.6.1.3 Overall System Design

The overall design of the assembled system—an integration of all the leaf-level designs

TABLE 8.43 Children of $FR_{21143132}$ and $DP_{21143132}$

	Functional Requirements (FRs)	Design Parameters (DPs)
21143132.1	Transmit force to carriage	Flexible belt
21143132.2	Transform torque to force	Drive pulley radius
21143132.3	Provide torque	CPS motor, gearbox, and amp
21143132.4	Control torque	BDS control software

developed in this section—is given in Figure 8.20a and b. The actual hardware is shown in Figure 8.21.

This case study gives the detailed process of designing the spindle head of a CMP machine. This detailed decomposition process is given to illustrate how products—in this case, a machine—can be designed so that there will not be any coupling.

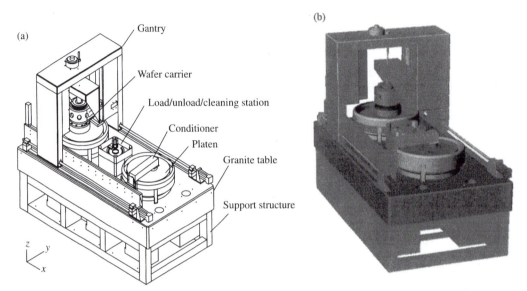

(a)

Gantry

Wafer carrier

Load/unload/cleaning station

Conditioner

Platen

Granite table

Support structure

(b)

Figure 8.20 (a) Assembly drawing of the CMP machine—an assemblage of all leaf-level DPs. (b) CAD rendering of the MIT CMP machine.

Figure 8.21 The mechanical part of the CMP machine built at MIT by four master's degree candidates (Jason Melvin, Kwang-duk Douglass Lee, Amir Torkaman, and Jamie Nam) under the supervision of Professor J. Chun, Dr. N. Saka, and Professor N. P. Suh. The machine shown is about 10 feet tall. The controller and the associated computer are not shown.

8.6.2 Axiomatic Development of CMP α Machine Control System[9]

The CMP machine shown in Figures 8.19 and 8.20 and described in the preceding section has great flexibility and high precision, and is designed to be superior to commercially available machines. The machine has one wafer carrier head and two platens, which enable multistep polishing. The wafer carrier is mounted on the bracket, and the bracket can move up and down along the gantry Z-axis. The gantry also provides X directional wafer carrier transportation. Loading, unloading, and cleaning of a wafer are all performed at the load/unload/cleaning station. All of these components are mounted on top of the granite table, which has a good vibrational damping characteristic. The table is then mounted on the steel supporting structure.

The machine control system needs to be designed to ensure the effectiveness of the CMP process and enhance the efficiency of the machine operation by integrating all the intended system functions. The control system development includes the following:

- The electrical wiring and interfacing of various actuators, sensors, and the control board.
- The design and implementation of individual control algorithms (or controllers) for each drive and actuation component.
- Creation of process steps, such as wafer loading, buffing, conditioning, and polishing, which are generated by chrono (time-wise) and conditional combinations of individual control action.
- Realization of machine operation, which allows a user to process sets of wafers to the desired specifications, using the various process parameters and sequences (recipe) in a fully automated environment.

Although not a single component of the control system is designed at this stage, based on the context and the demands from the machine design, a general idea of control system structure can be conceived as shown in Figure 8.22. The final system will differ from the one shown in the figure, but it is a good representation of what is required to control the machine.

The development (design and subsequent implementation) of a machine control system requires a sound knowledge of physical hardware, control technology, software engineering, and system integration, because of the scope involved. In addition, a systematic design tool is required to structure the whole system and to guide the design of its components. The development of various control systems has relied on human experience and trial and error, without any universal framework. As a result, some machine control systems with substantially high development time and cost lack interchangeability and reusability. Often the systems contain fatal errors undetected during the design phase, resulting in the endless upgrade and maintenance cycle, sometimes offsetting the very profitability of the systems.

Instead of system development based on experience and trial and error, axiomatic design is employed as a systematic design tool to guide the development of the overall CMP machine control system. Axiomatic design provides a concrete scientific methodology for system development, by identifying CAs first, setting up independent FRs, conceiving

---

[9] This section was prepared by Kwang-Duk Douglass Lee, a Research Assistant in the Department of Mechanical Engineering at MIT. This research was done as part of his M.S. thesis.

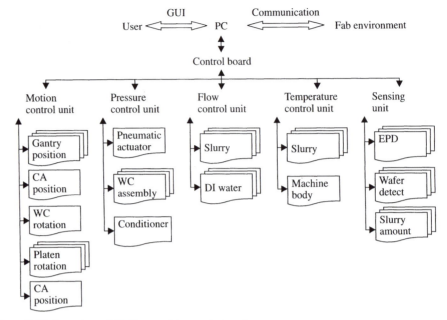

Figure 8.22 Overview of CMP α machine control system.

appropriate DPs, and also selecting PVs. Instead of relying on speculations and trial and error, axiomatic design ensures a coherent system development by clearly stating FRs and choosing the right DPs, and by showing their relationships in terms of design equations.

Axiomatic design is used throughout the entire development process, ranging from the highest level system organization to the lowest level of the specific procedural algorithm design. In this section, three design levels—system, application, and sequence—are presented as the examples of axiomatic machine control system development.

8.6.2.1 System Level

At the initial stage of a system development, it is often difficult to specify what is required for the system to function. In axiomatic design, these are in the realm of higher level decompositions. At this level, axiomatic design is used to organize and orient the overall system design by clearly stating FRs. FRs are much more important than DPs at this stage. Suppose we conceived the highest level FR as "control the machine in an automatic manner so that wafers can be processed in a preprogrammed way with minimal human intervention." The corresponding DP will simply be "auto mode." No specific detail about the auto mode needs to be known at this stage. The subsequent decompositions will detail the auto mode.

Stating FRs at an initial design stage serves the purpose of system definition and functionality clarification. Then a DP is a mere conceptualization of the corresponding FR, which enables further system development by the following decompositions. The strength of axiomatic design lies in the fact that it forces the designer to come up with a clearly defined set of FRs, which leads to a strategic placement of necessary system components.

Axiomatic design at this level identifies the major system components and transforms the once complex and poorly acknowledged original system to a hierarchical set of related subsystems that becomes much more easy to attack (design). The system level design of the CMP α machine control system will be given to illustrate this point.

Axiomatic decomposition often starts with a single objective or mission statement. In this case, the objective is clear: Integrate individual components to create an effective and efficient machine control system. The highest level FR—the mission statement—is given in Table 8.44.

As explained before, the equipment control system is a mere conceptualization of what is stated in the FR—a clear, yet extensive definition of the problem. It will take a few more decompositions to arrive at leaf-level DPs that have more concrete meaning.

At this stage, the following can be staged as the most fundamental customer attributes (CAs).

- Provide the required precision.
- Minimize the nonprocessing time.
- Enable a fully automated wafer processing.
- Ensure a flexible recipe editing and its implementation.
- Provide an equipment use and maintenance log.
- Provide ease of operation.
- Ensure safety of use.

These CAs can also be thought of as general guidelines or constraints throughout the overall system development process.

The design equation for the FRs and DPs is given in Table 8.45.

The equipment control system is essentially a set of software designed to achieve the goal of coordinated wafer processing, with appropriate interfaces to the machine and the user. However, we have to classify what needs to be done in the name of equipment control system to proceed to the next level decomposition. It is one thing to have a functional machine, i.e., all the machine components are working, and another to process wafers using those actions. Two functional requirements can be stated as in Table 8.45.

TABLE 8.44 Mission Statement (Highest Level FR and DP)

Functional Requirement(FR)	Design Parameter(DP)
Coordinate the individual components of the CMP α machine in a systematic manner so that it can process wafers effectively and efficiently to the desired specifications.	Equipment control system

TABLE 8.45 Decomposition of Mission Statement

Index		Control System Design	
		Functional Requirements (FRs)	**Design Parameters (DPs)**
	Parents	Coordinate the individual components of the CMP α machine in a systematic manner so that it can process wafers effectively and efficiently to the desired specifications	Equipment control system
1	Control	Generate control actions from each individual machine component	Machine-level control system
2	Process	Organize the controlled outputs for wafer processing	Process-level control system

$$\begin{Bmatrix} FR_1 \\ FR_2 \end{Bmatrix} = \begin{bmatrix} X & 0 \\ X & X \end{bmatrix} \begin{Bmatrix} DP_1 \\ DP_2 \end{Bmatrix} \tag{8.41}$$

The machine-level control system conceptualizes the FR of generating individual machine control actions, such as moving the gantry from position 1 to position 2, turning on and off the fluid valves, sensing reflectance signals, etc. It is highly biased toward the machine hardware. Because it deals with the real-time hardware control issues, it is programmed for and run on an external processing system (external real time computer).

However, to process wafers in a certain manner, there should be something more than just the machine-level system. A user should be able to command and supervise the machine operation. A type of software program is required to edit a set of processing data and sequence (recipe). A program should handle communication between the host PC and the machine-level processors. A sort of scheduler is necessary to dispatch a series of control actions to the machine side following the predefined sequence. All of these can be defined as the wafer processing-related issues. The process-level control system is a conceptualization which encompasses all those requirements. The process-level control software naturally resides on the host PC.

The design is decoupled, because the design of a process depends on what can be done on the machine side. Although not relevant to the current design, we can come up with a third design parameter: operation-level control system. The operation-level control system is required when the equipment is placed in a highly integrated and automated environment. The operation-level system will coordinate with other production equipment and supportive equipment, such as wafer-handling robots and the cleaning and inspection stations. The operation level will also be responsible for fabricating batches of wafers with different processes, reporting to a higher level production management system.

One more decomposition will clarify what is inside the machine-level control system. Table 8.46 shows the constituents of the system. A system can essentially be defined as a set of its members, which has inputs and outputs and its own unique processing on them. The I/O unit handles the input/output function of the machine-level control system. The control system communicates with both to the machine and to the process level. Other design parameters define the processing of the machine level.

The signal-processing unit converts the incoming signals to the desired units, filters the input signals, performs differentiation and integration, converts the outgoing signals, and saves the data internally. The controller unit is the essence of the machine-level

TABLE 8.46 Decomposition of Machine-Level Control System

Index 1.#		Control System Design	
	Functional Requirements (FRs)		**Design Parameters (DPs)**
Parents	Generate control actions from each individual machine component		Machine-level control system
1	Control	Receive inputs and send outputs	I/O unit
2	Control	Process incoming signals	Signal processing unit
3	Control	Produce controlled outputs	Controller unit
4	Control	Structure the machine-level control system	Machine-level overhead

system. It performs the control actions, including on/off and open-loop and closed-loop controllers. The machine-level overhead serves as an administrator at the machine level, organizing individual unit actions and coordinating time- and space-related issues within the level.

$$\begin{Bmatrix} FR_{11} \\ FR_{12} \\ FR_{13} \\ FR_{14} \end{Bmatrix} = \begin{bmatrix} X & 0 & 0 & 0 \\ X & X & 0 & 0 \\ X & X & X & 0 \\ X & X & X & X \end{bmatrix} \begin{Bmatrix} DP_{11} \\ DP_{12} \\ DP_{13} \\ DP_{14} \end{Bmatrix} \tag{8.42}$$

The design is highly decoupled. The input and output affect the way signals are processed. The design of the controller depends on what kinds of signal are available and how they are processed. The design of overhead is determined by what actions there are and how they need to be administered. The machine-level control system is logically placed between the machine hardware and the process-level control system, which is yet to be designed. Figure 8.23 shows the interaction of the system.

The DPs begin to have specific meanings and tasks. The I/O unit design, for example, will start to look at what kinds of signals are required to control the machine, then what types of I/O cards are required for the external processing system and what types of I/O variables are used for those cards, etc.

Before moving on to the next level, let us do a decomposition for the process-level control system. Having the machine control actions at its disposal, the process-level control system dictates the machine level in a concerted manner to achieve the desired wafer processing with the user interaction. The process-level control system is placed between the machine level and the user (and also the higher level control systems), and acts both as a messenger and a coordinator (and possibly as a manager).

The main function of the process-level control system is wafer processing, but to achieve that we may need a few supportive features. Table 8.47 shows the decomposition of the process level control system.

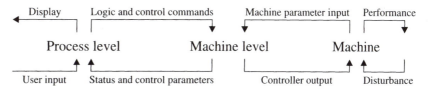

Figure 8.23 Interaction of machine-level control system with other systems. Logic commands: function buttons, initialization. Control commands: position, velocity, pressure, flow rate, valve and switch on/off, etc. State parameters: error, operative status. Control parameters: current position, velocity, sensor output, time, etc.

TABLE 8.47 Decomposition of Process-Level Control System

Index 2.#		**Control System Design**	
	Functional Requirements (FRs)		**Design Parameters (DPs)**
	Parents	Organize the controlled outputs for wafer processing	Process-level control system
1	Process	Provide supportive features for wafer processing	Supportive unit
2	Process	Process wafers	Process unit

The design equation is given by

$$\begin{Bmatrix} FR_{21} \\ FR_{22} \end{Bmatrix} = \begin{bmatrix} X & 0 \\ X & X \end{bmatrix} \begin{Bmatrix} DP_{21} \\ DP_{22} \end{Bmatrix} \tag{8.43}$$

The supportive unit is a collection of features designed to help wafer processing in the process unit. Although many items can be conceived, the following three listed in Table 8.48 seems to be the most essential:

• Recipe builder: required to process a wafer in an automatic fashion. The recipe builder defines a recipe and its file structure for storage, and provides a recipe editor for recipe handling. The recipe editor should be able to create, edit, and save a recipe, and retrieve the existing one. The process unit will have an adequate link to load the recipe data to its process parameter buffer. The design of a flexible and easy-to-use recipe editor is quite essential to ensure the effectiveness of the automatic wafer processing and to enhance the processing efficiency.

• Equipment database: can be classified into operation, which handles the data related to how the equipment is used, and maintenance, which logs the data concerning the equipment maintenance. The operation database records time, user, recipe, wafer ID, etc. and can be used to keep track of the equipment use history and the processed wafers. The operation database can provide useful data to improve the efficiency of the equipment operation and to optimize the process performed by the equipment.

• Setup mode: provides a nonprocess machine operation environment. Set points and calibration factors of sensors and actuators are adjusted and recorded in a configuration file. It also contains the machine initialization and the machine termination procedures to drive the machine to the on and off state. These procedures can be called by the process units to initialize and terminate the machine before and after the wafer processing.

The design equation is given by

$$\begin{Bmatrix} FR_{211} \\ FR_{212} \\ FR_{213} \end{Bmatrix} = \begin{bmatrix} X & 0 & 0 \\ 0 & X & 0 \\ 0 & 0 & X \end{bmatrix} \begin{Bmatrix} DP_{211} \\ DP_{212} \\ DP_{213} \end{Bmatrix} \tag{8.44}$$

A wafer is processed by the equipment under one of the modes in the process unit. At one extreme, the wafer can be treated in a purely manual manner. In this case, the user has to click the corresponding button at every sequence. At the other extreme, the user simply edits a recipe and loads it in a fully automated environment. Now the user only has to click the Run button to complete the entire recipe-based wafer processing.

TABLE 8.48 Decomposition of Supportive Unit

Index 21.#		Control System Design	
	Functional Requirements (FRs)		**Design Parameters (DPs)**
	Parents	Provide supportive features for wafer processing	Supportive unit
1	Process	Provide recipes for wafer processing	Recipe builder
2	Process	Provide a database for the machine use and maintenance	Equipment database
3	Process	Set up and calibrate the machine for wafer processing	Setup mode

Many different levels of automation can exist between the two extremes. However, two are the most fundamental and necessary—manual and auto. Many intermediate (or semiauto) modes can be designed by combining the manual and the auto mode in a certain proportion. The process unit is decomposed to the manual and the auto mode as shown in Table 8.49.

The design equation is given by

$$\left\{ \begin{array}{c} FR_{221} \\ FR_{222} \end{array} \right\} = \left[\begin{array}{cc} X & 0 \\ X & X \end{array} \right] \left\{ \begin{array}{c} DP_{221} \\ DP_{222} \end{array} \right\} \tag{8.45}$$

The design is decoupled because many manual mode procedures are reused in the auto mode. The auto mode simply replaces a user's decision making and command procedures in the manual mode. Once again, the design parameters become specific and now we need competent engineers to design the subsystems.

8.6.2.2 Application Level

An application is a subsystem of a parent system, which is identifiable as a nearly independent unit due to its clearly defined functionality, yet has a variety of functions compared to the procedure that has a single dedicated function. The auto mode of the equipment control system is a good example of the application. In this section, we will perform a thorough decomposition of the auto mode.

The FRs are well defined at this level, due to the foregoing decomposition at higher levels. More attention is paid to selecting appropriate DPs to satisfy the established FRs. DPs play a more important role at this level. From now on, the selection and design of each DP have technical flavors. For example, to design a graphic user interface, the programmer should design backend processes first to process data, next frontend window layout, then selection of programming languages, etc.

The application level usually has well-defined FRs, so the real difficulty is selecting and substantiating the right DPs. The designer is often required to have knowledge and experience related to the specific domain in which the design is being performed. However, axiomatic design is used to prevent the pitfalls of heuristic and trial-and-error approaches of experienced persons.

The auto mode is decomposed into five DPs as shown in Table 8.50. The auto mode overhead organizes the whole auto mode and links four other DPs. The recipe editor link provides access to the saved recipes and the recipe editor. The user interface handles the interaction between the user and the auto mode. The machine-level interface provides the communication channel between the auto mode and the machine. The process steps are

TABLE 8.49 Decomposition of Process Unit

Index 22.#	Control System Design		
	Functional Requirements (FRs)		**Design Parameters (DPs)**
	Parents	Process wafers	Process unit
1	Process	Process wafers manually	Manual mode
2	Process	Process wafers automatically	Auto mode

TABLE 8.50 Decomposition of Auto Mode

Index 222.#		Control System Design	
	Functional Requirements (FRs)		**Design Parameters (DPs)**
	Parents	Process wafers automatically	Auto mode
1	Process	Structure the auto mode software	Auto mode overhead
2	Process	Connect to the recipe editor	Recipe editor link
3	Process	Interact with a user	User interface
4	Process	Communicate with the machine level	Machine-level interface
5	Process	Designate distinct steps to process wafers in auto mode	Process steps

preprogrammed autonomous wafer processing stages to support the recipe-based automatic wafer processing.

The design equation is given by

$$
\begin{Bmatrix} FR_{2221} \\ FR_{2222} \\ FR_{2223} \\ FR_{2224} \\ FR_{2225} \end{Bmatrix} = \begin{bmatrix} X & 0 & 0 & 0 & 0 \\ X & X & 0 & 0 & 0 \\ X & X & X & 0 & 0 \\ X & 0 & 0 & X & 0 \\ X & 0 & 0 & 0 & X \end{bmatrix} \begin{Bmatrix} DP_{2221} \\ DP_{2222} \\ DP_{2223} \\ DP_{2224} \\ DP_{2225} \end{Bmatrix} \tag{8.46}
$$

Each design parameter will be decomposed down to one more level to show its constituents.

The design equation is given by

$$
\begin{Bmatrix} FR_{22211} \\ FR_{22212} \\ FR_{22213} \\ FR_{22214} \\ FR_{22215} \\ FR_{22216} \\ FR_{22217} \\ FR_{22218} \end{Bmatrix} = \begin{bmatrix} X & 0 & 0 & 0 & 0 & 0 & 0 & 0 \\ X & X & 0 & 0 & 0 & 0 & 0 & 0 \\ X & X & X & 0 & 0 & 0 & 0 & 0 \\ X & X & X & X & 0 & 0 & 0 & 0 \\ X & 0 & 0 & 0 & X & 0 & 0 & 0 \\ X & 0 & X & 0 & X & X & 0 & 0 \\ X & 0 & 0 & X & X & 0 & X & 0 \\ X & 0 & 0 & 0 & 0 & 0 & 0 & X \end{bmatrix} \begin{Bmatrix} DP_{22211} \\ DP_{22212} \\ DP_{22213} \\ DP_{22214} \\ DP_{22215} \\ DP_{22216} \\ DP_{22217} \\ DP_{22218} \end{Bmatrix} \tag{8.47}
$$

Decomposition of DP_{2221}: *Auto mode overhead.* Table 8.51 shows the decomposition of the auto mode overhead. It specifies the input and output frequencies of the auto mode, controls data flow between the various application components, and performs nonprocess routines such as initialization and termination.

DP_{22211}: *I/O frequency:*
      User input frequency: user-generated event
      Machine-level output frequency: user-generated event and 18-Hz PC timer
      Machine-level input frequency: 18-Hz PC timer
      User display output frequency: 18-Hz PC timer

DP_{22212}: *User input handler:* The user input handler extracts logic commands (run, stop, error reset, etc.) from the user interface and passes them to the machine-level output handler at the user input frequency.

DP_{22213}: *Recipe handler:* The recipe handler extracts the process commands from the

TABLE 8.51 Decomposition of Auto Mode Overhead

Index 2221.#		Control System Design	
	Functional Requirements (FRs)		**Design Parameters (DPs)**
	Parents	Structure the auto mode software	Auto mode overhead
1	Process	Designate how often the input to and the output from the auto mode are updated	I/O frequency
2	Process	Process user input	User input handler
3	Process	Process recipe data	Recipe handler
4	Process	Send outputs to the machine level	Machine-level output handler
5	Process	Update inputs from the machine level	Machine-level input handler
6	Process	Update the process information to the user	User display handler
7	Process	Initialize the auto mode	Auto mode initialization procedure
8	Process	Terminate the auto mode	Auto mode termination procedure

I/O buffer of the recipe editor link and places them in the data packets for the machine-level output handler.

DP_{22214}: *Machine-level output handler:* The output handler sends the data packets and logic commands to the machine-level interface, which sends them to the machine-level control system.

DP_{22215}: *Machine-level input handler:* The machine-level input handler receives the parameter and status inputs from the machine-level control system via the machine-level interface. The handler unpacks the inputs, analyzes them, and assigns them to the user display handler.

DP_{22216}: *User display handler:* The user display handler displays the machine-level inputs to the designated sections of the user interface window.

DP_{22217}: *Auto mode initialization procedure:* The auto mode initialization procedure is executed when the auto mode is loaded to the process-level control system. It simply assigns the constants, initializes the variables, calls the machine initialization routine of the setup mode if the machine was not initialized beforehand, and loads the user interface to the screen. Figure 8.24 illustrates the procedure.

DP_{22218}: *Auto mode termination procedure:* The auto mode termination procedure simply unloads the user interface from the screen. Optionally, it can perform the machine termination procedure. The machine termination procedure will move the machine components to the parking positions and then stop the ADwin program, or stop ADwin immediately if requested by an error set.

Decomposition of DP_{2222}: *Recipe editor link.* The recipe editor link provides an access to the recipe editor and the recipe files. It should be able to load and unload a saved recipe, and invoke the recipe editor to modify an existing recipe or to create a new recipe at run time without exiting the auto mode. Table 8.52 shows the decomposition of the recipe editor link.

DP_{22221}: *Load method:* Load method opens up a recipe file as read-only to prevent any accidental modification to the file. A recipe overhead object and various step objects are created based on the file header information and the recipe data are read by the object methods. Then the recipe data is loaded to the I/O buffer of the recipe editor link to be used for wafer processing.

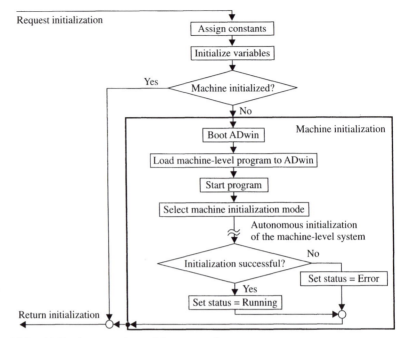

Figure 8.24 Initialization procedure of the auto mode.

TABLE 8.52 Decomposition of Recipe Editor Link

Index 2222.#		Control System Design	
	Functional Requirements (FRs)		**Design Parameters (DPs)**
	Parents	Connect to the recipe editor	Recipe editor link
1	Process	Load an existing recipe	Load method
2	Process	Unload the current recipe	Unload method
3	Process	Call the recipe editor to edit a recipe	Edit method

DP_{22222}: *Unload method:* Unload method simply empties the I/O buffer and destroys the created recipe objects.

DP_{22223}: *Edit method:* Edit method provides a direct link to the recipe editor if a user wants to create a new recipe at run time. The method calls the recipe editor and launches its screen on top of the auto mode user interface. Once the user finishes editing, he or she can go back to the auto mode user interface and load the just edited new recipe.

The design equation is given by

$$\left\{ \begin{array}{c} FR_{22221} \\ FR_{22222} \\ FR_{22223} \end{array} \right\} = \begin{bmatrix} X & 0 & 0 \\ 0 & X & 0 \\ 0 & 0 & X \end{bmatrix} \left\{ \begin{array}{c} DP_{22221} \\ DP_{22222} \\ DP_{22223} \end{array} \right\} \tag{8.48}$$

Decomposition of DP_{2223}: Auto mode user interface. An economical and logical design of the user interface is essential, because a great amount of information needs to be viewed in the auto mode. Table 8.53 shows the decomposition of the auto mode user interface.

TABLE 8.53 Decomposition of Auto Mode User Interface

Index 2223.#			Control System Design
	Functional Requirements (FRs)		**Design Parameters (DPs)**
	Parents	Interact with a user	User interface
1	Process	Show the administrative information	Administration display
2	Process	Show the wafer-related information	Wafer tracking
3	Process	Show the process-related information	Process information display
4	Process	Show the machine and operation statuses	Status display
5	Process	Provide means of process control	Function buttons
6	Process	Visualize machine movements	Graphic machine movement display
7	Process	Visualize process monitoring	Process monitoring graph
8	Process	Display current machine parameters	Parameter display
9	Process	Display current commands	Command display
10	Process	Allow the user to see the recipe	Recipe view
11	Process	Provide access to other modes	Navigation buttons

$DP_{2223.1}$: *Administration display:* The administrative information includes title, data and time, user name, etc. They will be displayed on the top portion of the interface screen.

$DP_{2223.2}$: *Wafer tracking:* In automated wafer processing, the interface should show the wafer-related information. The wafer tracking section will display the lot number, the wafer ID (if known), and the wafer count (number of wafers processed).

$DP_{2223.3}$: *Process information display:* Recipe name, current recipe step, and run time are examples of the process-related information.

$DP_{2223.4}$: *Status display:* The status display will show the equipment status (running, stopped, initializing, etc.) and the alarm status (normal, error, etc.).

$DP_{2223.5}$: *Function buttons:* Function buttons include run, pause, resume, abort, etc., which provide means of process control.

$DP_{2223.6}$: *Graphic machine movement display:* The graphic machine movement display is a small box that shows a simple animation of the machine movement. The motions of the wafer carrier and the conditioner, along with the wafer flow, will be displayed in real time.

$DP_{2223.7}$: *Process monitoring graph:* The process monitoring graph displays the parameters in graphic forms to catch the time-wise and the space-wise evolutions of the parameter signals. The graph can show different types of signals—wafer reflectance, strain gauge, current sensor, etc.

$DP_{2223.8}$: *Parameter display:* The parameter display shows the current machine parameters, such as wafer carrier positions and platen speeds.

$DP_{2223.9}$: *Command display:* The command display shows the process commands sent to the machine level at the current stage.

$DP_{2223.10}$: *Recipe view:* The recipe view allows the user to see the current recipe loaded, both the header and the process data.

$DP_{2223.11}$: *Navigation buttons:* Navigation buttons provide quick links to the other modes and utilities, such as the recipe editor and the database.

The design equation is given by

$$
\begin{Bmatrix}
FR_{2223.1} \\
FR_{2223.2} \\
FR_{2223.3} \\
FR_{2223.4} \\
FR_{2223.5} \\
FR_{2223.6} \\
FR_{2223.7} \\
FR_{2223.8} \\
FR_{2223.9} \\
FR_{2223.10} \\
FR_{2223.11}
\end{Bmatrix}
=
\begin{bmatrix}
X & 0 & 0 & 0 & 0 & 0 & 0 & 0 & 0 & 0 & 0 \\
0 & X & 0 & 0 & 0 & 0 & 0 & 0 & 0 & 0 & 0 \\
0 & 0 & X & 0 & 0 & 0 & 0 & 0 & 0 & 0 & 0 \\
0 & 0 & 0 & X & 0 & 0 & 0 & 0 & 0 & 0 & 0 \\
0 & 0 & 0 & 0 & X & 0 & 0 & 0 & 0 & 0 & 0 \\
0 & 0 & 0 & 0 & 0 & X & 0 & 0 & 0 & 0 & 0 \\
0 & 0 & 0 & 0 & 0 & 0 & X & 0 & 0 & 0 & 0 \\
0 & 0 & 0 & 0 & 0 & 0 & 0 & X & 0 & 0 & 0 \\
0 & 0 & 0 & 0 & 0 & 0 & 0 & 0 & X & 0 & 0 \\
0 & 0 & 0 & 0 & 0 & 0 & 0 & 0 & 0 & X & 0 \\
0 & 0 & 0 & 0 & 0 & 0 & 0 & 0 & 0 & 0 & X
\end{bmatrix}
\begin{Bmatrix}
DP_{2223.1} \\
DP_{2223.2} \\
DP_{2223.3} \\
DP_{2223.4} \\
DP_{2223.5} \\
DP_{2223.6} \\
DP_{2223.7} \\
DP_{2223.8} \\
DP_{2223.9} \\
DP_{2223.10} \\
DP_{2223.11}
\end{Bmatrix}
\quad (8.49)
$$

Figure 8.25 shows the preliminary layout of the auto mode user interface screen.

Description of DP_{2224}: *Machine-level interface.* The machine-level interface deals with the low-level serial communication between the auto mode and the machine-level control system. Table 8.54 shows the decomposition.

DP_{22241}: *Command output procedure:* Control commands (position, pressure, velocity, etc.) are placed in a command packet, and a library function is called to transmit the packet to the machine-level processor.

DP_{22242}: *Logic output procedure:* Logic commands (run, stop, etc.) are placed in single

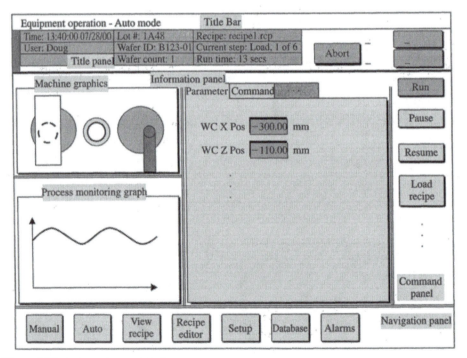

Figure 8.25 Auto mode user interface.

TABLE 8.54 Decomposition of Machine-Level Interface

Index 2224.#			Control System Design
	Functional Requirements (FRs)		**Design Parameters (DPs)**
	Parents	Communicate with the machine level	Machine-level interface
1	Process	Send command outputs	Command output procedure
2	Process	Send logic commands	Logic output procedure
3	Process	Receive parameter inputs	Parameter input procedure
4	Process	Receive status inputs	Status input procedure
5	Process	Receive data inputs	Data input procedure

variables and sent to the machine level after the control commands transfer, to ensure the update of control commands at the event of logic update.

DP_{22243}: *Parameter input procedure:* At each timer event, a library function is called that accesses the readable variables of the machine-level processor. The variables are preselected and contain machine parameters.

DP_{22244}: *Status input procedure:* The status input procedure is similar to the parameter input procedure. It extracts the machine status information from the machine level.

DP_{22245}: *Data input procedure:* The data input procedure receives a time series of process parameters such as motor torque output and reflectance sensing value. The procedure loads a batch of data into an array, which is accessible by the auto mode.

The design equation is given by

$$\begin{Bmatrix} FR_{22241} \\ FR_{22242} \\ FR_{22243} \\ FR_{22244} \\ FR_{22245} \end{Bmatrix} = \begin{bmatrix} X & 0 & 0 & 0 & 0 \\ X & X & 0 & 0 & 0 \\ 0 & 0 & X & 0 & 0 \\ 0 & 0 & 0 & X & 0 \\ 0 & 0 & 0 & X & X \end{bmatrix} \begin{Bmatrix} DP_{22241} \\ DP_{22242} \\ DP_{22243} \\ DP_{22244} \\ DP_{22245} \end{Bmatrix} \qquad (8.50)$$

Description of DP_{2225}: *Process steps.* Processing of a wafer demands that hundreds of individual control actions take place both in serial and parallel. Thus it is quite necessary to organize these actions into certain meaningful groups. Small sets of control actions can be organized as a "substep" to be used by a bigger "step." A process step is a sequence of control actions and substeps that can uniquely be distinguished in the stream of wafer processing. Then a user can construct his or her own process recipe by choosing the individual steps in a desired sequence, with the parameters edited for each step. Along with the substeps, seven different process steps can be conceived as shown in Table 8.55. They are the minimum set of steps to create every possible recipe configuration.

DP_{22251}: *Substeps:* A substep is written because it is repeatedly used by many steps or can hide many details within a single step. Although a further decomposition is necessary, we can group them into transport substeps, which deal with the wafer carrier and conditioner transportation; wafer polishing substeps, which contain polishing-related activities such as sweeping, buffing, and pressure

TABLE 8.55 Decomposition of Process Steps

Index 2225.#			Control System Design
	Functional Requirements (FRs)		**Design Parameters (DPs)**
	Parents	Designate distinct steps to process wafers in auto mode	Process steps
1	Process	Create sets of control actions repeatedly used in process steps	Substeps
2	Process	Provide a waiting step in automatic wafer processing	Step_LoadReady
3	Process	Condition polishing pads	Step_Condition
4	Process	Load a wafer	Step_Load
5	Process	Polish a wafer	Step_Polish
6	Process	Clean a wafer	Step_Clean
7	Process	Unload a wafer	Step_Unload
8	Process	Clean the conditioner head	Step_CondClean

profiling; and conditioning substeps, which house conditioner-related actions such as conditioner sweeping and condition zone parameter profiling.

DP_{22252}: *Step_LoadReady:* Step_LoadReady is a waiting step between the wafer unloading and loading in a continuous wafer processing. The ready step simply waits for the next wafer to be placed on the loading station. Either a run button click by the user or an automatic detection of wafer placement will trigger the run of the recipe-based automatic wafer polishing, beginning with Step_Load.

DP_{22253}: *Step_Condition:* Step_Condition performs the conditioning (regeneration) of the pad during (*in situ*) or after (*ex situ*) wafer polishing. It is a conditioner-related step and can be distinguished from most of the steps, which are wafer carrier-related ones. The condition step can cause coupling to other steps in the case of *in situ* condition, because the motions of the wafer carrier and the conditioner along with other machine parts need to be coordinated for dual purposes (parallel steps).

DP_{22254}: *Step_Load:* A wafer is picked up by the wafer carrier from the loading station in the loading step. The wafer carrier is transported to the designated position before the end of the step.

DP_{22255}: *Step_Polish:* This is the main and most important step in automatic wafer processing. It induces the desired physical change to the wafer. It will have a sequential algorithm to direct many stages involved in wafer polishing.

DP_{22256}: *Step_Clean:* A wafer is cleaned after polishing at the cleaning station by means of deionized (DI) water sprays. Then the wafer can be either unloaded or transferred to the subsequent polishing step.

DP_{22257}: *Step_Unload:* It simply unloads the wafer from the carrier to the unloading station. At the end of the step, the wafer is ready to be picked up either by a user or a wafer-handling robot.

DP_{22258}: *Step_CondClean:* It performs the cleaning of the conditioner. The conditioner often needs cleaning to rinse off slurry and pad wear particles.

The design equation is given by

$$
\begin{Bmatrix} FR_{22251} \\ FR_{22252} \\ FR_{22253} \\ FR_{22254} \\ FR_{22255} \\ FR_{22256} \\ FR_{22257} \\ FR_{22258} \end{Bmatrix} = \begin{bmatrix} X & 0 & 0 & 0 & 0 & 0 & 0 & 0 \\ X & X & 0 & 0 & 0 & 0 & 0 & 0 \\ X & 0 & X & 0 & 0 & 0 & 0 & 0 \\ X & 0 & X & X & 0 & 0 & 0 & 0 \\ X & 0 & X & 0 & X & 0 & 0 & 0 \\ X & 0 & X & 0 & 0 & X & 0 & 0 \\ X & 0 & X & 0 & 0 & 0 & X & 0 \\ X & 0 & 0 & 0 & 0 & 0 & 0 & X \end{bmatrix} \begin{Bmatrix} DP_{22251} \\ DP_{22252} \\ DP_{22253} \\ DP_{22254} \\ DP_{22255} \\ DP_{22256} \\ DP_{22257} \\ DP_{22258} \end{Bmatrix} \qquad (8.51)
$$

8.6.2.3 Sequence Level

It is often necessary to design a specific procedure or sequence while designing an application. The sequences usually comprise the lowest levels of decompositions (leaf modules in axiomatic design). Thus if an engineer designs the specific process sequences, he or she is near the completion of the whole system design. Axiomatic design is also useful in designing sequential procedures.

If the sequential stages are put into the form of design equations, they usually become a decoupled design because of their sequential (ordered) nature. But some DPs (stages) can be uncoupled from the previous stages, implying that they can be parallel procedures to them. The information contained in the design matrix can easily be exported to construct a sequential functional diagram, which is the graphic representation of a sequential algorithm.

Axiomatic design is a powerful tool in designing a logical and chronological sequence or algorithm. In this section, we will take a look at how axiomatic design can be employed to design sequences. Two examples from the previous application will be presented.

Step_Polish (DP_{22255}) requires a sequential algorithm to guide individual control actions to the common goal of coordinated wafer polishing. Table 8.56 shows the decomposition of the algorithm.

TABLE 8.56 Decomposition of Sequential Algorithm (Step_Polish)

Index 22255.#		Control System Design	
	Functional Requirements (FRs)		**Design Parameters (DPs)**
	Parents	Produce process commands for each stages in the step	Sequential algorithm
1	Process	Move WC to the nominal polishing position and CA to the home position	Substep Move_WC_CA
2	Process	Spin WC and platen at a touchdown speed	WC/Ptn spin procedure
3	Process	Supply slurry	Slurry supply procedure
4	Process	Move WC to the Z polish position	WC Z polish acquisition procedure
5	Process	Turn spray and drain valves on	Spray/drain valves on procedure
6	Process	Change AMA to positive (low) pressures	AMA pressurization procedure
7	Process	Spin WC and platen at a full polishing speed	WC/Ptn spin procedure
8	Process	Apply full polishing pressures to AMA	AMA pressurization procedure
9	Process	Set timer for polishing	Timer procedure
10	Process	Sweep WC if selected	Substep Sweep_WC
11	Process	Detect end point of the polishing	End-point detection procedure
12	Process	Turn sweep off	Sweep off procedure
13	Process	Turn slurry off	Slurry off procedure
14	Process	Buff wafer if selected	Substep Buff_Wafer

continued

TABLE 8.56 Continued

Index 22255.#			Control System Design
	Functional Requirements (FRs)		**Design Parameters (DPs)**
15	Process	Bring WC to the nominal X polish position	WC X move procedure
16	Process	Reduce WC and platen rotations to a lift up speed	WC/Ptn spin procedure
17	Process	Change AMA to vacuum to pick up wafer	AMA vacuuming procedure
18	Process	Move WC up to the Z clearance position for transportation	WC Z move procedure
19	Process	Turn spray and drain valves off	Spray/drain valves off procedure
20	Process	Move WC to the finish position	Substep Move_WC

Because each FR/DP set is self-explanatory, we will skip the detailed explanations of them. Logical requirements of the step are expressed as the functional requirements and the corresponding design parameters are chosen to realize the corresponding stages, which are mostly substeps and procedures. Each procedure usually has a single function and is coded in a few lines in the auto mode software. If a designer reached this level, he or she has arrived at the leaves of decomposition branches.

Equation 8.52 shows the design equation. The examination of causal relationship reveals some uncoupled elements in the lower diagonal, although most of them are decoupled. Coupling is natural, because in a sequential algorithm, later stages cannot be executed without the occurrence of the previous stages. However, the uncoupling means the corresponding stage can occur in parallel to its immediately preceding stage. For example, the design matrix indicates that $DP_{22255.3}$ can be a parallel process to $DP_{22255.2}$. Utilizing this information, the sequential functional diagram in Figure 8.26 can easily be constructed.

The design equation is given by

$$
\begin{Bmatrix}
FR_{22255.1} \\
FR_{22255.2} \\
FR_{22255.3} \\
FR_{22255.4} \\
FR_{22255.5} \\
FR_{22255.6} \\
FR_{22255.7} \\
FR_{22255.8} \\
FR_{22255.9} \\
FR_{22255.10} \\
FR_{22255.11} \\
FR_{22255.12} \\
FR_{22255.13} \\
FR_{22255.14} \\
FR_{22255.15} \\
FR_{22255.16} \\
FR_{22255.17} \\
FR_{22255.18} \\
FR_{22255.19} \\
FR_{22255.20}
\end{Bmatrix}
=
\begin{bmatrix}
X & 0 & 0 & 0 & 0 & 0 & 0 & 0 & 0 & 0 & 0 & 0 & 0 & 0 & 0 & 0 & 0 & 0 & 0 & 0 \\
X & X & 0 & 0 & 0 & 0 & 0 & 0 & 0 & 0 & 0 & 0 & 0 & 0 & 0 & 0 & 0 & 0 & 0 & 0 \\
X & 0 & X & 0 & 0 & 0 & 0 & 0 & 0 & 0 & 0 & 0 & 0 & 0 & 0 & 0 & 0 & 0 & 0 & 0 \\
X & X & X & 0 & 0 & 0 & 0 & 0 & 0 & 0 & 0 & 0 & 0 & 0 & 0 & 0 & 0 & 0 & 0 & 0 \\
X & X & X & 0 & X & 0 & 0 & 0 & 0 & 0 & 0 & 0 & 0 & 0 & 0 & 0 & 0 & 0 & 0 & 0 \\
X & X & X & X & X & X & 0 & 0 & 0 & 0 & 0 & 0 & 0 & 0 & 0 & 0 & 0 & 0 & 0 & 0 \\
X & X & X & X & X & X & X & 0 & 0 & 0 & 0 & 0 & 0 & 0 & 0 & 0 & 0 & 0 & 0 & 0 \\
X & X & X & X & X & X & 0 & X & 0 & 0 & 0 & 0 & 0 & 0 & 0 & 0 & 0 & 0 & 0 & 0 \\
X & X & X & X & X & X & X & X & X & 0 & 0 & 0 & 0 & 0 & 0 & 0 & 0 & 0 & 0 & 0 \\
X & X & X & X & X & X & X & X & X & 0 & 0 & 0 & 0 & 0 & 0 & 0 & 0 & 0 & 0 & 0 \\
X & X & X & X & X & X & X & X & X & X & 0 & 0 & 0 & 0 & 0 & 0 & 0 & 0 & 0 & 0 \\
X & X & X & X & X & X & X & X & X & X & X & 0 & 0 & 0 & 0 & 0 & 0 & 0 & 0 & 0 \\
X & X & X & X & X & X & X & X & X & X & X & 0 & X & 0 & 0 & 0 & 0 & 0 & 0 & 0 \\
X & X & X & X & X & X & X & X & X & X & X & X & X & 0 & 0 & 0 & 0 & 0 & 0 & 0 \\
X & X & X & X & X & X & X & X & X & X & X & X & X & X & 0 & 0 & 0 & 0 & 0 & 0 \\
X & X & X & X & X & X & X & X & X & X & X & X & X & X & X & 0 & 0 & 0 & 0 & 0 \\
X & X & X & X & X & X & X & X & X & X & X & X & X & X & X & X & 0 & 0 & 0 & 0 \\
X & X & X & X & X & X & X & X & X & X & X & X & X & X & X & X & X & 0 & 0 & 0 \\
X & X & X & X & X & X & X & X & X & X & X & X & X & X & X & X & X & 0 & X & 0 \\
X & X & X & X & X & X & X & X & X & X & X & X & X & X & X & X & X & X & X & X
\end{bmatrix}
\begin{Bmatrix}
DP_{22255.1} \\
DP_{22255.2} \\
DP_{22255.3} \\
DP_{22255.4} \\
DP_{22255.5} \\
DP_{22255.6} \\
DP_{22255.7} \\
DP_{22255.8} \\
DP_{22255.9} \\
DP_{22255.10} \\
DP_{22255.11} \\
DP_{22255.12} \\
DP_{22255.13} \\
DP_{22255.14} \\
DP_{22255.15} \\
DP_{22255.16} \\
DP_{22255.17} \\
DP_{22255.18} \\
DP_{22255.19} \\
DP_{22255.20}
\end{Bmatrix}
\quad (8.52)
$$

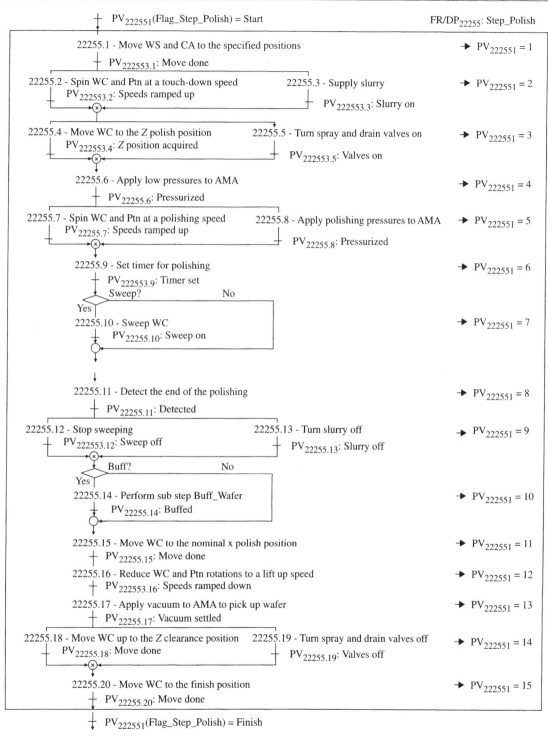

Figure 8.26 Sequential functional diagram of Step_Polish.

Figure 8.26 shows the sequential functional diagram constructed from the design matrix. Each FR/DP set corresponds to a stage in the sequential algorithm. The transition conditions between the stages are represented by the corresponding process variables. At the same time, the evolution or progress of the sequence is recorded by PV_{222551}, which is the status indicator of the polishing step. A higher level program determines the stage of the step and performs corresponding actions based on the status indicator monitoring. The logic flow is branched into parallel processes, where necessary, and joined again at an "$\otimes$ (AND)" junction.

Although comprehensive, the previous example is demanding because of its size. Let us take a look at a smaller example.

The substep Move_WC_CA (DP_{222551}) is a member of the transport substeps and transports both the wafer carrier and the conditioner in a coordinated manner to avoid collision. Similar to Step_Polish, it also has a sequential algorithm to guide the transportation. Table 8.57 is the decomposition of its sequential algorithm.

The design equation is given by

$$\begin{Bmatrix} FR_{222551} \\ FR_{222552} \\ FR_{222553} \\ FR_{222554} \\ FR_{222555} \\ FR_{222556} \\ FR_{222557} \end{Bmatrix} = \begin{bmatrix} X & 0 & 0 & 0 & 0 & 0 & 0 \\ X & X & 0 & 0 & 0 & 0 & 0 \\ X & 0 & X & 0 & 0 & 0 & 0 \\ X & X & X & X & 0 & 0 & 0 \\ X & X & X & X & X & 0 & 0 \\ X & X & X & X & 0 & X & 0 \\ X & X & X & X & X & X & X \end{bmatrix} \begin{Bmatrix} DP_{222551} \\ DP_{222552} \\ DP_{222553} \\ DP_{222554} \\ DP_{222555} \\ DP_{222556} \\ DP_{222557} \end{Bmatrix} \qquad (8.53)$$

The resulting sequential functional diagram is given Figure 8.27. The algorithm first checks the final positions of WC and CA to see if it is safe to move to the destinations. If either the X- or Z-axis relative distance between the two is less than the specified safety clearance, the collision check fails and the rest of the algorithms are not executed. The indicator returns an error to the calling step.

Once the check is successful, the wafer carrier and the conditioner are moved all the way up in parallel. Then the wafer carrier can clear the conditioner and the conditioner has enough Z clearance from the table objects. The conditioner is moved in the X-axis to its destination position. Once the X motion is complete, the conditioner is moved in its Z position, while the wafer carrier is in its X motion. Finally the wafer carrier settles in its Z position and the substep is finished. Figure 8.28 is the graphic display of the motions involved.

8.6.2.4 System Integration

Having finished the axiomatic decomposition from the system level down to the sequence level, it is time to assemble all the design parameters together in place and implement them. However, it is often necessary to provide a hardware base before an implementation. In a control system development, electrical wiring and component signal interfacing must be performed to enable further system development. Figure 8.29 shows the general outline of the hardware interface for the CMP α machine control system.

Also, it is a good practice to display all the design parameters in a single frame. If not possible because of the size, a few selected design parameters can still be shown. The design parameters can be arranged in the format of a DP tree diagram. It serves well as an

TABLE 8.57 Decomposition of Sequential Algorithm (Substep Move_WC_CA)

Index 2225113.#		Control System Design	
Functional Requirements (FRs)			**Design Parameters (DPs)**
Parents		Generate command values for WC and CA movements	Sequential algorithm
1	Process	Check the final points of WC and CA	Collision check procedure
2	Process	Move WC all the way up	WC up procedure
3	Process	Move CA to the up position	CA up procedure
4	Process	Move CA to the desired X position	CA X move procedure
5	Process	Move CA to the down position if specified	CA down procedure
6	Process	Move WC to the desired X position	WC X move procedure
7	Process	Move WC down to the desired Z position	WC down procedure

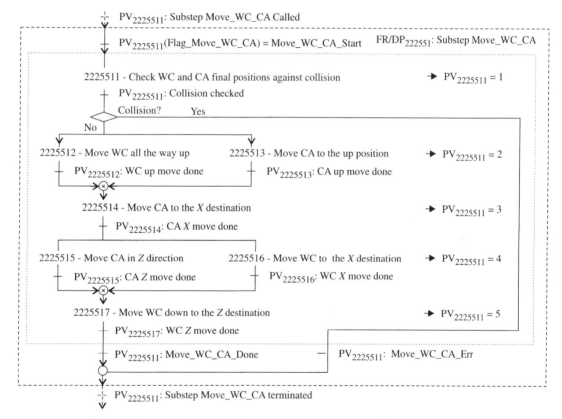

Figure 8.27 Sequential functional diagram of substep Move_WC_CA.

organization chart, which provides a quick overview of the system. Figures 8.30 and 8.31 show the DP tree diagram of the machine-level control system and the process-level control system, respectively. Note that only major DPs are shown because of space limitations.

Another method to display the design decomposition in a single frame is the design matrix. Each decomposition matrix can be entered as a subset of the parent design matrix.

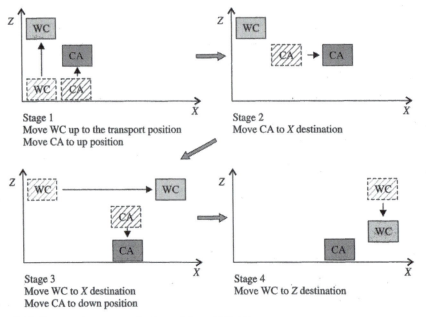

Figure 8.28 Four motion stages of substep Move_WC_CA.

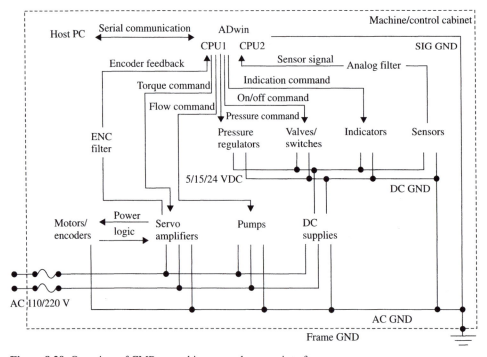

Figure 8.29 Overview of CMP α machine control system interface.

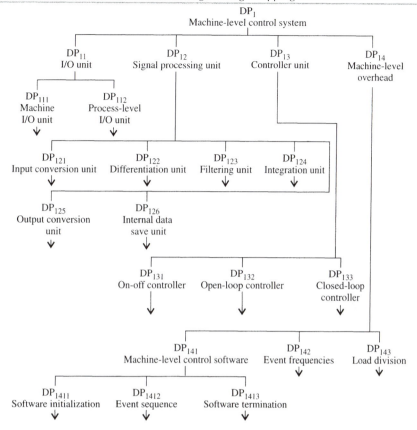

Figure 8.30 DP tree of machine-level control system.

Figure 8.32 shows the design matrix of the auto mode, which has been used as an example of application. The design matrix is useful to display causal relationships between different levels of FRs and DPs, because they all are displayed in a single table.

8.7 CONCURRENT ENGINEERING: MAPPING FROM FR TO DP TO PV

In Chapter 1, the basic requirement for concurrent engineering was given in terms of design matrices for product and process. This may be restated as

$$\{FR\} = [A]\{DP\} = [A][B]\{PV\} = [C]\{PV\} \tag{8.54}$$

For concurrent engineering to be possible, the Independence Axiom states that both the product design and the process design must satisfy functional independence, i.e., the matrix $[A]$ must be either diagonal or triangular and matrix $[B]$ must also be diagonal or triangular. Furthermore, when both of these matrices are fully triangular, either both of them must be lower triangular matrices or both of them must be upper triangular matrices. Otherwise, concurrent engineering cannot be done effectively.

To enable concurrent engineering, it may be necessary to change the manufacturing process if the condition for concurrent engineering cannot be satisfied with the existing

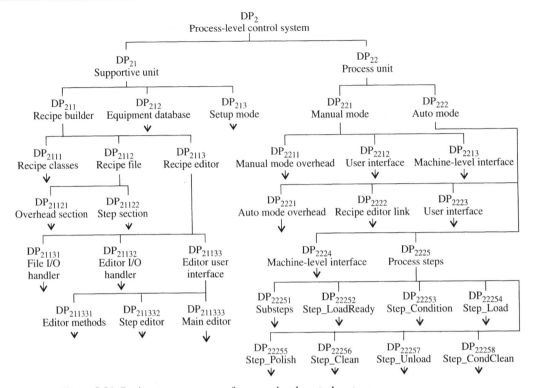

Figure 8.31 Design parameter tree of process-level control system.

processes. Sometimes, it may be easier to redesign the product rather than the manufacturing process. In *The Principles of Design* (Suh, 1990), the design and manufacture of a multilens plate were discussed. This is a good example of concurrent engineering where the product design had to be modified to be able to use an existing manufacturing process.

In many situations, the process variables are already fixed because the product must be manufactured using the available equipment. When this is the case, the choice of DPs is constrained by the given PVs.

8.8 PRODUCT SERVICE

Most products carry warranty and require service. The warranty cost can be more than 10% of a company's revenue, which is sometimes larger than profit. Therefore, if there is no way to eliminate the failure, it is important to develop a strategy for servicing a product.

There are two ways of servicing the product: regular preventive maintenance and service when the product fails. Regular preventive maintenance is often preferred because it will make customers happy and reduce the probability of having catastrophic failures that may be extremely expensive, both financially and in terms of the company's reputation. A company that sells large office copiers sends out its service people when its customers report machine failure. Each time it sends the service people out to the customer, no one wins. The customer cannot get its copies made while the copier is down, and the copy-machine

Design Parameters (DPs)

Functional Requirements (FRs)

			222																																		
			2221								2222			2223											2224					2225							
			22211	22212	22213	22214	22215	22216	22217	22218	22221	22222	22223	22231	22232	22233	22234	22235	22236	22237	22238	22239	2223A	2223B	22241	22242	22243	22244	22245	22251	22252	22253	22254	22255	22256	22257	22258
		22211	X	O	O	O	O	O	O	O	O	O	O	O	O	O	O	O	O	O	O	O	O	O	O	O	O	O	O	O	O	O	O	O	O	O	O
		22212	X	X	O	O	O	O	O	O	O	O	O	O	O	O	O	O	O	O	O	O	O	O	O	O	O	O	O	O	O	O	O	O	O	O	O
		22213	X	X	X	O	O	O	O	O	O	O	O	O	O	O	O	O	O	O	O	O	O	O	O	O	O	O	O	O	O	O	O	O	O	O	O
		22214	X	X	X	X	O	O	O	O	O	O	O	O	O	O	O	O	O	O	O	O	O	O	O	O	O	O	O	O	O	O	O	O	O	O	O
		22215	X	O	O	O	X	O	O	O	O	O	O	O	O	O	O	O	O	O	O	O	O	O	O	O	O	O	O	O	O	O	O	O	O	O	O
		22216	X	O	X	O	X	X	O	O	O	O	O	O	O	O	O	O	O	O	O	O	O	O	O	O	O	O	O	O	O	O	O	O	O	O	O
		22217	X	O	O	X	X	O	X	O	O	O	O	O	O	O	O	O	O	O	O	O	O	O	O	O	O	O	O	O	O	O	O	O	O	O	O
		22218	O	O	O	O	O	O	O	X	O	O	O	O	O	O	O	O	O	O	O	O	O	O	O	O	O	O	O	O	O	O	O	O	O	O	O
		22221	O	O	X	O	O	O	O	O	X	O	O	O	O	O	O	O	O	O	O	O	O	O	O	O	O	O	O	O	O	O	O	O	O	O	O
	2222	22222	O	O	O	O	O	O	O	O	O	X	O	O	O	O	O	O	O	O	O	O	O	O	O	O	O	O	O	O	O	O	O	O	O	O	O
		22223	O	O	O	O	O	O	O	O	O	O	X	O	O	O	O	O	O	O	O	O	O	O	O	O	O	O	O	O	O	O	O	O	O	O	O
		22231	X	O	O	O	O	O	O	O	O	O	O	X	O	O	O	O	O	O	O	O	O	O	O	O	O	O	O	O	O	O	O	O	O	O	O
		22232	X	O	O	O	O	O	O	O	O	O	O	O	X	O	O	O	O	O	O	O	O	O	O	O	O	O	O	O	O	O	O	O	O	O	O
		22233	X	O	X	O	O	X	O	O	X	X	O	O	O	X	O	O	O	O	O	O	O	O	O	O	O	O	O	O	O	O	O	O	O	O	O
		22234	X	O	O	O	O	X	O	O	O	O	O	O	O	O	X	O	O	O	O	O	O	O	O	O	O	O	O	O	O	O	O	O	O	O	O
		22235	X	X	O	X	O	O	O	O	O	O	O	O	O	O	O	X	O	O	O	O	O	O	O	O	O	O	O	O	O	O	O	O	O	O	O
222	2223	22236	X	O	O	O	O	X	O	O	O	O	O	O	O	O	O	O	X	O	O	O	O	O	O	O	O	O	O	O	O	O	O	O	O	O	O
		22237	X	O	O	O	O	X	O	O	O	O	O	O	O	O	O	O	O	X	O	O	O	O	O	O	O	O	O	O	O	O	O	O	O	O	O
		22238	X	O	O	O	O	X	O	O	O	O	O	O	O	O	O	O	O	O	X	O	O	O	O	O	O	O	O	O	O	O	O	O	O	O	O
		22239	X	O	X	O	O	O	O	O	X	O	O	O	O	O	O	O	O	O	O	X	O	O	O	O	O	O	O	O	O	O	O	O	O	O	O
		2223A	O	O	X	O	O	O	O	O	X	O	O	O	O	O	O	O	O	O	O	O	X	O	O	O	O	O	O	O	O	O	O	O	O	O	O
		2223B	O	O	O	O	O	O	O	O	O	O	X	O	O	O	O	O	O	O	O	O	O	X	O	O	O	O	O	O	O	O	O	O	O	O	O
		22241	X	X	X	X	O	O	O	O	O	O	O	O	O	O	O	X	O	O	O	O	O	O	X	O	O	O	O	O	O	O	O	O	O	O	O
		22242	X	X	O	X	O	O	O	O	O	O	O	O	O	O	O	X	O	O	O	O	O	O	X	X	O	O	O	O	O	O	O	O	O	O	O
	2224	22243	X	O	O	O	X	O	O	O	O	O	O	O	O	O	O	O	O	O	O	O	O	O	O	O	X	O	O	O	O	O	O	O	O	O	O
		22244	X	O	O	O	X	O	O	O	O	O	O	O	O	O	O	O	O	O	O	O	O	O	O	O	O	X	O	O	O	O	O	O	O	O	O
		22245	X	O	O	O	X	O	O	O	O	O	O	O	O	O	O	O	O	O	O	O	O	O	O	O	O	O	X	O	O	O	O	O	O	O	O
		22251	X	O	X	X	X	O	O	O	O	O	O	O	O	O	O	O	O	O	O	O	O	O	O	O	O	O	O	X	O	O	O	O	O	O	O
		22252	X	O	X	X	X	O	O	O	O	O	O	O	O	O	O	O	O	O	O	O	O	O	O	O	O	O	O	X	X	O	O	O	O	O	O
		22253	X	O	X	X	X	O	O	O	O	O	O	O	O	O	O	O	O	O	O	O	O	O	O	O	O	O	O	X	O	X	O	O	O	O	O
	2225	22254	X	O	X	X	X	O	O	O	O	O	O	O	O	O	O	O	O	O	O	O	O	O	O	O	O	O	O	X	O	X	X	O	O	O	O
		22255	X	O	X	X	X	O	O	O	O	O	O	O	O	O	O	O	O	O	O	O	O	O	O	O	O	O	O	X	O	X	O	X	O	O	O
		22256	X	O	X	X	X	O	O	O	O	O	O	O	O	O	O	O	O	O	O	O	O	O	O	O	O	O	O	X	O	X	O	O	X	O	O
		22257	X	O	X	X	X	O	O	O	O	O	O	O	O	O	O	O	O	O	O	O	O	O	O	O	O	O	O	X	O	X	O	O	O	X	O
		22258	X	O	X	X	X	O	O	O	O	O	O	O	O	O	O	O	O	O	O	O	O	O	O	O	O	O	O	X	O	O	O	O	O	O	X

Figure 8.32 Design matrix of auto mode.

manufacturer might lose its customer to a competitor. How often should you maintain the machine?

When companies think of failure, they often think of component failures. However, component failure is hard to monitor when there are thousands of parts in a given machine. What is much easier to monitor is the deterioration of functions. For example, when a copying machine starts making poor quality copies, we know that some of the highest FRs are not being satisfied. In this case, by sensing the contrast of image of the copies, we may identify all possible DPs that can cause the problem by examination of the system architecture or the function tree diagram. We can then narrow down the search for the failing component by checking other functions or even by intentionally causing failure of certain functions.

Once the system architecture is created, it is possible to create software or a table that can trace the failure to the component level by means of a suitable function analysis (Homework 8.7).

8.9 SYSTEM ARCHITECTURE

During the product design, the system architecture for the product—in the form of flow diagrams and the hierarchical trees of FRs and DPs—must be developed for several different reasons (Section 4.6). First, when a machine with many functional requirements is being designed, project coordination and project management can be done effectively if the system architecture is available so that everyone in the project team can have access to the necessary information. Second, the flow diagram will quickly identify the coupled designs. Third, the system architecture provides good documentation for the machine or system designed.

In Chapter 4, the system architecture for a machine with many functional requirements with tight tolerances was given to illustrate the construction of a flow diagram and the FR/DP/PV hierarchies. At one of the major automotive companies, a major project is being undertaken to construct the system architecture of its automatic transmission in order to reexamine the design. The company wants to make the most reliable and functional transmission. At the leading manufacturing firm of semiconductor equipment, the entire engineering practice is based on axiomatic design. The goal is to systematically design a sophisticated machine so as to eliminate ad hoc trial-and-error practices.

As is shown in Chapter 9, the probability of success decreases rapidly with an increase in the number of FRs when products are designed by ad hoc trial-and-error processes.

8.10 SUMMARY

In this chapter, the design of products is discussed. Successful products must be designed to satisfy clearly defined customer needs, functional requirements, and constraints. To develop competitive products, most appropriate DPs must be chosen to satisfy FRs and Cs. In engineered products, the quality of design and technology differentiates products' competitiveness. At the same time, it is necessary to put the importance of technology in a proper context. There are nontechnical factors that are equally important, such as the market size, marketing, purchasing, and human resources. Finally, to have a successful product, engineers, designers, and all those who have a stake in the success of the product must work together as a team.

The design process is illustrated by designing a power plant for automobiles in real time as this chapter was being written. This is done intentionally to illustrate the axiomatic design process for product design, not with the view of creating the best power plant in such a short period of time.

An industrial case study on the design of a depth charge is presented. The design of this product, which was based on the principles of axiomatic design, is much simpler and much more reliable than were the previous designs. This product is currently in production.

To illustrate how students can design complicated industrial-scale machines based on axiomatic design, the design and manufacture of a chemical-mechanical polishing (CMP) machine for semiconductor manufacturing is presented. This machine was designed and manufactured by four master's degree candidates in the Department of Mechanical

Engineering at MIT in two and a half years. They did a market study to determine the needs of industrial customers; designed the hardware, software, and control systems; manufactured some of the parts; used outside vendors to have key components manufactured according to their designs; and assembled them. The machine functioned as designed. The development cost of this machine was substantially less than what it would have cost in industrial firms.

The concept of error budgeting related to tolerances is discussed. It was pointed out that the traditional concept of error budgeting is done in the physical domain, which is incompatible with axiomatic design. In axiomatic design, we must deal with the design range in the functional domain and try to create a robust design in the physical domain by lowering the stiffness of the system.

The importance of system architecture in designing machines or systems with many FRs is again emphasized.

REFERENCES

Altshuller, G. *And Suddenly the Inventor Appeared*, Technical Innovation Center, Worcester, MA, 1996.

Clausing, D. *Total Quality Development*, ASME Press, New York, 1994.

Donaldson, R. R. "Error Budget," in *Technology of Machine Tools*, in *Machine Tool Accuracy*, Vol. 5, R. J. Hocken, ed., Machine Tool Task Force, 1980.

Galitello, K. A. *Two Stroke Cycle Engine*, U.S. Patent No. 4,876,991, October 31, 1989 (Galitello Research, Inc., P.O. Box 25, Torrington, CT 06790).

Nordlund, M. "An Information Framework for Engineering Design Based on Axiomatic Design," Doctoral Thesis, Royal Institute of Technology, Stockholm, Sweden, 1996.

Shiba, S., Graham, A., and Walden, D. *A New American TQM*, Productivity Press, Portland, OR, 1993.

Slocum, A. H. *Precision Engineering*, Prentice-Hall, Englewood Cliffs, NJ, 1992.

Suh, N. P. *Tribophysics,* Prentice-Hall, Englewood Cliffs, NJ, 1986.

Suh, N. P. *The Principles of Design,* Oxford University Press, New York, 1990.

Swenson, A. "Projektrapport I Axiomatic Design," Internal Document, Saab Missiles AB, Linkoping, Sweden, 1994.

Ulrich, K., and Eppinger, S. *Product Design and Development*, McGraw-Hill, New York, 1995.

Wood, K., and Otto, K. *Product Design*, Prentice-Hall, Englewood, NJ., 2000.

HOMEWORK

8.1 If there are six FRs and six DPs, what is the probability of finding the design solution if the design matrix is fully triangular?

8.2 Decompose the FRs and DPs of the free-floating piston engine shown in Figure 8.3 of Chapter 8. Draw detailed drawings that show your decomposed DPs.

8.3 Consider the design of a marketing organization that can sell automobiles through the Internet rather than through the usual dealership. The FRs are given by Equation (8.2) in Section 8.2. Develop a conceptual design and identify the DPs. Does your design satisfy the Independence Axiom?

8.4 Mechanical seals are used around shafts and flat surfaces to prevent oil from leaking out and to protect the device from abrasive particles penetrating into the device. Seal manufacturing is a $2 billion a year industry in the United States, but the cost in terms of loss of productivity and equipment resulting from poor seal performance is far greater. Seals are critical components in mechanical devices. Devices containing hundreds of seals are common and the failure of a single seal often has catastrophic consequences. Accordingly, significant effort has gone into improving seal design. At MIT, Professors Douglas Hart and Mary Boyce have been investigating the wear of face seals.

A conventional face seal is shown in Figure H.8.4.a. A face seal differs from the more common seal in that it contacts its bearing surface along a plane rather than along the perimeter of a shaft. The face seal assembly shown in the figure consists of the seal lip, stiffener ring, and load ring. The lip is manufactured from an elastomeric compound (e.g., filled polyurethane rubber) and provides the sealing action by contacting the moving part (i.e., shaft). The seal lip is bonded to a thermoplastic stiffener ring that distributes the force from the seal lip to the load ring. The rubber load ring positions the seal within the seal mount and allows axial movement between the bushing and pin without significantly altering the lip contact pressure. The outer edge of the seal is exposed to abrasive slurries and the inside of the seal is in contact with lubricant.

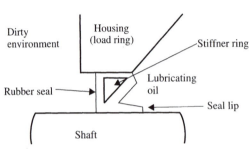

Figure H.8.4.a Schematic drawing of a rubber seal.

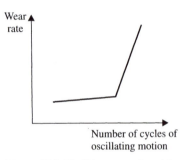

Figure H.8.4.b Wear rate of a rubber seal vs. the number of cycles of oscillating motion.

The seal undergoes a slow small oscillatory motion relative to the bushing. The wear of the seal consists of the "break-in" period and the aggressive wear period as shown in Figure H.8.4.b. The break-in period lasts about 7000 cycles. According to the finite-element analysis done by Professor Boyce and her student at MIT, the contact pressure is largest near the edge of the seal and decreases gradually.

Your design task is redesign the seal surface based on the following hypothetical wear model, which is an "educated" guess of how the seal wears.

Hypothesis:

1. Microscopic particles are generated at the interface between the seal and the shaft due to the rubbing action. In addition, small abrasive particles may also penetrate the interface from outside.
2. The particles at the interface agglomerate are under high pressure, forming larger particles.
3. The load at the interface is now carried by fewer but larger particles, which plow deeper into the seal surface, generating more particles.
4. Through this process, the seal wears.
5. When the seal wear reaches a critical stage, more abrasive particles from outside penetrate the interface rapidly, causing catastrophic failure.

8.5. A new diesel engine is being developed (see Figure H.8.5.a). This engine has many unique features: high power-to-weight ratio, low noise, lower emission of NO_x and particles than is the case for existing engines, and no cylinder head or gasket. The engine consists of a monoblock and an "engine housing" that contains the oil. The engine housing is attached to the engine through a cylindrical disk, which is a composite of two annular disks with rubber damper between them (see Figure H.8.5.b). A ball bearing is press-fit on this disk, which is in turn mounted on the crankshaft. When the engine is mounted on the car, the load is supported through the engine housing.

    a. State the FRs of the composite disk.
    b. What are the DPs?
    c. Write design equations.
    d. During a test run with a prototype engine, it was found that oil leaked between the disk and the engine housing, and also between the composite disk and the crankshaft. Propose a better design and justify your design.

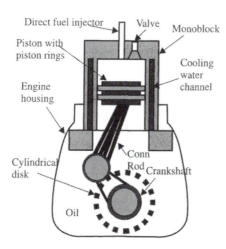

Figure H.8.5.a Schematic drawing of a mono-block diesel engine with direct fuel injector. The crankshaft is supported by ball bearings mounted on the engine block. The engine housing is held in place by a cylindrical disk, which is made of two annular disks with a rubber damper between them. Ball bearings are press-fit inside the inner annular disk.

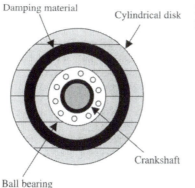

Figure H.8.5.b Cylindrical disk that connects the engine housing to the engine block through the crankshaft.

8.6   Silicon wafers used to make integrated circuit (IC) chips must be polished to control the waviness and roughness of the surface. It is done after growing oxides on the silicon wafer or after depositing metal or oxide layers on its surface. One of the functional requirements of the process is the creation of an oxide layer of uniform thickness. The thickness of the oxide layer should be within 20 Å. The polished silicon wafer must be flat within 25 Å and its roughness must be less than 20 Å. The wafer is 300 mm (12 inches) in diameter and about 1 mm thick.

One of the techniques used to polish the surface is called chemical and mechanical polishing (CMP). A schematic drawing of a CMP machine is shown in Figure H.8.6. A typical CMP machine consists of a polishing "cloth" (a relatively soft composite sheet consisting of polyurethane matrix and fibers) bonded on the top surface of a large rotating table (similar to a lapping table) and a spindle. The silicon wafer is placed on top of the polishing cloth and attached to the bottom flat plate of the spindle. Normal load is applied on the wafer by the rotating spindle, which can move perpendicular to the wafer surface along the spindle axis. During the polishing operation, silica slurry (i.e., fine silica particles dispersed in water) is continuously supplied to the polishing cloth. The interface temperature rise must be limited to a low temperature (about 1°C) to prevent warping of the wafer.

Figure H.8.6 Schematic drawing of a CMP machine.

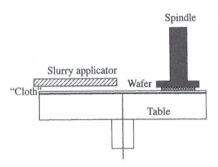

When a single spindle machine is used, the throughput rate (i.e., number of wafers processed per hour) is about 25. We must increase the throughput rate to about 100 wafers per hour to satisfy industrial needs. To increase the throughput rate, a proposal has been made to increase the number of spindles. One company has placed five spindles on top of the rotating plate. The idea is to place one wafer under each spindle to polish all five wafers at the same time and thus increase the throughout rate.

Each silicon wafer to be polished is slightly different from other wafers in thickness, waviness, and roughness. Also, the oxide layer thickness varies from wafer to wafer. Each spindle is also slightly different from the others. The polishing cloth is not perfectly uniform but is pliable.

The cost of ownership of these machines depends on many things. One of the important factors is the "footprint" of the machine (i.e., the floor space taken by each machine), as the cost of the semiconductor fabrication factory (the "fab") is very high. The 100 wafers-per-hour machine should not be more than twice the footprint of the current machine.

a. Is the single-spindle machine a good machine? How would you increase the throughput rate if you decide to use the single-spindle concept?

b. Is the proposed multispindle machine a good design? Please justify your answer by writing the design equations for the single-spindle machine and for the five-spindle machine.

c. How would you improve the design of the single-spindle CMP machine to satisfy the requirement of high throughput rate? How would you design a multispindle machine? How would you make the machine robust?

8.7 Study how your office copier works. Create the system architecture (either a flow diagram or the FR and DP trees). Based on the system architecture, design a preventive-maintenance strategy for the product.

8.8 Design a device that can satisfy FR_{13} and FR_{14} of the depth charge design discussed in Section 8.5.

8.9 Automobile doors are sealed using a weatherstrip to prevent dust and water from coming into the car and to insulate the inside compartment of the car from road noise. The weatherstrip design should not make it difficult to close the door. The weatherstrip must fill a gap between the door and car body, which can vary as much as 10 to 20 mm. It must last 10 years. The temperature of service may vary from $-30°C$ to $+50°C$. Design the weatherstrip.

8.10 A laptop computer manufacturer is interested in designing a mechanism that can hold the LCD display at any position relative to the keyboard of the computer, as shown in Figure H.8.10. It is important that the LCD part, which weighs 600 g, rotate smoothly without exerting much load. Design a mechanism that can achieve the desired functions. The mechanism must be durable and able to withstand 10,000 cycles of opening and closing motions.

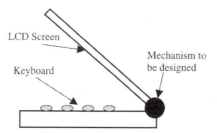

Figure H.8.10 Rotating hinge design for PCs.

8.11 Semiconductor manufacturing equipment must have a tightly controlled environmental chamber to control the temperature within $0.1°C$ and the relative humidity to within 0.5%. This is typically done by bringing in controlled air from outside the chamber. The operators of the equipment set the air temperature and the humidity of the chamber, which may change several times a day. The desired temperature typically ranges from $21°$ to $23°C$ and the desired relative humidity is in the range of 40 to 50%. Design an environmental chamber that is reliable, simple, and inexpensive to make. Write the design equation for your proposed design. Does it satisfy the Independence Axiom and the Information Axiom?

9.1 INTRODUCTION

On May 6, 1997, the *New York Times,* one of the leading newspapers in the United States, carried an article entitled "Researchers on Complexity Ponder What It's All About." The appearance of such an article in a daily newspaper indicates that the issue of complexity has reached the center stage of science and technology in the 1990s. The article stated that

> Some of the grandest phenomena, like the coursing of comets around the sun, are marvelously predictable. But some of the most mundane, like weather, are so convoluted that they continue to elude the most diligent forecasters. They are what scientists call complex systems. Though made up of relatively simple units—like the molecules in the atmosphere—the pieces interact to yield behavior that is full of surprise[s].

Past attempts to define complexity. In spite of all the efforts that have been made, mathematicians, scientists, and engineers have not even accepted a common definition of what is meant by complexity (Sohlenius, 1999). According to the author's colleague, Seth Lloyd,[1] there are about three-dozen different ways scientists use the word "complexity."

---

[1] This statement was attributed to Professor Seth Lloyd of MIT by the *New York Times* article.

Some definitions deal with the complexity of process; for instance, how much computing it would take to solve a problem (Cover, 1991). Complexity has also been equated with the scale of measure—how many bits of information it takes to describe an object or a message (Shannon and Weaver, 1949). Those in the field of manufacturing associate complexity with how much effort it would take to manufacture a product (Suh, 1990). Chaitin (1987) and others have come up with a concept called "algorithmic complexity." The basic idea is that simple tasks can be done by short computer programs whereas complex tasks require longer programs. According to this view, it should be possible to measure the complexity of a task by the length of its most compact description. The problem with this idea is that the length of even the shortest computer program depends on the design of the software as well as the coding.

Gell-Mann (Gell-Mann and Lloyd, 1996) proposed the idea of schema to identify the system's regularities as a means of defining complexity. He claimed that the length of the schema measures what he calls "effective complexity," which is roughly the length of a compact description of the identified regularities of an entity. In the case of language, the schema is its grammar. Gell-Mann and Lloyd (1996) also proposed the concept of total information, which is effective complexity plus an entropy term that measures the information required to describe the random aspects of the entity. Bennet (1985) has developed a different measure of complexity called "logical depth." The idea is to gauge how long it would plausibly take for a computer to go from a simple blueprint to the final product. Huberman and Hogg (1994) equate complexity with "a phase transition" between order and randomness. Lloyd and Pagels (1988) equated complexity to free energy. There are many other views (Yates, 1978, in Flood and Carson, 1993). All of these efforts are attempts to discover the basic absolute measure for complexity, which is contrary to the concept of information and complexity used in axiomatic design.

Science[2] devoted a special section to the topic of complex systems. The journal dealt with complex systems in many fields of science, including life sciences, chemistry, mathematics, biology, physiology, geology, meteorology, and economics. No attempt was made to present a unified definition of complexity. It is interesting to review the notion of complexity used by different authors to describe the complexity of their fields.

- In the introductory article by R. Gallagher and T. Appenzeller, a "complex system" is taken to be one whose properties are not fully explained by an understanding of its component parts.
- In their article entitled "Simple Lessons from Complexity," N. Goldenfield and L. P. Kadanoff state that "complexity means that we have structure with variations. Thus, a living organism is complex because it has many different working parts, each formed by variation in the working out of the same genetic coding."
- In their article on "Complexity in Chemistry," G. M. Whiteside and R. F. Ismagilov state: "a complex system is one whose evolution is very sensitive to initial conditions or to small perturbations, one in which the number of independent interacting components is large, or one in which there are multiple pathways by which the system can evolve. Analytical descriptions of such systems typically require nonlinear differential equations."

---

[2] These articles came out in *Science*, Vol. 284, No. 5411, April 2, 1999.

Their second characterization is more informal, that is, "the system is 'complicated' by some subjective judgement and is not amenable to exact description, analytical or otherwise."

- In the abstract of their article entitled "Complexity in Biological Signaling Systems," G. Weng, U. S. Bhalla, and R. Iyengar state: "Complexity arises from the large number of components, many with isoforms that have partially overlapping functions, from the connections among components, and from the spatial relationship between components."

Axiomatic design perspective of complexity and information. Many of the past ideas of complexity are not consistent with that defined in axiomatic design. In many of the past works, complexity was treated in terms of an absolute measure. In axiomatic design, information and complexity are defined only relative to what we are trying to achieve and/or want to know. Information is defined as a logarithmic function of the probability of achieving the specified functional requirements (Suh, 1990). Complexity is related to information.

Axiomatic design has other implications for complexity. According to the Independence Axiom, complexity arises because the task—decision making in design or the question we seek to answer in science—is an aggregation of individual elements or decisions that make up the hierarchy of the decomposed FRs and DPs. Some of these elements are related to each other serially in a given branch, whereas others are in parallel branches of the hierarchy that interact only when the branches merge at higher levels. Complexity arises when we are unable to deal with or understand the behavior of the aggregation. The information required to understand the behavior of the aggregation at the highest level is directly related to the number of interacting elements and the information associated with each interacting element. The complexity of aggregation or system increases with the information content required to make the aggregate behave as intended or answer the question posed in science.

To generalize the notion of complexity in the context of axiomatic design, we need to define the term complexity more precisely. In axiomatic design, complexity is defined only when specific functional requirements (or the exact nature of the query) are defined. Complexity is defined as a measure of uncertainty. Complexity is related to information, which is defined in terms of the probability of achieving the functional requirements (FRs). We should be able to specify the meaning of complexity in the following situations.

- What is the complexity in a bar of AISI 1020 mild steel that has to be machined to the dimensions of 1 m in length, 0.02 m in diameter, and 10 μm in surface finish?
- What is the complexity of a machine that has seven FRs?
- What is the complexity of a laser printer?
- What is the complexity of a manufacturing process designed to make nylon fibers with a microcellular structure consisting of 1-μm-diameter bubbles with a cell density of 10^{12} bubbles/cm^3?
- How complex is the job of being the weather person?

In this chapter, we will introduce two new concepts that classify and define complexity: *time-dependent complexity* and *time-independent complexity*. Time-independent complexity can further be divided into time-independent *real* complexity and time-independent *imaginary* complexity. Time-dependent complexity may also be divided into two kinds: time-dependent *combinatorial* complexity and time-dependent *periodic* complexity. These complexities are examined and defined further in this chapter.

9.2 COMPLEXITY, UNCERTAINTY, INFORMATION, AND PERIODICITY

9.2.1 Preliminary Remarks

In axiomatic design, the design process is described in terms of the mapping between domains. The design goals for a product (or software, systems, etc.) are described in terms of FRs in the functional domain. The design task is to achieve the set of specified FRs by mapping these FRs to DPs in the physical domain (see Figure 9.1). Thus, the selection of DPs determines the probability of satisfying the FRs. The Independence Axiom specifies that DPs must be chosen so that the resulting design is either uncoupled or decoupled, which are characterized by a diagonal and a triangular design matrix, respectively.

When the FR is defined, its desired target value FR_0 and its design range are specified, as shown in Figure 9.2. However, the actual probability density function (pdf) of the resulting design embodiment determines the system range, which may be different from the design range, as discussed in Chapters 1, 2, and 3. The portion of the design range overlapped by the system range is called the common range.

If the system pdf for a given FR_i is denoted $p_s(FR_i)$, then the probability P_i of satisfying FR_i is given by

$$P_i \left(dr^l \leq FR_i \leq dr^u \right) = \int_{dr^l}^{dr^u} p_s(FR_i) \, d(FR_i) \tag{9.1}$$

where dr^l and dr^u are the lower and upper limits of the design range, respectively. Information content I_i is defined in terms of the probability P_i of satisfying a given FR_i as

$$I_i = -\log_2 P_i \tag{9.2}$$

In the general case of m FRs, the information content for the entire system I_{sys} is

$$I_{sys} = -\log_2 P_{\{m\}} \tag{9.3}$$

where $P_{\{m\}}$ is the joint probability that all m FRs are satisfied.

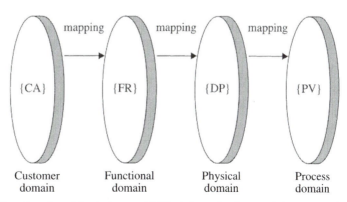

Figure 9.1 Four domains of the design world. The $\{x\}$ are the characteristic vectors of each domain. Design of products involves mapping from the functional domain to the physical domain. Design of processes involves mapping from the physical domain to the process domain.

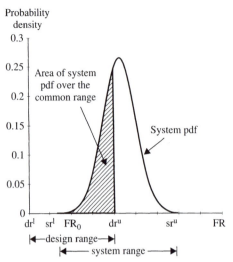

Figure 9.2. Design range, system range, and system pdf. The probability of satisfying the FR is given by the shaded area under the system pdf in the common range. dr^l and dr^u are the lower and upper bounds of the design range; sr^l and sr^u are the lower and upper bounds of the system range.

When all FRs are statistically independent, as is the case for an uncoupled design, $P_{\{m\}} = \Pi_i P_i$. Therefore I_{sys} may be expressed as

$$I_{sys} = \sum_{i=1}^{m} I_i = -\sum_{i=1}^{m} \log_2 P_i \tag{9.4}$$

When all FRs are not statistically independent, as is the case for a decoupled design,

$$I_{sys} = -\sum_{i=1}^{m} \log_2 P_{i|\{j\}} \qquad \{j\} = \{1, 2, \ldots, i-1\} \tag{9.5}$$

where $P_{i|\{j\}}$ is the conditional probability that FR_i is satisfied given that the other relevant (i.e., correlated) $\{FR_j\}_{j=1,\ldots,i-1}$ are also satisfied.

Because the system has a fixed number of FRs, complexity is unrelated to the number of FRs, but instead is the probability that a system will achieve all FRs. A system with low total I (high probability of satisfying all FRs) is less complex than a system with the same number of FRs and DPs, but with high total I (low probability of satisfying all FRs). This leads us to a specific definition of complexity.

9.2.2 Definition of Complexity

Complexity is defined as a measure of uncertainty in achieving the specified FRs.[3] Therefore, complexity is related to information content, which is defined as a logarithmic function of the probability of achieving the FRs. The greater the information required to achieve the FRs of a design, the greater is the information content, and thus the complexity.

---

[3] This definition is consistent with ideas such as a machine with many parts being more complex than a machine with fewer parts in many cases, as the uncertainty of achieving the functions of the machine increases with the number of parts. Uncertainty increases as the ability to predict the future outcome decreases.

In this section, we will discuss two new concepts of complexity: *time-dependent complexity* and *time-independent complexity*. Time-independent complexity consists of two orthogonal components: time-independent *real complexity* and time-independent *imaginary complexity*. Their vector sum is called *absolute complexity*. The real complexity of coupled designs is larger than that of uncoupled or decoupled designs. Imaginary complexity can be reduced when the design matrix is known. As an example of time-independent imaginary complexity, the design of a printing machine based on xerography will be discussed.

There are also two kinds of time-dependent complexity: time-dependent *combinatorial complexity* and time-dependent *periodic complexity*. Time-dependent combinatorial complexity can lead to a chaotic situation if the number of combinations continues to explode as a function of time. On the other hand, a time-dependent periodic complexity reduces the number of combinations to a finite set and reduces the complexity problem to a deterministic one. Using a robot-scheduling problem as an example, it is shown that a coupled design with a combinatorial complexity can be reduced to a decoupled design with periodic complexity. The introduction of periodicity simplifies the design by making it deterministic, which requires much less information. Whenever a combinatorial complexity is converted to a periodic complexity, uncertainty is reduced and design is simplified. These findings may have profound implications for engineering and other fields.

9.2.3 Time-Independent Complexities: Real Complexity, Imaginary Complexity, and Absolute Complexity

In axiomatic design, time-independent complexities—*real complexity* and *imaginary complexity*—are defined to deal with two kinds of uncertainties—*real* uncertainty and *imaginary* uncertainty. Imaginary complexity is not related to real complexity; that is, it is orthogonal to real complexity. *Absolute complexity* is defined as the vector sum of these two orthogonal components of time-independent complexity.

9.2.3.1 Real Complexity

Real complexity is defined as a measure of uncertainty when the probability of achieving the FR is less than 1.0 because the common range is not identical to the system range. In Figure 9.2, the uncertainty is given by the unshaded area of the system pdf. Real uncertainty in design exists because the actual embodiment of the design does not quite satisfy the desired FR at all times.

The probability of achieving a given FR is determined by the overlap between the design range and the system range, called the common range (Figure 9.2). Therefore, real uncertainty exists even when the Independence Axiom is satisfied, if the common range is not the same as the system range. Thus, real complexity can be related to the information content, which was defined in terms of the probability of success in achieving the desired set of functional requirements [see Equation (9.4)]. If we denote real complexity as C_R, then we will define real complexity to be equal to the information content as

$$C_R = I \qquad (9.6)$$

The information content is a measure of uncertainty and thus is related to real complexity.

Real complexity may be reduced when the design is either uncoupled or decoupled, i.e., when the design satisfies the Independence Axiom. For uncoupled designs, the system range for each FR can be shifted horizontally by changing the DPs until the information content is at a minimum, as other FRs are not affected by such a change. Therefore, the mean value of FR provided by the system can be determined by adjusting the corresponding DP until the information is at a minimum. For decoupled designs, the system range can be shifted to seek the minimum information point by changing the DPs in the sequence given by the design matrix. The best values of DPs can be obtained by finding where the value of real complexity reaches its minimum when the following two conditions are satisfied.

$$\sum_{j=1}^{n} \frac{\partial C_R}{\partial DP_j} = 0 \qquad (9.7)$$

$$\sum_{j=1}^{n} \frac{\partial^2 C_R}{\partial DP_j^2} > 0 \qquad (9.8)$$

When the design is uncoupled, the solution to Equations (9.7) and (9.8) can be obtained for each DP without regard to any of the other DPs, i.e., each term of the series must be equal to zero. In the case of decoupled designs, these equations must be evaluated in the sequence given by the design equation because the design matrix is triangular.

In the case of a coupled design, real complexity can also be changed, but the minimum information point for each FR is no longer meaningful, because when one of the DPs is changed in order to affect only one FR, all other FRs may change. Therefore, the minimum information point is defined only for the entire set of DPs where the information for the entire set of FRs is the minimum. This corresponds to an "optimum" point, which is often sought in operations research. However, this is a poor design solution because many of the FRs can be satisfied exactly in the design space if the Independence Axiom is satisfied. In many cases of coupled design, Equation (9.7) may never be satisfied. Coupled designs have larger real complexity than uncoupled or decoupled designs for the same set of FRs.[4]

9.2.3.2 Imaginary Complexity

Imaginary complexity is defined as uncertainty that is not real uncertainty, but arises because of the designer's lack of knowledge and understanding of a specific design itself. Even when the design is a good design, consistent with both the Independence Axiom and the Information Axiom, imaginary (or unreal) uncertainty exists when we are ignorant of what we have.

To understand the distinction between real and imaginary uncertainty, consider a decoupled design with m FRs and n DPs given by the triangular matrix in Equation (9.9) where $m = n$.

---

[4] This is consistent with Theorem 18 (Existence of an Uncoupled Design): There always exists an uncoupled design that has less information than a coupled design.

$$
\begin{Bmatrix} FR_1 \\ FR_2 \\ FR_3 \\ \dots \\ \dots \\ \dots \\ FR_m \end{Bmatrix} = \begin{bmatrix} X & 0 & 0 & 0 & \dots & 0 \\ X & X & 0 & 0 & \dots & 0 \\ X & X & X & 0 & \dots & 0 \\ \dots & \dots & \dots & \dots & \dots & 0 \\ \dots & \dots & \dots & \dots & \dots & 0 \\ \dots & \dots & \dots & \dots & \dots & 0 \\ X & X & X & X & \dots & X \end{bmatrix} \begin{Bmatrix} DP_1 \\ DP_2 \\ DP_3 \\ \dots \\ \dots \\ \dots \\ DP_n \end{Bmatrix} \tag{9.9}
$$

which may be generally written as

$$
\{FR\} = \left[A^{LT} \right] \{DP\} \tag{9.10}
$$

where $\left[A^{LT} \right]$ is a lower triangular matrix.

The design represented by Equation (9.9) satisfies the Independence Axiom. Thus the design can be implemented because there is no uncertainty associated with it if the DPs are changed in the order indicated in Equation (9.9) and if each of the system ranges is inside its associated design range. If the common range is the same as the system range for all FR_i, then the real complexity is equal to zero. If the common range is not the same as the system range for any of the FR_i, there is real uncertainty and real complexity. This real complexity cannot be removed unless the system range and the common range are made the same for all FRs by choosing new DPs or by making the design more robust so as to remove uncertainty.

The decoupled design given by Equation (9.9) can be a source of *imaginary complexity*, despite the fact that the design does satisfy the Independence Axiom. Imaginary complexity exists whenever the perceived complexity is not entirely due to real complexity. This imaginary uncertainty exists only in the mind of the designer because the designer does not know that the design represented by Equation (9.9) is a good design or when the designer does not write the design equation.

Suppose the designer does not recognize that the design is a decoupled design, although it can be represented by Equation (9.9), and thus does not know that the DPs must be changed in a proper order to make the design achieve the given set of m FRs. Then the designer resorts to trial-and-error methods of evaluation, trying many different sequences of DPs to satisfy the FRs. There are $n!$ distinct sequences of DPs, of which only one is correct. Then the probability of finding the right sequence of n DPs to satisfy the entire set of m FRs is given by[5]

$$
P = \frac{1}{n!} \tag{9.11}
$$

The probability of finding the right sequence through a random trial-and-error process goes down rapidly with an increase in the number of DPs as shown by Table 9.1. When n is 5, the probability of finding the right sequence is 0.008, which is a small number. Therefore, this design appears to be very complicated and one would say that this design is very complex because the uncertainty is large. However, it is not a case of real uncertainty;

---

[5] The actual probability of satisfying the FRs may be less than the probability of finding the right sequence, as the system range for each FR may be different from the design range. However, for large n, this probability of finding the right sequence is likely to dominate.

Table 9.1 Probability of Finding the Correct Decoupled Design
When There Are n FRs

n	$n!$	$P = 1/n!$
1	1	1
2	2	0.5
3	6	0.1667
4	24	0.04167
5	120	0.8333×10^{-2}
6	720	0.1389×10^{-2}
7	5,040	0.1984×10^{-3}
8	40,320	0.2480×10^{-4}

this uncertainty is artificially created as a result of the lack of understanding of the system designed. Therefore, this kind of uncertainty is defined as *imaginary uncertainty*. In many situations, this imaginary uncertainty leads to the erroneous conclusion that a design is complex—although it may not be—as a result of lack of fundamental understanding of axiomatic design theory.

If we denote imaginary complexity as C_I, then it may be related to the probability of finding the right sequence given by Equation (9.11) as

$$C_I = \log n! \tag{9.12}$$

For very large n (e.g., $n > 100$), Equation (9.12) may be written as

$$C_I \approx n(\log n - 1) \tag{9.13}$$

If the design matrix is such that there is more than one possible sequence of n DPs that can equally satisfy the m FRs, then the probability of finding an appropriate sequence is given by

$$P = z/n!$$

where z = the number of sequences that will satisfy the FRs. Therefore, as z increases, the design will appear to be less complex because the imaginary uncertainty decreases. However, the real uncertainty does not change with z.

EXAMPLE 9.1 Xerography-Based Printing Machine

HG Company, one of the leading printing-press manufacturers in the world, has just developed a commercial label-printing machine based on a xerography technique. This machine can quickly print commercial labels as soon as the original copy is inserted into the machine because it is based on the xerography principle. The design of the machine is schematically illustrated in Figure E.9.1.a.

The optical image of the label is transmitted to the surface of the selenium-coated aluminum cylinder using light. The cylinder rotates at a constant speed. When the charged section of the cylinder passes by the toner box, the oppositely charged liquid toner transfers to the charged part of the selenium surface. To control the thickness of the toner layer on the selenium drum, the wiper roll removes the extra-thick toner layer from the surface

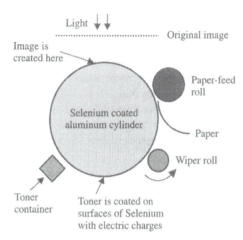

Light ↓ ↓

Original image

Image is
created here

Paper-feed
roll

Selenium coated
aluminum cylinder

Paper

Wiper roll

Toner
container

Toner is coated on
surfaces of Selenium
with electric charges

Figure E.9.1.a Schematic drawing of the xerography-based printing machine. The image is transmitted to the selenium-coated aluminum cylinder using light. When the charged section of the cylinder passes by the toner box, the oppositely charged toner liquid transfers to the charged part of the selenium surface. The wiper roll removes the extra-thick toner layer from the surface of the cylinder. Paper comes in contact with the selenium surface under the light pressure exerted on the paper by the paper-feed roll.

of the cylinder. Paper is fed into the gap between the main selenium cylinder and the paper-feed roll. When the paper comes in contact with the selenium surface under the light pressure exerted on the paper by the paper-feed roll, the image is firmly printed on the paper.

The Advanced Engineering Division of HG Co., which was developing this printing machine, ran into trouble. They found that sometimes the selenium coating is badly scratched, creating poor images and damaging the expensive selenium-coated rolls (about $4000 per cylinder, which was about 18 inches in diameter). Because the beta machine had to be shipped in a few months, they assigned many scientists and engineers to figure out the problem and solve it.

The scientists and engineers came to the conclusion that the scratch marks (in the form of lines) must have been the result of abrasive wear. They attributed the source of abrasion to unknown abrasive particles that somehow got into the toner tank. This reasoning received much internal support from everyone in the Advanced Engineering Division, as the machine (which was about 30 feet long) was being assembled at a corner of a large machine shop. They conjectured that tiny metal chips from the machining operation somehow got into the tank, occasionally scratching the selenium drum.

To make sure that the toner was free of any abrasive particles, they installed special filters that would remove all particles greater than a few microns and put a plastic sheet around the machine to create a clean environment around the machine. However, the despicable scratch marks would not go away! The high-level managers of the company became uneasy about the situation and decided to consult a tribologist at MIT about this abrasive wear problem. The tribologist told them to read a reference book on tribology to learn all about the things that affect abrasive wear.

After a few months, the tribologist received an urgent call from HG Co. They said that they had to ship the beta machine to a customer's factory in a week and yet the scratch marks were still there—apparently the reference book did not do any good! The tribologist was asked to hop on an airplane right way and visit the factory where the machine was being tested. So he went.

What do you think the tribologist found at the HG Company?

SOLUTION

The tribologist, who also knows something about axiomatic design, listened to the HG engineers and scientists who explained all the things they had done and their theory on the cause of the problem. They were sure that somehow devilish small particles were getting into the printing machine and the toner box and that it was these particles that caused the scratches on the surface. Indeed the examination of the surface and micrographs indicated that the scratch marks were typical scratches caused by abrasive particles. However, the tribologist was not convinced that the explanation given by the HG engineers and scientists was correct.

The functional requirements of the machine, assuming that abrasive particles somehow got into the toner box, may be chosen to be the following:

FR_1 = Create electrically charged images.
FR_2 = Coat the charged surface with toner.
FR_3 = Wipe off the excess toner.
FR_4 = Make sure that abrasive particles do not cause abrasion.
FR_5 = Feed the paper.
FR_6 = Transfer the toner to the paper.
FR_7 = Control throughput rate.

The tribologist reasoned that the DPs used by HG personnel (although they did not use axiomatic design) in their trial-and-error approach were:

DP_1 = Optical system with light on selenium surface
DP_2 = Electrostatic charges of the selenium surface and the toner
DP_3 = Wiper roller
DP_4 = Filter
DP_5 = Paper-feeding mechanism
DP_6 = Mechanical pressure
DP_7 = Speed of the cylinder

Because there are seven FRs and seven DPs, there are more than 5000 sequences of DPs to consider if they try to run the tests by trying different sequences of DPs. The probability of success using a trial-and-error method is quite small. Even if they devised an orthogonal array experiment, there are still too many tests to determine the cause. Furthermore, if the design is a decoupled design, simply identifying important DPs through the orthogonal array experiment will not yield the answer. Indeed their extensive tests did not yield any solution!

The design matrix that may represent the thinking of the HG engineers may be represented as follows:

	DP_1	DP_2	DP_3	DP_4	DP_5	DP_6	DP_7
FR1	X	0	0	0	0	0	0
FR2	X	X	0	0	0	0	0
FR3	0	0	X	0	0	0	0
FR4	0	0	X	X	X	0	0
FR5	0	0	0	0	X	0	0
FR6	0	0	0	0	0	X	0
FR7	0	0	0	0	X	0	X

According to the above design matrix, the order of FR_4 and FR_5 as well as DP_4 and DP_5 should be changed to obtain a triangular matrix. What the matrix is indicating is that if the paper-feeding mechanism or process creates particles, filtering the toner outside the machine will not do any good. The filter must also remove the particles generated by the paper-feeding mechanism. This is not easy to achieve.

Another solution is to prevent large particles from ever approaching the interface by means of controlling the fluid motion. For abrasion to occur, kinematic considerations indicate that somehow the abrasive particle, whatever it may be made of, must be stationary at the interface between the main cylinder and the wiper roll. If the particle goes through, then at most the selenium surface will be indented rather than scratched. Then FR_4 (make sure that abrasive particles do not cause abrasion) may be decomposed as

FR_{41} = Prevent the abrasive particle from being anchored at the interface between the main cylinder and the wiper roll.

FR_{42} = Prevent the particles from approaching the interface.

At this point, it is instructive to consider the kinematics and fluid mechanics of the toner motion near the entrance between the wiper roll and the main roll.[6] When the machine is first started, if the main cylinder rotates first before the wiper roller is rotated, the toner will be dragged along and any particle in the toner will anchor at the narrow section of the opening between the roller and the main cylinder. Furthermore, if the surface speed of the main cylinder is greater than that of the counterrotating wiper roller, the pressure at the narrow gap will be greater and the tendency to squeeze in the abrasive particle at the interface between the main cylinder and the wiper roller will be greater.

On the other hand, if the wiper roller starts turning first and if the surface speed of the wiper roller is greater than and opposite to the surface speed of the main cylinder (as indicated in Figure E.9.1.b), then the pressure at the entrance will be less. It will reduce the tendency for large particles to come into the narrow gap. Furthermore, the vortex motion in the toner will prevent the large particles from approaching the main cylinder/wiper interface as shown in Figure E.9.1.b.

Then DPs may be chosen as

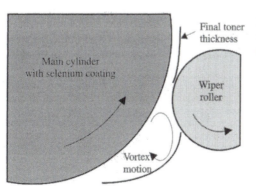

Figure E.9.1.b The vortex motion of the toner and the rotational direction of the main cylinder and the wiper roller.

---

[6] In selecting DPs, the designer's knowledge of associated physics and engineering is obviously indispensable. Axiomatic design cannot make up for lack of fundamental understanding of physics, mathematics, and other associated knowledge bases.

DP_{41} = The order of rotation of the wiper roller and the main cylinder (wiper roller rotates first)

DP_{42} = The surface speed of the wiper roller greater than and opposite to the surface speed of the main cylinder

The tribologist made the suggestion that DP_{41} and DP_{42} be implemented. The machine had a digital control system, and therefore DP_{41} and DP_{42} could be implemented immediately. He also asked the HG engineers to put abrasive particles into the toner intentionally. When the machine was turned on, there were no more scratch marks! The tribologist happily hopped on an airplane and returned to Boston. He had spent six working hours at HG Company to solve the problem, whereas many months using the trial-and-error approach prior to his visit produced no success.

The tribologist was a typical professor but not a good businessman. Instead of charging the company by the job done, he charged them for his service by the hour!

If a design is uncoupled with a diagonal design matrix and zero information content, both the real uncertainty and the imaginary uncertainty are equal to zero. In this case, both those who do and those who do not understand axiomatic design may come to the same conclusion on complexity and uncertainty.

9.2.3.3 Absolute Complexity

The absolute complexity C_A is defined as

$$C_A = C_R + C_I \qquad (9.14)$$

C_R, the real complexity, and C_I, the imaginary complexity, may be plotted in a two-dimensional complex plane, as shown in Figure 9.3. The vertical axis is the axis of imaginary complexity, i.e., the axis of "ignorance," as it is caused by lack of knowledge, which yields the perception that the design is more complex than it really is. The horizontal axis represents real uncertainty as a result of design and/or unknown behavior of nature. C_I and C_R are orthogonal to each other because the imaginary complexity has no relationship to the real complexity and vice versa. The absolute complexity C_A is shown as the vector sum of C_R and C_I because C_R and C_I are orthogonal to each other.

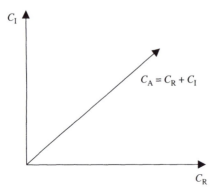

Figure 9.3 Real and imaginary components of complexity. The vertical axis is the axis of ignorance and the horizontal axis represents real uncertainty as a result of design and/or unknown behavior of nature.

It is difficult to predict the exact values of C_R and C_I a priori if the design is coupled or decoupled. However, a bound for C_I can be estimated if the design is a decoupled design using Equation (9.12). When the design is uncoupled, the imaginary component of complexity is equal to zero and only real complexity may exist if the system range is not inside the design range for all FRs. In the case of a coupled design, the magnitude of the imaginary complexity can be very large and dominate the real complexity.

Based on the foregoing discussion of absolute complexity, real complexity, and imaginary complexity, we may adopt the following definition of complexity:

> *Time-independent complexity* is a measure of uncertainty in achieving a given set of FRs and thus is related to information content. It consists of two orthogonal components—real complexity and imaginary complexity. Real complexity is defined as a measure of real uncertainty in achieving a set of FRs and thus is related to the information content given by Equation (9.5). Imaginary complexity—perceived uncertainty—is caused by the designer's lack of knowledge about the system designed.

9.2.4 Time-Dependent Complexity: Combinatorial Complexity and Periodic Complexity

The time-independent complexity discussed in the preceding section dealt with two different kinds of complexities involved in making design decisions: the real complexity associated with the uncertainties inherent in the system designed and the imaginary complexity associated with uncertainties caused by lack of design knowledge, i.e., ignorance.

There is another kind of uncertainty—time-dependent uncertainty—because future events occur in unpredictable ways and thus cannot be predicted. In this section, time-dependent complexity will be defined. In the next section, the means of reducing time-dependent complexity through the use of the Independence Axiom and the Information Axiom will be discussed.

Time-dependent complexity arises because in many situations, future events cannot be predicted a priori. Many of these problems are combinatorial problems that can grow complicated indefinitely as a function of time because the future events depend on the decisions made in the past, but in an unpredictable way. In some cases, this unpredictability is due to the violation of the Independence Axiom. An example is the problem associated with scheduling a job shop. Job shops are typically engaged in machining a variety of parts that are brought to them by their customers. In this case, the future scheduling—which parts are produced using which machines—is affected by the decisions made earlier and its complexity is a function of the decisions made over its past history. This type of time-dependent complexity will be defined as time-dependent *combinatorial complexity*.

There is another kind of time-dependent complexity, *periodic complexity.* Consider the problem of scheduling airline flights. Although airlines develop their flight schedules, uncertainties exist in actual flight departures and arrivals because of unexpected events such as bad weather or mechanical problems. The delayed departure or arrival of one airplane will affect many of the subsequent connecting flights and arrival times. However, because the airline schedule is periodic each day, all of the uncertainties introduced during the course of a day terminate at the end of a 24-hour cycle, and hence this combinatorial complexity does not extend to the following day. That is, each day the schedule starts all over again, i.e., it is periodic and thus uncertainties created during the prior period are irrelevant. However, during a given period there are uncertainties due to combinatorial and other complexities. This type of time-dependent complexity will be defined as time-dependent *periodic complexity*.

Both time-dependent combinatorial complexity and time-dependent periodic complexity are real complexities.

In the next section, it will be shown that time-dependent combinatorial complexity can be changed to time-dependent periodic complexity, greatly reducing information content, uncertainty, and, ultimately, complexity. This is done through redesign or by introducing decouplers.

9.3 REDUCTION OF UNCERTAINTY: CONVERSION OF A DESIGN WITH TIME-DEPENDENT COMBINATORIAL COMPLEXITY TO A DESIGN WITH TIME-DEPENDENT PERIODIC COMPLEXITY

The Independence Axiom and the Information Axiom can be used to reduce the information content of a design, to deal with time-dependent combinatorial complexity, and to convert a combinatorial complexity problem to a deterministic problem through the introduction of periodicity.

Example 9.2 is a beautiful case that shows how a coupled design was decoupled by applying the Independence Axiom so that the robot schedule would not affect the manufacturing process. This decoupling is achieved by adding decouplers (Black, 1991). This example shows how a time-dependent combinatorial complexity problem was reduced to a periodic complexity problem, thereby minimizing the information needed to make the system work and increasing the reliability of the system by removing uncertainty.

EXAMPLE 9.2 From Combinatorial Complexity to Periodic Complexity: Design of Fixed Manufacturing Systems for Identical Parts[7]

One of the processes involved in manufacturing semiconductors is the uniform coating and development of a photoresist material—a viscous substance that is light sensitive. When photoresist material is exposed to electromagnetic waves in a lithographic machine (i.e., a huge camera), the exposed photoresist material cross-links through chemical reaction. The unreacted photoresist material is removed to develop the image on the silicon wafer. Both the initial coating of the photoresist and development of the image are done in a "track" machine. A schematic diagram of a track machine is shown in Figure E.9.2.a.

In a track machine, a silicon wafer is picked up by a robot from a cassette station and then inserted into and removed from modules before and after various processes. The photoresist material is spread on the wafer surface uniformly by spraying it on the wafer

---

[7] From Oh and Lee (2000) and Lee (1999).

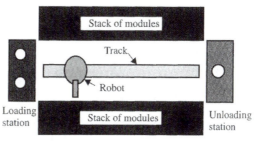

Figure E.9.2.a A schematic diagram of a "track" machine, which puts a thin layer (less than 1 μm thick within a tolerance of 10 Å) of a photoresist on a silicon wafer, typically by a spin-coating technique. The wafer is supplied to the track at the loading station on a periodic basis and a robot picks it up and inserts it into various modules. The wafer is subjected to many different processes before the coating process is completed. After the wafer is coated with the photoresist, it is unloaded by another robot that feeds it into a "lithography" tool—another semiconductor manufacturing equipment that is an ultraprecision camera that creates the desired circuit pattern on the photoresist-coated silicon wafer. The wafer is then brought back to the track to develop the image created by the lithography tool. There can be as many as 42 modules, most of which contain wafers. When the process is completed in each module, the wafer must be picked up and transported to the next module. A problem arises when several wafers are finished with their processes nearly at the same time, all waiting for the robot to transport them to the next module.

and spinning it. Prior to the application of the photoresist, the wafer is subjected to a solvent vapor and heated to improve the adhesion of the photoresist to the wafer. After the spin-coating operation, the wafer is cooled. The wafer then goes to a lithography machine, which prints the circuit pattern on the photoresist. This wafer is then returned to the track machine for development of the image. All of these operations are done in separate modules. Some of the modules and their functions are shown in Figure E.9.2.b. Throughout this operation, the main robot(s) moves the wafer from module to module.

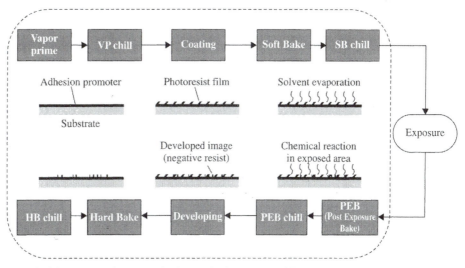

Figure E.9.2.b Some of the modules in a typical "track" machine.

The design task is to develop a schedule for the robot motion so as to maximize the manufacturing productivity of a track machine with many stations.

SOLUTION

Highest Level Design of a Fixed Manufacturing System

This is a simple manufacturing system for making identical parts. It processes wafers in a series of modules. The highest level FR may be stated as

FR_1 = Maximize the return on investment (ROI).

To maximize ROI, we have to produce the maximum number of coated wafers, sell them at the highest possible price, minimize the manufacturing cost, and minimize the capital investment. However, we will consider here only the task of maximizing the output of a dedicated, automated machine. Then the design parameter may be written as

DP_1 = Dedicated automated machine that can produce the desired part at the specified
     production rate

The FRs of the dedicated and automated machine may be written as

FR_{11} = Process wafers in various modules.
FR_{12} = Transport the wafers between modules, between the loading dock and mod-
     ules, between modules and unloading dock.

The corresponding DPs may be chosen as

DP_{11} = Process modules
DP_{12} = Robots

The constraints (Cs) are

C_1 = Throughput rate
C_2 = Manufacturing cost
C_3 = Quality of the product
C_4 = Yield (production of acceptable products divided by total output)

When a machine with a robot and process modules was designed, FR_{11} and FR_{12} were coupled. For example, sometimes the wafers from two or more different modules would be finished at nearly the same time and demand transport to the next module by the robot. However, because one robot cannot perform two functions at the same time, a decision had to be made as to which wafer was to be picked up first. This decision affected all subsequent decisions. The original design solved this problem by using the "if–then" type expert-system algorithm. Sometimes wrong decisions were made thereby delaying the operation, and the machine would come to a complete stop when there was no appropriate "if–then" rule. The problem with this design is that it is a coupled design. The design equation for this coupled design is given by

$$\begin{Bmatrix} FR_{11} \\ FR_{12} \end{Bmatrix} = \begin{bmatrix} X & X \\ X & X \end{bmatrix} \begin{Bmatrix} DP_{11} \\ DP_{12} \end{Bmatrix} \qquad \text{(a)}$$

Therefore, a decision was made to design the robot schedule based on axiomatic design so that the Independence Axiom is satisfied. The new design can be expressed using the design equation given by

$$\begin{Bmatrix} FR_{11} \\ FR_{12} \end{Bmatrix} = \begin{bmatrix} X & 0 \\ X & X \end{bmatrix} \begin{Bmatrix} DP_{11} \\ DP_{12} \end{Bmatrix} \tag{b}$$

Equation (b) expresses the fact that in the proposed design, DP_{11} (process modules) will affect FR_{12} (transport wafers), but DP_{12} (the robot) will not affect FR_{11} (process wafers). All subsequent design decisions as we decompose these FRs and DPs must be consistent with this decision. The design represented by Equation (b) states that given an arrangement of the modules, we must design a transportation system that will not affect the manufacturing process. This is a decoupled design.

Because this machine processes exactly identical parts, a "push" system may be designed, where the part will be supplied to the machine on a regular time interval T. T is equal to $(3600/m)$ seconds where m is the number of wafers supplied to the machine per hour. T is the cycle time during which the robot must pick up wafers from all modules at least once.

The number of modules needed for each process is related to the period T, because if the process time of a module is larger than T, it will take more than one module to be able to meet the required throughput rate. If the process time in Module i is denoted as t_{Pi} (seconds), the number n_i of modules for process i needed to meet the production requirement is given by

$$n_i = \text{Int} \uparrow \frac{t_{Pi}}{T} = \text{Int} \uparrow \left[\frac{t_{Pi}}{(3600/m)} \right] \tag{c}$$

where $\text{Int} \uparrow [x]$ is a function that rounds x up to the nearest integer. The total number of modules M required to process the wafers is given by

$$M = \sum_{i=1}^{N} n_i \tag{d}$$

where N is the number of tasks, i.e., processes. These process modules must be so arranged that the robots can serve all of these modules in the shortest possible time.

Within the cycle-time period T, the module for each process (or one of the modules when there is more than one module for a given process) completes its task. Therefore, within a given period T, the robot must pick up the wafers from the modules that have just completed their process cycles and transfer them to the next set of modules. The robot must also deliver a wafer from the supply cassette station to the first module as well as from the last module to the outgoing cassette station, all in a given period T. If it takes t_T for the robot to transport the wafer from one module to the next, then the number of moves the robot can make in time T is equal to T/t_T.

The sequence of the robot operation is as follows. The robot picks up a wafer from the supply cassette station and delivers it to Module 1 for Process 1. On completion of Process 1 in Module 1, the robot picks up the wafer from Module 1 and inserts it into Module 2 for Process 2, and then from Module 2 to Module 3, and so on. When it is again time to pick up another new wafer from the supply cassette after an elapse of time T, the robot goes back to the supply cassette and loads another unprocessed wafer from the cassette to the first module for Process 1. If the first Module 1 is still processing a wafer, this new wafer is loaded into the second Module 1. This sequence of wafer transfer continues until the entire process is completed. In one period T, the robot must move all the wafers that have just finished a prescribed process to the next module for another process [*Note*: there can be more than one module for each process as per Equation (c)]. In addition, the robot must load a new wafer from the supply cassette station and also deliver the finished wafer from the last module to the outgoing cassette station.

A conflict can arise in scheduling the robot motion if two processes are completed nearly at the same time (i.e., within the time required for a single robot motion), as the robot has to be at two different places at the same time. This coupling of functional requirements can cause a system failure. In the past, this problem was tackled using an "if–then" algorithm for deciding which wafer the robot should pick up next. An "if–then" type of approach is unreliable because the number of combinations increases continuously, as each subsequent decision depends on the decisions made earlier. The number of possible combinations increases to Πn_i where n_i is the number of modules available for process i. Furthermore, when there is no appropriate "if–then" rule, the system breaks down.

This problem can be solved rigorously by introducing decouplers, i.e., by redesigning the system! The coupling occurs when two or more wafers complete their prescribed processes nearly simultaneously (within the transport time of the robot). We can decouple the pick-up functions by introducing a "decoupler"—a device that stores the wafers until the robot becomes available. The role of decouplers is to decouple the functional requirements of the transport. The decouplers do not have to be separate physical devices. In this case, the modules can act as decouplers by letting the wafers stay in the modules longer. Decouplers provide queues between modules so that the wafers can be picked up in a predetermined sequence by the robot. The design task is to determine where the decouplers should be placed and how long their queue should be. Some modules cannot act as decouplers if the process time in the module is tightly controlled for chemical reasons.

When decouplers are introduced with queue q_i, the cycle time T_C increases. As T_C increases, the number of modules may increase, depending on the process time t_P of each module. Therefore, we have dual goals: decouple the process by means of the decouplers and minimize T_C by selecting the best set of q_i. Then FR_{12} (transport wafers) may be decomposed as:

$FR_{121} = $ Decouple the process times.
$FR_{122} = $ Minimize the number of modules M.

The corresponding DPs are

$DP_{121} = $ Decouplers with queues q_i
$DP_{122} = $ The minimum value of T_C

The design equation is given by

$$\left\{ \begin{matrix} FR_{121} \\ FR_{122} \end{matrix} \right\} = \left[\begin{matrix} X & 0 \\ X & X \end{matrix} \right] \left\{ \begin{matrix} DP_{121} \\ DP_{122} \end{matrix} \right\} \tag{e}$$

The minimum cycle time T_C results in a minimum number of modules M. To minimize M, we must satisfy the following two conditions:

$$\sum_{i=1}^{N} \frac{\partial M}{\partial q_i} = 0$$

$$\sum_{i=1}^{N} \frac{\partial^2 M}{\partial q_i^2} > 0$$

(f)

where N is the number of processes.

Analytical Solution for Queues in Decouplers[8]

Having designed the manufacturing system, we must replace each X with a mathematical expression if it can be modeled. In this section, the q_i will be determined through modeling and analysis to determine the exact relationship between FR_{121} (decouple the process times) and DP_{121} (decouplers with q_i).

If we denote the time the wafer has to be picked up on completion of process j in Module i as T_i, then T_i is the sum of the accumulated process time t_P and the accumulated transport time t_T, which may be expressed as

$$T_i = t_P + t_T$$

(g)

T_i, t_P, and t_T are normalized with respect to the sending period T, i.e., actual time divided by T. Therefore, throughout this analysis, all of the times will be dimensionless, i.e., the actual time divided by the period T.

Because the total process time t_P is the sum of the individual process times t_{Pi} and the transport time t_T is the sum of the all robot transport times t_{Ti}, Equation (g) may be expressed as

$$T_i = t_P + t_T = \sum_{j=1}^{i} t_{Pj} + \sum_{j=1}^{i-1} t_{Tj}$$

(h)

The number of pick-up moves the robot can make in a given period T is given by

$$n_R = \frac{T}{t_T}$$

(i)

If there are N process steps, there are M modules with wafers that have gone through their respective processes and are ready to be picked up within a given period T. Within this time period, the robot must pick up all these wafers from the modules that have completed their processes. The robot must pick up a wafer at time τ_i that is measured from the beginning of each period T, which may be expressed as

---

[8] This robot-scheduling problem comes from SVG, Inc., which hired many consultants to solve the problem without obtaining any satisfactory solution. While Dr. Larry Oh, SVG Vice President, and the author were waiting at an airport, the author suggested the use of decouplers and Dr. Oh (with the author's graduate student Tae-Sik Lee) came up with this elegant closed-form solution. A patent has been filed by SVG to protect this work.

$$\tau_i = T_i - \text{Int} \downarrow \left(\sum_{j=1}^{i} t_{Pj} + \sum_{j=1}^{i-1} t_{Tj} \right) = \sum_{j=1}^{i} t_{Pj} + \sum_{j=1}^{i-1} t_{Tj}$$

$$- \text{Int} \downarrow \left(\sum_{j=1}^{i} t_{Pj} + \sum_{j=1}^{i-1} t_{Tj} \right)$$

(j)

where $\text{Int} \downarrow (x)$ is a function that rounds x down to the nearest integer. However, if the pick-up times are coupled because two or more processes are finished nearly at the same time, the robot cannot implement the schedule given by Equation (j).

We must modify the design to decouple the process by adding decouplers with queues q_i. For example, if Process 1 in Module 1 and Process 3 in Module 3 are both finished within the transport time required t_{T1}, the robot cannot pick up both pieces at the same time. Therefore, in this case, we may add a "decoupler" to Module 1, which may be either a physically separate device or just a queue in Module 1 to keep the wafer there longer. In this case, T_1 given by Equation (j) is extended by q_1. Then the new time for pick-up T_1^* is given by

$$T_1^* - T_1 = q_1$$

(k)

Extending this to the more general case:

$$T_2^* - T_2 = q_1 + q_2$$

$$T_3^* - T_3 = q_1 + q_2 + q_3$$

etc.

Substituting these relationships into Equation (j), we obtain the modified actual pick-up time. If we denote this modified time as τ_i^*, then $(\tau_i^* - \tau_i)$ may be expressed approximately as

$$\tau_i^* - \tau_i = \sum_{j=1}^{i} a_{ij} q_j$$

(l)

where a_{ij} is defined as

$$a_{ij} = \begin{cases} 0 & \text{when} \quad i < j \\ 1 & \text{when} \quad i \geq j \end{cases}$$

We can determine τ_i^* approximately by solving Equation (j), by determining where the decouplers may be needed, and by approximating the values of queues.

Equation (l) may be expressed as

$$\{\Delta \tau\} = [a]\{q\}$$

(m)

where $\Delta \tau =$ vector of $\tau_i^* - \tau_i$
$[a] =$ triangular matrix with elements a_{ij}
$\{q\} =$ vector of q_j

Equation (m) may be solved for $\{q\}$ as

$${q} = [a]^{-1}{\Delta \tau} = \frac{1}{|a|}[A]{\Delta \tau} = [A]{\Delta \tau} \tag{n}$$

where $|a| \equiv$ determinant of $[a] = \prod_{i=1}^{N} a_{ii} = 1$

$[A] = \text{Adj } [a]'$

$A_{ji} = \text{cofactor of } a_{ij} \text{ in } [a] = (-1)^{i+j} M_{ij}$

$M_{ij} \equiv \text{minor of } a_{ij} \text{ in } [a]$

Equation (n) can be solved iteratively. To solve Equation (n), we need to know ${\Delta \tau}$, which can be approximated by estimating reasonable values for τ_i^* and by solving Equation (j) for τ_i. The value for τ_j^* can be estimated by adding transport time to τ_i^*, as $|\tau_i^* - \tau_j^*| > t_T$, for all js except $j = i$. The solution can be improved by successive substitution of the improved values of τ_i^*.

Because the best solution is the one that makes the total cycle time T_C a minimum, we must seek a set of values of q_i that yields a minimum value for the total queue, Σq_i. When precise control of processing time is critical, the queue q_i associated with the critical process should be set equal to zero.

To solve Equation (m) for the best set of queues q_i, Oh (1998) and Oh and Lee (2000) developed an optimization software program based on genetic algorithms. Multiplying these q_i by T, we can obtain the actual values of queues.

Determination of the Queues of a Fixed Manufacturing System That Processes Identical Parts

A manufacturing system is being developed for coating of wafers. To produce the final semiconductor product, wafers coated with photoresist must be subjected to various heating and cooling cycles at various temperatures for different durations before they can be shipped to the next operation. The manufacturing system is an integrated machine that consists of five process steps involving five different modules. A robot must place the wafers into these modules, then take them out of the modules, and transport them to the next process module according to a preset sequence. We want to maximize the throughput rate by using the robot and the modules most effectively. The desired throughput rate is 60 units an hour. A constraint is the use of a minimum number of modules. The time it takes for the robot to travel between the modules is 6 seconds. The wafers are processed through the following sequence:

Steps	Modules	Temperature (°C)	Duration ± Tolerance (seconds)
1	A	35	$50 + 25$
2	B	80	45 ± 0
3	C	10	$60 + 20$
4	D	50	$70 + 10$
5	E	68	35 ± 0

The process times in Modules B and E must be precise because of the critical nature of the process. The cycle time is assumed to be the process time plus the transport time both for placement and pick-up of the wafer.

The robot must pick up the wafer from a supply bin (load-lock) and deliver it to Module A, and when the process is finished, it must pick up the wafer from Module E and place it on a cassette. These operations take 6 seconds each.

Solution for q_i for Decouplers

The minimum number of modules is dependent on the cycle time T_C and the desired throughput rate. The required number of modules is as follows:

Modules	Number of Modules
A	2
B	1
C	2
D	2
E	1

Without any decouplers, there are simultaneous demands for the service of the robot as illustrated in Figure E.9.2.c, which shows the time the process is finished in each of the modules within a given period T. In this figure, the horizontal axis is dimensionless time—1.000 represents one period T. Because the transport time is equal to 6 seconds, i.e., $(T/10)$, the figure shows that Process 1 and Process 3 are completed so close to each other that the robot has a conflict. Similarly, Processes 2, 3, and 5 are all finished at nearly the same time.

The solution is obtained by solving Equation (m) using the software program developed by Oh and Lee (2000). The best solution was obtained by finding a set of values that gives the shortest cycle time T_C solving Equation (m) repeatedly and using a genetic algorithm. The solution yields the following values for q_i.

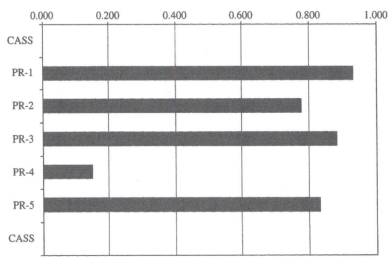

Figure E.9.2.c The pick-up schedule in a period T without decouplers. There are conflicts among the processes finished in Modules 1, 2, 3, and 5. This result is obtained using the software program developed by Oh (1998) and Oh and Lee (2000).

$q_A = 19.7$ seconds
$q_B = 0$ second
$q_C = 9.3$ seconds
$q_D = 9.3$ seconds
$q_E = 0$ second

The queues for B and E are zeros because the tolerance on these two modules is specified to be zero. Therefore, the queues of other modules have been adjusted to permit these two queues to be zero. The actual pick-up times at the completion of the processes of Modules A, B, C, D, and E are given in Figure E.9.2.d.

There are other possible sets of solutions for q_i, but they may not give the minimum M or minimum T_C and the minimum Σq_i.

One of the interesting results of this solution is that the number of combinations for part flow reduces from several thousands to a few, because the parts flow through the manufacturing system along deterministic paths. What the concept of decouplers has done is to change a combinatorial problem into a periodic function that repeats itself with a given cycle that is deterministic.

Summary of Example 9.2

What we showed through this example is how a coupled design was changed to a decoupled design by invoking the Independence Axiom and then how converting a combinatorial complexity problem into a periodic complexity problem further reduced the information content and simplified the design.

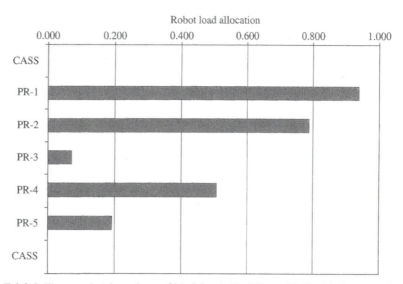

Figure E.9.2.d The actual pick-up times of Modules A, B, C, D, and E. PR-1 is Process 1 that takes place in Module A, PR-2 is for Module B, PR-3 is for Module C, PR-4 is for Module D, and PR-5 is for Module E. This solution is obtained using the software program developed by Oh (1998) and Oh and Lee (2000).

In the robot-scheduling problem discussed above, the Independence Axiom was violated by the original approach. Therefore, a decision was made to create a decoupled design. It was done by introducing decouplers that provided queues to the modules. By this means, the infinite time-dependent combinatorial problem was changed to a periodic problem where the cycle within the sending period was made to repeat itself by adding decouplers. The scheduling problem was thus changed from a combinatorial problem to a deterministic one, thereby immensely reducing the uncertainty and complexity. Furthermore, this is an important consequence of applying the Independence Axiom to these random events to create a "periodicity." This change of the task with an uncertain outcome to one with a definite outcome reduces uncertainty and makes the task much less complex. In other words, the introduction of decouplers has introduced periodicity and changed a combinatorial problem into a deterministic problem.

It is a very significant finding that this creation of "periodicity" reduces, if not eliminates, the uncertainty and therefore the complexity associated with an infinite combinatorial system. Moreover, an infinite time-dependent combinatorial system cannot be sustained because the uncertainty associated with its future events becomes too large. The system then becomes risky and unreliable. This means that even when it is not clear as to how a period can be defined, it is better to stop an event and restart with new initial values to reduce the uncertainty of future events, where the current decisions affect future events and probabilities. Nature forces this periodicity by giving a finite life to all living beings. These observations, which are extensions of the Independence Axiom and the Information Axiom, can be stated as a theorem:[9]

> *Theorem 26 (Conversion of a System with Infinite Time-Dependent Combinatorial Complexity to a System with Periodic Complexity)*
>
> *Uncertainty associated with a design (or a system) can be reduced significantly by changing the design from serial, time-dependent combinatorial complexity to periodic complexity.*

This relationship between complexity and periodicity has many important applications and implications.

■ 9.4 DISTINCTION BETWEEN TIME-INDEPENDENT AND TIME-DEPENDENT COMPLEXITIES

One of the interesting questions is whether there is any generalization that can be made about the relationship between time-independent and time-dependent complexity. Although no thorough investigation has been made of this issue, it seems that they are as distinct from each other as the elliptic partial differential equation is from the hyperbolic partial differential equation.

In the case of time-independent complexity, the complexity of a system is governed by the given set of FR and DP relationships. This is in contrast to the case of time-dependent complexity, which depends on the initial conditions, but unless the system goes back to the same set of initial conditions periodically, the distant future behavior is totally unpredictable.

---

[9] Other theorems can be found in Suh (1990).

That is, in the case of time-dependent combinatorial complexity, the initial condition has little control over the long-term behavior of a combinatorial complexity system. In the case of a system with time-dependent combinatorial complexity, the initial condition is not distinguishable from the state of the system at any other time in terms of its ability to control long-term behavior. However, in a system with a time-dependent periodic complexity, the initial condition at the beginning of each cycle determines the behavior of the system during the period and thus forever, as the period repeats itself.

9.5 OTHER IMPLICATIONS OF THE DESIGN AXIOMS AND PERIODIC COMPLEXITY: A SPECULATION[10]

9.5.1 Nature

One of the important discoveries this chapter has described is the power of changing a design with time-dependent combinatorial complexity to a design that has periodic complexity in order to reduce uncertainty. When uncertainty is large, the future outcome cannot be assured. The periodicity also renews the life cycle and increases the reliability of the system by restarting the system from the same initial conditions over and over again.

It is interesting to note that nature has known this fact all along. Many things in nature are periodic. Atomic structure is periodic. The animal life cycle is periodic. Most animals sleep daily to renew themselves. The life of all living beings is periodic and finite.

Weather is an interesting example of how a time-dependent periodic complexity problem is superimposed on a combinatorial complexity problem. Because the earth revolves around the sun and the intensity of the energy it receives from the sun undergoes a periodic change depending on the orientation of the earth relative to the sun—the period being approximately 365 days—the weather undergoes a periodic change on a yearly basis. However, within a given year, the weather pattern is affected by a number of events that occur at a given locale, which then, in turn, affect subsequent events that occur throughout the globe. Thus, the weather prediction at a given instant is a combinatorial complexity problem. It should not take much imagination to understand the implication of the periodic nature of the earth revolving around the sun—otherwise, the earth would not be able to support the life system of all living beings.

If we extend this speculation one step further, nature may deal with the continuing level of environmental pollution by replacing the current expansion of combinatorial complexity of nature with a periodic one, on a grand scale. This will happen if the earth cannot support the current form of living beings without starting all over again. It can be concluded that to prevent this unpleasant event from happening, human beings must discover a means of renewing nature through less pollution.

9.5.2 Biological Systems and Living Beings

One of the interesting questions is the design of biological systems by nature. Are biological systems a coupled design, a decoupled design, or an uncoupled design? It appears that

---

[10]The topics discussed in this section may be classified as intellectual speculations, as they are not supported by any proof and/or experimental confirmation. Some of the speculative ideas are given as food for thought.

Table 9.2 Design Matrix for Biological Systems

	ATP or Carbohydrates	Oxygen (Reactants)	Catalysts/ Enzymes	Reproduction of Cells	Chemicals for Control of Equilibrium State
Supply fuel (energy)	X	0	0	0	0
Supply oxidizer (reactant)	0	X	0	0	0
Induce reaction of fuel/oxidizer	X	X	X	0	0
Reproduce cells	X	0	0	X	0
Control reaction rate (growth)	0	0	0	X	X

many biological systems—at a molecular level, a cellular level, and a physiological level—are a decoupled system. For example, five of the basic functions of biological systems—supply fuel, supply oxidizer, react the fuel and oxidizer, reproduce themselves, and regulate growth—are satisfied by supply of carbohydrates, supply of oxygen or other elements that react with the fuel, catalysts, reproduction of cells, and chemical equilibrium. If, for instance, these were a correct set of FRs and DPs, it appears that it is a decoupled system as shown in the matrix in Table 9.2.

There are several implications:

- Catalysts and chemicals for equilibrium can be supplied without any sequence.
- Fuel (carbohydrates) must be supplied first, followed by oxidizer and cell reproduction.
- If any one of the DPs is eliminated, we have a coupled system and the cells cannot survive.
- Living biological systems may be decoupled systems, and any event that changes a decoupled biological system may cause death of the cell.

Nature sustains life by renewing itself periodically. If living beings were to live forever, they would go through mutations and other changes that can neither be predicted nor controlled—a combinatorial complexity problem. Therefore, all living beings stop living when these unanticipated events occur. They sustain and renew themselves by reproduction through the combination of the fundamental building blocks from ground zero periodically. Living beings represent a design—Nature's design—with periodic complexity, where the period is the life span of the living being.

9.5.3 Artificial Systems

Theorem 26 also has important implications for political and societal systems. A kingdom or a country that is ruled by a dictator without any possibility of renewal is a system with time-dependent combinatorial complexity rather than one with periodic complexity. Therefore, such a political system can undergo unexpected mutations, as the future outcome cannot be controlled or predicted. Hence, the system can corrupt and deteriorate in a totally unexpected manner. There are certainly many historical examples of such systems, one of the most recent ones being the Soviet Union.

We must introduce periodicity even to political systems so that they can reduce uncertainty and renew themselves. Possible renewal mechanisms are periodic elections, a periodic setting of budgets, and periodic auditing.

Universities must also be designed to have periodic complexity. The existing mechanisms are academic semesters, fixed periods of study, and the academic tenure systems for faculty. Contrary to popular view, the tenure system at leading universities guarantees the renewal of academic life. It provides a means of renewing faculty on a regular cycle (typically 6 to 8 years), as many who are judged to be less than the best end up leaving the institution, although wrong decisions are sometimes made in this context just like many other human decisions. For the tenured faculty, the university depends on the retirement system for periodic renewal. The fact that universities no longer have a mandatory retirement system is not good in terms of periodic complexity.

9.6 COMPLEXITY OF NATURAL PHENOMENA

Many of the complexity issues discussed in the articles of *Science*[11] are different from the cases discussed in Sections 9.3 and 9.4 in that they deal with preexisting natural systems with many FRs and many DPs in a "self-selecting" and "self-organizing" system. The system is governed by natural forces, geometric constraints, and chemical and biological interactions. FRs are the questions scientists want to have answered so as to understand certain phenomena.[12] Once the questions are posed, we have to determine DPs. Then we have to zigzag and decompose FRs and DPs until we can answer the question posed. The lowest level FRs (leaves) may be many. The issue here is closely associated with the issues discussed in Section 4.5.4 on large flexible systems, but with additional constraints.

The problem may be formulated as follows: The lowest level FR, which is one of the fundamental questions the scientists want to know, can be related to several DPs, all of which can affect the FR. FR_1 could be "how does a crack nucleate in sliding wear." Candidates for DP_1 could be many: "plastic deformation," "stress around second-phase particles," "twinning," etc. This knowledge base can be expressed as

$$FR_1 \ \$ \ \left(DP_1^a, DP_1^b, \ldots DP_1^r\right) \tag{9.15}$$

Similarly, the knowledge base for other FRs can be structured as

$$FR_2 \ \$ \ \left(DP_2^a, DP_2^b, \ldots DP_2^q\right)$$

$$FR_3 \ \$ \ \left(DP_3^a, DP_3^b, \ldots DP_3^w\right)$$

$$\ldots \tag{9.16}$$

$$\ldots$$

$$FR_m \ \$ \ \left(DP_m^a, DP_m^b, \ldots DP_m^s\right)$$

Equation (9.15) simply states that FR_1 can be satisfied (indicated by \$) by DP_1^a, DP_1^b,

---

[11] *Science*, Vol. 284, No. 5411, April 2, 1999.

[12] For example, the FR may be stated as "we want to know what determines the number of cells nucleated in microcellular plastics." There can be many DPs. One DP could be "cell nucleation mechanism." The lower-level FRs must then be determined and the process should continue.

DP_1^r, etc. Similarly, FR_m is satisfied by DP_m^a, DP_m^s, etc. Equations (9.15) and (9.16) represent the *knowledge base* or the *database* for a large flexible system. Adding more DPs to these equations is equivalent to expanding the knowledge base or the database. The richness of the system is defined by the size and quality of the database, as the larger the number of available DPs, the greater is the likelihood for better designs.

As we search for DPs in the physical domain that will enable us to satisfy the FRs, we may find that there is more than one DP_i that can satisfy a given FR_i. Equation (9.16) does not indicate which DP_3^i, for example, is the best choice for FR_3. Furthermore, because all or a subset of the FRs must be satisfied at any given instant, it is not possible to say a priori which DP_3 is the best solution without considering the other FRs that must be satisfied at the same time. That is, the choice of DP_3 may be different depending on the chosen subset of FRs.

In natural science, we want to know how certain phenomena occur. Even in nature, there are situations in which several FRs must be satisfied at the same time. If the natural system is an uncoupled system, then it can be treated as a set of uncoupled FRs. In such a system, the information content can be minimized by eliminating the bias of the system by adjusting the value of DP and by reducing variance. Because the information content is analogous to the negative entropy of the system, the minimization of information content may be equivalent to maximizing entropy. In the case of no chemical affinity, seeking a system with the minimum information content may be equivalent to finding a system with the maximum free energy of formation of the system.

For example, in the case of wear of materials under the load exerted by the counter surface that slides over another surface, there are many questions (i.e., FRs) we must answer at the same time. For example, how is the applied load distributed, how and where does a crack initiate and propagate, how is a loose wear particle generated, and what determines the tangential force? These are possible FRs. When all these phenomena occur at the same time, only a specific set of DPs can operate to yield the FRs. Furthermore, the set of phenomena of interest may change over a period of time.

9.7 SUMMARY

This chapter examined the issue of complexity, information, and uncertainty based on the Independence Axiom and the Information Axiom. It was shown that there are many different kinds of complexities.

In the time-independent situation, it was shown that there are two kinds of complexities: real complexity and imaginary complexity, which are orthogonal to each other. Absolute complexity is defined to be the vector sum of the real and the imaginary complexities.

In the time-dependent complexity arena, it was shown that there are two different kinds of complexities: combinatorial complexity and periodic complexity. In a system that is subject to combinatorial complexity, the uncertainty of the future outcome continues to grow as a function of time, and as a result, the system cannot have long-term stability and reliability. In the case of systems with periodic complexity, the system is deterministic and can renew itself with each period. Therefore, a stable and reliable system must be periodic. Starting from the application of the Independence Axiom, it was shown how a coupled system was decoupled through design changes and how a combinatorial complexity problem could be changed into a periodic complexity problem.

A case study was presented to show how complexity could be reduced by redesign and by replacement of a combinatorial complexity problem with a periodic complexity one.

Finally, the consistency between nature and Theorem 26 (Conversion of a System with Infinite Time-Dependent Combinatorial Complexity to a System with Periodic Complexity) was discussed. It was shown that many things in nature are periodic, consistent with the need to change a combinatorial complexity design to one of periodic complexity to reduce uncertainty. It was argued that the theorem may apply to political and societal systems as well. Periodic renewal of political systems and societal systems is essential for long-term sustainability of the system.

REFERENCES

Bennet, C. H. *Emerging Syntheses in Science: Proceedings of the Founding Workshops of the Santa Fe Institute*, The Institute, Santa Fe, NM, 1985.

Black, J. T. *The Design of the Factory with a Future,* McGraw-Hill, New York, 1991.

Chaitin, G. J. *Algorithmic Information Theory*, Cambridge University Press, Cambridge, 1987.

Cover, T. M., and Thomas, J. A. *Elements of Information Theory,* Wiley, New York, 1991.

Flood, R. L., and Carson, E. R. *Dealing with Complexity: An Introduction to the Theory and Application of Systems Science*, 2nd ed., Plenum Press, New York, 1993.

Gell-Mann, M., and Lloyd, S. "Information Measures, Effective Complexity, and Total Information," *Complexity*, Vol. 2, no. 1, pp. 44–52, 1996.

Huberman, B. A., and Hogg, T. In *Märk världen. En bok om vetenskap och intuition*, T. Nørretranders, ed., Bonnier Alba., Stockholm, 1994.

Lee, T.-S. "The System Architecture Concept in Axiomatic Design Theory: Hypothesis Generation and Case Study Validation," M. S. Thesis, Department of Mechanical Engineering, MIT, 1999.

Lloyd, S., and Pagels, H. "Complexity as Thermodynamic Depth," *Annals of Physics*, Vol. 188, pp 186–213, 1988.

Oh, H. L. "Synchronizing Wafer Flow to Achieve Quality in a Single-Wafer Cluster Tool," Unpublished Report, Silicon Valley Group, Inc., 1998 (Patent Pending, 1998).

Oh, H. L., and Lee, T.-S. "Synchronous Algorithm to Reduce Complexity in Wafer Flow," presented at the First International Conference on Axiomatic Design, MIT, Cambridge, MA, June 2000.

Shannon C. E., and Weaver, W. *The Mathematical Theory of Communication*, University of Illinois Press, Urbana, 1949.

Sohlenius, G. "Notes on Complexity, Difficulty and Axiomatic Design," unpublished note, 1999.

Suh, N. P. *The Principles of Design,* Oxford University Press, New York, 1990.

| HOMEWORK |

9.1 Consider the design and manufacture of a car. Estimate the order of magnitude of real and imaginary complexities that the engineers at Toyota Motor Co. may be dealing with by analyzing how they design the vehicle.

9.2 In Chapter 1, the design of LCD holders was analyzed. Estimate the real and the imaginary complexity an engineer must deal with when the products are evaluated by a trial-and-error method.

9.3 State the conditions under which a combinatorial complexity can be changed into a periodic complexity.

9.4 Compare the quantitative definition of complexity provided by axiomatic design to other quantitative definitions of complexity (for example, Kolmogorov complexity and Shannon's information entropy). What is the fundamental difference?

Index